Applied Mathematical Sciences
Volume 133

Springer Science+Business Media, LLC

Applied Mathematical Sciences

1. *John:* Partial Differential Equations, 4th ed.
2. *Sirovich:* Techniques of Asymptotic Analysis.
3. *Hale:* Theory of Functional Differential Equations, 2nd ed.
4. *Percus:* Combinatorial Methods.
5. *von Mises/Friedrichs:* Fluid Dynamics.
6. *Freiberger/Grenander:* A Short Course in Computational Probability and Statistics.
7. *Pipkin:* Lectures on Viscoelasticity Theory.
8. *Giacoglia:* Perturbation Methods in Non-linear Systems.
9. *Friedrichs:* Spectral Theory of Operators in Hilbert Space.
10. *Stroud:* Numerical Quadrature and Solution of Ordinary Differential Equations.
11. *Wolovich:* Linear Multivariable Systems.
12. *Berkovitz:* Optimal Control Theory.
13. *Bluman/Cole:* Similarity Methods for Differential Equations.
14. *Yoshizawa:* Stability Theory and the Existence of Periodic Solution and Almost Periodic Solutions.
15. *Braun:* Differential Equations and Their Applications, 3rd ed.
16. *Lefschetz:* Applications of Algebraic Topology.
17. *Collatz/Wetterling:* Optimization Problems.
18. *Grenander:* Pattern Synthesis: Lectures in Pattern Theory, Vol. I.
19. *Marsden/McCracken:* Hopf Bifurcation and Its Applications.
20. *Driver:* Ordinary and Delay Differential Equations.
21. *Courant/Friedrichs:* Supersonic Flow and Shock Waves.
22. *Rouche/Habets/Laloy:* Stability Theory by Liapunov's Direct Method.
23. *Lamperti:* Stochastic Processes: A Survey of the Mathematical Theory.
24. *Grenander:* Pattern Analysis: Lectures in Pattern Theory, Vol. II.
25. *Davies:* Integral Transforms and Their Applications, 2nd ed.
26. *Kushner/Clark:* Stochastic Approximation Methods for Constrained and Unconstrained Systems.
27. *de Boor:* A Practical Guide to Splines.
28. *Keilson:* Markov Chain Models—Rarity and Exponentiality.
29. *de Veubeke:* A Course in Elasticity.
30. *Shiatycki:* Geometric Quantization and Quantum Mechanics.
31. *Reid:* Sturmian Theory for Ordinary Differential Equations.
32. *Meis/Markowitz:* Numerical Solution of Partial Differential Equations.
33. *Grenander:* Regular Structures: Lectures in Pattern Theory, Vol. III.
34. *Kevorkian/Cole:* Perturbation Methods in Applied Mathematics.
35. *Carr:* Applications of Centre Manifold Theory.
36. *Bengtsson/Ghil/Källén:* Dynamic Meteorology: Data Assimilation Methods.
37. *Saperstone:* Semidynamical Systems in Infinite Dimensional Spaces.
38. *Lichtenberg/Lieberman:* Regular and Chaotic Dynamics, 2nd ed.
39. *Piccini/Stampacchia/Vidossich:* Ordinary Differential Equations in $\mathbf{R}^n$.
40. *Naylor/Sell:* Linear Operator Theory in Engineering and Science.
41. *Sparrow:* The Lorenz Equations: Bifurcations, Chaos, and Strange Attractors.
42. *Guckenheimer/Holmes:* Nonlinear Oscillations, Dynamical Systems, and Bifurcations of Vector Fields.
43. *Ockendon/Taylor:* Inviscid Fluid Flows.
44. *Pazy:* Semigroups of Linear Operators and Applications to Partial Differential Equations.
45. *Glashoff/Gustafson:* Linear Operations and Approximation: An Introduction to the Theoretical Analysis and Numerical Treatment of Semi-Infinite Programs.
46. *Wilcox:* Scattering Theory for Diffraction Gratings.
47. *Hale et al:* An Introduction to Infinite Dimensional Dynamical Systems—Geometric Theory.
48. *Murray:* Asymptotic Analysis.
49. *Ladyzhenskaya:* The Boundary-Value Problems of Mathematical Physics.
50. *Wilcox:* Sound Propagation in Stratified Fluids.
51. *Golubitsky/Schaeffer:* Bifurcation and Groups in Bifurcation Theory, Vol. I.
52. *Chipot:* Variational Inequalities and Flow in Porous Media.
53. *Majda:* Compressible Fluid Flow and System of Conservation Laws in Several Space Variables.
54. *Wasow:* Linear Turning Point Theory.
55. *Yosida:* Operational Calculus: A Theory of Hyperfunctions.
56. *Chang/Howes:* Nonlinear Singular Perturbation Phenomena: Theory and Applications.
57. *Reinhardt:* Analysis of Approximation Methods for Differential and Integral Equations.
58. *Dwoyer/Hussaini/Voigt (eds):* Theoretical Approaches to Turbulence.
59. *Sanders/Verhulst:* Averaging Methods in Nonlinear Dynamical Systems.
60. *Ghil/Childress:* Topics in Geophysical Dynamics: Atmospheric Dynamics, Dynamo Theory and Climate Dynamics.

(continued following index)

I.I. Vorovich

Nonlinear Theory of Shallow Shells

English Edition edited by L.P. Lebedev
Translated by Michael Grinfeld

With 46 Illustrations

Springer

I.I. Vorovich
Research Institute of Mechanics
and Applied Mathematics
Rostov State University
ul. Stachki 200/1
Rostov on Don, 344090
Russia

English Edition by
L.P. Lebedev
P.O. Box 4105
ul. Sodruzhestva 37
Kv 16
Rostov on Don, 344103
Russia

Translated by
Michael Grinfeld
Department of Mathematics
University of Strathclyde
26 Richmond Street
Glasgow, Scotland, G1 1XH
UK

Editors

Mathematics Subject Classification (1991): 73K15, 35J50, 73HXX, 73C35, 73VXX

Library of Congress Cataloging-in-Publication Data
Vorovich, Iosif Izrailevich, 1920–
Nonlinear theory of shallow shells /Isoif I. Vorovich.
p. cm. — (Applied mathematical sciences ; 133)
Includes bibliographical references and index.
ISBN 978-1-4757-7288-3 ISBN 978-0-387-22776-4 (eBook)
DOI 10.1007/978-0-387-22776-4
1. Shells (Engineering) 2. Nonlinear theories. 3. Boundary value problems. I. Title. II. Series: Applied mathematical sciences (Springer-Verlag New York, Inc.) : v. 133.
QA1.A647 vol. 133
[TA660.S5]
510 s—dc21
[624.1'7762] 98-17535

Printed on acid-free paper.

Production managed by Timothy Taylor; manufacturing supervised by Jacqui Ashri.
Photocomposed copy prepared from the author's TEX files.

9 8 7 6 5 4 3 2 1

ISBN 978-1-4757-7288-3 SPIN 10645080

Preface to the English Edition

The author is deeply grateful to Prof. S.S. Antman for assistance in the translation of the monograph. Since the publication of the Russian version of the book, there have appeared a number of works that are directly relevant to the problems we are considering. Therefore, the English edition is supplemented by references, some of which I found in S.S. Antman's monograph *Nonlinear Problems of Elasticity* [396].

An important part in the preparation of the translation was played by Prof. L.P. Lebedev, who made many useful corrections and additions to the text of the translation, and by the translator, Dr. M. Grinfeld. I would like to express my appreciation of their efforts.

I consider it a pleasant duty to express my gratitude to Springer-Verlag for their attention and encouragement in this project.

Rostov on Don, Russia *I.I. Vorovich*

Preface to the Russian Edition

The nonlinear theory of shallow shells originates in the work of I.G. Bubnov and T. von Kármán. Its present state owes a considerable amount to ideas of L. Donnell, K. Marguerre, Kh.M. Mushtari, V.Z. Vlasov, V.V. Novozhilov, Chien Wei-Zhang, and others. The intensive development of this branch of the mathematical theory of elasticity is first and foremost connected to its wide applicability, since it turns out that the problem of stability of thin-walled constructions can only be solved on the basis of nonlinear boundary value problems.

At the same time, nonlinear shell theory can be considered as a wide generalization of the classical Plateau problem, and this is its great importance for the natural sciences. Indeed, the Plateau problem deals with surfaces with a well defined deformation law: The density of the potential energy of deformation is proportional to the change in the area of the element. On the other hand, in shell theory one considers surfaces in which the density of potential energy of deformation is a scalar function of the deformation tensor, which makes the problem not only much more difficult, but also widely applicable and of great interest in the natural sciences. There is a voluminous amount of work in which particular examples of nonlinear shell theory are being studied. However, there is no problem in this theory in which a solution is obtainable in closed form. Therefore, a wide range of computer-based approximation techniques are in use. This makes a mathematically rigorous analysis of the underlying class of nonlinear problems all the more necessary. Let us also add that mechanical phenomena of practical interest do not allow an analysis based on weakly nonlinear assumptions; they are related to essential "deep" nonlinearities.

Therefore, the construction of the mathematical theory of nonlinear problems requires the utilization of a wide range of mathematical techniques, and this has to a large extent determined the structure of the present monograph.

Chapter I deals with the formulation of boundary value problems of nonlinear shallow shell theory. In the course of that chapter we analyze in detail the concept of shallowness itself; it is of a complicated physical and geometrical nature. This concept had been developed in the work of Mushtari, Vlasov, Galimov, Novozhilov, and others. A unified criterion for "shallowness" of a shell is given. The main boundary value problems are formulated in general nonorthogonal coordinates, both in displacements and with an Airy stress function.

In Chapter II of the monograph we review Sobolev spaces and state a number of functional-analytic theorems that lay the foundation for the subsequent considerations. Here we also present an important topological invariant, the degree of a mapping, and study its properties. A coercivity lemma, which is widely used in the sequel, is proved. In Sections 11 and 12 we present the basic function spaces $H_{t\kappa}$, H_t, H_κ, which serve as a setting for the generalized formulation of the boundary value problems. Since we are dealing with essentially nonlinear problems, there exist a number of ways in which the passage to generalized solutions can be effected. The author has chosen generalized solutions that follow immediately from the variational principles of Lagrange and Alumyae. This also clarifies the mechanical meaning of generalized solutions.

In Sections 13–16 of Chapter III we develop a topological approach to proofs of solvability of the main boundary value problems in displacements. It is based on the computation of the winding number of a certain completely continuous vector field on spheres of large radius in a Hilbert space H, for which we use ideas of homotopy that require an a priori estimate of the solution. This estimate is obtained here under very general conditions on the clamping of the shell under the action of both transverse loads and of loads tangent to the midsurface of the shell.

Since obtaining the a priori estimate is one of the crucial steps in the proof, we observe that it is based on subdividing the sphere in the energy space H into two parts, one of which does not contain the weak closure of zero. For this part of the sphere the proof is based on the analysis of the energy of stretching of the shell, while for the remaining part of the sphere the corresponding inequalities are obtained by estimating the bending energy. The computation of the winding number of the vector field gives us not only the solvability theorem, but a tool for analyzing the number of solutions as well. In particular, in some cases this approach allows us to prove nonuniqueness of solutions. In Sections 17–19 similar arguments are given for the case of a boundary value problem with an Airy stress function. Here we also compute the winding number of the corresponding vector field and prove solvability theorems.

In Section 20 we study in detail differentiability properties of generalized solutions. In particular, we establish conditions for the existence of classical solutions. These results are used later on in the estimates of rate of convergence of approximation methods.

In Chapter V we use the variational approach to study boundary value problems of nonlinear shallow shell theory. Though the principal results here are again solvability theorems, the solutions obtained here are substantially different from those of Sections 16 and 19, as they characterize extremal states of the system.

A characteristic feature of the arguments of Chapters III–IV is that the methods used there give nonlocal results. In other words, the boundary value problems of nonlinear shell theory are analyzed without any assumptions on smallness of nonlinear terms, loading parameters, curvature, etc.

In Chapters VI–VII we study a wide range of methods frequently used currently in the numerical solution of boundary value problems of nonlinear shallow shell theory. First, we analyze local methods (small parameter, successive approximations, the Newton–Kantorovich method). Limits of their applicability are determined, and recommendations to increase their effectiveness are given. In Chapter VII we present a complete justification of the methods of Bubnov–Galerkin and Ritz in their most widely used forms, due to Papkovich, Mushtari, and Vlasov. The analysis here again does not use any local considerations, based on smallness of the defining factors.

Let us note an important feature of the problems under consideration: Justification of approximation methods is done under conditions of possible nonuniqueness, which makes the results quite general. The technique suggested in Chapter VII to justify direct methods to solve boundary value problems of nonlinear shallow shell theory is applicable to an analysis of finite-difference and finite-element methods. Here we can formulate a general principle for these methods: The error made in looking for an approximate solution of the boundary value problem by these methods is, as a rule, asymptotically equivalent to the error of direct approximation of the solution by the suggested method. Exceptions to this rule may be provided only by solutions lying on "folds," that is, in boundary regions in parameter space, in which multiplicity of solutions of the nonlinear system changes.

Questions of stability of shells "in the large" are considered in Chapters VIII–IX. We assume that in a given mechanical problem the solution to the problem of stability should single out the most realistic equilibrium configuration under the conditions of their nonuniqueness. Thus, first of all, we should study the question of multiplicity of equilibrium configurations under a given load. Therefore in Sections 29–31 we study questions of uniqueness of solutions of nonlinear boundary value problems of shallow shell theory. Here we have to distinguish between global uniqueness (in the entire function space) and local uniqueness (in a small neighborhood of zero in the function space). In connection with the question of global uniqueness we introduce the concept of a rigid shell. It is shown that stiffness of a shell is determined by a new invariant quantity of both mechanical and geometrical nature. In Section 31 of Chapter VIII we determine well-posedness classes, in which a small change of the strains of a shell is guaranteed. An important part in issues of stability of thin-walled constructions is played by the question of a lower critical number. It can be said that its approximate computation is an invariable constituent of any stability analysis. Therefore, in Section 33 we present an existence theorem for the lower critical number for a wide range of problems. The entire Chapter IX is devoted to the derivation of relations among the lower and the upper critical numbers, points of spectrum of the corresponding linearized problem, and the Euler characteristic, the first eigenvalue of that problem.

Chapter X deals with the first part of the stability problem, that is, estimating the number of equilibrium configurations of a shell under various loads. In Section 38 we make an attempt to analyze the second part of that problem, that is, to understand how realistic are the different equilibrium configurations in the case of nonuniqueness. Hence the working of a shell is described taking into account stochastic factors. It turns out that it is possible to describe the probability of a shell being in a particular equilibrium configuration. In a particular case, as the probability measure we can take the level of the potential energy of the shell. This, in effect, completes the planned study of stability of shallow shells. The author hopes that the book will prove useful to scientists and engineers working with thin-walled constructions, as well as to mathematicians interested in nonlinear problems of continuum mechanics.

L.P. Lebedev has thoroughly read and edited the manuscript, making many valuable suggestions that led to its improvement. N.F. Morozov has made a number of remarks in the process of refereeing the monograph, which I also took into account. I am very grateful to both of them.

Rostov on Don, Russia *I.I. Vorovich*

From the Editor

The Western reader is presented here with a translation of the fundamental monograph by the outstanding Russian scientist I.I. Vorovich. This work brings together the results of more than thirty years of scientific activity in the field of mathematical analysis of shallow shell theory. The first publications of Prof. Vorovich in this area appeared in 1953. In 1957 he obtained his doctorate from the Leningrad State University.

I first encountered Prof. Vorovich's doctoral thesis as a student, when he, as my scientific advisor, suggested I should study mathematical questions in mechanics. I remember well the impression this bulky substantial work made on me; I did not at the time understand completely its results, which were obtained by methods not to be found in textbooks. All the results were presented with complete proofs. Amazingly, all the mathematical arguments and computations had a clear mechanical subtext. This work showed that in complex mathematical questions of mechanics significant results could be obtained only by mastering the mechanical nature of the problem.

After the defense of the thesis, it was suggested to I.I. Vorovich that he should publish it as a monograph in the Leningrad State University Press. This, however, was not done. Essentially all the results of the thesis appeared in the prestigious journal *Doklady of the USSR Academy of Sciences*. As this journal publishes only brief summaries of research, an extended presentation of the majority of the results have appeared only in the Russian edition of the present monograph, in 1989. This may be the reason why the many results in the theory of shallow shells obtained by I.I. Vorovich remained unknown to readers in the West for so long. Furthermore, all the while there kept appearing in the West numerous papers rediscovering particular cases of the results that saw light long before, in Vorovich's doctoral thesis. This applies to solvability theorems for problems in displacements in the geometrically nonlinear theory of plates, and for the von

Kármán equations. In fact, Vorovich's thesis dealt in great detail with these and other questions of the mathematical theory of the corresponding boundary value problems of shell theory, which includes the theory of plates as a particular case. In particular, the thesis also contains a rigorous analysis of numerical methods for the solution of nonlinear problems. We should also mention here Vorovich's results in the qualitative theory of stability of nonlinear shells. Actually, the reader can see for himself the scope of ideas and methods developed by I.I. Vorovich in the 1950s, since the present work is a significantly expanded and revised version of his 1957 thesis. The main difference is that in the monograph before us, the theory is developed for shallow shells defined in general curvilinear coordinates, while in the thesis Vorovich considered only the case of orthogonal coordinate systems. In the nonlinear problems under consideration, such a generalization does not result in a simple change of coordinates, but rather requires substantially modified mathematical tools. In addition, the present monograph includes numerous results obtained by I.I. Vorovich and his students subsequently. The new mathematical results allow us a deeper understanding of the mechanical contents of the equations of the theory and give us a better idea of the domain of its possible applications.

I think the Western reader will find much of interest (and some surprises!) in the present book. It has to be said, however, that the changes connected with the introduction of general coordinate systems have not resulted in a simplification from the point of view of understanding the tools being employed. I think a pure mathematician would have enjoyed the 1957 manuscript, which does not use tensor notation.

Reading Prof. Vorovich's monograph requires some knowledge in the area of applications of functional analysis in problems of mechanics. This material can be found in the recent textbook of L.P. Lebedev, I.I. Vorovich, and G.M.L. Gladwell, *Functional Analysis. Applications in Mechanics and Inverse Problems* (Kluwer Academic Press, 1996).

Finally, the English translation of I.I. Vorovich's differs in certain respects from the Russian original. Since the book makes extensive use of results from diverse areas of mathematics, explanations of fundamental concepts and terms employed were added to make it accessible to a wider readership. Numerous typographical errors of the Russian version were corrected, an index and a list of notations were added; the system of equation numbering was rationalized. Changes in the notation were made with a view to imparting to the terms in English the mechanical meaning they carried in the Russian version.

The translator, Dr. Michael Grinfeld, and Prof. S.S. Antman have made a valuable contribution to these changes.

Rostov on Don, Russia *L.P. Lebedev*

Contents

CHAPTER I

The Main Boundary Value Problems in the Nonlinear Theory of Shallow Shells

1. Results from the Theory of Surfaces

1.1. Let a surface S in R^3 be parametrized by (α^1, α^2), so that every point of the surface is defined by a vector

$$\rho = \rho(\alpha^1, \alpha^2), \tag{1.1}$$

which can be written in Cartesian coordinates as

$$x_i = x_i(\alpha^1, \alpha^2), \; i = 1, \ldots, 3.$$

We shall interpret the numbers α^i as the Cartesian coordinates in the plane, taking their values in some planar domain Ω. We shall also assume that (1.1) defines a homeomorphism between Ω and S, that is, a continuous bijection with a continuous inverse. We note that in the sequel we shall allow S to be multiply connected, in which case Ω will also be multiply connected (see Figure 1.1).

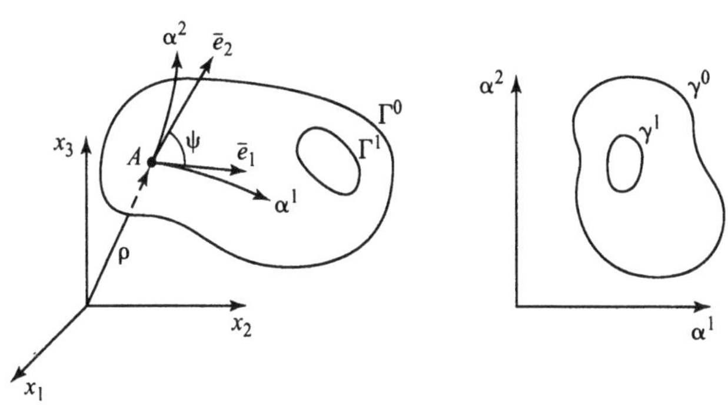

FIGURE 1.1.

We shall also assume that $\rho \in C^2_{\Omega}$, that is, all the second partial derivatives of ρ are continuous in $\overline{\Omega}$. Furthermore, let the following condition be satisfied everywhere in $\overline{\Omega}$:

$$D^2 = \left|\frac{\partial(x_1, x_2)}{\partial(\alpha^1, \alpha^2)}\right|^2 + \left|\frac{\partial(x_1, x_3)}{\partial(\alpha^1, \alpha^2)}\right|^2 + \left|\frac{\partial(x_2, x_3)}{\partial(\alpha^1, \alpha^2)}\right|^2 \geq \alpha > 0, \qquad (1.2)$$

where α is some positive constant and where by $\partial(x_1, x_2)/\partial(\alpha^1, \alpha^2)$, etc., we denote the Jacobian matrix of the corresponding planar mapping. Condition (1.2) by itself guarantees only that S and Ω are locally homeomorphic. We shall say that a surface $S \in C^2_{\Omega}$, for which the relation (1.2) holds is a *regular surface*. Below we present some results from the theory of surfaces; more details can be found in any of the standard sources [410, 71, 79, 128, 218, 220, 243, 254].

In the sequel we shall denote by the subscript α_k the corresponding partial derivative. The vectors $\mathbf{e}_k = \rho_{\alpha^k}$, $k = 1, 2$, tangent to the coordinate lines of the surface at a point, we shall call the *tangent basis*. It is defined at each point of S. At any point $\mathcal{A}$ of the surface (see Figure 1.1) we can define the first and the second quadratic forms with the coefficients

$$A_{ij} = \rho_{\alpha^i} \cdot \rho_{\alpha^j} \quad B_{ij} = [\rho_{\alpha^i \alpha^j} \cdot \rho_{\alpha^i} \cdot \rho_{\alpha^j}] D^{-1}; \quad i, j = 1, 2. \qquad (1.3)$$

It is easy to prove the equality

$$D^2 = A_{11}A_{22} - A_{12}^2 = |\mathbf{e}_1 \times \mathbf{e}_2|^2. \qquad (1.4)$$

The arc-length element ds defined by the "vector" $(d\alpha^1, d\alpha^2)$ on the surface at a point is defined by the relation

$$ds^2 = A_{ij} d\alpha^i \, d\alpha^j,$$

where, as everywhere below, we are summing over repeated indices. Let us note in addition that if an expression involves the lower index α^k, that denoting differentiation with respect to α^k, then the summation over k is performed as if instead of α^k we had the lower index k.

Let us also introduce the unit normal to S at a point $\mathbf{n}$:

$$\mathbf{n} = \mathbf{e}_3 = D^{-1}(\mathbf{e}_1 \times \mathbf{e}_2).$$

Let us assume that the vectors $\mathbf{e}_1$, $\mathbf{e}_2$, $\mathbf{n}$ form a right trihedron. We set $\mathbf{e}^3 = \mathbf{n}$.

Let us now introduce another basis for the three-dimensional Euclidean space, the so-called reciprocal basis $\mathbf{e}^k$, which satisfies

$$\mathbf{e}^k \cdot \mathbf{e}_l = \delta^k_l,$$

where δ^k_l is the Kronecker delta, which equals one if the indices are the same and zero otherwise; the dot denotes the usual scalar product. It can be shown [71, 79, 128, 218] that such a basis exists. It is easily seen that $\mathbf{e}^3 = \mathbf{n}$. The angle ψ between $\mathbf{e}_1$ and $\mathbf{e}_2$ is given by the relation

$$\sin\psi = D(A_{11}A_{22})^{-\frac{1}{2}}.$$

The vectors of the original basis can be expressed in terms of the reciprocal basis:

$$\mathbf{e}^k = a^{kl}\mathbf{e}_l.$$

In courses in differential geometry [71, 79, 128, 218] it is shown that the matrices A_{kl} and A^{kl} are inverses of each other. We say that the A_{kl} are covariant components of the metric tensor, which is a tensor of rank two, while the A^{kl} are its contravariant components. They satisfy the relations

$$A^{11} = A_{22}D^{-2}, \quad A^{12} = A^{21} = -A_{12}D^{-2}, \quad A^{22} = A_{11}D^{-2}.$$

Let us remind the reader of the definition of a tensor of rank two. An ordered pair of vectors $\mathbf{xy}$ is called a tensor product of these vectors (or a dyad, for short). Let us note that in the general case $\mathbf{xy} \neq \mathbf{yx}$. Let us introduce the operation of multiplying a dyad by a scalar number a, denoted by $a\mathbf{xy}$. The product of a dyad and a number has properties similar to those of a usual product, namely

$$a\mathbf{xy} = (a\mathbf{x})\mathbf{y} = \mathbf{x}(a\mathbf{y}).$$

Introducing now the operation of adding two dyads, $a_1\mathbf{x}_1\mathbf{y}_1 + a_2\mathbf{x}_2\mathbf{y}_2$, with the properties

$$\begin{aligned} \mathbf{x}(a_1\mathbf{y}_1 + a_2\mathbf{y}_2) &= a_1\mathbf{xy}_1 + a_2\mathbf{xy}_2, \\ (a_1\mathbf{x}_1 + a_2\mathbf{x}_2)\mathbf{y} &= a_1\mathbf{x}_1\mathbf{y} + a_2\mathbf{x}_2\mathbf{y}, \end{aligned} \tag{1.5}$$

we obtain a linear space. If this construction is based on the Euclidean spaces E^2 or E^3, the resulting spaces are, respectively, four- and nine-dimensional. An element of these spaces is called a tensor of rank two. Let $\mathbf{e}^k$, $k = 1, 2, \dots, n$, be a basis of the underlying space. Then the set of all dyads $\mathbf{e}^k\mathbf{e}^l$ is a basis for the corresponding space of tensors. Any tensor can be uniquely represented as an algebraic sum of tensor dyads. the coefficients of this representation, as in the case of vectors, are called the coordinates of the tensor in the corresponding basis.

Using the formulae (1.5), we can, as in the underlying Euclidean space, perform a change of basis. Under such a change, the coordinates of a tensor defined in any basis uniquely determine its coordinates in any other basis. Let $g^{ij} = \mathbf{e}^i \cdot \mathbf{e}^j$ and $g_{ij} = \mathbf{e}_i \cdot \mathbf{e}_j$, where $\mathbf{e}_i$ is a vector of the reciprocal basis. A particular case of this relation is given in (1.3). This set of coefficients defines an important tensor, called the *metric tensor* given by $g_{kl}\mathbf{e}^k\mathbf{e}^l$. It can be verified that

$$g_{kl}\mathbf{e}^k\mathbf{e}^l = g^{kl}\mathbf{e}_k\mathbf{e}_l.$$

This means that the quantities A_{kl} and A^{kl}, called above the covariant and the contravariant components of the metric tensor, are indeed the components of the same tensor in two different bases, the principal and the reciprocal ones. It turns out that the following relations hold as well:

$$g_{kl}\mathbf{e}^k\mathbf{e}^l = g_l^k\mathbf{e}_k\mathbf{e}^l = g_l^k\mathbf{e}^l\mathbf{e}_k, \quad g_i^k = \delta_i^k.$$

In this way we have introduced the so-called mixed components of the metric tensor. The coefficients of a tensor in a basis can be written in the form of an

$n \times n$ matrix. The corresponding matrix composed of the mixed components of the metric tensor is the identity matrix, which is why it is said that the metric tensor corresponds to the identity matrix.

We have already introduced above the coefficients of the second quadratic form of the surface B_{ij}. This collection of coefficients B_{kl} can also be considered as the covariant components of some tensor of rank two. Since $B_{kl} = B_{lk}$, such a tensor is said to be symmetric. Its contravariant and mixed components are given by

$$B^{ij} = A^{it}A^{jk}B_{kt}, \quad B_i^j = A^{jk}B_{ki}.$$

We see that the connections among the components of different types of the tensor corresponding to the second quadratic form are defined using the coefficients of the metric tensor. This type of relations among components of different types holds also for any symmetric tensor of rank two, $c_{kl}\mathbf{e}^k\mathbf{e}^l$, when we start the construction with the n-dimensional Euclidean space; namely,

$$c^{ij} = g^{ik}g^{jl}c_{kl}, \; c_{ij} = g_{ik}g_{jl}c^{kl}, \; c_j^i = g^{ik}c_{kj} = g_{jl}c^{il}.$$

The mixed components of a symmetric tensor do not depend on the order in which the upper and the lower indices are written.

We can introduce the operation of convolution for dyads and vectors: $(\mathbf{ab})\cdot\mathbf{c} = (\mathbf{b}\cdot\mathbf{c})\mathbf{a}$, $\mathbf{c}\cdot(\mathbf{ab}) = (\mathbf{c}\cdot\mathbf{a})\mathbf{b}$, $(\mathbf{ab})\cdot(\mathbf{cd}) = (\mathbf{b}\cdot\mathbf{c})\mathbf{ad}$, where by $(\mathbf{b}\cdot\mathbf{c})$ we denote the usual scalar product of the vectors $\mathbf{b}$ and $\mathbf{c}$. Convolution on dyads and vectors indexes the corresponding operation on arbitrary tensors and vectors. The convolution operation is constructed in such a way that in an orthonormal basis the matrix of convolution of two tensors is obtained by multiplying the corresponding matrices of components of these tensors, while the matrix of convolution of a tensor and a vector is obtained as the usual product of the matrix of the tensor and the column vector of the components of the vector. Further information on tensor calculus can be found in [71, 79, 128, 218].

Let us now state some properties of surfaces of class C^2_Ω.

Lemma 1.1. *There are positive constants m and M such that in $\overline{\Omega}$ we have the inequalities*

$$m \le |A_{ii}| \le M, \; i = 1, 2, \tag{1.6}$$
$$|A_{12}| \le M, \tag{1.7}$$
$$D \le M. \tag{1.8}$$

The right inequality of (1.6) and the inequality (1.7) follow from the fact that $S \in C^2_\Omega$, while the left inequality of (1.6) follows from the second one and the relation

$$A_{11}A_{22} > A_{12}^2 + \alpha^2,$$

which comes from (1.2) and (1.4). Inequality (1.8) follows from (1.6).

We remark that in the following we shall frequently introduce various constants; in cases where it is the existence of these constants that will be important, and not their numerical values, these will always be denoted by m and M, since only a finite number of such constants are introduced in the book.

Lemma 1.2. *Let $\mathcal{A}$, $\mathcal{B}$ be arbitrary points of S, with a, b their images in Ω. Furthermore, let L be an arbitrary curve in S connecting $\mathcal{A}$ and $\mathcal{B}$, and let d be its image in Ω. Then*

$$m \leq \frac{\mathcal{D}_L}{\mathcal{D}_d} \leq M, \tag{1.9}$$

where $\mathcal{D}_L$, $\mathcal{D}_d$ are the lengths of the corresponding curves.

Indeed, let ds and $d\sigma$ be, respectively, arc-length elements of L and d:

$$\left(\frac{ds}{d\sigma}\right)^2 = A_{ij}\frac{d\alpha^i}{d\sigma}\frac{d\alpha^j}{d\sigma}.$$

Since by (1.2) and (1.4) A_{ij} is a positive definite form with respect to $d\alpha^1/d\sigma$ and

$$\left(\frac{d\alpha^1}{d\sigma}\right)^2 + \left(\frac{d\alpha^2}{d\sigma}\right)^2 = 1,$$

we have

$$m \leq \frac{ds}{d\sigma} \leq M, \tag{1.10}$$

and (1.9) follows from (1.10).

Lemma 1.3. *Let $\alpha_i^1(\sigma)$ and $\alpha_i^2(\sigma)$ be two curves in $\overline{\Omega}$ that originate at the same point. Let $\rho_1(s)$, $\rho_2(s)$ be their images on $\overline{S}$. Then if the angle between these curves in $\overline{\Omega}$ is different from 0 and π, then the angle between ρ_1 and ρ_2 is also different from 0 and π, and vice versa.*

Lemma 1.3 follows from Lemma 1.2.

Lemma 1.4. *Two coordinate lines of one family on S do not touch or intersect.*

Lemma 1.4 follows from the homeomorphism of $\overline{S}$ and $\overline{\Omega}$.

Furthermore, let the surface S be bounded by a closed curve Γ. We shall say that Γ is of class C^1 if at each point $\mathcal{A}$ of Γ there is a neighborhood such that in that neighborhood the equation of the curve can be represented in the form

$$\rho_\Gamma = \rho_\Gamma(s),$$

where ρ_Γ is a continuously differentiable vector and the entire curve Γ is covered by a finite number of such neighborhoods. Now let γ be the closed curve in the plane α^1, α^2 corresponding to the curve Γ. Then, obviously, $\Gamma \in C_\Gamma^1$ if and only if $\gamma \in C_\gamma^1$. Let us assume now that Γ consists of a finite number of arcs $\Gamma_i \in C_{\Gamma_i}^1$. In this case we shall say that Γ is a *piecewise smooth curve* (PSC) of class C_Γ^1. Clearly, if Γ is a C_Γ^1 PSC, then γ is a C_γ^1 PSC.

Lemma 1.5. *Let Γ be a C^1_Γ PSC. Then angles of Γ are images of angles of γ. If Γ has no entrance angles equal to 0 and exit angles equal to π, then the same is true for γ.*

Lemma 1.5 follows from Lemma 1.3.

Lemma 1.6. *Coordinate lines on S have no self-tangencies or self-intersections.*

Lemma 1.7. *Coordinate lines from different families on S intersect in at most at one point.*

Lemmas 1.6 and 1.7 follow from the homeomorphism of $\overline{S}$ and $\overline{\Omega}$.

1.2. In a number of cases the surface S is such that it is impossible for $\overline{S}$ to be homeomorphic to a bounded connected domain $\overline{\Omega}$ in R^2. For example, a practically important case is of S a closed surface homeomorphic to a sphere, or of S a piece of a cylindrical surface homeomorphic to a circular cylinder (see Figure 1.2). In this case we resort to surgery [3] (Figure 1.3). Let us cut S by a closed curve $\mathcal{L}$ into two parts: $S = S_1 + S_2$. Then S_1 will be homeomorphic to a domain Ω_1 with boundary γ^1, while S_2 will be homeomorphic to Ω_2 with boundary γ^2. This procedure automatically generates a homeomorphism between the curves γ^i: We say that a point on γ^1 maps into a point on γ^2 if both correspond to the same point on $\mathcal{L}$. In each of the domains Ω_1, Ω_2 we can introduce its own parametrization α_1^i, α_2^i with the necessary compatibility condition

$$\rho(\alpha_1^i)\,|_{\mathcal{L}} = \rho(\alpha_2^i)\,|_{\mathcal{L}},$$

and furthermore, at each point of $\mathcal{L}$ the surfaces S_1 and S_2 must have the same tangent planes, equal values of the principal curvature, and the same directions of lines of curvature. If we now introduce smooth functions or vector fields on S, we must also satisfy compatibility conditions on $\mathcal{L}$, which include equality of the elements themselves and of some of their derivatives (depending on the smoothness of the field).

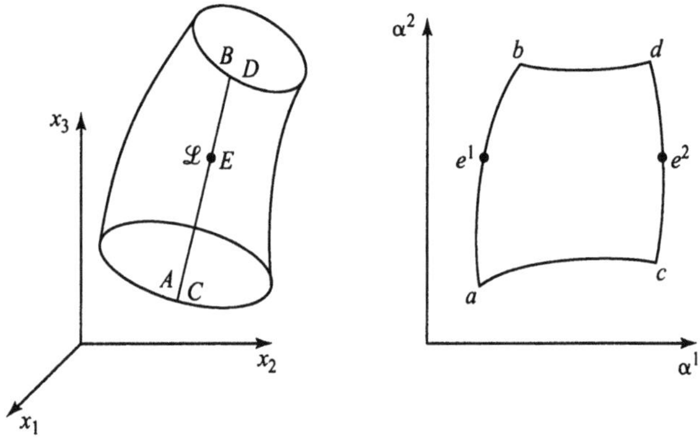

FIGURE 1.2.

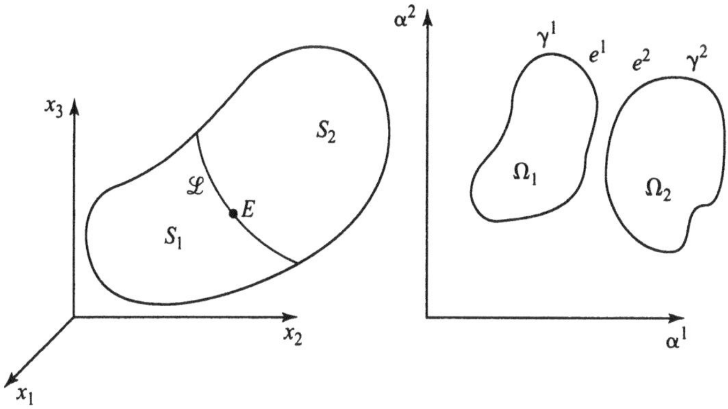

FIGURE 1.3.

If the surface S is homeomorphic to a piece of a cylinder (Figure 1.2), we have that the cut $\mathcal{L}$ has to be along the generating curve, and then the resulting cut cylinder is homeomorphic to the rectangle $abdc$ in which we have identified the corresponding points of ab and cd. The compatibility conditions of the parametrization $\rho(\alpha^i)$ and of the corresponding scalar and vector fields must then be satisfied at corresponding points. If the surface S is homeomorphic to a torus (Figure 1.4), we find a homeomorphism with a planar domain Ω as follows: We cut the torus along a generating curve $\mathcal{L}$ and unfold the obtained surface into an annulus in the plane α^1, α^2. Turning this surface into an annulus can clearly be done in such a way that the corresponding points have the same angular coordinate (Figure 1.4).

Finally, let us explain how to construct the parametrization α^1, α^2 and the corresponding homeomorphism for a surface Σ homeomorphic to a sphere with two handles (Figure 1.5). Let us cut the surface with a plane that passes through the two handles and let us unfold the resulting surfaces S_1, S_2 in the plane α^1, α^2. The resulting domains Ω_1, Ω_2 will be triply connected. In this case we have to identify the corresponding points of the closed curves γ^1 and γ^2, γ^3 and γ^4, γ^5 and γ^6, and choose the parametrization of $\rho(\alpha^i)$ in such a way that compatibility conditions hold at corresponding points of the boundary curves. Any function and vector

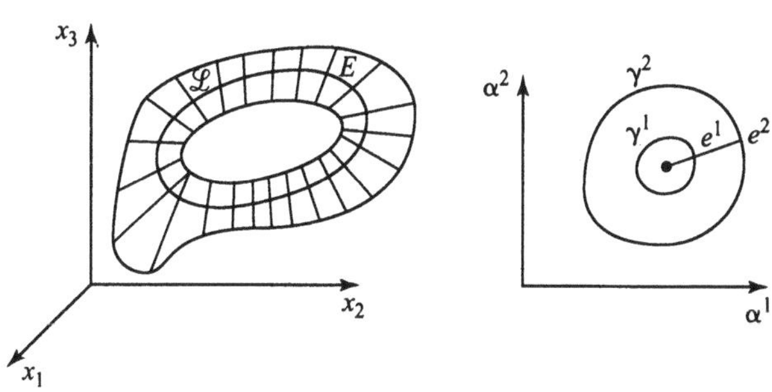

FIGURE 1.4.

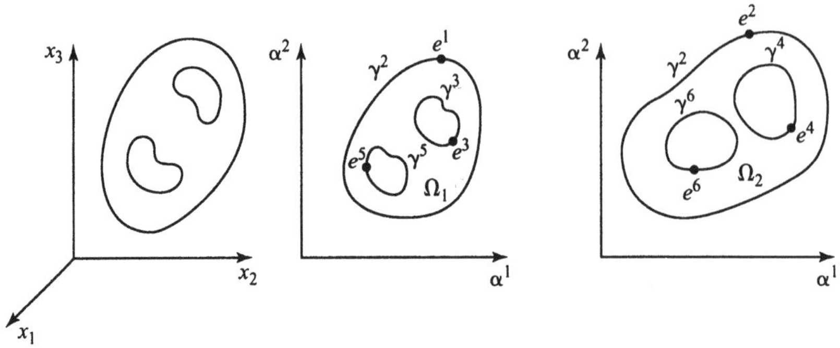

FIGURE 1.5.

fields we introduce on S_1 and S_2 must also satisfy the appropriate compatibility conditions.

In view of the above, it is completely clear that for any real shell we can define a homeomorphism of S and a multiply connected domain Ω if we identify points of components of the boundary γ. An important class of problems consists of the cases in which Ω is a multiply connected domain without identification of any points of γ. Let us also note that practical methods of surface parametrization are of importance. Moreover, it is important to find a parametrization that would correspond best to the structure of the surface S and would lead to simple analytical and computational algorithms in the subsequent analysis of the problems [95, 226].

1.3. Let a curve $\mathcal{L}$ be defined on a surface S parametrically,

$$\alpha^i = \alpha^i(t), \quad \rho = \rho(\alpha^i(t)), \quad 0 \le t \le T.$$

We shall assume that $\mathcal{L}$ is a PSC of class $C^1_{\mathcal{L}}$. Clearly, if s is the arc length on $\mathcal{L}$ and σ is the arc length of its image in the coordinate plane α^1, α^2, then

$$\frac{ds}{dt} = \sqrt{A_{ij}(t)\frac{d\alpha^i}{dt}\frac{d\alpha^j}{dt}},$$

$$\frac{d\sigma}{dt} = \sqrt{\left(\frac{d\alpha^1}{dt}\right)^2 + \left(\frac{d\alpha^2}{dt}\right)^2},$$

where we take s, σ to be increasing in t.

Next, let $\boldsymbol{\tau}$ be the unit tangent vector to $\mathcal{M}$, and let $\mathbf{m}$ be the unit normal vector to $\mathcal{M}$ contained in S. Let $\mathbf{n} = \mathbf{e}_3 = \mathbf{e}^3 = D^{-1}(\mathbf{e}_1 \times \mathbf{e}_2)$ be the unit normal vector to S (see Figure 1.6). We have

$$\boldsymbol{\tau} = \frac{d\rho}{ds} = \tau^k \mathbf{e}_k = \tau_k \mathbf{e}^k, \quad \tau^k = \frac{d\alpha^k}{ds} = \frac{d\alpha^k}{dt}\left(A_{ij}(t)\frac{d\alpha^i}{dt}\frac{d\alpha^j}{dt}\right)^{-\frac{1}{2}},$$

$$\tau_k = A_{kp}\tau^p = A_{kp}\frac{d\alpha^p}{dt}\left(A_{ij}(t)\frac{d\alpha^i}{dt}\frac{d\alpha^j}{dt}\right)^{-\frac{1}{2}}.$$

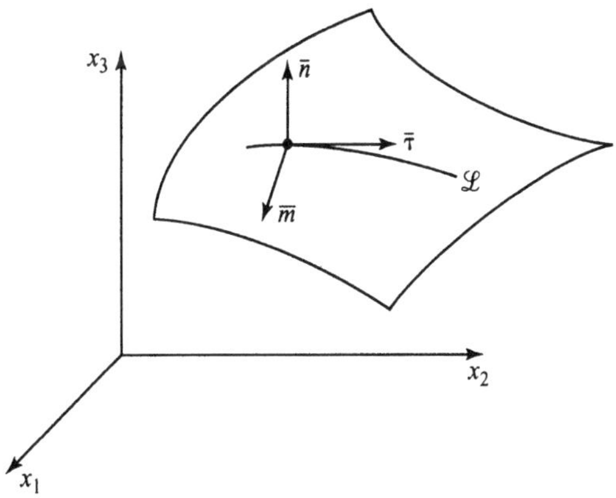

FIGURE 1.6.

To determine the components of the vector $\mathbf{m}$, we note that

$$\mathbf{m} = m^k \mathbf{e}_k, \quad m^k = \frac{d\alpha^k}{dm},$$

and furthermore,

$$\mathbf{m}^2 = 1, \quad \mathbf{m} \cdot \boldsymbol{\tau} = 0,$$

so that

$$A^{ij} m_i m_j = 1, \quad m_k \tau^k = 0,$$

whence

$$m_1 = -D\frac{d\alpha^2}{ds} = -D\tau^2, \quad m_2 = D\frac{d\alpha^1}{ds} = D\tau^1. \tag{1.11}$$

Let us introduce the discriminant tensor of the surface S, defined by the relations

$$C_{12} = -C_{21} = -D, \quad C^{12} = -C^{21} = -D^{-1}, \quad C^{ii} = C_{ii} = 0, \ i = 1, 2. \tag{1.12}$$

Then we obtain

$$m_k = C_{kj}\frac{d\alpha^j}{ds} = C_{kj}\tau^j,$$

and thus

$$m^k = A^{kp} C_{pj}\frac{d\alpha^j}{ds} = C^k_j \frac{d\alpha^j}{ds}, \quad m^1 = \frac{d\alpha^1}{dm} = -\frac{\tau_2}{D}, \quad m^2 = \frac{d\alpha^2}{dm} = \frac{\tau_1}{D}. \tag{1.13}$$

Let us note some useful formulae that follow from (1.13):

$$m^k = C^k_j \frac{d\alpha^j}{ds} = C^k_j \tau^j = C^{kj}\tau_j, \quad \tau^k = C^{kj} m_j,$$

$$\frac{\partial s}{\partial \alpha^k} = \tau_k = A_{kp}\frac{d\alpha^p}{ds}, \quad \frac{\partial m}{\partial \alpha^k} = m_k = C_{kj}\frac{d\alpha^j}{ds}.$$

If on a component the curve Γ and its image γ in Ω are given by the equation

$$\Phi(\alpha^1, \alpha^2) = 0,$$

then we have

$$\frac{d\alpha^1}{ds} = \frac{d\alpha^i}{\sqrt{A_{ij}d\alpha^i d\alpha^j}} = \pm\frac{1}{\sqrt{A_{11} + 2A_{12}\frac{d\alpha^2}{d\alpha^1} + A_{22}\left(\frac{d\alpha^2}{d\alpha^1}\right)^2}}$$
$$= \pm\frac{\Phi_{\alpha^2}}{\sqrt{A_{11}\Phi^2_{\alpha^2} - 2A_{12}\Phi_{\alpha^1}\Phi_{\alpha^2} + A_{22}\Phi^2_{\alpha^1}}}, \quad (1.14)$$

$$\frac{d\alpha^2}{ds} = \frac{d\alpha^2}{\sqrt{A_{ij}d\alpha^i d\alpha^j}} = \pm\frac{1}{\sqrt{A_{11} + 2A_{12}\frac{d\alpha^2}{d\alpha^1} + A_{22}\left(\frac{d\alpha^2}{d\alpha^1}\right)^2}}$$
$$= \pm\frac{\Phi_{\alpha^1}}{\sqrt{A_{11}\Phi^2_{\alpha^2} - 2A_{12}\Phi_{\alpha^1}\Phi_{\alpha^2} + A_{22}\Phi^2_{\alpha^1}}}. \quad (1.15)$$

The $\pm$ signs in (1.14), (1.15) must be the same as the signs of $d\alpha^k/ds$ and Φ_{α^k}.

It is well known that the vectors $\boldsymbol{\tau}, \mathbf{m}, \mathbf{n}$ satisfy on $\mathcal{L}$ the Frenet–Serret differential equations [34, 79, 128, 254],

$$\frac{d\boldsymbol{\tau}}{ds} = \kappa_1\mathbf{n} - \kappa_2\mathbf{m}, \quad \frac{d\mathbf{m}}{ds} = \kappa_2\boldsymbol{\tau} - \kappa_3\mathbf{n}, \quad \frac{d\mathbf{n}}{ds} = \kappa_3\mathbf{m} - \kappa_1\boldsymbol{\tau},$$

where κ_1 is the normal sectional curvature of S in the direction of $\boldsymbol{\tau}$ (the curvature of the geodesic), κ_2 is the geodesic curvature of $\mathcal{L}$ on S, κ_3 is the geodesic torsion determined by the relations [71, 79, 254]

$$\kappa_1 = B_{ij}\frac{d\alpha^i}{ds}\frac{d\alpha^j}{ds}, \quad \kappa_2 = -C_{ij}\frac{d\alpha^j}{ds}\left(\frac{d^2\alpha^i}{ds^2} + G^i_{\beta\gamma}\frac{d\alpha^\beta}{ds}\frac{d\alpha^\gamma}{ds}\right),$$
$$\kappa_3 = -C_{ij}B^i_\mu\frac{d\alpha^\mu}{ds}\frac{d\alpha^j}{ds}; \quad B^i_\mu = A^{it}B_{\mu t}. \quad (1.16)$$

$G^i_{\beta\gamma}$ will be defined by (1.21).

Let us quote the relations that connect the derivatives u_{α^i}, u_s, u_m of a function u defined on $\mathcal{L}$. We have

$$\frac{\partial u}{\partial s} = u_{\alpha^k}\frac{d\alpha^k}{ds} = u_{\alpha^k}\tau^k = u_{\alpha^k}C^{kj}m_j = D^{-1}\left(u_{\alpha^1}m_2 - u_{\alpha^2}m_1\right), \quad (1.17)$$
$$\frac{\partial u}{\partial m} = u_{\alpha^k}\frac{d\alpha^k}{dm} = u_{\alpha^k}m^k = u_{\alpha^k}C^{kj}\tau_j = D^{-1}\left(-u_{\alpha^1}\tau_2 + u_{\alpha^2}\tau_1\right), \quad (1.18)$$
$$u_{\alpha^1} = \frac{\partial u}{\partial s}\tau_1 + \frac{\partial u}{\partial m}m_1, \quad u_{\alpha^2} = \frac{\partial u}{\partial s}\tau_2 + \frac{\partial u}{\partial m}m_2. \quad (1.19)$$

To conclude, we note that all the formulae of this section apply verbatim in the case when $\mathcal{L}$ is the boundary curve Γ.

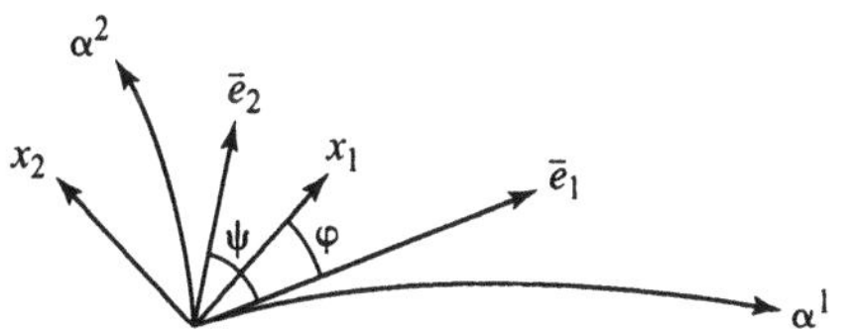

FIGURE 1.7.

1.4. Let us assume that at point $\mathcal{A}$ of the surface S there is an orthogonal system of coordinates $\mathbf{x}_1, \mathbf{x}_2$ oriented in such a way that the directions $\mathbf{e}_1, \mathbf{x}_1$ form an angle φ (see Figure 1.7). Elementary computations show that

$$\frac{d\alpha^1}{dx_1} = \frac{\sin(\psi-\varphi)}{\sqrt{A_{11}}\sin\psi}, \quad \frac{d\alpha^1}{dx_2} = -\frac{\cos(\psi-\varphi)}{\sqrt{A_{11}}\sin\psi},$$
$$\frac{d\alpha^2}{dx_1} = \frac{\sin\varphi}{\sqrt{A_{22}}\sin\psi}, \quad \frac{d\alpha^2}{dx_2} = \frac{\cos\varphi}{\sqrt{A_{22}}\sin\psi};$$
$$\frac{\partial\alpha^k}{\partial x_1} = C^{pk}\frac{\partial x_2}{\partial\alpha^p}, \quad \frac{\partial\alpha^k}{\partial x_2} = -C^{pk}\frac{\partial x_1}{\partial\alpha^p},$$
$$\frac{\partial x_1}{\partial\alpha^l} = C_{kl}\frac{\partial\alpha^k}{\partial x_2}, \quad \frac{\partial x_2}{\partial\alpha^l} = C_{kl}\frac{\partial\alpha^k}{\partial x_1}.$$

Let u be a differentiable function of α^i defined on the surface S. Then

$$u_{x_1} = u_{\alpha^1}\frac{\sin(\psi-\varphi)}{\sqrt{A_{11}}\sin\psi} + u_{\alpha^2}\frac{\sin\varphi}{\sqrt{A_{22}}\sin\varphi},$$
$$u_{x_2} = -u_{\alpha^1}\frac{\cos(\psi-\varphi)}{\sqrt{A_{11}}\sin\psi} + u_{\alpha^2}\frac{\cos\varphi}{\sqrt{A_{22}}\sin\psi},$$
$$\frac{1}{\sqrt{A_{11}}}u_{\alpha^1} = u_{x_1}\cos\varphi - u_{x_2}\sin\varphi,$$
$$\frac{1}{\sqrt{A_{22}}}u_{\alpha^2} = u_{x_1}\cos(\psi-\varphi) + u_{x_2}\sin(\psi-\varphi).$$

1.5. Let us consider now the problem of differentiating an arbitrary vector defined on the surface S.

Let $\mathbf{u} = \sum_{p=1}^{3} u^p(\alpha^1, \alpha^2)\mathbf{e}_p$ be an arbitrary differentiable vector on S. The usual rules of differentiating vectors imply that

$$\mathbf{u}_{\alpha^k} = \frac{\partial\mathbf{u}}{\partial\alpha^k} = \sum_{p=1}^{3}\left(\frac{u^p(\alpha^1, \alpha^2)}{\partial\alpha^k}\mathbf{e}_p + u^p(\alpha^1, \alpha^2)\frac{\partial\mathbf{e}_p}{\partial\alpha^k}\right).$$

Thus, differentiation formulae involve derivatives of basis vectors. Since the $\mathbf{e}^k$ form a basis of the space E^3, these derivatives can be expanded in this basis. The corresponding coefficients of the expansion, denoted by G^i_{jk}, are called Christoffel symbols of the second kind. In [71, 79, 128, 218, 243, 254] the reader can find the following differentiation formulae:

$$\mathbf{e}_{k\alpha^p} = G^q_{kp}\mathbf{e}_q + B_{kp}\mathbf{n}, \quad \mathbf{e}^k_{\alpha^p} = -G^k_{pq}\mathbf{e}^q + B^k_p\mathbf{n},$$
$$\mathbf{e}_{3\alpha^p} = \mathbf{e}^3_{\alpha^p} = -B^q_p\mathbf{e}_q = B_{pq}\mathbf{e}^q; \; k,\ p,\ q = 1,\ 2. \tag{1.20}$$

In (1.20) G^q_{kp} are the Christoffel symbols of the second kind, defined by the relations

$$G^1_{11} = \frac{A_{11\alpha^1}A_{22} + A_{11\alpha^2}A_{12} - 2A_{12}A_{12\alpha^1}}{2D^2},$$
$$G^1_{22} = \frac{A_{22}(2A_{12\alpha^2} - A_{12}A_{22\alpha^2}) - A_{12}A_{22\alpha^2}}{2D^2} \tag{1.21}$$
$$G^1_{12} = G^1_{21} = \frac{A_{22}A_{11\alpha^2} - A_{12}A_{22\alpha^1}}{2D^2}, \quad 1 \rightleftharpoons 2.$$

Next, let $\mathbf{u}$ be an arbitrary differentiable vector on S. We denote the coordinates of the vector $\mathbf{u}$ in the basis $(\mathbf{e}^1,\ \mathbf{e}^2,\ \mathbf{n})$ by $u_1,\ u_2,\ u_3$ (or by $u_1\ u_2,\ u$). In the basis $(\mathbf{e}_1,\ \mathbf{e}_2,\ \mathbf{n})$ they will be denoted by $u^1,\ u^2,\ u^3$ (or u^1, u^2, u, since $u^3 = u_3 = u$). The dependence of a vector on coordinates will be denoted by $\mathbf{u}(u_1,\ u_2,\ u)$. Sometimes, to denote the same fact we shall use an alternative (not completely correct) notation $\mathbf{u} = (u_1,\ u_2,\ u)$. Frequently, we construct using the components of a vector $\mathbf{u}$ a new vector $\omega = u_1\mathbf{e}^1 + u_2\mathbf{e}^2$; in such cases we shall use the notation $\mathbf{u}(\omega,\ u)$. All the above methods of notation will be considered interchangeable. Then

$$\mathbf{u}_{\alpha^k} = \nabla_k u^p\mathbf{e}_p = \nabla_k u_p\mathbf{e}^p,$$
$$\nabla_k u^p = u^p_{\alpha^k} + G^p_{ks}u^s - B^p_k u^3;$$
$$\nabla_k u_p = u_{p\alpha^k} + G^s_{pk}u_s - B_{kp}u^3,$$
$$\nabla_k u_3 = \nabla_k u^3 = u^3_{\alpha^k} + B_{ks}u^s = u^3_{\alpha^k} + B^s_k u_s; \; k,\ p,\ s = 1,\ 2. \tag{1.22}$$

Thus the formulae of the usual coordinate-wise differentiation of vectors in a Cartesian basis are replaced by more complicated ones, obtained by assuming that the basis vectors now depend on the coordinates α^i. The corresponding formulae (1.22) are called *covariant differentiation formulae.*

We shall need the formulae (1.21) in some specific coordinate systems. First let us consider orthogonal coordinates: $A_{12} = 0$,

$$G^1_{11} = \frac{A_{11\alpha^1}}{2A_{11}}, \quad G^1_{12} = G^1_{21} = \frac{A_{11\alpha^2}}{2A_{11}}, \quad G^1_{22} = \frac{A_{22\alpha^1}}{2A_{11}},$$
$$G^2_{11} = -\frac{A_{11\alpha^2}}{2A_{22}}, \quad G^2_{12} = G^2_{21} = \frac{A_{22\alpha^1}}{2A_{22}}, \quad G^2_{22} = \frac{A_{22\alpha^2}}{2A_{22}}.$$

Next, we assume that α^i is an isothermal (conformal) parametrization:

$$A_{11} = A_{22} = \Lambda(\alpha^1,\ \alpha^2), \; A_{12} = 0.$$

Then we have

$$G^1_{11} = \frac{\Lambda_{\alpha^1}}{2\Lambda}, \quad G^1_{12} = G^1_{21} = \frac{\Lambda_{\alpha^2}}{2\Lambda}, \quad G^1_{22} = \frac{\Lambda_{\alpha^1}}{2\Lambda},$$

$$G^2_{11} = -\frac{\Lambda_{\alpha^2}}{2\Lambda}, \; G^2_{12} = G^2_{21} = \frac{\Lambda_{\alpha^1}}{2\Lambda}, \; G^2_{22} = \frac{\Lambda_{\alpha^2}}{2\Lambda}. \tag{1.23}$$

Conditions for the existence of an isothermal parametrization are given by the following theorem.

Theorem 1.1 **([325]).** *Let a parametrization α^k of a surface S be such that $\rho_{\alpha^i\alpha^j}$ satisfy in Ω Hölder continuity conditions with an exponent $0 < \lambda \leq 1$. Then we can parametrize S by a global isothermal parametrization*

$$ds^2 = \Lambda[(d\alpha^1)^2 + (d\alpha^2)^2],$$

where Λ satisfies a Hölder condition with exponent λ^1 such that $\lambda^1 = \lambda$ if $0 < \lambda < 1$ and λ^1 can be taken to be arbitrarily close to (but less than) 1 if $\lambda = 1$.

Theorem 1.2 **([325]).** *Assume that the conditions of Theorem 1.1 hold and that furthermore,*

$$B_{11}B_{22} - B^2_{12} = 0,$$

that is, S is a surface with zero Gaussian curvature. Then there is a global Euclidean parametrization of S, α^k, for which

$$ds^2 = (d\alpha^1)^2 + (d\alpha^2)^2, \; A_{11} = A_{22} = 1, \; A_{12} = 0.$$

1.6. Below we shall be using formulae for second derivatives on S. We have [96]

$$\tau^k\tau^l\nabla_{kl}\Psi = \frac{\partial^2\Psi}{\partial s^2} + \kappa_2\frac{\partial\Psi}{\partial m}, \tag{1.24}$$

$$\tau^k m^l\nabla_{kl}\Psi = \frac{\partial}{\partial s}\frac{\partial\Psi}{\partial m} - \kappa_2\frac{\partial\Psi}{\partial s}. \tag{1.25}$$

In (1.24), (1.25),

$$\nabla_{kl}\Psi = \nabla_{lk}\Psi = \Psi_{\alpha^k\alpha^l} - G^t_{kl}\Psi_{\alpha^t}. \tag{1.26}$$

1.7. Let us also quote the relations due to Gauss and Peterson–Codazzi [34, 71, 79, 128, 218, 220, 254]:

$$\begin{aligned} B_{11}B_{22} - B^2_{12} &= A_{12\alpha^1\alpha^2} - \frac{1}{2}(A_{11\alpha^2\alpha^2} + A_{22\alpha^1\alpha^1}) + \\ &\quad + A_{\gamma\delta}G^\gamma_{12}G^\delta_{12} - A_{\alpha\beta}G^\alpha_{11}G^\beta_{22}, \end{aligned} \tag{1.27}$$

$$B_{i1\alpha^2} - B_{i2\alpha^1} = G^k_{i2}B_{k1} - G^k_{i1}B_{k2}.$$

We shall also need some less familiar relations for the derivatives G^k_{ij}. We have

$$\rho_{\alpha^i\alpha^j} = G^k_{ij}\mathbf{e}_k + B_{ij}\mathbf{n}, \tag{1.28}$$

whence

$$G^p_{i1} = \mathbf{e}_{i\alpha^1}\cdot\mathbf{e}^p = -\mathbf{e}_i\cdot\mathbf{e}^p_{\alpha^1}, \tag{1.29}$$

$$G^p_{i2} = \mathbf{e}_{i\alpha^2}\cdot\mathbf{e}^p = -\mathbf{e}_i\cdot\mathbf{e}^p_{\alpha^2}. \tag{1.30}$$

From (1.29), (1.31) it follows that

$$G^p_{i1\alpha^2} - G^p_{i2\alpha^1} = -\mathbf{e}_{i\alpha^2} \cdot \mathbf{e}^p_{\alpha^1} - \mathbf{e}_i \cdot \mathbf{e}^p_{\alpha^1\alpha^2} + \mathbf{e}_{i\alpha^1} \cdot \mathbf{e}^p_{\alpha^2} + \mathbf{e}_i \cdot \mathbf{e}^p_{\alpha^1\alpha^2}$$
$$= -\mathbf{e}_{i\alpha^2} \cdot \mathbf{e}^p_{\alpha^1} + \mathbf{e}_{i\alpha^1} \cdot \mathbf{e}^p_{\alpha^2} = -(G^k_{i2}\mathbf{e}_k + B_{i2}\mathbf{n}) \cdot \mathbf{e}^p_{\alpha^1}$$
$$+(G^k_{i1}\mathbf{e}_k + B_{i1}\mathbf{n}) \cdot \mathbf{e}^p_{\alpha^2}. \tag{1.31}$$

From (1.30), (1.31) we have [34, pp. 77–80]

$$G^p_{i1\alpha^2} - G^p_{i2\alpha^1} = -G^k_{i1}G^p_{2k} + G^k_{i2}G^p_{1k} + B_{i1}B^p_2 - B_{i2}B^p_1. \tag{1.32}$$

The relations (1.32) are analogous to the Voss–Weyl relations [128]

$$D^{-1}_{\alpha^i} + G^k_{ik}D^{-1} = 0, \tag{1.33}$$

which are easily proved by a direct verification using (1.4), (1.21).

1.8. Let $A(\alpha^1, \alpha^2)$, $B(\alpha^1, \alpha^2)$ be two arbitrary differentiable functions on S. We shall be using the obvious relations

$$\int_\Omega AB_{\alpha^1}D\,d\alpha^1 d\alpha^2 = \int_\Gamma ABD\,d\alpha^2 - \int_\Omega (AD)_{\alpha^1}B\,d\alpha^1 d\alpha^2$$
$$= \int_\Gamma ABD\frac{d\alpha^2}{ds}ds - \int_\Omega (AD)_{\alpha^1}B\,d\alpha^1 d\alpha^2, \tag{1.34}$$

$$\int_\Omega AB_{\alpha^2}D\,d\alpha^1 d\alpha^2 = -\int_\Gamma ABD\,d\alpha^1 - \int_\Omega (AD)_{\alpha^2}B\,d\alpha^1 d\alpha^2$$
$$= -\int_\Gamma ABD\frac{d\alpha^1}{ds}ds - \int_\Omega (AD)_{\alpha^2}B\,d\alpha^1 d\alpha^2. \tag{1.35}$$

In (1.34), (1.35) while integrating with respect to s, Γ is traversed in such a way that Ω is always on our left-hand side. The formulae (1.34), (1.35) hold also when Ω is a multiply connected domain.

2. S-Coordinates in Space. Formation of a Shell. Components of Finite Deformation in S-Coordinates and Their Simplification

2.1. The position of an arbitrary point $\mathcal{B}$ in a neighborhood of S will be described by the vector

$$\mathbf{r}(\alpha^1, \alpha^2, \alpha^3) = \rho(\alpha^1, \alpha^2) + \alpha^3\mathbf{n}(\alpha^1, \alpha^2). \tag{2.1}$$

(Here the parentheses reflect the dependence on spatial coordinates.) Thus, to define the coordinates of the point $\mathcal{B}$, we have to drop from it a perpendicular onto S (see Figure 2.1). The coordinates α^i of the point $\mathcal{A}$ of its intersection with S and the signed distance α^3 from $\mathcal{B}$ to $\mathcal{A}$ along the normal will be called be S-coordinates of the point $\mathcal{B}$ [90, 88, 91]. It is clear that if $S \in C^2_\Omega$, then S-coordinates can be defined at least in some neighborhood of S.

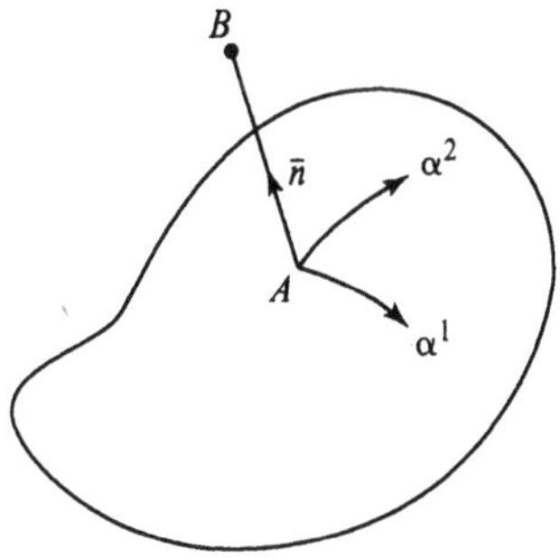

FIGURE 2.1.

The S-parametrization (2.1) generates at each point of the three-dimensional space the tetrahedron

$$\gamma_k = \mathbf{r}_{\alpha^k}, \ k = 1, 2, 3.$$

However, in shell theory it makes more sense to use a different basis, obtained by parallel transport to the point $\mathcal{B}$ along $\mathbf{n}$ of the tetrahedron $\mathbf{e}_1, \mathbf{e}_2, \mathbf{n} = \mathbf{e}_3$. Below we shall call this tetrahedron the $\mathbf{e}$-tetrahedron. Since $\mathbf{n}$ depends on α^i, in general $\gamma_1 \neq \mathbf{e}_1, \gamma_2 \neq \mathbf{e}_2$. For an arc-length element we have in the $\mathbf{e}$-tetrahedron

$$ds^2 = A_{ji} d\alpha^i d\alpha^j + (d\alpha^3)^2, \ i, \ j = 1, 2.$$

We obtain the formulae for differentiation of the vectors in the $\mathbf{e}$-tetrahedron by supplementing (1.22) by the relations

$$\mathbf{e}_{k\alpha^3} = \mathbf{e}^k_{\alpha^3} = 0; \ k = 1, 2, 3.$$

Then we shall have

$$\mathbf{e}_{k\alpha^p} = \Gamma^s_{kp}\mathbf{e}_s, \ \mathbf{e}^k_{\alpha^p} = G^s_{kp}\mathbf{e}^s, \tag{2.2}$$

where

$$\Gamma^q_{kp} = G^q_{kp}, \ \Gamma^q_{k3} = -B^q_k, \ \Gamma^3_{kp} = B_{kp}, \ \Gamma^3_{k3} = 0, \ q, \ k, \ p = 1, 2.$$

In accordance with (2.2), for an arbitrary vector $\mathbf{u}$ we have

$$\mathbf{u}_{\alpha^k} = \nabla_k u^p \mathbf{e}_p = \nabla_k u_p \mathbf{e}^p,$$

where

$$\nabla_k u^p = u^p_{\alpha^k} + G^p_{kq}u^q - B^p_k u^3; \ \nabla_k u^3 = u^3_{\alpha^k} + B_{kq}u^q,$$
$$\nabla_3 u^t = u^t_{\alpha^3}; \ p, \ k, \ q = 1, 2; \ t = 1, 2, 3; \tag{2.3}$$
$$\nabla_k u_p = u_{p\alpha^k} - G^q_{kp}u_q - B_{kp}u_3; \ \nabla_k u_3 = u_{3\alpha^k} + B^q_k u^q,$$
$$\nabla_3 u_t = u_{t\alpha^3}. \tag{2.4}$$

It must be remembered that the relation $\Gamma^k_{ij} = \Gamma^k_{ji}$ does not always hold in the $\mathbf{e}$-tetrahedron; for example, $\Gamma^1_{13} \neq \Gamma^1_{31}$. This is an indication of the fact that there is no α-parametrization of the three-dimensional space that corresponds to the e-tetrahedron. In other words, we are dealing here with a parametrization of the affine connection with torsion.

2.2. To create a shell, let us mark sufficiently small segments $h^{\pm}(\alpha^1, \alpha^2)$ at a point $\mathcal{A}$ along the normal to S. If now $\mathcal{A}$ runs through the whole of the surface S, the normal with the marked segments $h^{+}(\alpha^1, \alpha^2)$, $h^{-}(\alpha^1, \alpha^2)$ will trace out a volume V. In this way we obtain a geometrical shell. If the volume V is filled by a deformable body, we obtain a physical shell. The boundary surfaces of the shell, defined by the relations

$$\mathbf{r}(\alpha^1, \alpha^2, \alpha^3) = \rho(\alpha^1, \alpha^2) \pm h^{\pm}(\alpha^1, \alpha^2)\mathbf{n}(\alpha^1, \alpha^2),\ \alpha^3 = h^{\pm}(\alpha^1, \alpha^2),$$

will be denoted below by $S^{\pm}$. Finally, we shall denote the lateral surface of the shell V, defined by the relation $\Gamma \times (-h^{-}(\alpha^1, \alpha^2) \le \alpha^3 \le h^{+}(\alpha^1, \alpha^2))$, by S^0 (see Figure 5.1). We shall assume that the volume V is immersed into the domain of validity of S-parametrization. The surface S is called the *middle surface*.

In this section we shall obtain the formulae for the components of the finite deformation tensor in the e-tetrahedron. Here we shall also develop a successive scheme for simplifying the main relations, which takes into account certain characteristic properties of the work of a thin shell. For finite deformation we have [90, 133, 174, 184, 221, 219, 270, 237]

$$\begin{aligned} 2\epsilon_{ij} &= 2e_{ij} + \mathbf{u}_{\alpha^i} \cdot \mathbf{u}_{\alpha^j},\ 2e_{ij} = \mathbf{r}_{\alpha^i} \cdot \mathbf{u}_{\alpha^j} + \mathbf{r}_{\alpha^j} \cdot \mathbf{u}_{\alpha^i} \\ &= \mathbf{e}_i \nabla_j \cdot u_k \mathbf{e}^k + \mathbf{e}_j \nabla_i \cdot u_k \mathbf{e}^k = \nabla_j u_i + \nabla_i u_j,\ i,\ j,\ k = 1,\ 2,\ 3. \end{aligned} \tag{2.5}$$

In (2.5), $\mathbf{u} = (u_1, u_2, u_3)$ is the vector of displacements of a point of the shell V.

Let us consider a fiber $\alpha^k(s)$, $k = 1, 2, 3$ originating from an arbitrary point $\mathcal{B}$. For its relative extension ϵ we have [90, 133, 174, 184, 221, 270]

$$(1 + \epsilon)^2 = 1 + 2\epsilon_{ij} \frac{d\alpha^i}{ds} \frac{\alpha^j}{ds}. \tag{2.6}$$

Below we shall consider the case of small strains, for which we shall introduce the following asssumption

Assumption 1.

$$|\epsilon| \ll 1. \tag{2.7}$$

From (2.6), (2.7) we obtain

$$\epsilon \approx \epsilon_{ij} \frac{d\alpha^i}{ds} \frac{\alpha^j}{ds}. \tag{2.8}$$

Since in (2.8) the derivatives $d\alpha^k/ds$ are related by the relation

$$\sum_{k=1}^{3} \left(\frac{d\alpha^k}{ds} \right)^2 = 1$$

but are otherwise arbitrary, it follows from (2.8) that

$$|\epsilon_{ij}| \ll 1. \tag{2.9}$$

An easy computation shows that from (2.9) it follows that

$$|\delta_{ij}| \ll 1,$$

where δ_{ij} is the distortion of the angles between $\mathbf{e}_i$ and $\mathbf{e}_j$ due to the deformation.

2.4. To simplify the equations further, let us introduce ω_i, which characterizes the rotation angles of $\mathbf{e}_i$ under the deformation of the shell,

$$\omega_i = \frac{\mathbf{e}_i \times (\mathbf{e}_i + \mathbf{u}_{\alpha^i})}{\mathbf{e}_i \cdot (\mathbf{e}_i + \mathbf{u}_{\alpha^i})} = \frac{\mathbf{e}_i \times \mathbf{u}_{\alpha^i}}{A_{ii} + e_{ii}}.$$

(Recall that summation is performed only over indices placed at different positions: upper and lower ones.) It is clear that the vector ω_i is perpendicular to both the initial and the deformed states of $\mathbf{e}_i$, while $|\omega_i|$ is the tangent of the rotation angle of $\mathbf{e}_i$ due to the deformation.

Using the well-known relation [128, 141] of vector calculus

$$(\mathbf{a} \times \mathbf{b}) \cdot (\mathbf{c} \times \mathbf{d}) = (\mathbf{a} \cdot \mathbf{c})(\mathbf{b} \cdot \mathbf{d}) - (\mathbf{b} \cdot \mathbf{c})(\mathbf{a} \cdot \mathbf{d}),$$

we obtain

$$|\omega_i|^2 = \frac{A_{ii}\,|\mathbf{u}_{\alpha^i}|^2 - e_{ii}^2}{(A_{ii} + e_{ii})^2}. \tag{2.10}$$

Furthermore, from (2.5) we have

$$\epsilon_{ii} = e_{ii} + \frac{1}{2}\,|\mathbf{u}_{\alpha^i}|^2, \; 1 + \frac{2\epsilon_{ii}}{A_{ii}} = 1 + \frac{2e_{ii}}{A_{ii}} + \frac{1}{A_{ii}}\,|\mathbf{u}_{\alpha^i}|^2. \tag{2.11}$$

Solving (2.11) for $|\mathbf{u}_{\alpha^i}|^2$ and substituting into (2.10), we obtain

$$1 + \frac{2\epsilon_{ii}}{A_{ii}} = (1 + |\omega_i|^2)\left(1 + \frac{e_{ii}}{A_{ii}}\right)^2. \tag{2.12}$$

See also [190]. The relation (2.12) shows that the condition of smallness of deformation (2.7) does not by itself entail smallness of the rotation angles and of e_{ii}, and that therefore additional hypotheses on $|\omega_i|$ and e_{ii} are required.

The simplest of these reduces to the assumption

$$|\omega_i| \sim \epsilon \ll 1.$$

If we explore this assumption systematically, we obtain linear shell theory. The next class of problems can be characterized by the following relation:

Assumption 2.

$$|\omega_i|^2 \sim \epsilon \ll 1. \tag{2.13}$$

This can be naturally called the *moderate bending assumption* (deformation with moderate rotation angles). This singles out a large class of deformations of three-dimensional continua, which is widely applicable, at least in shell theory.

Under Assumption 2, we obtain from (2.12)

$$1 + \frac{2e_{ii}}{A_{ii}} \approx \sqrt{1 + \frac{2\epsilon_{ii}}{A_{ii}}}, \quad e_{ii} \sim \epsilon \ll 1, \tag{2.14}$$

so that

$$\omega_i \approx \frac{\mathbf{e}_i \times \mathbf{u}_{\alpha^i}}{A_{ii}}. \tag{2.15}$$

Furthermore, let ω_{ik} be the covariant components of the vector ω_i. From (2.15) we have

$$\omega_{11} = 0, \ \omega_{12} = -\nabla_1 u^3 \frac{D}{A_{11}}, \ \omega_{13} = \nabla_1 u^1 \frac{D}{A_{11}}, \tag{2.16}$$

$$\omega_{21} = \nabla_2 u^3 \frac{D}{A_{22}}, \ \omega_{22} = 0, \ \omega_{23} = -\nabla_2 u^1 \frac{D}{A_{22}}, \tag{2.17}$$

$$\omega_{31} = -\nabla_3 u^2 \frac{D}{A_{33}}, \ \omega_{32} = \nabla_3 u^1 \frac{D}{A_{33}}, \ \omega_{33} = 0. \tag{2.18}$$

The relations (2.16)–(2.18) clarify the geometrical meaning of the quantities $\nabla_p u^k$. From these formulae it can be seen that $\nabla_p u^k$ is approximately the value of the projection of the rotation angle of the $\mathbf{e}_p$-axis onto the $\mathbf{e}_k$-axis. This argument indicates the way to further simplifications.

2.5. As it is our intention to apply these considerations to an analysis of deformation of shells, let us observe distinguishing features of the situation. Since the shell is thin, rotation around $\mathbf{e}_1$, $\mathbf{e}_2$ axes, which lie in a plane tangent to the middle surface, will be much larger than the rotation around the $\mathbf{e}_3 = \mathbf{n}$ axis, if, of course, we neglect rigid rotations. In view of this, it makes sense to consider the following assumption concerning the relations among the ω_{ik}:

Assumption 3.

$$\omega_{13}, \ \omega_{23} \sim \omega_{12}^2, \ \omega_{21}^2, \ \omega_{31}^2, \ \omega_{32}^2 \sim \epsilon \ll 1. \tag{2.19}$$

From (2.16)–(2.19) it follows that

$$\nabla_1 u^2, \ \nabla_2 u^1 \sim (\nabla_1 u^3)^2, \ (\nabla_2 u^3)^2; (\nabla_3 u^2)^2, \ (\nabla_3 u^1)^2 \sim \epsilon \ll 1. \tag{2.20}$$

These relations hold throughout the entire volume of the shell.

In order to determine the order of $\nabla_i u^i$, $i = 1, 2$, we take into account the relations (2.14), from which we obtain

$$e_{ii} = \mathbf{e}_i \cdot (\nabla_i u^k \mathbf{e}_k) = A_{ik} \nabla_i u^k, \tag{2.21}$$

whence

$$\nabla_i u^i \sim \epsilon \ll 1, \ i = 1, 2. \tag{2.22}$$

Relations (2.21)–(2.22) allow us to simplify (2.5) further. We have

$$\begin{aligned}\epsilon_{ij} &= e_{ij} + \frac{1}{2}\mathbf{u}_{\alpha^i}\cdot\mathbf{u}_{\alpha^j} = e_{ij} + \frac{1}{2}(\nabla_i u^k \mathbf{e}_k)\cdot(\nabla_j u^p \mathbf{e}_p)\\ &= e_{ij} + \frac{1}{2}A_{pk}\cdot\nabla_i u^k \nabla_j u^p + \frac{1}{2}\nabla_i u^3\cdot\nabla_j u^3;\ i,\ j = 1,\ 2;\\ \epsilon_{ij} &\approx e_{ij} + \frac{1}{2}\nabla_i u^3\cdot\nabla_j u^3. \qquad (2.23)\end{aligned}$$

We are not going to simplify ϵ_{33}, as this component is excluded from consideration in the construction of shell theory. As for the magnitudes of ϵ_{i3}, it has to be noted that the simplification scheme used above does not lead to any conclusion at this stage. Indeed,

$$\epsilon_{i3} = e_{i3} + \frac{1}{2}\nabla_i u^t \mathbf{e}_t\cdot\nabla_3 u^\gamma \mathbf{e}_\gamma = e_{i3} + \frac{1}{2}A_{t\gamma}\nabla_i u^t\nabla_3 u^\gamma + \nabla_i u^3\nabla_3 u^3, \qquad (2.24)$$

and the second term in (2.24) is of order $\epsilon^{\frac{3}{2}}$, which follows from (2.20), while the last term cannot be estimated at all at this stage. Further simplification of ϵ_{i3} is possible by appealing to additional hypotheses; this will be done in Section 3.

Let us note that the above simplification scheme can be continued if instead of (2.13) we use the assumption

$$|\omega_i|^m \sim \epsilon \ll 1.$$

A number of boundary value problems that do not use the assumption (2.13) can be found in [5, 148, 234, 235, 236, 237, 238, 256, 257, 258, 260, 267, 278, 279]. Naturally, in such cases both the theoretical and the numerical analysis of the problems are much harder.

3. The Kirchhoff–Love Hypotheses. Their Mathematical and Mechanical Content. Computation of Deformations of a Shallow Shell Using the Kirchhoff–Love Hypotheses

3.1. Below we shall use the *Kirchhoff–Love* hypotheses in the following form:

Assumption 4. *The vector* $\mathbf{u}(u_1(\alpha^i),\ u_2(\alpha^i),\ u_3(\alpha^i))$ *of displacements of the points of the shell has the structure*

$$\begin{aligned}\mathbf{u} &\equiv (u_1(\alpha^i),\ u_2(\alpha^i),\ u_3(\alpha^i))\\ &= \mathbf{a}_0(w_1(\alpha^1,\ \alpha^2),\ w_2(\alpha^1,\ \alpha^2),\ w(\alpha^1,\ \alpha^2))\\ &\quad + \alpha^3\mathbf{a}_1(v_1(\alpha^1,\ \alpha^2),\ v_2(\alpha^1,\ \alpha^2),\ v(\alpha^1,\ \alpha^2)); \qquad (3.1)\end{aligned}$$

that is,

$$u_i(\alpha^1,\ \alpha^2,\ \alpha^3) = w_i(\alpha^1,\ \alpha^2) + \alpha^3 v_i(\alpha^1,\ \alpha^2);\, i = 1,\ 2;\ u_3 = w(\alpha^1,\ \alpha^2). \quad (3.2)$$

Assumption 4 means that the deformation in the interior of the shell is completely determined by the deformation of the middle surface. Furthermore, the distribution of displacements over the thickness of the shell (coordinate α^3) is linear. Note that on the middle surface $\mathbf{a}_0 = \mathbf{u}$. We shall take this fact into account in the notation used below for the vector of displacements: In the theory we are considering, all the quantities that characterize the state of stress and strain are indeed expressed in terms of the values of the vector of displacements of the middle surface.

Assumption 5. *Transverse displacements ϵ_{i3}, $i = 1, 2$, and transverse deformations are small in comparison with all the other components of the deformation tensor and are taken to be zero.*

The following assumption has to do with the distribution of stresses inside a shell.

Assumption 6. *The stress σ_{33} is negligibly small in comparison with all the other components of the stress tensor and is taken to be zero.*

The Kirchhoff–Love hypotheses are a generalization of the well-known Bernoulli–Euler hypothesis to the case of a shell. Ever since the appearance of the theory of plates and shells, its mathematical justification invariably attracted the attention of scientists. Here we find the names of Cauchy and Poisson, Saint-Venant, Galerkin and Krauss, Vlasov, Kil'chevskii, Donnell, Lur'e, Marguerre, Mushtari, Novozhilov and others. The history of this period in the development of the problem is treated in more detail in [376, 353, 142].

A new stage in the problem of justifying the Kirchhoff–Love hypotheses was begun in the years after the Second World War, starting with the work of Friedrichs and Dressler [82]. A general approach to the problem of justifying the Kirchhoff–Love hypotheses was developed in the work of Gol'denvaizer and his students [97]. A different general approach was suggested by Vorovich; it was developed in the papers of Akstentyan and Vorovich and of Bazarenko and Vorovich. This approach was found to be efficient also from the point of view of the numerical analysis of the problem. In all the papers of this group we find the construction of algorithms that uncover the asymptotic nature of solutions of three-dimensional elastic thin bodies. In parallel, the problem was being studied from a purely mathematical viewpoint. Among the recent papers in this direction, we note the work of Destuynder and Ciarlet [65].

3.2. Let us explain the main conclusions concerning the character of states of stress and strain for thin isotropic bodies, i.e., for isotropic shells, that have been obtained using the mathematical considerations described above. The geometry of a sufficiently thin shell is characterized by two dimensionless parameters, $\lambda = h/R$, $\mu = h/\mathcal{L}$, where h is its thickness, R is the characteristic radius of curvature, and $\mathcal{L}$ is the characteristic length of the middle surface S (it can also be the minimal radius of curvature of the boundary curve).

(1) For $h \to 0$ and fixed R and $\mathcal{L}$ we have a qualitatively different character of the state of stress and strain of the shell close to the boundary S^0 and inside the volume V.

In the interior points of V the strain state changes smoothly and is given by the so-called first iteration process [97]. In other words, any property θ of the stress-deformed state of the shell inside V is given by

$$\theta = \theta_0 + \mu\theta_1 + \mu^2\theta_2 + \cdots, \tag{3.3}$$

where θ_0 is given by the solution of the corresponding theory of shells based on the Kirchhoff–Love hypotheses. Thus, in interior subdomains of V, corrections to the Kirchhoff–Love hypotheses are $O(\mu)$, $\mu = h/\mathcal{L}$.

A different situation obtains in a neighborhood of the boundary S^0. Here, in general, the stress-deformed state changes rapidly, and we shall describe it in a very general fashion. It is composed of the so-called boundary layer states. These can be subdivided into two groups. In the first group of boundary layers belong strain states of the form

$$\theta \sim M \exp\left(-\sqrt{\frac{R}{h}}\kappa_i\right), \quad i = 1, 2, 3, 4. \tag{3.4}$$

These stress-deformed boundary layer states also arise in theories based on the Kirchhoff–Love hypotheses. There are only four of them.

(2) Boundary layers of the second group are essentially different components of three-dimensional stress-deformed states that are not detected by Kirchhoff–Love theories. Here any characteristic of the state contains components of the form

$$\theta_i \sim M \exp\left(-\frac{\mathcal{L}}{h}\kappa_i\right), \quad i = 1, 2, \ldots \infty, \tag{3.5}$$

where κ_i are roots of a transcendental equation defined by the conditions of clamping of the shell at the boundary surfaces and $\mathcal{L}$ is the characteristic radius of curvature of Γ. In the general case, these enter the stress-deformed state of a shell in a neighborhood of S^0; they are absent in applied theories. Though these boundary layers decay rapidly, faster than the boundary layers (3.4), they can nonetheless exert a decisive influence on the stress-deformed state in a neighbourhood of S^0. In particular, they can exert a decisive influence on the concentration of stresses. Methods for computing a three-dimensional deformation state as a sum of states of the form (3.3)–(3.5) have been developed in [2, 18, 19, 326, 327, 352, 355, 363, 376, 353, 367, 368, 369, 370, 364].

(3) Let us note a principal difference of the asymptotics of the stress-deformed state of a thin shell in a neighborhood of S^0 from the well-known asymptotic solutions due to Lyusternik and Vishik, namely, the presence of a countably infinite number of boundary layers. This creates considerable difficulties, both from the computational point of view and for the problem of mathematical justification.

(4) To characterize the mechanical meaning of the Kirchhoff–Love hypotheses, let us present results dealing with stress concentration on S^0 itself. It can be shown

that on S^0 any characteristic of the strained state of the shell is given by the asymptotic expansion

$$\theta = \theta_0 + \mu\theta_1 + \mu^2\theta_2 + \cdots, \tag{3.6}$$

and here it would have been reasonable to expect that the principal term of the expansion θ_0 should be determined using the Kirchhoff–Love hypotheses. However, a detailed analysis of the passage to the limit from a three-dimensional problem of elasticity theory to a two-dimensional one, constructed using the Kirchhoff–Love hypotheses, shows that this is far from being always the case; this is the main reason behind the present study. For example, if the shell is fixed on S^0, that is, if the conditions

$$\mathbf{a}\,|_{S^0} = 0$$

hold, then the expansion (3.6) for σ_s, the stresses on the surface S^0 tangent to the midsurface (see Figure 3.1), gives us

$$\sigma_s|_{S^0} = \sigma_{s0} + \mu\sigma_{s1} + \mu^2\sigma_{s2} + \cdots,$$

where σ_{s0} is not the same as the value given by the Kirchhoff–Love theory. On the other hand, if the edge S^0 is free, then σ_{s0} coincides with the value obtained from the Kirchhoff–Love theory. However, already the transverse shear stress τ_{nm}(see Figure 3.1), defined by the expansion

$$\tau_{nm}|_{S^0} = \tau_{nm0} + \mu\tau_{nm1} + \mu^2\tau_{nm2} + \cdots$$

is asymptotically incorrect. In other words, τ_{nm0} is not the same as the value given by the Kirchhoff–Love theory.

The above examples clarify, if only partially, the mechanical content of the Kirchhoff–Love hypotheses. A more detailed analysis of these questions is to be found in the already cited works [2, 18, 19, 20, 326, 327, 352, 355, 363, 376, 353, 367, 368, 369, 370, 364].

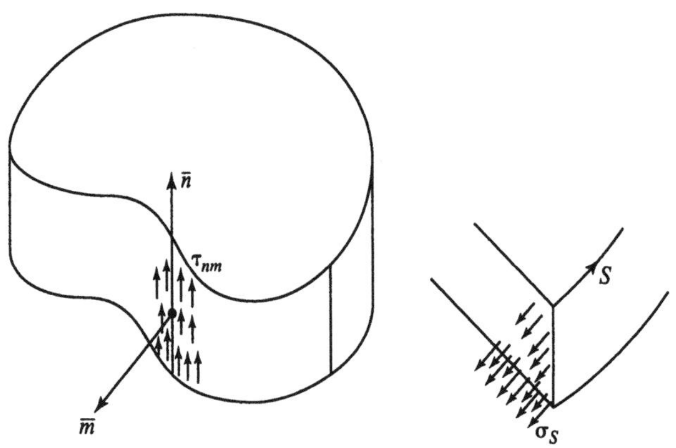

FIGURE 3.1.

3.3. Let us compute the components of finite deformation of the shell under the assumptions (1)–(6) of the Kirchhoff–Love hypotheses. Let us start with ϵ_{13}. From (3.1), (3.2), and (2.5) we have

$$e_{i3} = \overset{0}{e}_{i3} + \alpha^3 \overset{1}{e}_{i3}, \quad i = 1,\ 2,$$

where

$$\overset{0}{e}_{i3} = \nabla_i w + \nabla_3 w_i = \nabla_i w + v_i, \quad \overset{1}{e}_{i3} = 0.$$

Furthermore, from (2.24) we have

$$\epsilon_{i3} = \overset{0}{e}_{i3} = \frac{1}{2} A_{kj} \nabla_i w^k \nabla_3 w^j + \nabla_i w \nabla_3 w.$$

Using (2.3) and (2.4) we deduce that

$$\begin{aligned} \epsilon_{i3} &= v_i + \nabla_i w + \frac{1}{2} A_{kj} \nabla_j w^k \nabla_3 w^j = v_i + \nabla_i w + \frac{1}{2} v_j \nabla_i w^j \\ &= v_i + \nabla_i w + \frac{1}{2} v_j \nabla_j w^j, \quad i = 1,\ 2. \end{aligned} \tag{3.7}$$

Next, by Assumption 5 of the Kirchhoff–Love hypotheses ($\epsilon_{i3} = 0,\ i = 1,\ 2$), from (3.7) we obtain the two equations

$$\begin{aligned} v_1 \left(1 + \frac{1}{2} \nabla_1 w^1\right) + \frac{1}{2} v_2 \nabla_1 w^2 &= -\nabla_1 w, \\ v_1 \frac{1}{2} \nabla_2 w^1 + v_2 \left(1 + \frac{1}{2} \nabla_2 w^2\right) &= -\nabla_2 w. \end{aligned} \tag{3.8}$$

Taking into account the estimates (2.20), we obtain from (3.8) the important expressions

$$v_i = -\nabla_i w. \tag{3.9}$$

Let us consider next the expressions for ϵ_{ij}, $i,\ j = 1,\ 2$, that characterize the angles of rotation of the normal to the midsurface. From (2.23) it follows that

$$\epsilon_{ij} = \overset{0}{e}_{ij} + \alpha^3 \overset{1}{e}_{ij} + \frac{1}{2} \nabla_i u^3 \nabla_j u^3, \tag{3.10}$$

where

$$2\overset{0}{e}_{ij} = \nabla_i w_j + \nabla_j w_i = w_{i\alpha^j} + w_{j\alpha^i} - 2B_{ij} w - 2G^k_{ij} w_k, \tag{3.11}$$

$$2\overset{1}{e}_{ij} = \nabla_i v_j + \nabla_j v_i. \tag{3.12}$$

Moreover, from (2.3) and (3.9) we have

$$\begin{aligned} \nabla_i u^3 = \nabla_i w &= w_{\alpha^i} + B_{iq} u^q = w_{\alpha^i} + B_{iq}(w^q + \alpha^3 v^q) \\ &= w_{\alpha^i} + B_{iq}(w^q - \alpha^3 \nabla_q w) = \nabla_i w - \alpha^3 B_{iq} \nabla_q w. \end{aligned}$$

Let us now introduce *Assumption* 7, that the shell has thin walls:

Assumption 7.

$$|h^{\pm} B_{is}| \ll 1. \tag{3.13}$$

Using this we obtain

$$\nabla_i u^3 \approx \nabla_i w, \quad i = 1, 2.$$

Finally, substituting (3.9) into (3.12) we have

$$\overset{1}{e}_{ij} = -\nabla_{ij} w. \tag{3.14}$$

The next stage of simplifications is done under *Assumption* 8, that the shell is shallow.

Assumption 8.

$$\nabla_i w = w_{\alpha^i} + B_{ik} w^k \approx w_{\alpha^i}, \tag{3.15}$$

We can take this assumption in a wide range of problems, substituting (3.11), (3.12), (3.14) into (3.10), we have in final form

$$\epsilon_{ij} = \overset{0}{\epsilon}_{ij} + \alpha^3 \overset{1}{\epsilon}_{ij}, \quad 2\overset{0}{\epsilon}_{ij}$$
$$= 2\overset{0}{e}_{ij} + \nabla_i w \cdot \nabla_j w = \nabla_i w_j + \nabla_j w_i + \nabla_i w \cdot \nabla_j w$$
$$= w_{j\alpha^i} + w_{i\alpha^j} - 2B_{ij} w - 2G^k_{ij} w_k + w_{\alpha^i} w_{\alpha^j}; \tag{3.16}$$

$$\overset{1}{\epsilon}_{ij} = \overset{1}{e}_{ij} = -\nabla_{ij} w. \tag{3.17}$$

3.4. Let us return now to a more detailed explanation of the concept of shallowness, defined by the simplification (3.15), which is equivalent to the assumption

$$|w_{\alpha^i}| \gg \left|B_{ij}\right| \cdot |w|. \tag{3.18}$$

Relation (3.18) will hold under the following two conditions: the curvature B_{ij} is small enough, and the growth of w under differentiation is sufficiently fast. In general, both these factors cooperate. Different strategies can be used to give (3.15), (3.18) a quantitative interpretation. In some works, e.g., in [97], the so-called variability exponent for the stress-deformed state of the shell is introduced: A characteristic ratio $|w| : |w_{\alpha^i}| \sim \delta$ is considered, and a number m is found such that $\delta \sim (h/\mathcal{L})^m$, where $\mathcal{L}$ is the characteristic size of the middle surface and h is the thickness of the shell. The exponent m describes the rate of growth of w under differentiation. However, in our view the complex nature of condition (3.18) is better clarified as follows: Any sufficiently smooth function $w(\alpha^1, \alpha^2)$ in Ω can be be extended by zero outside of Ω, and we can compute the Fourier transform of the extended function,

$$\widetilde{w}(\lambda_1, \lambda_2) = \frac{1}{4\pi^2} \int_{-\infty}^{\infty} w(\alpha^1, \alpha^2) \exp i(\lambda_1 \alpha^1 + \lambda_2 \alpha^2) d\alpha^1, \, d\alpha^2.$$

As a rule, $\widetilde{w}$ will be essentially nonzero only for a certain domain of the plane of the parameters λ_1, λ_2. Let a typical (central) point of that domain be $\overset{0}{\boldsymbol{\lambda}} = (\overset{0}{\lambda}_1, \overset{0}{\lambda}_2)$. In other words, let the support of $\widetilde{w}$ be concentrated in a neighborhood of $\overset{0}{\boldsymbol{\lambda}}$. Then clearly,

$$|w_{\alpha^k}| \sim |\overset{0}{\boldsymbol{\lambda}}| \cdot |w|,$$

and the shallowness condition (3.18) assumes the form

$$\frac{|\overset{0}{\boldsymbol{\lambda}}|}{|B_{is}|} \gg 1. \tag{3.19}$$

In our opinion, the relation (3.19), which is, by the way, written in dimensionless form, reveals explicitly the interdependence of the smallness of curvature of the shell and the characteristic pattern of its stress-deformed state, which are united in the fundamental concept of "shallowness."

3.5. Let us state the final form of the relations for the components of finite deformation obtained using the Kirchhoff–Love hypotheses. We have

$$\overset{0}{\epsilon}_{11} = w_{1\alpha^1} - B_{11}w - G_{11}^k w_k + \frac{1}{2}w_{\alpha^1}^2, \tag{3.20}$$

$$\overset{0}{\epsilon}_{22} = w_{2\alpha^2} - B_{22}w - G_{22}^k w_k + \frac{1}{2}w_{\alpha^2}^2, \tag{3.21}$$

$$\overset{0}{\epsilon}_{12} = \frac{1}{2}(w_{1\alpha^2} + w_{2\alpha^1}) - B_{12}w - G_{12}^k w_k + \frac{1}{2}w_{\alpha^1}w_{\alpha^2}, \tag{3.22}$$

$$\overset{0}{\epsilon}_{13} \equiv \overset{0}{\epsilon}_{23} \equiv 0. \tag{3.23}$$

Instead of $\overset{0}{\epsilon}_{ij}$ we shall be frequently using the quantities $\overset{0}{\gamma}_{ij}$ defined by the relations

$$\overset{0}{\gamma}_{11} = \overset{0}{\epsilon}_{11} = w_{1\alpha^1} - B_{11}w - G_{11}^k w_k + \frac{1}{2}w_{\alpha^1}^2, \tag{3.24}$$

$$\overset{0}{\gamma}_{22} = \overset{0}{\epsilon}_{22} = w_{2\alpha^2} - B_{22}w - G_{22}^k w_k + \frac{1}{2}w_{\alpha^2}^2, \tag{3.25}$$

$$\overset{0}{\gamma}_{12} = 2\overset{0}{\epsilon}_{12} = w_{1\alpha^2} + w_{2\alpha^1} - 2B_{12}w - 2G_{12}^k w_k + w_{\alpha^1}w_{\alpha^2}. \tag{3.26}$$

Next,

$$\begin{aligned}
\overset{1}{\epsilon}_{11} &= \overset{1}{e}_{11} = -\nabla_{11}w = -w_{\alpha^1\alpha^1} + G_{11}^k w_{\alpha^k}, \\
\overset{1}{\epsilon}_{22} &= \overset{1}{e}_{22} = -\nabla_{22}w = -w_{\alpha^2\alpha^2} + G_{22}^k w_{\alpha^k}, \\
\overset{1}{\epsilon}_{12} &= \overset{1}{e}_{12} = -\nabla_{12}w = -w_{\alpha^1\alpha^2} + G_{12}^k w_{\alpha^k},
\end{aligned} \tag{3.27}$$

and finally, we define $\overset{1}{\gamma}_{ij}$ by

$$\overset{1}{\gamma}_{11} = \overset{1}{\epsilon}_{11} = \overset{1}{e}_{11} = -\nabla_{11} w = -w_{\alpha^1\alpha^1} + G^k_{11} w_{\alpha^k},$$
$$\overset{1}{\gamma}_{22} = \overset{1}{\epsilon}_{22} = \overset{1}{e}_{22} = -\nabla_{22} w = -w_{\alpha^2\alpha^2} + G^k_{22} w_{\alpha^k},$$
$$\overset{1}{\gamma}_{22} = 2\overset{1}{\epsilon}_{12} = 2\overset{1}{e}_{12} = -2\nabla_{12} w = 2\left(-w_{\alpha^1\alpha^2} + G^k_{12} w_{\alpha^k}\right).$$

4. Potential Energy of Deformation of a Shallow Shell

4.1. Let us describe in more detail the mechanical properties of the material of the shell. We shall assume that the shell is orthotropic, and that one of the axes of orthotropy is α^3. Furthermore, assume that the middle surface S admits an orthogonal parametrization $x^1 = \alpha^1$, $x^2 = \alpha^2$ that is an orthotropy parametrization. In other words, at each point of S the axes of orthotropy coincide with x^1, x^2. On equidistant surfaces $\alpha^3 = \text{const}$, the axes of orthotropy are obtained by parallel transport along the normal $\mathbf{n}$ of the orthotropy axes from S at the corresponding points. In principal orthotropy axes, Hooke's law [8, 176] has the form

$$\epsilon_{11} = \frac{\sigma_{11}}{E_1} - \frac{\nu_{12}\sigma_{22}}{E_2} - \frac{\nu_{13}\sigma_{33}}{E_3}, \quad \epsilon_{12} = \frac{\sigma_{12}}{2G_{12}} \quad (1 \to 2 \to 3 \to 1).$$

Taking the shell to be nonhomogeneous, we shall assume that the elastic constants $E(\mathbf{x}, \alpha^3)$, $G(\mathbf{x}, \alpha^3)$ depend on $\alpha^1, \alpha^2, \alpha^3$. We shall also assume that the following conditions are satisfied:

Assumption 9.

(1) E, G *are piecewise-smooth functions of* $\alpha^1, \alpha^2, \alpha^3$;

(2) $E_1\nu_{12} = E_2\nu_{21}$ $(1 \to 2 \to 3 \to 1)$; (4.1)

(3) *the quadratic form*

$$\begin{aligned} 2\Pi &= \left(\frac{\sigma_{11}}{E_1} - \frac{\nu_{12}\sigma_{22}}{E_2} - \frac{\nu_{13}\sigma_{33}}{E_3}\right)\sigma_{11} + \left(\frac{\sigma_{22}}{E_2} - \frac{\nu_{21}\sigma_{11}}{E_1} - \frac{\nu_{23}\sigma_{33}}{E_3}\right)\sigma_{22} \\ &+ \left(\frac{\sigma_{33}}{E_3} - \frac{\nu_{31}\sigma_{11}}{E_1} - \frac{\nu_{32}\sigma_{22}}{E_2}\right)\sigma_{33} + \frac{\sigma_{12}^2}{2G_{12}} + \frac{\sigma_{13}^2}{2G_{13}} + \frac{\sigma_{23}^2}{2G_{23}} \\ &= \frac{\sigma_{11}^2}{E_1} + \frac{\sigma_{22}^2}{E_2} + \frac{\sigma_{33}^2}{E_3} - \left(\frac{\nu_{12}}{E_2} + \frac{\nu_{21}}{E_1}\right)\sigma_{11}\sigma_{22} - \left(\frac{\nu_{13}}{E_3} + \frac{\nu_{31}}{E_1}\right)\sigma_{11}\sigma_{33} \\ &- \left(\frac{\nu_{23}}{E_3} + \frac{\nu_{32}}{E_2}\right)\sigma_{22}\sigma_{33} + \frac{\sigma_{12}^2}{2G_{12}} + \frac{\sigma_{13}^2}{2G_{13}} + \frac{\sigma_{23}^2}{2G_{23}} \end{aligned} \quad (4.2)$$

is uniformly positive definite over the whole of the volume V occupied by the shell; that is, we have the inequality [8, 97, 176, 184, 218]

$$2\Pi \geq m(\sigma_{11}^2 + \sigma_{22}^2 + \sigma_{12}^2 + \sigma_{13}^2 + \sigma_{23}^2 + \sigma_{33}^2). \tag{4.3}$$

We shall call the conditions (1) and (3) the *material regularity conditions*. Let us also note the useful relations

$$G_{12} = \frac{E_1 E_2}{E_1 + E_2 + E_1 \nu_{12} + E_2 \nu_{21}} = \frac{E_1 E_2}{E_1 + E_2 + 2\nu_{12} E_1} = \frac{E_1 E_2}{E_1 + E_2 + 2\nu_{21} E_2}.$$

By the Kirchhoff–Love hypotheses we have

$$\sigma_{11} = \frac{E_1}{1 - \nu_{12}\nu_{21}}(\epsilon_{11} + \nu_{12}\epsilon_{22}); \quad \sigma_{12} = 2G_{12}\epsilon_{12} \ (1 \rightleftharpoons 2), \tag{4.4}$$

where the notation $1 \rightleftharpoons 2$ means that the corresponding formulae for σ_{22} and σ_{21} are obtained by replacing all occurrences of the subscript 1 by 2 and vice versa.

For the volume density of potential energy Π of deformation of the shell we have

$$2\Pi = \sigma_{11}\epsilon_{11} + \sigma_{22}\epsilon_{22} + 2\sigma_{12}\epsilon_{12} = \sigma_{11}\gamma_{11} + \sigma_{22}\gamma_{22} + \sigma_{12}\gamma_{12},$$

where $\gamma_{11} = \epsilon_{11}$, $\gamma_{12} = 2\epsilon_{12}$ $(1 \to 2 \to 3 \to 1)$. Taking (4.4) into account, we have for 2Π the relation

$$2\Pi = B^{\lambda\mu qs}(\mathbf{x}, \alpha^3)\gamma_{\lambda\mu}(\mathbf{x}, \alpha^3)\gamma_{qs}(\mathbf{x}, \alpha^3), \ \lambda \leq \mu, \ q \leq s,$$

where $B^{\lambda\mu qs}$ are the elastic constants (moduli) in orthogonal orthotropy axes defined by the relations

$$B^{1111}(\mathbf{x}, \alpha^3) = \frac{E_1}{1 - \nu_{12}\nu_{21}}, \quad B^{1122}(\mathbf{x}, \alpha^3) = \frac{E_1 \nu_{12}}{1 - \nu_{12}\nu_{21}} = \frac{E_1 \nu_{21}}{1 - \nu_{12}\nu_{21}}, \tag{4.5}$$

while $\gamma_{\lambda\mu}(\mathbf{x}, \alpha^3)$ are the components of the strain tensor in the same coordinates. By (3.10) we can write

$$\gamma_{\lambda\mu}(\mathbf{x}, \alpha^3) = \overset{0}{\gamma}_{\lambda\mu}(\mathbf{x}, \alpha^3) + \alpha^3 \overset{1}{\gamma}_{\lambda\mu}(\mathbf{x}, \alpha^3),$$

which gives the following expression for Π:

$$\Pi = \Pi_s + \Pi_f + \Pi^*, \tag{4.6}$$

where

$$2\Pi_s = B^{\lambda\mu qs}(\mathbf{x}, \alpha^3)\overset{0}{\gamma}_{\lambda\mu}(\mathbf{x})\overset{0}{\gamma}_{qs}(\mathbf{x}),$$

$$2\Pi_f = (\alpha^3)^2 B^{\lambda\mu qs}(\mathbf{x}, \alpha^3)\overset{1}{\gamma}_{\lambda\mu}(\mathbf{x})\overset{1}{\gamma}_{qs}(\mathbf{x}),$$

$$2\Pi^* = \alpha^3 B^{\lambda\mu qs}(\mathbf{x}, \alpha^3)\left[\overset{0}{\gamma}_{\lambda\mu}(\mathbf{x})\overset{1}{\gamma}_{qs}(\mathbf{x}) + \overset{1}{\gamma}_{\lambda\mu}(\mathbf{x})\overset{0}{\gamma}_{qs}(\mathbf{x})\right].$$

For the potential energy U stored in the volume V of the shell we shall have

$$U = \int_V \Pi D^*(\mathbf{x}, \alpha^3) d\alpha^1, d\alpha^2, d\alpha^3, \tag{4.7}$$

where $D^*(\mathbf{x}\,\alpha^3)$ is the Jacobian in the coordinates $(\alpha^1, \alpha^2, \alpha^3)$. If we take into account the fact (3.13) that the shell is thin, we can take

$$D^*(\mathbf{x}, \alpha^3) \approx D^*(\mathbf{x}, 0) = \sqrt{A_{11}A_{22} - A_{12}^2}. = D$$

4.2. In the subsequent construction of the theory we shall have to make an assumption, which states that the shell is symmetric with respect to width.

Assumption 10. *The elastic constants $E_i(\alpha^1, \alpha^2, \alpha^3)$, $G_{ij}(\alpha^1, \alpha^2, \alpha^3)$ are even functions of α^3, and furthermore,*

$$h^+ = h = h^-.$$

Then we have from (4.6), (4.7),

$$U = U_s + U_f, \tag{4.8}$$

where

$$2U_s = \int_V 2\Pi_s D d\alpha^1\, d\alpha^2\, d\alpha^3$$
$$= \int_S \int_{-h}^{h} B^{\lambda\mu q s}(\mathbf{x}\,\alpha^3) \overset{0}{\gamma}_{\lambda\mu}(\mathbf{x}) \overset{0}{\gamma}_{qs}(\mathbf{x}) D d\alpha^1\, d\alpha^2\, d\alpha^3 = \int_S Q_s D d\alpha^1\, d\alpha^2 \tag{4.9}$$

and

$$Q_s = D_s^{\lambda\mu q s}(\mathbf{x}) \overset{0}{\gamma}_{\lambda\mu}(\mathbf{x}) \overset{0}{\gamma}_{qs}(\mathbf{x}), \quad D_s^{\lambda\mu q s}(\mathbf{x}) = \int_{-h}^{h} B^{\lambda\mu q s}(\mathbf{x}, \alpha^3) d\alpha^3. \tag{4.10}$$

For U_f we have

$$2U_f = \int_V 2\Pi_f D d\alpha^1\, d\alpha^2\, d\alpha^3$$
$$= \int_S \int_{-h}^{h} (\alpha^3)^2 B^{\lambda\mu q s}(\mathbf{x}, \alpha^3) \overset{1}{\gamma}_{\lambda\mu}(\mathbf{x}) \overset{1}{\gamma}_{qs}(\mathbf{x}) D d\alpha^1\, d\alpha^2\, d\alpha^3 = \int_S Q_f D d\alpha^1\, d\alpha^2,$$

$$Q_f = D_f^{\lambda\mu q s}(\mathbf{x}) \overset{1}{\gamma}_{\lambda\mu}(\mathbf{x}) \overset{1}{\gamma}_{qs}(\mathbf{x}), \quad D_f^{\lambda\mu q s}(\mathbf{x}) = \int_{-h}^{h} (\alpha^3)^2 B^{\lambda\mu q s}(\mathbf{x}, \alpha^3) d\alpha^3. \tag{4.11}$$

Let us now obtain expressions for the potential energy in arbitrary S-coordinates $\alpha^1, \alpha^2, \alpha^3$. Clearly,

$$2U_s = \int_S Q_s D d\alpha^1 d\alpha^2, \quad 2U_f = \int_S Q_f D d\alpha^1 d\alpha^2, \tag{4.12}$$

where

$$Q_{\mathrm{s}} = D_{\mathrm{s}}^{\lambda\mu qs}(\alpha)\overset{0}{\gamma}_{\lambda\mu}(\alpha)\overset{0}{\gamma}_{qs}(\alpha), \quad Q_{\mathrm{f}} = D_{\mathrm{f}}^{\lambda\mu qs}(\alpha)\overset{1}{\gamma}_{\lambda\mu}(\alpha)\overset{1}{\gamma}_{qs}(\alpha), \quad \alpha = (\alpha^1, \alpha^2),$$

and $D_{\mathrm{s,f}}^{\lambda\mu qs}(\alpha)$ (the subscript "s, f" means that these formulae hold for both "s" and "f") are expressed in terms of $D_{\mathrm{s,f}}^{\lambda\mu qs}(\mathbf{x})$ by

$$D_{\mathrm{s,f}}^{1111}(\alpha) = \frac{1}{A_{11}^2 \sin^4\psi}\{D_{\mathrm{s,f}}^{1111}(\mathbf{x})\sin^4(\psi-\varphi) + D_{\mathrm{s,f}}^{2222}(\mathbf{x})\cos^4(\psi-\varphi)$$
$$+ 2\sin^2(\psi-\varphi)\cos^2(\psi-\varphi)(2D_{\mathrm{s,f}}^{1212}(\mathbf{x}) + D_{\mathrm{s,f}}^{1122}(\mathbf{x}))\},$$

$$D_{\mathrm{s,f}}^{2222}(\alpha) = \frac{1}{A_{22}^2 \sin^4\psi}\{D_{\mathrm{s,f}}^{1111}(\mathbf{x})\sin^4\varphi + D_{\mathrm{s,f}}^{2222}(\mathbf{x})\cos^4\varphi$$
$$+ 2\sin^2\varphi\cos^2\varphi(2D_{\mathrm{s,f}}^{1212}(\mathbf{x}) + D_{\mathrm{s,f}}^{1122}(\mathbf{x}))\},$$

$$D_{\mathrm{s,f}}^{1212}(\alpha) = \frac{1}{A_{11}A_{22}\sin^4\psi}\{D_{\mathrm{s,f}}^{1111}(\mathbf{x})\sin^2(\psi-\varphi)\sin^2\varphi +$$
$$+ D_{\mathrm{s,f}}^{2222}(\mathbf{x})\cos^2\varphi\cos^2(\psi-\varphi) + D_{\mathrm{s,f}}^{1212}(\mathbf{x})\sin^2(\psi-2\varphi)$$
$$- D_{\mathrm{s,f}}^{1122}(\mathbf{x})\sin 2\varphi\sin(\psi-\varphi)\cos(\psi-\varphi)\},$$

$$D_{\mathrm{s,f}}^{1122}(\alpha) = \frac{1}{A_{11}A_{22}\sin^4\psi}\{D_{\mathrm{s,f}}^{1111}(\mathbf{x})\sin^2(\psi-\varphi)\sin^2\varphi$$
$$+ D_{\mathrm{s,f}}^{2222}(\mathbf{x})\cos^2\varphi\cos^2(\psi-\varphi) - D_{\mathrm{s,f}}^{1212}(\mathbf{x})\sin 2\varphi\sin 2(\psi-\varphi)$$
$$+ D_{\mathrm{s,f}}^{1122}(\mathbf{x})[\sin^2(\psi-\varphi)\cos^2\varphi + \cos^2(\psi-\varphi)\sin^2\varphi]\},$$

$$D_{\mathrm{s,f}}^{1112}(\alpha) = \frac{1}{A_{11}^{3/2}A_{22}^{1/2}\sin^4\psi}$$
$$\times\left\{D_{\mathrm{s,f}}^{1111}(\mathbf{x})\sin^3(\psi-\varphi)\sin\varphi - D_{\mathrm{s,f}}^{2222}(\mathbf{x})\cos^3(\psi-\varphi)\cos\varphi\right.$$
$$\left.+ (D_{\mathrm{s,f}}^{1122}(\mathbf{x}) + 2D_{\mathrm{s,f}}^{1212}(\mathbf{x}))\sin(\psi-\varphi)\cos(\psi-\varphi)\sin(2\varphi-\psi)\right\},$$

$$D_{\mathrm{s,f}}^{2212}(\alpha) = \frac{1}{A_{11}^{1/2}A_{22}^{3/2}\sin^4\psi}$$
$$\times\left\{D_{\mathrm{s,f}}^{1111}(\mathbf{x})\sin(\psi-\varphi)\sin^3\varphi - D_{\mathrm{s,f}}^{2222}(\mathbf{x})\cos(\psi-\varphi)\cos^3\varphi\right.$$
$$\left.+ (D_{\mathrm{s,f}}^{1122}(\mathbf{x}) + 2D_{\mathrm{s,f}}^{1212}(\mathbf{x}))\sin\varphi\cos\varphi\sin(\psi-2\varphi)\right\},$$

and also

$$2Q_{\mathrm{s}} = D_{\mathrm{s}}^{1111}(\mathbf{x})\overset{0}{\gamma}{}_{11}^{2}(\mathbf{x}) + D_{\mathrm{s}}^{2222}(\mathbf{x})\overset{0}{\gamma}{}_{22}^{2}(\mathbf{x}) + D_{\mathrm{s}}^{1212}(\mathbf{x})\overset{0}{\gamma}{}_{12}^{2}(\mathbf{x})$$
$$+ 2D_{\mathrm{s}}^{1122}(\mathbf{x})\overset{0}{\gamma}_{11}(\mathbf{x})\overset{0}{\gamma}_{22}(\mathbf{x}) = D_{\mathrm{s}}^{1111}(\alpha)\overset{0}{\gamma}{}_{11}^{2}(\alpha) + D_{\mathrm{s}}^{2222}(\alpha)\overset{0}{\gamma}{}_{22}^{2}(\alpha)$$
$$+ D_{\mathrm{s}}^{1212}(\alpha)\overset{0}{\gamma}{}_{12}^{2}(\alpha) + 2D_{\mathrm{s}}^{1122}(\alpha)\overset{0}{\gamma}_{11}(\alpha)\overset{0}{\gamma}_{22}(\alpha)$$
$$+ 2D_{\mathrm{s}}^{1112}(\alpha)\overset{0}{\gamma}_{11}(\alpha)\overset{0}{\gamma}_{12}(\alpha) + 2D_{\mathrm{s}}^{2212}(\alpha)\overset{0}{\gamma}_{22}(\alpha)\overset{0}{\gamma}_{12}(\alpha), \tag{4.13}$$

$$2Q_{\mathrm{f}} = D_{\mathrm{f}}^{1111}(\mathbf{x})\overset{1}{\gamma}{}_{11}^{2}(\mathbf{x}) + D_{\mathrm{f}}^{2222}(\mathbf{x})\overset{1}{\gamma}{}_{22}^{2}(\mathbf{x}) + D_{\mathrm{f}}^{1212}(\mathbf{x})\overset{1}{\gamma}{}_{12}^{2}(\mathbf{x})$$
$$+ 2D_{\mathrm{f}}^{1112}(\mathbf{x})\overset{1}{\gamma}_{11}(\mathbf{x})\overset{1}{\gamma}_{22}(\mathbf{x}) = D_{\mathrm{f}}^{1111}(\alpha)\overset{1}{\gamma}{}_{11}^{2}(\alpha) + D_{\mathrm{f}}^{2222}(\alpha)\overset{1}{\gamma}{}_{22}^{2}(\alpha)$$
$$+ D_{\mathrm{f}}^{1212}(\alpha)\overset{1}{\gamma}{}_{12}^{2}(\alpha) + 2D_{\mathrm{f}}^{1122}(\alpha)\overset{1}{\gamma}_{11}(\alpha)\overset{1}{\gamma}_{22}(\alpha)$$
$$+ 2D_{\mathrm{f}}^{1112}(\alpha)\overset{1}{\gamma}_{11}(\alpha)\overset{1}{\gamma}_{12}(\alpha) + 2D_{\mathrm{f}}^{2212}(\alpha)\overset{1}{\gamma}_{22}(\alpha)\overset{1}{\gamma}_{12}(\alpha).$$

4.3. Let us derive expressions for the potential energy dual to (4.12). For that we introduce the stresses T^{ij} and the moments M^{ij} by

$$T^{ij} = \frac{\partial Q_{\mathrm{s}}}{\partial \overset{0}{\gamma}_{ij}} = D_{\mathrm{s}}^{ijkl}\overset{0}{\gamma}_{kl}, \quad M^{ij} = \frac{\partial Q_{\mathrm{s}}}{\partial \overset{1}{\gamma}_{ij}} = D_{\mathrm{f}}^{ijkl}\overset{1}{\gamma}_{kl}. \tag{4.14}$$

The expressions (4.14) can be written in the following form:

$$T^{ij} = D_{\mathrm{s}}^{ijkl}\overset{0}{\epsilon}_{kl}, \quad M^{ij} = D_{\mathrm{f}}^{ijkl}\overset{1}{\epsilon}_{kl}. \tag{4.15}$$

However, in (4.14) if $k \neq l$, then $k = 1, l = 2$. At the same time, in (4.15) we necessarily have $k = 1, l = 2$ or $k = 2, l = 1$. From (4.14) we find that

$$\overset{0}{\gamma}_{kl} = C_{klij,\mathrm{s}}T^{ij}, \quad \overset{1}{\gamma}_{kl} = C_{klij,\mathrm{f}}M^{ij}. \tag{4.16}$$

Solubility of (4.14) with respect to $\overset{0}{\gamma}_{kl}, \overset{1}{\gamma}_{kl}$ is guaranteed by the inequality (4.3). Using (4.16), we obtain the following relations for Q_{s} and Q_{f}:

$$Q_{\mathrm{s}} = C_{\mathrm{s}}^{\lambda\mu qs}T_{\lambda\mu}T_{qs}, \quad Q_{\mathrm{f}} = C_{\mathrm{f}}^{\lambda\mu qs}M_{\lambda\mu}M_{qs}. \tag{4.17}$$

By positive-definiteness of Π (the inequality (4.3)) we have the inequalities

$$\begin{aligned} &Q_{\mathrm{s}} \geq m(\overset{0}{\gamma}{}_{11}^{2} + \overset{0}{\gamma}_{22}{}^{2} + \overset{0}{\gamma}_{12}{}^{2}), \quad Q_{\mathrm{f}} \geq m(\overset{1}{\gamma}_{11}{}^{2} + \overset{1}{\gamma}{}_{22}^{2} + \overset{1}{\gamma}{}_{12}^{2}), \\ &Q_{\mathrm{s}} \geq m(T_{11}^{2} + T_{22}^{2} + T_{12}^{2}), \quad Q_{\mathrm{f}} \geq m(M_{11}^{2} + M_{22}^{2} + M_{12}^{2}). \end{aligned} \tag{4.18}$$

Therefore, the tensors D_{s}^{ijkl}, D_{f}^{ijkl} and C_{s}^{ijkl}, C_{f}^{ijkl} are positive definite, which follows from the material regularity condition (4.3).

4.4. Let us consider the case of an isotropic shell in more detail. From (4.5) in this case we have

$$B^{1111}(\mathbf{x}, \alpha^3) = \frac{E}{1-\nu^2}, \quad B^{1122}(\mathbf{x}, \alpha^3) = \frac{E\nu}{1-\nu^2}, \quad B^{1212}(\mathbf{x}, \alpha^3) = \frac{E}{2(1+\nu)}, \tag{4.19}$$

where we have

$$\begin{aligned} 2Q_s &= D_{s1}\overset{0}{I}{}_1^2 + 2\frac{D_{s2}}{D^2}\left(\frac{\overset{0}{\gamma}{}_{12}^2}{4} - \overset{0}{\gamma}_{11}\overset{0}{\gamma}_{22}\right), \\ 2Q_f &= D_{f1}\overset{1}{I}{}_1^2 + 2\frac{D_{f2}}{D^2}\left(\frac{\overset{0}{\gamma}{}_{12}^2}{4} - \overset{0}{\gamma}_{11}\overset{0}{\gamma}_{22}\right); \end{aligned} \tag{4.20}$$

$$\begin{aligned} D_{s1} &= \int_{-h}^{h} \frac{E d\alpha^3}{1-\nu^2}, \quad D_{s2} = \int_{-h}^{h} \frac{E d\alpha^3}{1+\nu}, \\ D_{f1} &= \int_{-h}^{h} \frac{E(\alpha^3)^2 d\alpha^3}{1-\nu^2}, \quad D_{f2} = \int_{-h}^{h} \frac{E(\alpha^3)^2 d\alpha^3}{1+\nu}. \end{aligned} \tag{4.21}$$

The invariants $\overset{0}{I}_1$, $\overset{1}{I}_1$ in (4.20) are given by

$$\begin{aligned} \overset{0}{I}_1 &= A^{ij}\overset{0}{\gamma}_{ij} = A^{11}\overset{0}{\gamma}_{11} + A^{22}\overset{0}{\gamma}_{22} + A^{12}\overset{0}{\gamma}_{12}, \\ \overset{1}{I}_1 &= A^{ij}\overset{1}{\gamma}_{ij} = A^{11}\overset{1}{\gamma}_{11} + A^{22}\overset{1}{\gamma}_{22} + A^{12}\overset{1}{\gamma}_{12}. \end{aligned} \tag{4.22}$$

The relations (4.19)–(4.22) can be considered in an arbitrary parametrization α^1, α^2, and then we obtain from them

$$\begin{aligned} T^{11} &= D_{s1}\overset{0}{I}_1 A^{11} - \frac{D_{s2}}{D^2}\overset{0}{\gamma}_{22} = D_{s3}\overset{0}{I}_1 A^{11} + D_{s2}\overset{0}{\epsilon}_{11}, \\ T^{12} &= D_{s1}\overset{0}{I}_1 A^{12} - \frac{D_{s2}}{2D^2}\overset{0}{\gamma}_{12} = D_{s3}\overset{0}{I}_1 A^{12} + D_{s2}\overset{0}{\epsilon}_{12}, \end{aligned} \tag{4.23}$$

where

$$D_{s3} = \int_{-h}^{h} \frac{E\nu}{1-\nu^2}\, d\alpha^3 = D_{s1} - D_{s2}. \tag{4.24}$$

The relations (4.23) correspond to the following representation of the tensors:

$$\begin{aligned} D_s^{1111} &= D_{s1}(A^{11})^2 = D^{-4}D_{s1}(A^{22})^2, \\ D_s^{1122} &= D_{s1}A^{11}A^{22} - D_{s2}D^{-2} = D^{-4}(D_{s3}A_{11}A_{22} + D_{s2}A_{12}^2), \\ D_s^{1112} &= D_{s1}A^{11}A^{12} = -D^4 D_{s1}A_{22}A_{12}; \; D_s^{1222} = D_{s1}A^{22}A^{12} \\ &= -D^{-4}D_{s1}A_{11}A_{12}, \\ D_s^{1212} &= D_{s1}A^{12}A^{12} + \frac{D_{s2}}{2}D^{-2} = \left(D_{s1}A_{12}^2 + \frac{D_{s2}}{2}D^2\right)D^{-4} \; (1 \rightleftharpoons 2). \end{aligned} \tag{4.25}$$

Similarly, for an isotropic shell we have

$$
\begin{aligned}
M^{11} &= D_{\mathrm{f1}}\overset{1}{I}_1 A^{11} - \frac{D_{\mathrm{f2}}}{D^2}\overset{1}{\gamma}_{22} = D_{\mathrm{f3}}\overset{1}{I}_1 A^{11} + D_{\mathrm{f2}}\overset{1}{\epsilon}{}^{11} \ (1 \rightleftharpoons 2),\\
M^{12} &= D_{\mathrm{f1}}\overset{1}{I}_1 A^{12} - \frac{D_{\mathrm{f2}}}{D^2}\overset{1}{\gamma}_{12} = D_{\mathrm{f3}}\overset{1}{I}_1 A^{12} + D_{\mathrm{f2}}\overset{1}{\epsilon}{}^{12}.
\end{aligned}
\tag{4.26}
$$

For D_{f}^{ijkl} we have the expressions (4.25), in which the subscript "s" must be replaced by "f."

Finally, from (4.25), (4.26) we can obtain

$$
\begin{aligned}
D_{\mathrm{s2}}\overset{0}{\epsilon}{}^{ij} &= T^{ij} - \frac{D_{\mathrm{s3}}A^{ij}}{2D_{\mathrm{s1}} - D_{\mathrm{s2}}}T, \quad T = T^{ij}A_{ij} = T^{11}A_{11} + T^{22}A_{22} + 2T^{12}A_{12},\\
D_{\mathrm{f2}}\overset{1}{\epsilon}{}^{ij} &= M^{ij} - \frac{D_{\mathrm{s3}}}{2D_{\mathrm{s1}} - D_{\mathrm{s2}}}M, \quad M = M^{ij}A_{ij}\\
&= M^{11}A_{11} + M^{22}A_{22} + 2M^{12}A_{12}.
\end{aligned}
$$

Similarly, for the potential energy we have

$$
\begin{aligned}
Q_{\mathrm{s}} &= \frac{D_{\mathrm{s1}}T^2}{2D_{\mathrm{s2}}(2D_{\mathrm{s1}} - D_{\mathrm{s2}})} + \frac{D^2}{D_{\mathrm{s2}}}(T^{12}T^{21} - T^{11}T^{22}),\\
Q_{\mathrm{f}} &= \frac{D_{\mathrm{f1}}M^2}{2D_{\mathrm{f2}}(2D_{\mathrm{f1}} - D_{\mathrm{f2}})} + \frac{D^2}{D_{\mathrm{f2}}}(M^{12}M^{21} - M^{11}M^{22}).
\end{aligned}
\tag{4.27}
$$

To the expressions in (4.27) correspond the following values of $C_{\mathrm{s,f}}^{ijkl}$:

$$
\begin{aligned}
C_{1111\mathrm{s}} &= \frac{D_{\mathrm{s1}}A_{11}^2}{D_{\mathrm{s2}}(2D_{\mathrm{s1}} - D_{\mathrm{s2}})}, \quad C_{1122\mathrm{s}} = \frac{D_{\mathrm{s1}}A_{11}A_{22}}{D_{\mathrm{s2}}(2D_{\mathrm{s1}} - D_{\mathrm{s2}})} - \frac{D^2}{D_{\mathrm{s}}^2},\\
C_{1112\mathrm{s}} &= \frac{D_{\mathrm{s1}}A_{11}A_{12}}{D_{\mathrm{s2}}(2D_{\mathrm{s1}} - D_{\mathrm{s2}})} \ (1 \rightleftharpoons 2).
\end{aligned}
$$

Similar relations also hold for $C_{ijkl,\mathrm{f}}$ if we replace the subscript "s" by "f."

If the elastic properties of the shell do not depend on α^3, then

$$
\begin{aligned}
Q_{\mathrm{s}} &= \frac{1}{4Eh}[T^2 + 2D^2(1+\nu)(T^{12}T^{21} - T^{11}T^{22})],\\
Q_{\mathrm{f}} &= \frac{1}{4Eh^3}[M^2 + 2D^2(1+\nu)(M^{12}M^{21} - M^{11}M^{22})];
\end{aligned}
$$

$$
\begin{aligned}
C_{1111\mathrm{s}} &= \frac{A_{11}^2}{2Eh}, \quad C_{1122\mathrm{s}} = \frac{1}{2Eh}(A_{12}^2 - D^2\nu),\\
C_{1212\mathrm{s}} &= \frac{1}{Eh}(2A_{12}^2 + D^2(1+\nu));\\
C_{1111\mathrm{f}} &= \frac{3A_{11}^2}{2Eh^3}, \quad C_{1122\mathrm{f}} = \frac{3}{2Eh}(A_{12}^2 - D^2\nu),\\
C_{1112\mathrm{f}} &= \frac{3A_{11}A_{12}}{2Eh}, \quad C_{1212,\mathrm{f}} = \frac{3}{Eh}(2A_{12}^2 + D^2(1+\nu)).
\end{aligned}
$$

5. Independent Displacements, Generalized Stresses and the Work of External Forces Under the Kirchhoff–Love Hypotheses

5.1. Under the conditions of the Kirchhoff–Love hypotheses, by (3.1), (3.2), (3.9), the strained state of a shell can be expressed in terms of $w^k(\alpha^1, \alpha^2)$, $k = 1, 2, 3$, displacements of points of the middle surface S. On the boundary S^0 the displacements, due to (3.1), (3.2), (3.9), cannot be specified arbitrarily. If the displacement vector $\mathbf{u}\,|_{S^0}$ does not have the structure (3.1), (3.2), (3.9), then in order to solve for it in the framework of the Kirchhoff–Love hypotheses, it is necessary to approximate it optimally by a relation of the form of (3.1), (3.2), (3.9). However, it is easily seen that not all the quantities $w^k(\alpha^1, \alpha^2)$, $k = 1, 2, 3$, $v_i = -\nabla_i w$ are independent on Γ. Indeed, we can specify $w^k(s)$ independently. But then by (1.17), $w^3_{\alpha^1}$ and $w^3_{\alpha^2}$ will be linearly related, and in general, we can specify in an arbitrary fashion only one of the two quantities $w^3_{\alpha^i}$. Thus, if the Kirchhoff–Love hypotheses hold, we have on Γ three independent displacements $w^k(s)$, 1, 2, 3, and one of the quantities $w^3_{\alpha^i}$. However, from physical and geometrical considerations it makes more sense to introduce on Γ the angle of the normal to the deformed middle surface S at a boundary point of S^0, which approximately equals $\partial w/\partial m$. Then (1.17), (1.18) determine $w^3_{\alpha^i}$. Thus, under the Kirchhoff–Love conditions, displacements on S^0 are determined by the four parameters independent on Γ, $w^k(s)$, $k = 1, 2, 3$, $(w^3(s) = w_3(s) = w(s))$ and $\partial w/\partial m = w_4(s)$.

5.2. Let us compute the work of external forces applied to the shell on displacements allowed under the conditions of the Kirchhoff–Love hypotheses, (3.1), (3.2), (3.9), (3.13). We shall assume that stresses $\mathbf{F}^{\pm}(\alpha^1, \alpha^2)$ are applied on the edges $S^{\pm}$. Furthermore, let body forces $\mathbf{F}(\alpha^1, \alpha^2, \alpha^3)$ act on the shell, and finally, let surface forces $\mathbf{F}(s, \alpha^3)$ act on S^0 (see Figure 5.1).

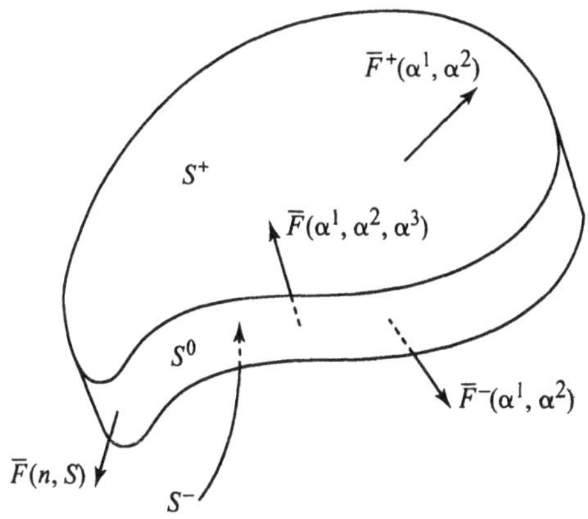

FIGURE 5.1.

For the elementary work δA_1 of body forces and forces applied to $S^{\pm}$ performed on virtual displacements $\delta \mathbf{u} = \delta \mathbf{a}_0 + \alpha^3 \delta \mathbf{a}_1$, we have

$$\begin{aligned}\delta A_1 = &\int_{S^+} \mathbf{F}^+(\alpha^1, \alpha^2) \cdot \delta \mathbf{u}(\alpha^1, \alpha^2)\,|_{S^+}\, D^+ d\alpha^1 d\alpha^2 \\ &+ \int_{S^-} \mathbf{F}^-(\alpha^1, \alpha^2) \cdot \delta \mathbf{u}(\alpha^1, \alpha^2)\,|_{S^-}\, D^- d\alpha^1 d\alpha^2 \\ &+ \int_V \mathbf{F}(\alpha^1, \alpha^2, \alpha^3) \cdot \delta \mathbf{u}(\alpha^1, \alpha^2, \alpha^3) D^* d\alpha^1 d\alpha^2 d\alpha^3.\end{aligned} \tag{5.1}$$

In (5.1) D^+, D^- are area elements on S^+ and S^-, while D^* is a volume element in S-coordinates α^1, α^2, α^3. It is easy to show that if Assumption 7 (3.13) holds, we have

$$D^+ \approx D^- \approx D^* \approx D,$$

and then, taking into account (3.1),

$$\begin{aligned}\delta A_1 = \int_S \Big[&\mathbf{F}^+(\alpha^1, \alpha^2) \cdot (\delta \mathbf{a}_0 + h \delta \mathbf{a}_2) + \mathbf{F}^-(\alpha^1, \alpha^2)(\delta \mathbf{a}_0 - h \delta \mathbf{a}_2) \\ &+ \int_{-h}^{h} \mathbf{F}(\alpha^1, \alpha^2, \alpha^3) \cdot (\delta \mathbf{a}_0 + \alpha^3 \delta \mathbf{a}_1)\, d\alpha^3 \Big] d\Omega,\end{aligned}$$

and furthermore,

$$\begin{aligned}\delta A_1 = &\int_S \Big[\mathbf{F}^+(\alpha^1, \alpha^2) + \mathbf{F}^-(\alpha^1, \alpha^2) + \int_{-h}^{h} \mathbf{F}(\alpha^1, \alpha^2, \alpha^3)\, d\alpha^3\Big] \cdot \delta \mathbf{a}_0\, d\Omega \\ &+ \int_S \Big[\mathbf{F}^+(\alpha^1, \alpha^2) h(\alpha^1, \alpha^2) - \mathbf{F}^-(\alpha^1, \alpha^2) h(\alpha^1, \alpha^2) \\ &+ \int_{-h}^{h} \mathbf{F}(\alpha^1, \alpha^2, \alpha^3) \alpha^3\, d\alpha^3\Big] \cdot \delta \mathbf{a}_1\, d\Omega.\end{aligned} \tag{5.2}$$

Transforming (5.2) and using (1.34), (1.35), and (3.9), we obtain

$$\delta A_1 = \int_S R^k \delta w_k\,|_{k=1,2,3}\, d\Omega + \int_\Gamma \mathcal{Z}_1 \delta w\, ds, \tag{5.3}$$

where

$$R^i = F^{+i} + F^{-i} + \int_{-h}^{h} F^i(\alpha^1, \alpha^2, \alpha^3)\, d\alpha^3, \quad i = 1, 2, \tag{5.4}$$

$$R^3 = F^{+3} + F^{-3} + \int_{-h}^{h} F^3(\alpha^1, \alpha^2, \alpha^3)\, d\alpha^3 + D^{-1}(\mathcal{Z}^i D)_{\alpha^i}, \tag{5.5}$$

$$\mathcal{Z}^i = (F^{+i} - F^{-i})h - \int_{-h}^{h} F^i(\alpha^1, \alpha^2, \alpha^3) \alpha^3\, d\alpha^3,$$

$$\mathcal{Z}_1 = \Big[(F^{+2} - F^{-2})h + \int_{-h}^{h} F^2 \alpha^3\, d\alpha^3\Big] m_2 + \Big[(F^{+1} - F^{-1})h$$

$$+\int_{-h}^{h} F^1\alpha^3\,d\alpha^3\Big]m_1 = \Big[(\mathbf{F}^+ - \mathbf{F}^-)h + \int_{-h}^{h} \mathbf{F}\alpha^3\,d\alpha^3\Big]\cdot \mathbf{m}. \quad (5.6)$$

In obtaining (5.3) we assumed that $\delta \mathbf{a}_1$ is a sufficiently smooth function, which can always be done.

5.3. Let us consider the work A_2 of stresses applied to S^0. We have

$$\delta A_2 = \int_{S^0} \mathbf{F}\cdot\delta\mathbf{u}\,d\alpha^3\,ds = \int_\Gamma\int_{-h}^{h} \mathbf{F}\cdot(\delta\mathbf{a}_0 + \alpha^3\delta\mathbf{a}_1)\,d\alpha^3\,ds. \quad (5.7)$$

From (5.7) we have

$$\delta A_2 = \int_\Gamma\int_{-h}^{h} \mathbf{F}d\alpha^3\cdot(\delta w_m\mathbf{m} + \delta w_\tau\boldsymbol{\tau} + \delta w\mathbf{n})\,ds = \int_\Gamma\int_{-h}^{h} \mathbf{F}\alpha^3 d\alpha^3\cdot\delta v_i\mathbf{e}^i\,ds. \quad (5.8)$$

In (5.8),

$$\delta w_m = \delta w^k m_k,\ \delta w_\tau = \delta w^k\tau_k,\ \delta v_i = -\delta w_{\alpha^i}.$$

Transforming (5.8) using integration by parts, and taking δw to be a sufficiently smooth function, we have

$$\delta A_2 = \int_\Gamma \Big(\widetilde{T}^m\delta w_m + \widetilde{T}^\tau\delta w_\tau + \mathcal{Z}_2\delta w + \widetilde{M}^m\cdot\delta\frac{\partial w}{\partial m}\Big)\,ds$$
$$= \int_\Gamma \big(\widetilde{T}^k\delta w_k + \mathcal{Z}_2\cdot\delta w + \widetilde{M}^m\cdot\delta w_4\big)\,ds,$$

where

$$\widetilde{T}^m = \int_{-h}^{h}\mathbf{F}\cdot\mathbf{m}\,d\alpha^3 = \int_{-h}^{h} F^i\mathbf{e}_i\cdot m_k\mathbf{e}^k\,d\alpha^3 = \int_{-h}^{h} F^k d\alpha^3 m_k$$
$$= \widetilde{T}^k m_k, \quad (5.9)$$

$$\widetilde{T}^k = \int_{-h}^{h} F^k d\alpha^3;$$

$$\widetilde{T}^\tau = \int_{-h}^{h}\mathbf{F}\cdot\boldsymbol{\tau}\,d\alpha^3 = \int_{-h}^{h} F^i\mathbf{e}_i\cdot\tau_k\mathbf{e}^k\,d\alpha^3 = \int_{-h}^{h} F^k d\alpha^3\tau_k$$
$$= \widetilde{T}^k\tau_k, \quad (5.10)$$

$$\widetilde{M}^m = \int_{-h}^{h}\mathbf{F}\alpha^3\,d\alpha^3\cdot\mathbf{m} = \int_{-h}^{h} F^k\alpha^3\,d\alpha^3 m_k; \quad (5.11)$$

$$\mathcal{Z}_2 = \int_{-h}^{h} F^3\,d\alpha^3 + \frac{\partial\widetilde{M}^s}{\partial s},\ \widetilde{M}^s = \int_{-h}^{h}\mathbf{F}\alpha^3\,d\alpha^3\cdot\boldsymbol{\tau}$$
$$= \int_{-h}^{h} F^k\alpha^3\,d\alpha^3\tau_k. \quad (5.12)$$

5.4. Let us write down a general expression for the total work of external forces on virtual displacements $\delta\mathbf{u}$ of a shell. We have

$$\begin{aligned}\delta A &= \delta A_1 + \delta A_2 \\ &= \int_S R^k \delta w_k \,|_{k=1,2,3}\, d\Omega + \int_\Gamma (\widetilde{T}^m \delta w_m + \widetilde{T}^s \delta w_s + \widetilde{Q}\,\delta w + \widetilde{M}^m \delta w_4)\, ds \\ &= \int_S R^k \delta w_k \,|_{k=1,2,3}\, d\Omega + \int_\Gamma (\widetilde{T}^i \delta w_i + \widetilde{Q}\,\delta w + \widetilde{M}^m \delta w_4)\, ds.\end{aligned} \tag{5.13}$$

Here R^k are defined by the relations (5.4), (5.5), while $\widetilde{T}^m$, $\widetilde{T}^s$, $\widetilde{M}^m$ are given by (5.9)–(5.11). Furthermore,

$$\widetilde{Q} = \mathcal{Z}_1 + \mathcal{Z}_2,$$

where $\mathcal{Z}_1$, $\mathcal{Z}_2$ are defined by (5.6), (5.12), respectively.

6. Boundary Value Problems in Displacements of the Moderate Bending Theory for Shallow Shells

6.1. To write down the boundary conditions for a shell, we assume that the boundary curve Γ admits two decompositions,

$$\Gamma = \Gamma_1 + \Gamma_2 + \Gamma_3 + \Gamma_4,$$

which is used to specify boundary conditions on the transverse displacements w and its normal derivative, and

$$\Gamma = \Gamma_5 + \Gamma_6 + \Gamma_7 + \Gamma_8,$$

used to specify boundary conditions on the tangential components w_1 and w_2 and their derivatives.

Here the Γ_i can be disconnected sets, but they must always consist of a finite number of components.

For the first decomposition of Γ, we distinguish four types of boundary conditions on w and its normal derivative.

(1) On Γ_1 there is "hard clamping":

$$w|_{\Gamma_1} = \widetilde{w}, \tag{6.1}$$

$$\left.\frac{\partial w}{\partial m}\right|_{\Gamma_1} = \widetilde{w}_4. \tag{6.2}$$

(2) On Γ_2 the transverse displacement w is fixed:

$$w|_{\Gamma_2} = \widetilde{w}, \tag{6.3}$$

while the rotation $\partial w/\partial m$ is elastically resisted by the support, so that this rotation on Γ_2 contributes

$$U_{\mathrm{supp}}\big|_{\Gamma_2} = \frac{1}{2}\int_{\Gamma_2} k_{\mathrm{f}}^{44} w_4^2 \, ds, \quad w_4 = \frac{\partial w}{\partial m} \tag{6.4}$$

to the energy of deformation. In (6.4) $k_{\mathrm{f}}^{44}(s) \geq 0$ is a piecewise-continuous elastic coefficient for the support, and w_4 is the angle of rotation.

(3) On Γ_3 the rotation $\partial w/\partial m$ is fixed,

$$w_4\big|_{\Gamma_3} \equiv \left.\frac{\partial w}{\partial m}\right|_{\Gamma_3} = \widetilde{w}_4, \tag{6.5}$$

while the transverse displacement w is elastically resisted by the support, so that this displacement on Γ_3 contributes

$$U_{\mathrm{supp}}\big|_{\Gamma_3} = \frac{1}{2}\int_{\Gamma_3} k_{\mathrm{f}}^{33} w^2 \, ds \tag{6.6}$$

to the energy of deformation. Here k_{f}^{33} is a piecewise-continuous nonnegative function of s.

(4) On Γ_4 both the translation w and the rotation $\partial w/\partial m$ are elastically resisted by the support, so that these deformations on Γ_4 contribute

$$U_{\mathrm{supp}}\big|_{\Gamma_4} = \frac{1}{2}\int_{\Gamma_4} k_{\mathrm{f}}^{ij} w_i \, w_j\big|_{i,j=3,4} \, ds \tag{6.7}$$

to the energy of deformation. The matrix $k_{\mathrm{f}}^{ij}(s)$ is assumed to be piecewise continuous and at least positive semidefinite, that is, for all w_i, w_j,

$$k_{\mathrm{f}}^{ij}(s) w_i \, w_j\big|_{i,j=3,4} \geq 0.$$

For the second decomposition of Γ, we distinguish four kinds of boundary conditions on w_1, w_2 and their derivatives:

(1) On Γ_5 the tangential components w_1, w_2 of the displacement are fixed:

$$w_1\big|_{\Gamma_5} = \widetilde{w}_1, \quad w_2\big|_{\Gamma_5} = \widetilde{w}_2. \tag{6.8}$$

(2) On Γ_6 the normal component w_m of the tangential displacement is fixed:

$$w_m\big|_{\Gamma_6} = w_k m^k\big|_{\Gamma_6} = w^k m_k\big|_{\Gamma_6} = \mathbf{w}\cdot\mathbf{m}\big|_{\Gamma_6} = \widetilde{\mathbf{w}}_m, \tag{6.9}$$

while the displacement w_τ tangent to Γ_6 is elastically resisted by the support, so that this displacement on Γ_6 contributes

$$U_{\mathrm{supp}}\big|_{\Gamma_6} = \frac{1}{2}\int_{\Gamma_6} k_{\mathrm{s}}^{\tau\tau} w_\tau^2 \, ds \tag{6.10}$$

to the energy of deformation. Note that below in formulae containing $k_{\mathrm{s}}^{\tau\tau}$, $k_{\mathrm{s}}^{\tau m}$, k_{s}^{mm}, there is no summation over indices.

Next, we have

$$w_\tau = w_k \tau^k = w^k \tau_k. \tag{6.11}$$

In (6.9), (6.10), (6.11) w_k, w^k are co- and contravariant components of the tangential displacement ω; m_k and m^k, τ_k and τ^k are the co- and contravariant components of the unit normal to Γ contained in S and of the unit tangent vector $\boldsymbol{\tau}$ to Γ, respectively. The coefficient $k_{\mathrm{s}}^{\tau\tau}(s)$ is assumed to be piecewise continuous and nonnegative.

(3) On Γ_7 the tangential displacement w_τ is fixed,

$$w_\tau|_{\Gamma_7} = \widetilde{w}_\tau, \tag{6.12}$$

while the normal component w_m is elastically resisted by the support, so that this displacement on Γ_7 contributes

$$U_{\text{supp}}\big|_{\Gamma_7} = \frac{1}{2}\int_{\Gamma_7} k_{\mathrm{s}}^{mm} w_m^2 \, ds \tag{6.13}$$

to the energy of deformation. In (6.13) $k_{\mathrm{s}}^{mm}(s)$ is nonnegative and piecewise continuous; there is no summation on m.

(4) On Γ_8 both components w_1 and w_2 (or, equivalently, w_m and w_τ) are elastically resisted by the support, so that these deformations contribute

$$U_{\text{supp}}\big|_{\Gamma_8} = \frac{1}{2}\int_{\Gamma_8} k_{\mathrm{s}}^{ij}(s) w_i \, w_j\big|_{i,j=1,2} \, ds \tag{6.14}$$

to the energy of deformation. The support matrix k_{s}^{ij} on Γ_8 is positive semidefinite and piecewise continuous.

6.2. To derive the equilibrium equations for the shell, we use the Lagrange variational principle, according to which the change in the potential energy of the shell (or, what is the same, the work of internal forces of the system on virtual displacements) equals the work of all the external forces on the same displacements. According to the principle of Lagrange,

$$\delta U_{\mathrm{s}} + \delta U_{\mathrm{f}} + \delta U_{\text{supp}} = \delta A_1 + \delta A_2, \tag{6.15}$$

where U_{s}, U_{f} are given by (4.9), (4.11), respectively. U_{supp}, depending on the boundary conditions, is a sum consisting of the contributions to the energy of deformation of the elastic resistance of supports to deformation, while δA_1, δA_2 are given by (5.13). As test functions in (6.15) we take all sufficiently smooth vectors (w_1, w_2, w) that satisfy the geometric boundary conditions (6.1)–(6.3), (6.5), (6.8), (6.9), (6.12).

Interchanging the order of integration and variation in (6.15), we obtain an integro-differential equation for $\mathbf{u}$, the vector of displacements, with arbitrary (sufficiently smooth) functions $\delta\mathbf{u} = (\delta w^1, \delta w^2, \delta w)$. Such equations occur frequently in the calculus of variations. Using traditional methods of the calculus of variations, we integrate by parts the expressions for the internal energy (transferring derivatives from the functions δw^1, δw^2, δw). Collecting terms, we obtain an expression of the form

$$\int_\Omega [(\cdots)\delta w^1 + (\cdots)\delta w^2 + (\cdots)\delta w] \, d\Omega + \int_\Gamma (\cdots) \, ds, \tag{6.16}$$

from which, using the fact that the functions δw^1, δw^2, δw are arbitrary, we obtain the equilibrium equations inside Ω,

$$\widetilde{\nabla}_j(DT^{ij}) + DR^i = 0,\ \widetilde{\nabla}_j(DT^{ij}) = (DT^{ij})_{\alpha^j} + DT^{st}G^i_{st}, \tag{6.17}$$

where

$$T^{ij} = \partial U_{\mathrm{s}}/\partial \overset{0}{\epsilon}_{ij},$$
$$\widetilde{\nabla}_{ij}(DM^{ij}) + \left(DT^{ij}w_{\alpha^i}\right)_{\alpha^j} + DT^{ij}B_{ij} + DR^3 = 0, \tag{6.18}$$

where

$$M^{ij} = \partial U_{\mathrm{f}}/\partial \overset{1}{\epsilon}_{ij},\ \widetilde{\nabla}_{ij}(DM^{ij}) = (DM^{ij})_{\alpha^i\alpha^j} + (DM^{ij}G^s_{ij})_{\alpha^s}. \tag{6.19}$$

The contour integral part of (6.16) gives us the so-called natural boundary conditions, the static boundary conditions of the form

$$-D_{\mathrm{f}}^{ijkl}\nabla_{kl}wm_i\, m_j\big|_{\Gamma_2} = (k_{\mathrm{f}}^{44}w + \widetilde{M}^m)\big|_{\Gamma_2}, \tag{6.20}$$

$$\begin{aligned}&\left\{\frac{\partial}{\partial s}\left(DD_{\mathrm{f}}^{ijkl}\nabla_{kl}w\right)\tau_j m_i D^{-1} + \frac{\partial}{\partial m}\left(DD_{\mathrm{f}}^{ijkl}\nabla_{kl}w\right)m_j m_i D^{-1}\right.\\ &\left.+\frac{\partial}{\partial s}\left(D_{\mathrm{f}}^{ijkl}\nabla_{kl}w\right)\tau_j m_i + D_{\mathrm{f}}^{ijkl}\nabla_{kl}wG^s_{ij}m_s - \frac{\partial}{\partial s}\left(D_{\mathrm{s}}^{ijkl}\overset{0}{\epsilon}_{kl}m_i\tau_j\right)\right\}\Big|_{\Gamma_3}\\ &= (k^{3j}w_j + \widetilde{Q})\big|_{\Gamma_3},\end{aligned} \tag{6.21}$$

$$-D_{\mathrm{f}}^{ijkl}\nabla_{ij}wm_k\, m_l|_{\Gamma_4} = \left(k_{\mathrm{f}}^{4j}\ w_j\big|_{j=3,4} + \widetilde{M}^m\right)\Big|_{\Gamma_4}, \tag{6.22}$$

$$\begin{aligned}&\left\{\frac{\partial}{\partial s}\left(DD_{\mathrm{f}}^{ijkl}\nabla_{ij}w\right)\tau_k m_l D^{-1} + \frac{\partial}{\partial m}\left(DD_{\mathrm{f}}^{ijkl}\nabla_{ij}w\right)m_k m_l D^{-1}\right.\\ &\left.+\frac{\partial}{\partial s}\left(D_{\mathrm{f}}^{ijkl}\nabla_{ij}w\tau_l m_k\right)D_{\mathrm{f}}^{ijkl}\nabla_{ij}wG^s_{kl}m_s - \frac{\partial}{\partial s}\left(D_{\mathrm{s}}^{ijkl}\overset{0}{\epsilon}_{kl}m_i\tau_j\right)\right\}\Big|_{\Gamma_5}\\ &= (k^{3j}w_j + \widetilde{Q})\big|_{\Gamma_5},\end{aligned} \tag{6.23}$$

$$T^\tau|_{\Gamma_6} = T^{ij}m_i\tau_j\big|_{\Gamma_6} = D_{\mathrm{s}}^{ijkl}\overset{0}{\gamma}_{kl}m_i\ \tau_j\big|_{\Gamma_6} = (k_{\mathrm{s}}^{\tau\tau}w_\tau + \widetilde{T}^s)\big|_{\Gamma_6}, \tag{6.24}$$

$$T^m\big|_{\Gamma_7} = T^{ij}m_i m_j\big|_{\Gamma_7} = D_{\mathrm{s}}^{ijkl}\overset{0}{\gamma}_{kl}m_i\ m_j\big|_{\Gamma_7} = (k_{\mathrm{s}}^{mm}w_m + \widetilde{T}^m)\big|_{\Gamma_7}, \tag{6.25}$$

$$T^\tau|_{\Gamma_8} = T^{ij}m_i\tau_j\big|_{\Gamma_8} = D_{\mathrm{s}}^{ijkl}\overset{0}{\gamma}_{kl}m_i\ \tau_j\big|_{\Gamma_8} = (k_{\mathrm{s}}^{\tau\tau}w_\tau + k_{\mathrm{s}}^{\tau m}w_m + \widetilde{T}^s)\big|_{\Gamma_8}, \tag{6.26}$$

$$T^m\big|_{\Gamma_8} = T^{ij}m_i m_j\big|_{\Gamma_8} = D_{\mathrm{s}}^{ijkl}\overset{0}{\gamma}_{kl}m_i\ m_j\big|_{\Gamma_8} = (k_{\mathrm{s}}^{m\tau}w_\tau k_{\mathrm{s}}^{mm}w_m + \widetilde{T}^m)\big|_{\Gamma_8}. \tag{6.27}$$

In relations (6.20)–(6.27), $\widetilde{M}^m$, $\widetilde{Q}$ are the bending moment and the transverse force, applied, respectively, to Γ_2, Γ_3, Γ_4; $\widetilde{T}^\tau$, $\widetilde{T}^m$ are the external tangential and normal forces, applied, respectively, to Γ_6, Γ_7, Γ_8. They are given by the relations (5.9). In view of the use to which we will put them, we rewrite the expressions (6.24)–(6.27) in the form

$$\left.\left(D_{\mathrm{s}}^{ijkl}\overset{0}{e}_{kl}m_i\tau_j - k^{\tau\tau}w_\tau\right)\right|_{\Gamma_6} = \left.\left(-\frac{1}{2}m_i\tau_j D_{\mathrm{s}}^{ijkl}w_{\alpha^k}w_{\alpha^l} + \widetilde{T}^\tau\right)\right|_{\Gamma_6}, \tag{6.28}$$

$$\left.\left(D_{\mathrm{s}}^{ijkl}\overset{0}{e}_{kl}m_i m_j - k^{mm}w_m\right)\right|_{\Gamma_7} = \left.\left(-\frac{1}{2}m_i m_j D_{\mathrm{s}}^{ijkl}w_{\alpha^k}w_{\alpha^l} + \widetilde{T}^m\right)\right|_{\Gamma_7}, \tag{6.29}$$

$$\left.\left(D_{\mathrm{s}}^{ijkl}\overset{0}{e}_{kl}m_i\tau_j - k^{\tau\tau}w_\tau - k^{\tau m}w_m\right)\right|_{\Gamma_8} = \left.\left(-\frac{1}{2}m_i\tau_j D_{\mathrm{s}}^{ijkl}w_{\alpha^k}w_{\alpha^l} + \widetilde{T}^\tau\right)\right|_{\Gamma_8}, \tag{6.30}$$

$$\left.\left(D_{\mathrm{s}}^{ijkl}\overset{0}{e}_{kl}m_i m_j - k^{m\tau}w_\tau - k^{mm}w_m\right)\right|_{\Gamma_8} = \left.\left(-\frac{1}{2}m_i m_j D_{\mathrm{s}}^{ijkl}w_{\alpha^k}w_{\alpha^l} + \widetilde{T}^m\right)\right|_{\Gamma_8}. \tag{6.31}$$

Next, taking into account (6.17) we can rewrite (6.18). Here we have

$$\begin{aligned}\left(DT^{ij}w_{\alpha^i}\right)_{\alpha^j} &= \left(DT^{ij}\right)_{\alpha^j}w_{\alpha^i} + DT^{ij}w_{\alpha^i\alpha^j}\\ &= -D\left(T^{st}G^i_{st} + R^i\right)w_{\alpha^i} + DT^{ij}w_{\alpha^i\alpha^j}.\end{aligned} \tag{6.32}$$

Substituting (6.32) into (6.18) we obtain

$$\widetilde{\nabla}_{ij}\left(DM^{ij}\right) + DT^{ij}\left(B_{ij} + w_{\alpha^i\alpha^j}\right) - DT^{ij}G^s_{ij}w_{\alpha^s} + D\left(R^3 - R^s w_{\alpha^s}\right) = 0,$$

and moreover,

$$\widetilde{\nabla}_{ij}\left(DM^{ij}\right) + DT^{ij}\left(B_{ij} + \nabla_{ij}w\right) + D\left(R^3 - R^s w_{\alpha^s}\right) = 0. \tag{6.33}$$

In equations (6.17), (6.18), (6.33) all the force terms M^{ij}, T^{ij} can be expressed in terms of w_1, w_2, w (by using (3.20)–(3.22), (3.27), (4.14)). As a result we obtain a system of three differential equations in three unknown functions w_1, w_2, w:

$$\widetilde{\nabla}_j\left(DD_{\mathrm{s}}^{ijkl}\nabla_k w_l\right) = DG^i_{st}D_{\mathrm{s}}^{stkl}\left(B_{kl}w - \frac{1}{2}w_{\alpha^k}w_{\alpha^l}\right) + \left[DD_{\mathrm{s}}^{ijkl}\left(B_{kl}w - \frac{1}{2}w_{\alpha^k}w_{\alpha^l}\right)\right]_{\alpha^j} - DR^i \tag{6.34}$$

$$= \widetilde{\nabla}_j \Big[D D_{\mathrm{s}}^{ijkl} \Big(B_{kl} w - \frac{1}{2} w_{\alpha^k} w_{\alpha^l} \Big) \Big] - D R^i ,$$

$$\nabla_{ij} \Big(D D_{\mathrm{f}}^{ijkl} \nabla_{kl} w \Big) = D D_{\mathrm{s}}^{ijkl} \overset{0}{\epsilon}_{kl} \left(B_{ij} + \nabla_{ij} w \right) + D(R^3 - R^s w_{\alpha^s}). \quad (6.35)$$

The system (6.34), (6.35) supplemented by the geometric boundary conditions (6.1), (6.3), (6.5), (6.8), (6.9), (6.12) and the static boundary conditions (6.20)–(6.23), (6.28)–(6.31) describes a wide range of problems in the theory of thin-walled structures. Their distinctive peculiarity is contained in the essential nonlinearity, which conceals all the most important mechanical effects. At the same time, it has to be noted that the system (6.34) is linear in w_1, w_2.

For convenience in the analysis below, we rewrite the system (6.34)–(6.35) in the following form:

$$\mathcal{P}^i(w_1, w_2) = f_0^i(w) + f_{\alpha^j}^{ij}(w) - D R^i = f^i(w) - D R^i , \quad (6.36)$$

where

$$\begin{aligned}
\mathcal{P}^i(w_1, w_2) &= \widetilde{\nabla}_j \left(D D_{\mathrm{s}}^{ijkl} \nabla_k w_l \right), \\
f^i(w) &= D G_{st}^i D_{\mathrm{s}}^{stkl} \Big(B_{kl} w - \frac{1}{2} w_{\alpha^k} w_{\alpha^l} \Big) \\
&\quad + \Big[D D_{\mathrm{s}}^{ijkl} \Big(B_{kl} w - \frac{1}{2} w_{\alpha^k} w_{\alpha^l} \Big) \Big]_{\alpha^j} \\
&= \widetilde{\nabla}_j \Big[D D_{\mathrm{s}}^{ijkl} \Big(B_{kl} w - \frac{1}{2} w_{\alpha^k} w_{\alpha^l} \Big) \Big], \\
f_0^i(w) &= D G_{st}^i D_{\mathrm{s}}^{stkl} \Big(B_{kl} w - \frac{1}{2} w_{\alpha^k} w_{\alpha^l} \Big), \\
f^{ij}(w) &= D D_{\mathrm{s}}^{ijkl} \Big(B_{kl} w - \frac{1}{2} w_{\alpha^k} w_{\alpha^l} \Big).
\end{aligned}$$

Next, let us introduce the notation

$$\mathcal{P}^3(w) = \widetilde{\nabla}_{ij} \left(D D_{\mathrm{f}}^{ijkl} \nabla_{kl} w \right),$$

$$f^3(w_1, w_2, w) = D D_{\mathrm{s}}^{ijkl} \overset{0}{\epsilon}_{kl} \left(B_{ij} + \nabla_{ij} w \right).$$

Then equation (6.35) can be written in the form

$$\mathcal{P}^3(w) = f^3(w_1, w_2, w) + D \left(R^3 - R^s w_{\alpha^s} \right). \quad (6.37)$$

6.3. In the formulation of general boundary value problems for the system (6.34)–(6.35) we can combine any form of bending boundary conditions and any form of tangential boundary conditions. Therefore, in our reasonably general setting we have 16 boundary value problems for (6.34), (6.35). Thus we shall distinguish the Problems $t\kappa$, $\kappa = 1, 2, 3, 4$, $t = 5, 6, 7, 8$. For example, problem 25 consists in finding the vector $\mathbf{a} = (w_1, w_2, w)$ from (6.34), (6.35) under the boundary conditions of type 2 on Γ_2, (6.3), and (6.20) and the boundary condition of type 1 on Γ_5, (6.8). Of course, we shall also consider the cases when some of the parts of the boundary are absent.

In the analysis of the boundary value problems formulated above an important part will be played by the total energy functional of the system shell-external forces. By the considerations above, we have

$$
\begin{aligned}
\mathcal{I}_{t\kappa}(\mathbf{a}) = \frac{1}{2}\Bigg\{ & \int_{\Omega} \left[D_{\mathrm{f}}^{ijkl} \overset{1}{\epsilon}_{ij}(w) \overset{1}{\epsilon}_{kl}(w) + D_{\mathrm{s}}^{ijkl} \overset{0}{\epsilon}_{ij}(\omega) \overset{0}{\epsilon}_{kl}(\omega) \right] D d\alpha^1 d\alpha^2 \\
& + \int_{\Gamma_2} k_{\mathrm{f}}^{44} w_4^2 ds + \int_{\Gamma_3} k_{\mathrm{f}}^{33} w^2 ds + \int_{\Gamma_4} k_{\mathrm{f}}^{ij}\, w_i w_j \big|_{i,j=3,4} ds + \int_{\Gamma_6} k_{\mathrm{s}}^{\tau\tau} w_\tau^2 ds \\
& + \int_{\Gamma_7} k_{\mathrm{s}}^{mm} w_m^2 ds + \int_{\Gamma_8} k_{\mathrm{s}}^{ij}\, w_i w_j \big|_{i,j=1,2} ds \Bigg\} - \int_{\Gamma_2} \widetilde{M}^m w_4 ds \\
& - \int_{\Gamma_3} \widetilde{Q} w ds - \int_{\Gamma_4} (\widetilde{M}^m w_4 + \widetilde{Q} w) ds - \int_{\Gamma_6} \widetilde{T}^\tau w_\tau ds - \int_{\Gamma_7} \widetilde{T}^m w_m ds \\
& - \int_{\Gamma_8} \left(\widetilde{T}^m w_m ds + \widetilde{T}^\tau w_\tau \right) ds - \int_{\Omega} \left(R^3 w + R^i w_i \right) D\, d\alpha^1 d\alpha^2 .
\end{aligned}
\tag{6.38}
$$

In (6.38) $\mathbf{a} = (w_1,\ w_2,\ w)$ is the total displacement vector for points in the middle surface S, and $\omega = (w_1,\ w_2)$ is the vector of tangential displacements of points of the middle surface S.

6.4. Let us discuss in more detail the classification of strained states and structure of shells that was introduced in the course of the above analysis. First of all, we distinguish a class of strained states for which (2.13) holds, and which we have called the case of moderate bending. Further, under the assumption of moderate bending, we single out a class of strained states of a shell in which the condition (3.15) holds; therefore, (3.18) and hence (3.19) hold as well. Such shells were called *shallow*. Recall that the criterion of shallowness is a composite one: It involves both the curvatures of the shell, B_{ij}, and the parameter $\overset{0}{\lambda}$, which describes the variability of the strained state of the shell. If (3.19) is satisfied due to a large value of $\overset{0}{\lambda}$, it is natural to talk about *physically shallow shells*. If (3.19) is satisfied due to large values of $|B_{ij}|$, that is, for shells of small curvature, it is natural to talk about *geometrically shallow shells*. The boundary value problems are, of course, the same for the various cases, and their analysis is the main theme of the present book. In the class of shallow shells we distinguish *developable shells*, that is, shells the middle surface of which is a developable surface (a cone, a cylinder, etc.). In the same class we also place shells whose middle surface is close to a developable one. In this case we will identify the geometry of the middle surface with the geometry of the corresponding developable surface. Finally, we distinguish the class of *properly shallow shells* (PSS), for which the middle surface is close to a plane. This is precisely the class of shells studied by the founders of shallow shell theory.

Let us consider the theory of PSS in more detail. Let the equations of the middle surface be

$$z = f(\alpha^1, \alpha^2).$$

Then for the curvatures we have

$$B_{ij} = \frac{f_{\alpha^i\alpha^j}}{\left(1 + f^2_{\alpha^1} + f^2_{\alpha^1}\right)}.$$

In the theory of properly shallow shells we take

$$B_{ij} = f_{\alpha^i\alpha^j},$$

and therefore

$$f^2_{\alpha^i} \ll 1. \tag{6.39}$$

Furthermore, from the last inequality it follows that

$$A_{11} \approx A_{22} \approx 1, \quad A_{12} \approx 0. \tag{6.40}$$

Next, in the framework of this theory we set

$$G^k_{ij} = 0. \tag{6.41}$$

As a result, we have

$$\overset{0}{\epsilon}_{ij} = \frac{1}{2}\left(w^1_{\alpha^i} + w^2_{\alpha^j}\right) - f_{\alpha^i\alpha^j} w + \frac{1}{2} w_{\alpha^i} w_{\alpha^j}, \tag{6.42}$$

$$\overset{1}{\gamma}_{11} = \overset{1}{\epsilon}_{11} = -w_{\alpha^1\alpha^1}; \; \overset{1}{\gamma}_{22} = \overset{1}{\epsilon}_{22} = -w_{\alpha^2\alpha^2};$$
$$\overset{1}{\gamma}_{12} = 2\overset{1}{\epsilon}_{12} = -2w_{\alpha^1\alpha^2}. \tag{6.43}$$

Stresses in the theory of PSS are given by (4.14), in which $\overset{0}{\epsilon}_{ij}$, $\overset{1}{\epsilon}_{ij}$ are to be taken from (6.42), (6.43). Since now D is taken to be equal to one, equations (6.34), (6.35) now assume the form

$$\left(D^{ijkl}_{\mathsf{s}} w^l_{\alpha^k}\right)_{\alpha^j} = \left[D^{ijkl}_{\mathsf{s}}\left(f_{\alpha^k\alpha^l} w - \frac{1}{2} w_{\alpha^k} w_{\alpha^l}\right)\right]_{\alpha^j} - R^i, \; i = 1, 2, \tag{6.44}$$

$$\left(D^{ijkl}_{\mathsf{f}} w_{\alpha^k\alpha^l}\right)_{\alpha^i\alpha^j} = D^{ijkl}_{\mathsf{s}} \overset{0}{\epsilon}_{kl} \left(f_{\alpha^i\alpha^j} + w_{\alpha^i\alpha^j}\right) + R^3 - R^s w_{\alpha^s}. \tag{6.45}$$

In the boundary conditions (6.20)–(6.27) we should also put

$$D = 1, G^k_{ij} = 0.$$

In the case of properly shallow shells, the functional $\mathcal{I}_{t\kappa}$ in (6.38) is also simplified as a result of (6.39)–(6.43). Therefore, the boundary value problems for the theory of PSS are simpler than in the general theory of shallow shells. At the same time, the mathematical results in the theory of PSS are not, strictly speaking, direct corollaries of theorems of the shallow shell theory, though they are obtained by similar methods. Therefore, we adopt the following approach: We shall prove in detail theorems dealing with the general theory of shallow shells, and will quote the corresponding facts for PSS without proof.

7. Boundary Value Problems with Airy Stress Function in the Moderate Bending Theory for Shallow Shells

7.1. Let us consider first the homogeneous system (6.17) by putting $R^i \equiv 0$, $i = 1, 2$. It can be immediately verified that it will be approximately solved if we put [93, 212]

$$T^{ij} = C^{ik} C^{jl} \nabla_{kl} \widetilde{\Psi}, \tag{7.1}$$

where the C^{ik} are the contravariant components of the discriminant tensor defined by the relation (1.12) and $\widetilde{\Psi}$ is some function of the variables α^i. Expanding (7.1), we obtain

$$\begin{aligned} D^2 T^{11} &= \nabla_{22} \widetilde{\Psi} = \widetilde{\Psi}_{\alpha^2 \alpha^2} - G^s_{22} \widetilde{\Psi}_{\alpha^s}, \\ D^2 T^{12} &= -\nabla_{12} \widetilde{\Psi} = -\widetilde{\Psi}_{\alpha^1 \alpha^2} + G^s_{12} \widetilde{\Psi}_{\alpha^s} \quad (1 \rightleftharpoons 2). \end{aligned} \tag{7.2}$$

Substituting (7.1), (7.2) into the homogeneous equations (6.17), we obtain

$$\left(D T^{kt} \right)_{\alpha^t} + D T^{ij} G^k_{ij} = K D \widetilde{\Psi}_{\alpha^k}, \tag{7.3}$$

where K is the Gaussian curvature of S:

$$K = (B_{11} B_{22} - B_{12}^2) D^{-2}. \tag{7.4}$$

In the derivation of (7.3) we used the formulae (1.27), (1.32), (1.33). The relations (7.3) show that if the surface S is a developable surface, that is, if $K = 0$, then (7.2) is an exact solution of the homogeneous system (6.17). On the other hand, if $K \neq 0$, then in a number of cases (7.1) gives us an approximate solution of the system (6.17).

In order to characterize the domain of validity of (7.1), it makes sense to use the arguments of the conclusion of Section 3. Let $\widetilde{\widetilde{\Psi}}(\lambda_1, \lambda_2)$ be the Fourier transform of $\widetilde{\Psi}(\alpha^1, \alpha^2)$. In naturally occurring situations, the domain where $\widetilde{\widetilde{\Psi}}(\lambda_1, \lambda_2)$ is nonnegligible is concentrated in some region of the λ_1, λ_2 plane; let $\overset{0}{\boldsymbol{\lambda}} = (\overset{0}{\lambda}_1, \overset{0}{\lambda}_2)$ be a typical (central) point of that region. Then the domain of validity of (7.1) can be characterized by the relation

$$\left| \overset{0}{\boldsymbol{\lambda}} \right| / K \gg 1. \tag{7.5}$$

In practice, one usually acts as follows: The metric on S is taken to be the metric of a developable surface that is close to S. For example, if S is close to the plane, then the metric of the plane is chosen for S. In this approach (7.1), (7.2) give us the exact solution of the homogeneous equilibrium equations on a developable surface that is close to S. In the nonhomogeneous cases we introduce a particular solution T_{p}^{ij} of the system (6.17), and then

$$T^{ij} = C^{ik} C^{jl} \nabla_{kl} \widetilde{\Psi} + T_{\mathrm{p}}^{ij}. \tag{7.6}$$

Introduction of the Airy stress function $\widetilde{\Psi}$ allows us to reduce the boundary value problem of the theory of shallow shells to two equations in the unknown functions

w, $\widetilde{\Psi}$. As is natural, doing this we somewhat restrict the domain of rational applicability of the obtained boundary value problems. Namely, we consider the case when along the entire boundary Γ we are given stresses $\widetilde{T}^m$, $\widetilde{T}^s$:

$$T^m\big|_\Gamma = T^{ij} m_i m_j\big|_\Gamma = \widetilde{T}^m, \quad T^s\big|_\Gamma = T^{ij} m_i \tau_j\big|_\Gamma = \widetilde{T}^s. \tag{7.7}$$

For the transverse conditions we shall take

$$\Gamma = \Gamma_1 + \Gamma_2 \tag{7.8}$$

so that on the entire curve Γ,

$$w|_\Gamma = \widetilde{w}. \tag{7.9}$$

In accordance with (7.8), boundary conditions (6.2), (6.20) must be satisfied.

Let us consider boundary conditions for $\widetilde{\Psi}$. By (7.6), (7.7) we have

$$T^m\big|_\Gamma = (\tau^k \tau^l \nabla_{kl} \widetilde{\Psi})\big|_\Gamma + \overset{0}{N}, \quad \overset{0}{N} = T_{\mathrm{p}}^{sk} m_s m_t, \tag{7.10}$$

$$T^\tau|_\Gamma = (m^k \tau^l \nabla_{kl} \widetilde{\Psi})\big|_\Gamma + \overset{0}{S}, \quad \overset{0}{S} = T_{\mathrm{p}}^{sk} m_s \tau_t. \tag{7.11}$$

Using (1.25), (1.26) we write (7.10), (7.11) in the following form:

$$T^m\big|_\Gamma = \left(\frac{\partial^2 \widetilde{\Psi}}{\partial s^2} + \kappa_2 \frac{\partial \widetilde{\Psi}}{\partial m} + \overset{0}{N}\right)\bigg|_\Gamma = \widetilde{T}^m, \tag{7.12}$$

$$T^s\big|_\Gamma = \left(-\frac{\partial}{\partial s}\frac{\partial \widetilde{\Psi}}{\partial m} + \kappa_2 \frac{\partial \widetilde{\Psi}}{\partial s} + \overset{0}{S}\right)\bigg|_\Gamma = \widetilde{T}^s. \tag{7.13}$$

We remind the reader that in (7.12), (7.13) κ_2 is the geodesic curvature of Γ defined by (1.16). The formulae (7.12), (7.13) allow us to compute $\widetilde{\Psi}$, $\partial\widetilde{\Psi}/\partial s$, $\partial\widetilde{\Psi}/\partial m$ on Γ. Indeed,

$$\left(\frac{\partial^2 \widetilde{\Psi}}{\partial s^2} + \kappa_2 \frac{\partial \widetilde{\Psi}}{\partial m}\right)\bigg|_\Gamma = -\overset{0}{N} + \widetilde{T}^m,$$

$$\left(\frac{\partial}{\partial s}\frac{\partial \widetilde{\Psi}}{\partial m} - \kappa_2 \frac{\partial \widetilde{\Psi}}{\partial s}\right)\bigg|_\Gamma = \overset{0}{S} - \widetilde{T}^s.$$

Multiplying the second of these equations by i ($i^2 = -1$) and adding together the two equations, we obtain

$$\left\{\frac{\partial}{\partial s}\left(\frac{\partial \widetilde{\Psi}}{\partial s} + i\frac{\partial \widetilde{\Psi}}{\partial m}\right) - i\kappa_2\left(\frac{\partial \widetilde{\Psi}}{\partial s} + i\frac{\partial \widetilde{\Psi}}{\partial m}\right)\right\}\bigg|_\Gamma = \widetilde{T}^m - \overset{0}{N} - i\left(\widetilde{T}^s - \overset{0}{S}\right).$$

Considering this relation as an ordinary differential equation in $(\partial\widetilde{\Psi}/\partial s + i\partial\widetilde{\Psi}/\partial m)$, we see that its solution is

$$\left(\frac{\partial \widetilde{\Psi}}{\partial s} + i\frac{\partial \widetilde{\Psi}}{\partial m}\right)\bigg|_\Gamma = \int_0^s \left[\widetilde{T}^m - \overset{0}{N} - i\left(\widetilde{T}^s - \overset{0}{S}\right)\right] \exp i \int_0^\mu \kappa_2(\sigma) d\sigma\, d\mu. \tag{7.14}$$

Separating the real and the imaginary parts in (7.14), we have

$$\frac{\partial \widetilde{\Psi}}{\partial s}\bigg|_\Gamma = \int_0^s \left[\left(\widetilde{T}^m - \overset{0}{N}\right) \cos \int_0^\mu \kappa_2(\sigma) d\sigma\right.$$

$$+\left(\widetilde{T}^{s}-\overset{0}{S}\right)\sin\int_{0}^{\mu}\kappa_{2}(\sigma)d\sigma\Bigg]d\mu, \tag{7.15}$$

$$\left.\frac{\partial\widetilde{\Psi}}{\partial m}\right|_{\Gamma}=\int_{0}^{s}\Bigg[\left(\widetilde{T}^{m}-\overset{0}{N}\right)\sin\int_{0}^{\mu}\kappa_{2}(\sigma)d\sigma$$

$$-\left(\widetilde{T}^{s}-\overset{0}{S}\right)\cos\int_{0}^{\mu}\kappa_{2}(\sigma)d\sigma\Bigg]d\mu, \tag{7.16}$$

and finally,

$$\widetilde{\Psi}=\int_{0}^{s}\frac{\partial\widetilde{\Psi}(\sigma)}{\partial\sigma}d\sigma. \tag{7.17}$$

Let us note that this construction is similar to the construction of the stress function for a multiply connected domain in the problem of the theory of elasticity in the plane. Thus we have obtained the values of $\widetilde{\Psi}$, $\partial\widetilde{\Psi}/\partial m$, and $\partial\widetilde{\Psi}/\partial s$ on Γ.

Two problems have to be discussed in more detail. The first of these has to do with the fact that the right-hand sides of (7.15)–(7.17) will not always be single-valued functions on the simply connected curve Γ. This will be the case if and only if the following conditions are satisfied:

$$\int_{0}^{L}\left[\left(\widetilde{T}^{m}-\overset{0}{N}\right)\cos\int_{0}^{\mu}\kappa_{2}(\sigma)d\sigma+\left(\widetilde{T}^{\tau}-\overset{0}{S}\right)\sin\int_{0}^{\mu}\kappa_{2}(\sigma)d\sigma\right]d\mu$$
$$=0, \tag{7.18}$$

$$\int_{0}^{L}\left[\left(\widetilde{T}^{m}-\overset{0}{N}\right)\sin\int_{0}^{\mu}\kappa_{2}(\sigma)d\sigma-\left(\widetilde{T}^{\tau}-\overset{0}{S}\right)\cos\int_{0}^{\mu}\kappa_{2}(\sigma)d\sigma\right]d\mu$$
$$=0, \tag{7.19}$$

$$\int_{0}^{L}\int_{0}^{s}\left[\left(\widetilde{T}^{m}-\overset{0}{N}\right)\cos\int_{0}^{\mu}\kappa_{2}(\sigma)d\sigma+\left(\widetilde{T}^{\tau}-\overset{0}{S}\right)\sin\int_{0}^{\mu}\kappa_{2}(\sigma)d\sigma\right]d\mu\,ds$$
$$=0. \tag{7.20}$$

In (7.18)–(7.20) L is the length of the boundary curve. The relations (7.18)–(7.20) reflect self-equilibration of loads applied to the shell. If S is simply connected, then it is the self-equilibration of the loads that determines single-valuedness of the boundary values of $\widetilde{\Psi}$, $\partial\widetilde{\Psi}/\partial m$, $\partial\widetilde{\Psi}/\partial s$ on Γ. If the shell is multiply connected and $\Gamma=\cup_{i=1}^{N}\Gamma^{i}$, the problems in which the loads on each boundary curve are not equilibrated also make physical sense, under the condition that they are all equilibrated over the whole shell. To explain this case, we note that here $\widetilde{\Psi}$, $\partial\widetilde{\Psi}/\partial m$, $\partial\widetilde{\Psi}/\partial s$ will not be single-valued on each of the curves Γ^{i}. Let us construct a function $\widetilde{\Psi}_{i}$ that is single-valued on all the boundary curves apart from Γ^{i} and such that $\widetilde{\Psi}-\widetilde{\Psi}_{i}$ and $\partial(\widetilde{\Psi}-\widetilde{\Psi}_{1})/\partial m$ are single-valued on Γ^{i}. We shall not show how to construct such a function in practice. Let us introduce a function $\widetilde{\Psi}_{0}$ by the relation

$$\widetilde{\Psi}=\sum_{i=1}^{N}\widetilde{\Psi}_{i}+\widetilde{\Psi}_{0}, \tag{7.21}$$

where $\widetilde{\Psi}_0$ already is single-valued on the whole of Γ. Then evidently,

$$\widetilde{\Psi}_0(s)\big|_\Gamma = \Phi_0(s), \quad \left.\frac{\partial \widetilde{\Psi}_0(s)}{\partial m}\right|_\Gamma = \Phi_1(s). \tag{7.22}$$

Finally, choosing and fixing a function $\widetilde{\Psi}_{00}$ that satisfies (7.22) and introducing the function Ψ by the relation

$$\widetilde{\Psi}_0 = \Psi + \widetilde{\Psi}_{00}, \tag{7.23}$$

we obtain for Ψ the homogeneous boundary conditions

$$\Psi|_\Gamma = \left.\frac{\partial \Psi}{\partial m}\right|_\Gamma = 0. \tag{7.24}$$

Here it follows from (7.1), (7.21), (7.23) that

$$T^{ij} = C^{ik}C^{jl}\nabla_{kl}\Psi + T^{ij}_{\mathrm{p}} + S^{ij},$$

$$S^{ij} = C^{ik}C^{jl}\nabla_{kl}\left(\sum_{i=1}^{N}\widetilde{\Psi}_i + \widetilde{\Psi}_{00}\right),$$

$$T^{ij} = C^{ik}C^{jl}\nabla_{kl}\Psi + \widetilde{T}^{ij}, \quad \widetilde{T}^{ij} = \widetilde{T}^{ij}_{\mathrm{p}} + S^{ij}.$$

Recall that the C^{ij} are given by (1.12).

The second problem has to do with the arbitrariness in the boundary conditions for $\widetilde{\Psi}$ and Ψ. From (7.12), (7.13) it follows that for homogeneous boundary conditions we have the relations

$$\frac{\partial}{\partial s}\left(\frac{\partial \Psi}{\partial s} + i\frac{\partial \Psi}{\partial m}\right) - i\kappa_2\left(\frac{\partial \Psi}{\partial s} + i\frac{\partial \Psi}{\partial m}\right) = 0.$$

Considering this relation as an ordinary differential equation in $(\partial\Psi/\partial s + i\partial\Psi/\partial m)$, we obtain its solution,

$$\frac{\partial \Psi}{\partial s} + i\frac{\partial \Psi}{\partial m} = (\gamma_1 + i\gamma_2)\exp\left(-\int_0^s \kappa_2(\sigma)d\sigma\right).$$

Thus in boundary conditions for $\partial\Psi/\partial m$ and $\partial\Psi/\partial s$ there are two arbitrary constants, γ_1 and γ_2. When we find Ψ from (7.17), we pick up a third constant. Thus we have three arbitrary constants, which were chosen to be equal to zero in (7.15)–(7.17).

To simplify the technical aspects of this question, we shall henceforth make the following assumption: $\widetilde{T}^m = \overset{0}{N}$, $\widetilde{T}^s = \overset{0}{S}$. Then we have from (7.15)–(7.17),

$$\left.\frac{\partial \widetilde{\Psi}}{\partial s}\right|_\Gamma = \left.\frac{\partial \widetilde{\Psi}}{\partial m}\right|_\Gamma = \widetilde{\Psi}\big|_\Gamma = 0.$$

Clearly, we also have $\widetilde{\Psi}_i \equiv 0$ and $\widetilde{\Psi} \equiv \widetilde{\Psi}_0$, and since in (7.22) $\Phi_0(s) \equiv \Phi_1(s) \equiv 0$, we have $\widetilde{\Phi}_{00} = 0$ and $\widetilde{\Psi} = \Psi$, $S^{ij} \equiv 0$, $\widetilde{T}^{ij} = \widetilde{T}^{ij}_{\mathrm{p}}$.

7.2. To derive the equations that couple w and $\widetilde{\Psi}$, we shall use the mixed variational principle due to Alumyae [6, 7]. In accordance with this principle, a pair of functions w, Ψ describes a real stress-deformed state of a shell if and only if it is an extremum of the functional

$$\mathcal{I} = \frac{1}{2}\Big\{ \int_\Omega \Big[D_{\mathrm{f}}^{ijkl} \nabla_{ij} w \nabla_{kl} w - 2w \left(C^{ik} C^{jl} \nabla_{kl} \Psi + \widetilde{T}_{\mathrm{p}}^{ij} \right) B_{ij} + \left(C^{ik} C^{jl} \nabla_{kl} \Psi + \widetilde{T}_{\mathrm{p}}^{ij} \right) w_{\alpha^i} w_{\alpha^j} - C_{ijkls} \left(C^{k\lambda} C^{l\mu} \nabla_{\lambda\mu} \Psi + \widetilde{T}_{\mathrm{p}}^{kl} \right) \times \left(C^{is} C^{jt} \nabla_{st} \Psi + \widetilde{T}_{\mathrm{p}}^{ij} \right) \Big] d\Omega + \int_{\Gamma_2} k_{\mathrm{f}}^{44} w_4^2 ds \Big\} - \int_{\Gamma_2} \widetilde{M}^m w_4 ds - \int_\Omega R^3 w d\Omega. \tag{7.25}$$

As test functions in the functional $\mathcal{I}$ we take all the functions w that satisfy the geometric boundary conditions (7.9) and (6.2), (6.3), and functions Ψ for which $\Psi|_\Gamma$, $\frac{\partial \Psi}{\partial m}\big|_\Gamma$ are given by (7.24). Thus for the variations δw, $\delta \Psi$ we have

$$\delta\, w|_\Gamma = \left.\frac{\partial \delta w}{\partial m}\right|_{\Gamma_1} = \delta\, \Psi|_\Gamma = \left.\frac{\partial \delta \Psi}{\partial m}\right|_\Gamma = 0.$$

Before we derive the equilibrium and compatibility equations for w and Ψ in the usual way, let us state some auxiliary results.

Lemma 7.1. *Let $S \in C^3_\Omega$, $a(\alpha^1, \alpha^2)$, $b(\alpha^1, \alpha^2) \in C^2_\Omega$ and the conditions*

$$b|_\Gamma = \left.\frac{\partial b}{\partial m}\right|_\Gamma = 0 \tag{7.26}$$

hold. Then

$$\int_\Omega a B_{ij} C^{ik} C^{jl} \nabla_{kl} b d\Omega = \int_\Omega b B_{ij} C^{ik} C^{jl} \nabla_{kl} a d\Omega. \tag{7.27}$$

To prove the claim, we note that

$$\begin{aligned} \int_\Omega a B_{ij} C^{ik} C^{jl} \nabla_{kl} b d\Omega &= \int_\Omega (B_{11} \nabla_{22} b + B_{22} \nabla_{11} b - 2B_{12} \nabla_{12} b) a D^{-1} d\alpha^1 d\alpha^2 \\ &= \int_\Omega \Big[B_{11} \left(b_{\alpha^2\alpha^2} - G^s_{22} b_{\alpha^s} \right) + B_{22} \left(b_{\alpha^1\alpha^1} - G^s_{11} b_{\alpha^s} \right) \\ &\qquad - 2B_{12} \left(b_{\alpha^1\alpha^2} - G^s_{12} b_{\alpha^s} \right) \Big] a D^{-1} d\alpha^1 d\alpha^2 \\ &= \int_\Omega (\lambda_1 b_{\alpha^1} + \lambda_2 b_{\alpha^2}) d\alpha^1 d\alpha^2, \end{aligned} \tag{7.28}$$

where

$$\lambda_1 = -aD \cdot C^{ik} C^{jl} G^2_{ij} B_{kl} - (B_{22} a D^{-1})_{\alpha^1} + (B_{12} a D^{-1})_{\alpha^2} \ (1 \rightleftharpoons 2). \tag{7.29}$$

Next, taking into account the Peterson–Codazzi relations (1.27) and the Voss–Weyl formulae (1.33), we obtain from (7.29)

$$\lambda_1 = D^{-1}(a_{\alpha^2} B_{12} - a_{\alpha^1} B_{22}) \ (1 \rightleftharpoons 2).$$

Moreover, using again (1.27), (1.33) and taking (7.26) into account, we obtain from (7.28)

$$\begin{aligned}
&\int_\Omega a B_{ij} C^{ik} C^{jl} \nabla_{kl} b d\Omega \\
&= \int_\Omega (\lambda_1 b_{\alpha^1} + \lambda_2 b_{\alpha^2}) d\alpha^1 d\alpha^2 \\
&= \int_\Omega D^{-1} \big[(a_{\alpha^2} B_{12} - a_{\alpha^1} B_{22}) b_{\alpha^1} + (a_{\alpha^1} B_{12} - a_{\alpha^2} B_{11}) b_{\alpha^2} \big] d\alpha^1 d\alpha^2 \\
&= -\int_\Omega b \Big\{ \Big[D^{-1} (a_{\alpha^2} B_{12} - a_{\alpha^1} B_{22}) \Big]_{\alpha^1} + \Big[D^{-1} (a_{\alpha^1} B_{12} - a_{\alpha^2} B_{11}) \Big]_{\alpha^2} \Big\} d\alpha^1 d\alpha^2 \\
&= -\int_\Omega b \Big[D^{-1}_{\alpha^1} (a_{\alpha^2} B_{12} - a_{\alpha^1} B_{22}) + D^{-1}_{\alpha^2} (a_{\alpha^1} B_{12} - a_{\alpha^2} B_{11}) \\
&\qquad + D^{-1} (2 a_{\alpha^1 \alpha^2} B_{12} - a_{\alpha^2 \alpha^1} B_{22} - a_{\alpha^2 \alpha^2} B_{11}) \\
&\qquad + D^{-1} a_{\alpha^2} (B_{12\alpha^1} - B_{11\alpha^2}) + D^{-1} a_{\alpha^1} (B_{12\alpha^2} - B_{22\alpha^1}) \Big] d\alpha^1 d\alpha^2 \\
&= -\int_\Omega b \Big[(G^1_{11} + G^2_{12}) D^{-1} \times (a_{\alpha^1} B_{22} - a_{\alpha^2} B_{12}) \\
&\qquad + (G^1_{21} + G^2_{22}) (a_{\alpha^2} B_{11} - a_{\alpha^1} B_{12}) \\
&\qquad + D^{-1} (2 a_{\alpha^1 \alpha^2} B_{12} - a_{\alpha^1 \alpha^2} B_{22} - a_{\alpha^2 \alpha^2} B_{11}) \\
&\qquad + D^{-1} a_{\alpha^2} (G^k_{11} B_{k2} - G^k_{12} B_{k1}) \\
&\qquad + D^{-1} a_{\alpha^1} (G^k_{22} B_{k1} - G^k_{21} B_{k2}) \Big] d\alpha^1 d\alpha^2 \\
&= \int_\Omega b (B_{22} \nabla_{11} a + B_{11} \nabla_{22} a - 2 B_{12} \nabla_{12} a) D^{-1} d\alpha^1 d\alpha^2,
\end{aligned}$$

which is the same as (7.27). Lemma 7.1 is proved.

For arbitrary functions a, b, c let us introduce the notation

$$(a,\ b,\ c) = \int_\Omega C^{ik} C^{jl} a_{\alpha^i} b_{\alpha^j} \nabla_{kl} c \, d\Omega \tag{7.30}$$

and

$$[a,\ b,\ c] = \int_\Omega a C^{ik} C^{jl} \nabla_{ij} b \nabla_{kl} c \, d\Omega. \tag{7.31}$$

Lemma 7.2. *Let S, a, $b \in C^3_\Omega$, $c \in C^2_\Omega$, and suppose c satisfies the conditions* (7.26). *Then*

$$(a,\, b,\, c) = (c,\, a,\, b) = (b,\, a,\, c) = (c,\, b,\, a). \tag{7.32}$$

To prove the first of the relations (7.32) note that

$$\begin{aligned}
(a,\, b,\, c)& \\
&= \int_\Omega \big(\nabla_{11} c a_{\alpha^2} b_{\alpha^2} + \nabla_{22} c a_{\alpha^1} b_{\alpha^1} - \nabla_{12} c a_{\alpha^1} b_{\alpha^2} - \nabla_{21} c a_{\alpha^2} b_{\alpha^1}\big) D^{-1} d\alpha^1 d\alpha^2 \\
&= \int_\Omega \Big\{ c_{\alpha^1} \Big[-\big(a_{\alpha^2} b_{\alpha^2} D^{-1}\big)_{\alpha^1} - G^1_{11} a_{\alpha^2} b_{\alpha^2} D^{-1} - G^1_{22} a_{\alpha^1} b_{\alpha^1} D^{-1} \\
&\qquad + \big(a_{\alpha^1} b_{\alpha^1} D^{-1}\big)_{\alpha^2} + G^1_{22} a_{\alpha^1} b_{\alpha^2} D^{-1} + G^1_{21} a_{\alpha^2} b_{\alpha^1} D^{-1} \Big] \\
&\qquad + c_{\alpha^2} \Big[\big(-a_{\alpha^1} b_{\alpha^1} D^{-1}\big)_{\alpha^2} - G^2_{22} a_{\alpha^1} b_{\alpha^1} D^{-1} - G^2_{11} a_{\alpha^2} b_{\alpha^2} D^{-1} \\
&\qquad + \big(a_{\alpha^2} b_{\alpha^2} D^{-1}\big)_{\alpha^1} + G^2_{12} a_{\alpha^2} b_{\alpha^1} D^{-1} + G^2_{12} a_{\alpha^1} b_{\alpha^2} D^{-1} \Big] \Big\} d\alpha^1 d\alpha^2 \\
&= \int_\Omega (c_{\alpha^1} \mu_1 + c_{\alpha^2} \mu_2) d\alpha^1 d\alpha^2.
\end{aligned} \tag{7.33}$$

Using (1.33) we shall have

$$\begin{aligned}
\mu_1 = &- \big(a_{\alpha^2} b_{\alpha^2} D^{-1}\big)_{\alpha^1} - G^1_{11} a_{\alpha^2} b_{\alpha^2} D^{-1} - G^1_{22} a_{\alpha^1} b_{\alpha^1} D^{-1} + \big(a_{\alpha^1} b_{\alpha^2} D^{-1}\big)_{\alpha^2} \\
&+ G^1_{22} a_{\alpha^1} b_{\alpha^2} D^{-1} + G^1_{21} a_{\alpha^2} b_{\alpha^1} D^{-1} = D^{-1} \left(a_{\alpha^1} \nabla_{22} b - a_{\alpha^2} \nabla_{12} b\right),
\end{aligned} \tag{7.34}$$

and similarly,

$$\mu_2 = D^{-1} \left(a_{\alpha^2} \nabla_{11} b - a_{\alpha^1} \nabla_{21} b\right). \tag{7.35}$$

From (7.30), (7.33)–(7.35) we obtain

$$\begin{aligned}
(a,\, b,\, c) &= \int_\Omega \big[c_{\alpha^1} a_{\alpha^1} \nabla_{22} b + c_{\alpha^2} a_{\alpha^2} \nabla_{11} b - (c_{\alpha^1} a_{\alpha^2} + c_{\alpha^2} a_{\alpha^1}) \nabla_{12} b\big] D^{-1} d\alpha^1 d\alpha^2 \\
&= \int_\Omega C^{ik} C^{jl} c_{\alpha^i} a_{\alpha^j} \nabla_{kl} b \, d\Omega = (c,\, a,\, b).
\end{aligned}$$

The second and third relations in (7.32) follow from the symmetry of $(a,\, b,\, c)$ with respect to a, b.

Lemma 7.3. *Let $a \in C^1_\Omega$ satisfy the first of the relations* (7.26), *and let $b \in C^2_\Omega$ and S, $c \in C^3_\Omega$. Then*

$$(a,\, b,\, c) = -[a,\, b,\, c] + \int_\Omega a K b_{\alpha^t} c_{\alpha^k} A^{tk} d\Omega, \tag{7.36}$$

where K is the curvature of S determined by (7.4).

To prove (7.36) we have

$$
\begin{aligned}
&\int_\Omega C^{ik}C^{jl}a_{\alpha^i}b_{\alpha^j}\nabla_{kl}c\,d\Omega \\
&= \int_\Omega D^{-1}\big(a_{\alpha^1}b_{\alpha^1}\nabla_{22}c + a_{\alpha^2}b_{\alpha^2}\nabla_{11}c - a_{\alpha^1}b_{\alpha^2}\nabla_{12}c \\
&\qquad - a_{\alpha^2}b_{\alpha^1}\nabla_{21}c\big)d\alpha^1 d\alpha^2 \\
&= \int_\Omega \Big[a_{\alpha^1}D^{-1}\left(b_{\alpha^1}\nabla_{22}c - b_{\alpha^2}\nabla_{11}c\right) \\
&\qquad + a_{\alpha^2}D^{-1}\left(b_{\alpha^2}\nabla_{11}c - b_{\alpha^1}\nabla_{21}c\right)\Big]d\alpha^1 d\alpha^2 \\
&= -\int_\Omega a\left(\mathcal{L}_{1\alpha^1} + \mathcal{L}_{2\alpha^2}\right)d\alpha^1 d\alpha^2,
\end{aligned}
\tag{7.37}
$$

where

$$
\begin{aligned}
\mathcal{L}_1 &= D^{-1}\left(b_{\alpha^1}\nabla_{22}c - b_{\alpha^2}\nabla_{12}c\right), \\
\mathcal{L}_2 &= D^{-1}\left(b_{\alpha^2}\nabla_{11}c - b_{\alpha^1}\nabla_{21}c\right),
\end{aligned}
$$

and so

$$
\begin{aligned}
\mathcal{L}_{1\alpha^1} &= D^{-1}\Big[\big(b_{\alpha^1\alpha^1} - G_{1k}^k b_{\alpha^1}\big)\nabla_{22}c - c\big(b_{\alpha^1\alpha^2} - G_{1k}^k b_{\alpha^2}\big)\nabla_{12}c\Big] \\
&\quad + D^{-1}b_{\alpha^1}\left(c_{\alpha^1\alpha^2\alpha^2} - G_{22\alpha^1}^k c_{\alpha^k} - G_{22}^k c_{\alpha^k\alpha^1}\right) \\
&\quad - D^{-1}b_{\alpha^2}\left(c_{\alpha^1\alpha^1\alpha^2} - G_{12\alpha^1}^k c_{\alpha^k} - G_{12}^k c_{\alpha^k\alpha^1}\right)\ (1 \rightleftharpoons 2).
\end{aligned}
\tag{7.38}
$$

In the derivation of (7.38) we used the formulae (1.32).

From (7.38) it follows that

$$
\begin{aligned}
\mathcal{L}_{1\alpha^1} + \mathcal{L}_{2\alpha^2} &= D^{-1}\Big[\big(b_{\alpha^2\alpha^2} - G_{2k}^k b_{\alpha^2}\big)\nabla_{11}c + \big(b_{\alpha^1\alpha^1} - G_{1k}^k b_{\alpha^1}\big)\nabla_{22}c \\
&\qquad + \big(-2b_{\alpha^1\alpha^2} + G_{1k}^k b_{\alpha^2} + G_{2k}^k b_{\alpha^1}\big)\nabla_{12}c\Big] \\
&\quad + D^{-1}b_{\alpha^1}\Big[-G_{22}^k c_{k\alpha^1} + G_{12}^k c_{k\alpha^2} + \big(G_{12\alpha^2}^k - G_{22\alpha^1}^k\big)c_{\alpha^k} \\
&\quad + D^{-1}b_{\alpha^2}\Big[-G_{11}^k c_{k\alpha^2} + G_{12}^k c_{k\alpha^1} + \big(G_{12\alpha^1}^k - G_{11\alpha^2}^k\big)c_{\alpha^k}\Big].
\end{aligned}
\tag{7.39}
$$

Using the formulae for the derivatives of G^k_{ij} (1.32), we obtain from (7.39)

$$\begin{aligned}
\mathcal{L}_{1\alpha^1} + \mathcal{L}_{2\alpha^2} = {} & DC^{ik}C^{jl}\nabla_{ij}c\nabla_{kl}b \\
& + D^{-1}\Big\{\big(G^1_{22}b_{\alpha^1} - G^1_{21}b_{\alpha^2}\big)\nabla_{11}c + \big(G^2_{11}b_{\alpha^2} - G^2_{12}b_{\alpha^1}\big)\nabla_{22}c \\
& \qquad + \Big[\big(G^1_{11} - G^2_{21}\big)b_{\alpha^2} - \big(G^1_{12} - G^2_{22}\big)b_{\alpha^1}\Big]\nabla_{12}c\Big\} \\
& + D^{-1}b_{\alpha^1}\Big[-G^k_{22}c_{\alpha^1\alpha^k} + G^k_{12}c_{\alpha^2\alpha^k} + \big(-G^t_{21}G^k_{2t} + G^t_{22}G^k_{1t}\big)c_{\alpha^k} \\
& \qquad + \big(B^2_{21} - B_{11}B_{22}\big)A^{1k}c_{\alpha^k}\Big] \\
& + D^{-1}b_{\alpha^2}\Big[-G^k_{11}c_{\alpha^2\alpha^k} + G^k_{12}c_{\alpha^1\alpha^k} + \big(G^t_{11}G^k_{2t} - G^t_{12}G^k_{1t}\big)c_{\alpha^k} \\
& \qquad + \big(B^2_{12} - B_{11}B_{22}\big)A^{2k}c_{\alpha^k}\Big].
\end{aligned} \tag{7.40}$$

By elementary simplifications we obtain from (7.40)

$$\mathcal{L}_{1\alpha^1} + \mathcal{L}_{2\alpha^2} = DC^{ik}C^{jl}\nabla_{ij}c\nabla_{kl}b + D^{-1}b_{\alpha^t}A^{tk}c_{\alpha^k}(B^2_{12} - B_{11}B_{22}), \tag{7.41}$$

while from (7.37), (7.41) we have

$$\int_\Omega (\mathcal{L}_{1\alpha^1} + \mathcal{L}_{1\alpha^1})a\,d\alpha^1 d\alpha^2 = \int_\Omega aC^{ik}C^{jl}\nabla_{ij}c\nabla_{kl}b\,d\Omega - \int_\Omega aKb_{\alpha^t}c_{\alpha^k}A^{tk}d\Omega,$$

and furthermore,

$$(a,\,b,\,c) = -[a,\,b,\,c] + \int_\Omega aKb_{\alpha^t}c_{\alpha^k}A^{tk}d\Omega,$$

which is the same as (7.36). Lemma 7.3 is proved.

7.3. Let us return to a variational analysis of the functional $\mathcal{I}$. We have

$$\delta\mathcal{I} = \delta\mathcal{I}_1 + \delta\mathcal{I}_2,$$

where

$$\begin{aligned}
\delta\mathcal{I}_1 = {} & \frac{1}{2}\int_\Omega \big[D^{ikjl}_{\mathrm{f}}(\nabla_{ij}w\nabla_{kl}\delta w + \nabla_{ij}\delta w\nabla_{kl}w) - 2\delta w(C^{ik}C^{jl}\nabla_{kl}\Psi + \widetilde{T}^{ij}_{\mathrm{p}}B_{ij}) \\
& \qquad + (C^{ik}C^{jl}\nabla_{kl}\Psi + \widetilde{T}^{ij}_{\mathrm{p}})(\delta w_{\alpha^i}w_{\alpha^j} + w_{\alpha^i}\delta w_{\alpha^j})\big]d\Omega \\
& + \int_{\Gamma_2} k^{44}_{\mathrm{f}}w_4\delta w_4 ds - \int_{\Gamma_2}\widetilde{M}^m\delta w_4 dS - \int_\Omega R^3\delta w\,d\Omega,
\end{aligned} \tag{7.42}$$

$$\begin{aligned}
\delta\mathcal{I}_2 = {} & \frac{1}{2}\int_\Omega \big[-2wC^{ik}C^{jl}\nabla_{kl}\delta\Psi B_{ij} + C^{ik}C^{jl}\nabla_{kl}\delta\Psi w_{\alpha^i}w_{\alpha^j} \\
& - C_{ijkls}C^{k\lambda}C^{l\mu}\nabla_{\lambda\mu}\delta\Psi(C^{is}C^{jt}\nabla_{st}\Psi + \widetilde{T}^{ij}_{\mathrm{p}}) \\
& - C_{ijkls}(C^{k\lambda}C^{l\mu}\nabla_{\lambda\mu}\Psi + \widetilde{T}^{kl}_{\mathrm{p}})C^{is}C^{jt}\nabla_{st}\delta\Psi\big]d\Omega.
\end{aligned} \tag{7.43}$$

Transforming the right-hand side of (7.42) in the usual fashion by transferring differentiation to the other terms, we obtain

$$\begin{aligned}\frac{1}{2}&\int_\Omega D_{\mathrm{f}}^{ikjl}(\nabla_{ij}w\nabla_{kl}\delta w+\nabla_{ij}\delta w\nabla_{kl}w)d\Omega\\&=\int_\Omega D_{\mathrm{f}}^{ikjl}\nabla_{ij}w\nabla_{kl}\delta w d\Omega=\int_\Omega D_{\mathrm{f}}^{ikjl}\nabla_{ij}\delta w\nabla_{kl}w d\Omega\\&=\int_\Omega\widetilde{\nabla}_{ij}(DD_{\mathrm{f}}^{ijkl}\nabla_{kl}w)\delta w d\Omega+\int_{\Gamma_2}D_{\mathrm{f}}^{ijkl}\nabla_{kl}wm_im_j\delta w_4 ds.\end{aligned}\tag{7.44}$$

The operator $\widetilde{\nabla}_{ij}$ on the right-hand side of (7.44) is defined by the relation

$$\widetilde{\nabla}_{ij}\Theta=\Theta_{\alpha^i\alpha^j}+\left(\Theta G_{ij}^s\right)_{\alpha^s}.$$

Furthermore, if in (7.36) we set

$$a=\delta w,\ b=w,\ c=\Psi,$$

we then have

$$\begin{aligned}&\int_\Omega C^{ik}C^{jl}\delta w_{\alpha^i}w_{\alpha^j}\nabla_{kl}\Psi d\Omega\\&\quad=-\int_\Omega C^{ik}C^{jl}\nabla_{kl}\Psi\nabla_{ij}w\delta w d\Omega+\int_\Omega Kw_{\alpha^k}\Psi_{\alpha^t}A^{tk}\delta w d\Omega.\end{aligned}\tag{7.45}$$

Taking now into account the shallowness assumption (7.5) that was already used earlier, we obtain from (7.45)

$$\int_\Omega C^{ik}C^{jl}\delta w_{\alpha^i}w_{\alpha^j}\nabla_{kl}\Psi d\Omega\approx-\int_\Omega C^{ik}C^{jl}\nabla_{kl}\Psi\nabla_{ij}w\delta w\, d\Omega.\tag{7.46}$$

Now, since $\widetilde{T}_{\mathrm{p}}^{ij}$ satisfies (6.17), we have the formulae

$$\begin{aligned}\int_\Omega\widetilde{T}_{\mathrm{p}}^{ij}\delta w_{\alpha^i}w_{\alpha^j}d\Omega&=\int_\Omega\widetilde{T}_{\mathrm{p}}^{ij}\delta w_{\alpha^j}w_{\alpha^i}d\Omega=-\int_\Omega\left(\widetilde{T}_{\mathrm{p}}^{ij}Dw_{\alpha^i}\right)_{\alpha^j}\delta w d\alpha^1 d\alpha^2\\&=-\int_\Omega\left(\widetilde{T}_{\mathrm{p}}^{ij}D\right)_{\alpha^j}w_{\alpha^i}\delta w d\alpha^1 d\alpha^2\\&\quad-\int_\Omega\widetilde{T}_{\mathrm{p}}^{ij}Dw_{\alpha^i\alpha^j}\delta w d\alpha^1 d\alpha^2.\end{aligned}\tag{7.47}$$

Finally, from (7.47) we have

$$\begin{aligned}\int_\Omega\widetilde{T}_{\mathrm{p}}^{ij}\delta w_{\alpha^i}w_{\alpha^j}d\Omega&=\int_\Omega D\widetilde{T}_{\mathrm{p}}^{ij}G_{ij}^k w_{\alpha^k}\delta w d\alpha^1 d\alpha^2-\int_\Omega\widetilde{T}_{\mathrm{p}}^{ij}Dw_{\alpha^i\alpha^j}\delta w d\alpha^1 d\alpha^2\\&\quad+\int_\Omega R^s w_{\alpha^s}w d\Omega\\&=-\int_\Omega\left(\widetilde{T}_{\mathrm{p}}^{ij}\nabla_{ij}w-R^s w_{\alpha^s}\right)\delta w d\Omega.\end{aligned}\tag{7.48}$$

Substituting (7.44), (7.46)–(7.48) into (7.42), we have

$$\delta\mathcal{I}_1 = \int_\Omega \Big[D^{-1}\widetilde{\nabla}_{ij}(DD_{\rm s}^{ijkl}\nabla_{kl}w) - (C^{ik}C^{jl}\nabla_{kl}\Psi + \widetilde{T}_{\rm p}^{ij})(B_{ij} + \nabla_{ij}w) + R^s w_{\alpha^s} - R^3\Big]\delta w\, d\Omega + \int_{\Gamma_2}\left(D_{\rm f}^{ijkl}\nabla_{kl}w m_i m_j + k_{\rm s}^{44}w_{\rm p} - \widetilde{M}^m\right)\delta w\, ds. \tag{7.49}$$

By the Alumyae variational principle [6, 7],

$$\delta\mathcal{I}_1 = 0 \tag{7.50}$$

for any δw that satisfy the homogeneous geometric boundary conditions on Γ_1 and Γ_2. Therefore, from (7.49), (7.50) it follows that

$$\widetilde{\nabla}_{ij}(DD_{\rm f}^{ijkl}\nabla_{kl}w) = D\Big[(C^{ik}C^{jl}\nabla_{kl}\Psi + \widetilde{T}_{\rm p}^{ij})(B_{ij} + \nabla_{ij}w) + R^3 - R^s w_{\alpha^s}\Big]. \tag{7.51}$$

The static boundary condition of (7.49) is obtained in the form (6.20). Equation (7.51) is the first equation of the nonlinear theory of shallow shells with an Airy stress function Ψ.

To obtain the second equation, let us transform the right-hand side of (7.43) by moving the derivatives of $\delta\Psi$ onto the adjacent terms. Setting in (7.27)

$$b = \delta\Psi,\ a = w,$$

we obtain

$$\int_\Omega wC^{ik}C^{jl}\nabla_{kl}\delta\Psi B_{ij}d\Omega = \int_\Omega C^{ik}C^{jl}\nabla_{kl}wB_{ij}\delta\Psi d\Omega. \tag{7.52}$$

Next, if in (7.32) we set

$$c = \delta\Psi,\ a = b = w,$$

then we shall have

$$\int_\Omega C^{ik}C^{jl}w_{\alpha^i}w_{\alpha^j}\nabla_{kl}\delta\Psi d\Omega = \int_\Omega C^{ik}C^{jl}\delta\Psi_{\alpha^i}w_{\alpha^j}\nabla_{kl}wd\Omega, \tag{7.53}$$

and if we now use (7.36) setting

$$a = \delta\Psi,\ b = c = w,$$

then for the right-hand side of (7.53) we shall have

$$\int_\Omega C^{ik}C^{jl}\delta\Psi_{\alpha^i}w_{\alpha^j}\nabla_{kl}wd\Omega = -\int_\Omega \delta\Psi C^{ik}C^{jl}\nabla_{ij}w\nabla_{kl}wd\Omega + \int_\Omega \delta\Psi K w_{\alpha^l}w_{\alpha^k}A^{lk}d\Omega. \tag{7.54}$$

Again using the shallowness assumption (7.5) we can write (7.54) in the form

$$\int_\Omega C^{ik}C^{jl}\delta\Psi_{\alpha^i\alpha^j}\nabla_{kl}wd\Omega = -\int_\Omega \delta\Psi C^{ik}C^{jl}\nabla_{ij}w\nabla_{kl}wd\Omega. \tag{7.55}$$

Finally,

$$\int_\Omega C_{ijkls}C^{k\lambda}C^{l\mu}C^{is}C^{jt}\nabla_{\lambda\mu}\delta\Psi\nabla_{st}\Psi d\Omega = \int_\Omega \widetilde{\nabla}_{\lambda\mu}DC_*^{\lambda\mu st}\nabla_{st}\Psi\delta\Psi d\alpha^1 d\alpha^2,$$

$$C_*^{\lambda\mu st} = C_{ijkls}C^{k\lambda}C^{l\mu}C^{is}C^{jt}, \tag{7.56}$$

$$\int_\Omega C_{ijkls}C^{k\lambda}C^{l\mu}\nabla_{\lambda\mu}\delta\Psi\widetilde{T}_{\mathrm{p}}^{ij}d\Omega = \int_\Omega \widetilde{\nabla}_{\lambda\mu}(C_{*ij}^{\lambda\mu}D\widetilde{T}_{\mathrm{p}}^{ij})\delta\Psi d\alpha^1 d\alpha^2. \tag{7.57}$$

Substituting (7.52), (7.53), (7.55)–(7.57) into (7.43) we shall have

$$\begin{aligned}\delta\mathcal{I}_2 = \int_\Omega \Big[&- DC^{ik}C^{jl}B_{ij}\nabla_{kl}w - \frac{1}{2}C^{ik}C^{jl}\nabla_{ij}w\nabla_{kl}wD \\ &- \widetilde{\nabla}_{\lambda\mu}DC_*^{\lambda\mu st}\nabla_{st}\Psi - \widetilde{\nabla}_{\lambda\mu}(C_{*ij}^{\lambda\mu}D\widetilde{T}_{\mathrm{p}}^{ij})\Big]\delta\Psi d\alpha^1 d\alpha^2,\end{aligned} \tag{7.58}$$

and by the Alumyae mixed variational principle,

$$\delta\mathcal{I}_2 = 0 \tag{7.59}$$

for any $\delta\Psi$ that satisfies (7.24). From (7.59) we obtain

$$\widetilde{\nabla}_{\lambda\mu}DC_*^{\lambda\mu st}\nabla_{st}\Psi = -DC^{ik}C^{jl}\nabla_{kl}w\left(B_{ij} + \frac{1}{2}\nabla_{ij}w\right) - \widetilde{\nabla}_{\lambda\mu}(C_{*ij}^{\lambda\mu}D\widetilde{T}_{\mathrm{p}}^{ij}). \tag{7.60}$$

Equation (7.60) is the second equation of the nonlinear theory of shallow shells with an Airy stress function Ψ.

Thus the system (7.51), (7.60) with the boundary conditions (7.9), (6.2), (7.24) defines the boundary value problem of the shallow shell theory with an Airy function Ψ.

Let us note two points. Equation (7.60) with the boundary conditions (7.24) is linear in Ψ, which simplifies both the analysis of the problem and its approximate solution. Furthermore, it is necessary to realize that the boundary value problem obtained above does not quite correspond to the Alumyae mixed variational principle, since we repeatedly used the shell shallowness condition (7.5). Total correspondence will hold if S is a developable surface, that is, if $K \equiv 0$, or if we can introduce on it the metric of a developable surface to any required precision.

7.4. In the case of a properly shallow shell (PSS), the defining equations with a stress function have the form

$$T^{11} = -\widetilde{\Psi}_{\alpha^2\alpha^2} + T_{\mathrm{p}}^{11};\ T^{12} = -\widetilde{\Psi}_{\alpha^1\alpha^2} + T_{\mathrm{p}}^{12};\ T^{22} = -\widetilde{\Psi}_{\alpha^1\alpha^1} + T_{\mathrm{p}}^{22}.$$

The functional $\mathcal{I}$ of (7.25) becomes

$$\begin{aligned}\mathcal{I} = \frac{1}{2}\Big\{ \int_\Omega \Big[&D_{\mathrm{f}}^{ijkl}w_{\alpha^i\alpha^j}w_{\alpha^k\alpha^l} - \left(C^{ik}C^{jl}\widetilde{\Psi}_{\alpha^k\alpha^l} + T_{\mathrm{p}}^{ij}\right)(2wf_{\alpha^i\alpha^j} - w_{\alpha^i}w_{\alpha^j}) \\ &- C_{ijkl\,\mathrm{s}}\left(C^{ki}C^{lj}\widetilde{\Psi}_{\alpha^i\alpha^j} + T_{\mathrm{p}}^{kl}\right)\left(C^{is}C^{jt}\widetilde{\Psi}_{\alpha^s\alpha^t} + T_{\mathrm{p}}^{ij}\right)\Big] d\alpha^1 d\alpha^2\Big\} \\ &+ \int_{\Gamma_2} k_{\mathrm{f}}^{44}\left(\frac{\partial w}{\partial m}\right)^2 ds - \int_{\Gamma_2}\widetilde{M}^m\left(\frac{\partial w}{\partial m}\right) ds - \int_\Omega R^3 w d\alpha^1 d\alpha^2.\end{aligned} \tag{7.61}$$

In passing from (7.25) to (7.61) we took into account that for properly shallow shells,

$$
\begin{aligned}
&A_{11} = A_{22} = 1, \; A_{12} = 0; \; D = 1; \\
&G_{ij}^k = 0, \; C^{11} = C^{22} = 0, \; C^{12} = -C^{21} = -1.
\end{aligned}
\tag{7.62}
$$

Let us state now the main equations with a stress function for the theory of properly shallow shells. Taking into consideration (7.62), we obtain from (7.51), (7.60)

$$
\left(D_{\mathrm{f}}^{ijkl} w_{\alpha^k\alpha^l}\right)_{\alpha^i\alpha^j} = \left(C^{is}C^{jt}\Psi_{\alpha^s\alpha^t} + \widetilde{T}_{\mathrm{p}}^{ij}\right)(f_{\alpha^i\alpha^j} + w_{\alpha^i\alpha^j}) - R^3 - R^s w_{\alpha^s}, \tag{7.63}
$$

$$
\left(C_{\mathrm{s}}^{ijst}\Psi_{\alpha^s\alpha^t}\right)_{\alpha^i\alpha^j} = -C^{is}C^{jt} w_{\alpha^s\alpha^t}(f_{\alpha^i\alpha^j} + w_{\alpha^i\alpha^j}) - \left(C_{\mathrm{s}kl}^{ij}\widetilde{T}_p^{kl}\right)_{\alpha^i\alpha^j}. \tag{7.64}
$$

The boundary conditions (7.15)–(7.17) remain the same.

8. Some Remarks on Nonlinear Shallow Shell Theory. A Historical Survey

8.1. Let us summarize all the assumptions used in the derivation of the main boundary value problems (6.17), (6.33)–(6.35), (6.1)–(6.3), (6.5), (6.8), (6.9), (6.11)–(6.19), (6.20)–(6.23), (6.28)–(6.31) and (7.51), (7.60), (7.24), (7.1), (7.2), (7.7), (7.9).

Assumption 1: Smallness of strains (2.7).

Assumption 2: "Moderate bending" (2.13).

Assumption 3: On the prevailing value of the angles of rotation ω_{12}, ω_{21}, ω_{31}, ω_{32} (2.19).

Assumption 4: The Kirchhoff hypotheses (3.1), (3.2).

Assumption 5: The Kirchhoff hypotheses: smallness of transverse displacements ϵ_{i3}, $i = 1, 2$ in comparison with other components of the deformation tensor ϵ_{ij}, $i, j = 1, 2$.

Assumption 6: The Kirchhoff hypotheses: the stress σ_{33} is negligibly small in comparison with other components of the stress tensor.

Assumption 7: The shell is thin-walled (3.13).

Assumption 8: The shell is shallow (3.15), (3.18).

Assumption 9: The material of the shell is regular (4.3).

Assumption 10: The shell is symmetrical with respect to the α^3 axis: elastic moduli are even in α^3; $h^+ = h^- = h$.

In the process of solving a particular problem, the question of applicability of the theory of shallow shells arises naturally. It can be conceivably resolved in two different ways. One could solve the problem at hand using a more exact theory that does not make the assumptions of the shallow shell theory and then compare the results. Of course, this approach is very labor-intensive and can only be put into practice in a number of benchmark cases.

A different alternative is to use the results obtained in the framework of the shallow shell theory to estimate the magnitude of the approximations made. These estimates can be obtained in parallel with the solution of the problem proper; it involves a large amount of additional computational work, is quite robust, but not completely convincing from the point of view of a purely logical analysis of the problem. Some reflections on this question are presented in Chapter III.

8.2. Let us note an important feature of the basic relation of shallow shell theory: Assumption (2.13) concerning moderate bending is invariant neither with respect to small displacements nor with respect to finite ones. Indeed, imposing on the shell a rigid body rotation, rotation angles ω_i can assume any value. Therefore, strictly speaking, relations (2.13), (2.19) make no sense, since relative extensions are invariant with respect to rigid body motions of the shell. This is also related to the fact that formulae (3.20)–(3.26) are invariant neither with respect to small displacements, which property holds for the formulae for e_{ij}, nor with respect to finite rotations, which is true for the exact relations (2.5) for ϵ_{ij}. Therefore, relations (2.13), (2.19) are to be understood in a very particular sense. Namely, from the very beginning we must assume that the shell is operating in conditions that preclude its rigid body rotations. Conversely, if the shell can move as a rigid body, relations (2.13), (2.19) are to be understood in the sense that there exists a configuration of the deformed shell for which rotation angles ω_i have values that correspond to (2.13), (2.19).

8.3. It seems useful to discuss further the concept of a "shallow shell" itself. We shall do it first for problems $t\kappa$ in displacements. The shallowness criterion (3.19) suggested in Section 3 appears to be sufficiently universal and adequate. From it, it can be seen that the relations of shallow shell theory can hold for large values of $|\overset{0}{\lambda}|$ even if the values of B_{ij}, and thus the curvature, are large. Here the metric A_{ij} on the middle surface S does not have to be Euclidean. In fact, it can be significantly different from a Euclidean metric. Such shells can be collected in a separate class and can be justifiably called physically shallow shells.

On the other hand, relations (3.19) can hold for sufficiently small values of $|B_{ij}|$. This fact can be interpreted also in the sense that the middle surface of the shell S has to be close to a plane. In particular, Vlasov [332, 333], one of the founders of the theory of shallow shells, suggested calling a shell shallow if the ratio of its elevation $\mathcal{H}$ to its linear size $\mathcal{L}$ is of order $\frac{1}{10}$ to $\frac{1}{15}$. Of course, he had in mind shells whose curvature changes smoothly, which is the case for many types of construction structures. In the general case the ratio $\mathcal{H}/\mathcal{L}$ cannot provide an exhaustive criterion of the shallowness of the shell, and to solve the problem completely, one needs to estimate either $|B_{ij}|$ or (3.19). Chien [51, 52] suggests considering a shell shallow if the ratio of the size of its projection to the average radius of curvature is small. Here it can also be said that this criterion is applicable only if the curvature changes slowly, but in that case the criteria of Vlasov and Chien coincide.

Below, shells for which (3.19) holds for small B_{ij} will be called geometrically shallow.

Let us also note that an important factor in the use of a version of shell theory is the ability to introduce a Euclidean metric. Conditions to be able to do this are given in Theorem 1.2. The metric on S can also be taken to be Euclidean if that surface is close to a developable one. In that case equations (7.51), (7.60) are a consequence of the mixed variational principle of Alumyae [6, 7], as shown in Section 7. Clearly, such shallow shells should be called developable. As a criterion of a shell being in the class of developable shells we could use the inequality

$$A_{12\alpha^1\alpha^2} - \frac{1}{2}(A_{11\alpha^2\alpha^2} + A_{22\alpha^1\alpha^1}) + A_{\gamma\delta}G^{\gamma}_{12}G^{\delta}_{12} - A_{\alpha\beta}G^{\alpha}_{11}G^{\beta}_{22} \gg B_{11}B_{22} - B^2_{12}, \tag{8.1}$$

which follows from Gauss's equation (1.27).

Shells that are shallow in the sense of Vlasov and Chien, for which the middle surface is close to the plane, can be called properly shallow. Properly shallow shells are a particular case of developable ones. The theory of such shells is being widely used.

8.4. The development of the theory of shallow shells displays a number of observable stages. Its foundations lie in the work of Bubnov [44] and von Kármán [131]. Bubnov was the first to formulate the problem of buckling of a curved plate and introduced the term "buckling" itself. Von Kármán was the first to derive the "moderate bending" equations for a plate. He explicitly stated assumption (2.13) and introduced the stress function Ψ for the plate. This work was the source of much subsequent technical work. The work of Bubnov and von Kármán comprises the first stage in the development of shallow shell theory.

Later, in the 1930s, we find in the work of Donnell [67, 68] the first formulation of the idea of shallowness, expressed as the assumption of negligibility of shear forces in the equations governing the stresses tangent to the middle surface of the shell. The latter condition is clearly equivalent to Assumption 8, defined by relations (3.18), (3.19). In the papers of Mushtari [209, 210, 211, 212] dating from 1938, this idea was extensively developed, and using it, many problems of stability of shells were solved. In the work of Donnell and Mushtari the idea of shallowness was used only in linear problems of shell theory.

The third stage is connected with the names of Marguerre, Vlasov, Chien, Feodos'ev, and other authors. The principal work of Marguerre [191] appeared in 1939. In it, he extended the idea of von Kármán to the case of properly shallow shells; Marguerre's equations are written in Cartesian coordinates in the plane. The mid 1940s saw the publication of the work of Vlasov [332, 333] and Chien [51, 52]. There the boundary value problems of the theory of properly shallow shells were written in general curvilinear coordinates in the plane. In 1946 Feodos'ev proposed a version of nonlinear theory of shallow shells of revolution [78]. Feodos'ev equations were widely used in solving practical problems.

The work of Marguerre, Vlasov, Chien, and Feodos'ev made use of Assumption 2 of "moderate bending" and Assumption 8 of shallowness (3.15), (3.18).

An important part in the development of the nonlinear theory of shallow shells was played by the monograph of Novozhilov [221], which contains a consistent derivation of von Kármán's equations, together with an analysis of their errors and limits of applicability.

Galimov [88, 89, 90, 91, 92, 93, 94, 95], using his work on general nonlinear three-dimensional theory of elasticity, used extensively methods of modern differential geometry and tensor analysis.

A new stage in the development of the theory is connected with the work of Reissner [256, 257, 258, 259, 260] and Alumyae [5], in which the main boundary value problems were derived without the assumption of "moderate bending," while angles of rotation were considered to be finite. In the mid-sixties appeared the work in this direction of Koiter [148], Sanders [267], Simmonds, and Danielson [278, 279], developing a nonlinear shell theory free of the "moderate bending" assumption. In this connection, let us also note an important series of papers by Pietraszkiewicz [234, 235, 236, 237, 238], where the problem is considered in great generality. A review of the work of Pietraszkiewicz is contained in [237]. An asymptotic derivation of the boundary value problems of nonlinear plate theory is to be found in the work of Destuynder and Ciarlet [58, 65]. Let us also note the work of Shapovalov [277].

It is to be expected that in the near future workers in this area will try to construct a nonlinear shell theory based only on assumption 7 (3.13), of the shell being thin-walled. The scheme of a consistent analysis of the main assumptions of nonlinear shell theory and derivation of boundary value problems is due to the author. In a different notation, the same equations can be found in the work of Koiter [148].

CHAPTER II

General Mathematical Questions

9. Some General Mathematical Results

9.1. Let us remind the reader of the necessary facts and definitions from the theory of real Banach spaces. A more detailed presentation can be found in any textbook of functional analysis [156, 408].

Let X be a real vector space of elements ϕ. X is called a *normed* space if each element ϕ can be a associated with a real number $\|\phi\|$, called the *norm*, having the following properties (the norm axioms):

(1) $\|\phi\| \geq 0$, and $\|\phi\| = 0$ if and only if $\phi = 0$;
(2) $\|c\phi\| = |c|\|\phi\|$;
(3) $\|\phi + \psi\| \leq \|\phi\| + \|\psi\|$,

which hold for all $\phi,\ \psi \in X$ and any real number c.

Frequently, it is convenient to use different norms in the same space. Two norms $\|\cdot\|_1$ and $\|\cdot\|_2$ defined on the same normed space are said to be *equivalent* if there are positive constants m and M such that for any element $\phi \in X$ we have

$$m\|\phi\|_2 \leq \|\phi\|_1 \leq M\|\phi\|_2.$$

Equivalent norms do not change the nature of convergence of sequences of elements of the space.

A sequence of elements $\phi_n \in X$ is said to *converge* to an element $\phi_0 \in X$ if $\lim_{n\to\infty} \|\phi_n - \phi_0\| = 0$.

A sequence of elements $\phi_n \in X$ is called a *Cauchy sequence* if $\lim \|\phi_n - \phi_m\| = 0$ as $n,\ m \to \infty$.

A space X is called a Banach space if each Cauchy sequence has a limit in the space.

If we can define in a Banach space a bilinear function (ϕ, ψ), the *scalar product* of the elements, with the properties

(1) $(\phi, \phi) = \|\phi\|^2$;
(2) $(\phi, \psi) = (\psi, \phi)$;
(3) $(c_1\phi_1 + c_2\phi_2, \psi) = (c_1\phi_1, \psi) + (c_2\phi_2, \psi)$,

which hold for all elements $\phi, \phi_k, \psi \in X$ and all real numbers c_1, c_2, we call X a *Hilbert space.*

Throughout the book we will frequently construct various Banach spaces using certain energy norms obtained from one or another part of the potential energy of the shell. As a rule, these spaces are constructed by taking the closure of a set of smooth (vector) functions in the energy norm, in which these sets do not form a Banach space. The closure operation consists, roughly speaking, of adjoining to the original set of functions all the possible limits of sequences of elements of the set that do not belong to the original set. This operation can be justified using the so-called completion theorem. A different approach is to use ideas of the theory of functions of a real variable. The two methods are equivalent.

9.2. The set of elements φ of a Banach space B that satisfy the inequality

$$\|\varphi - \varphi_0\|_B \leq R \tag{9.1}$$

is called a closed ball of radius R with center at φ_0. If in (9.1) we have strict inequality, then we have an open ball.

If all elements of a set in the Banach space are contained in a ball, then we say that the set is bounded.

An operator $\mathbf{G}$ taking elements of a Banach space B_1 into elements of a Banach space B_2 is called linear if for any two elements $\varphi_1, \varphi_2 \in B_1$ and any two real numbers c_1, c_2 we have

$$\mathbf{G}(c_1\varphi_1 + c_2\varphi_2) = c_1\mathbf{G}\varphi_1 + c_2\mathbf{G}\varphi_2.$$

An operator $\mathbf{G}$ (not necessarily linear) is said to be bounded if it maps any bounded set in B_1 into a bounded set in B_2.

Lemma 9.1. *A linear operator* $\mathbf{G}$ *is bounded if and only if there is a constant* m *such that*

$$\|\mathbf{G}\varphi\|_{B_2} \leq m\,\|\varphi\|_{B_1} \tag{9.2}$$

for any $\varphi \in B_1$.

Indeed, sufficiency of (9.2) for boundedness of $\mathbf{G}$ is obvious. Assume now that $\mathbf{G}$ is a bounded operator and that (9.2) does not hold. Then there must exist a sequence such that

$$\|\varphi_n\|_{B_1} = 1$$

$$\|\mathbf{G}\varphi_n\|_{B_2} \to \infty,$$

which contradicts the boundedness of $\mathbf{G}$.

The lowest bound of all values of m in (9.2) is called the norm of the linear operator **G** and will be denoted by $\|G\|$.

An operator **G** is said to be continuous if from

$$\varphi_n \to \varphi \text{ in } B_1$$

it follows that

$$\mathbf{G}\varphi_n \to \mathbf{G}\varphi \text{ in } B_2.$$

Here and everywhere below an arrow indicates strong convergence of a sequence. A bounded linear operator is continuous at every point of B_1.

In the particular case of B_2 being the Euclidean space R^1, we obtain real functionals, to which therefore we can apply all the above definitions and results.

Theorem (Riesz). *Every linear bounded functional Φ in a Hilbert space can be represented in the form*

$$\Phi(\varphi) = (\varphi_\Phi \cdot \varphi),$$

where φ_Φ is some fixed element of the space, which is uniquely defined by the functional Φ.

A sequence of elements φ_n of a Banach space is said to *converge weakly* to an element φ_0 if for each bounded linear functional Φ we have

$$\Phi(\varphi_n) \to \Phi(\varphi_0).$$

For weak convergence of sequences φ_n we shall use the symbol $\varphi_n \rightharpoonup \varphi_0$. It is easy to see that if $\varphi_n \to \varphi_0$, then $\varphi_n \rightharpoonup \varphi_0$. The converse is false.

A set of elements of a Banach space B is said to be *strongly (weakly) compact* if every infinite sequence of the set contains a strongly (weakly) convergent subsequence.

An operator **G** is called *compact* if it maps bounded sets in B_1 into compact sets in B_2.

We shall say that an operator **G** is completely continuous if it is compact and continuous.

An operator **G** will be called strongly continuous if it maps weakly convergent sequences in a Hilbert space into strongly convergent ones.

A set of elements of a Banach space is called strongly *(weakly) closed* if it contains all the strong (weak) limits of its elements.

Definition (9.1) is equivalent to the one given here.

9.3. Let us formulate several theorems.

Theorem 9.1. *Every bounded infinite set in a Hilbert space is weakly compact.*

Theorem 9.2. *A closed ball in a Hilbert space is weakly compact and weakly closed.*

Theorem 9.3. *Assume that the following conditions hold for a sequence of elements φ_n in a Hilbert space:*

$$\varphi_n \rightharpoonup \varphi_0, \quad \|\varphi_n\| \to \|\varphi_0\| .$$

Then

$$\varphi_n \to \varphi_0 .$$

Theorem 9.4. *Every weakly continuous functional attains its maximal value on any weakly closed set.*

Theorem 9.5. *Let* **G** *be a strongly continuous operator from a Banach space B_1 into a Banach space B_2. Assume also that the closed ball in B_1 is weakly compact. Then* **G** *is a completely continuous operator.*

To prove this theorem, note that if **G** is strongly continuous, then it is necessarily continuous. We shall demonstrate that it is also compact. Let ω be a bounded set in B_1. Let us show that the set $\mathbf{G}\omega$ is compact. Consider an infinite subset $\mathbf{G}\widetilde{\omega}$ of $\mathbf{G}\omega$. Clearly, since $\overline{\omega} \subset \omega$, it is also bounded and therefore, by the condition on B_1, contains a sequence $\omega_1, \omega_2, \ldots, \omega_n, \ldots$ that converges weakly in B_1. But then $\mathbf{G}\omega_1, \ldots, \mathbf{G}\omega_n, \ldots$ is a strongly convergent sequence in B_2. Thus $\mathbf{G}\widetilde{\omega}$ contains a strongly convergent sequence. Theorem 9.5 is proved.

Theorem 9.6. *Let* **G** *be a completely continuous operator from a reflexive Banach space B_1 into a reflexive Banach space B_2. Then* **G** *is a strongly continuous operator, which means that from*

$$\omega_n \rightharpoonup \omega_0 \text{ in } B_1$$

it follows that

$$\mathbf{G}\omega_n \to \mathbf{G}\omega_0 \text{ in } B_2 .$$

We do not prove this theorem here.

Note that a Hilbert space is reflexive, and thus a completely continuous operator acting in a Hilbert space is strongly continuous as well.

The last theorem has important applications. In particular, from it, it follows that in all the Sobolev space embedding theorems the embedding operator is not only completely continuous, as is usually stated, but also strongly continuous. This fact is a consequence of reflexivity of Sobolev spaces and of the linearity of the embedding operator.

We have presented the concepts and result of functional analysis required in the sequel. For more details, the reader may consult [130, 156, 283, 287, 310, 403, 415].

9.4. Let us move on to present an important topological concept, that of the winding number of a vector field. It will be frequently used below. Let us start with the simplest case, that of a vector field in the plane. Suppose that at every point of a planar domain Σ we define a vector $\mathbf{\Pi}$, which depends continuously on the coordinates x_1, x_2 (see Figure 9.1). In this case we shall say that a continuous

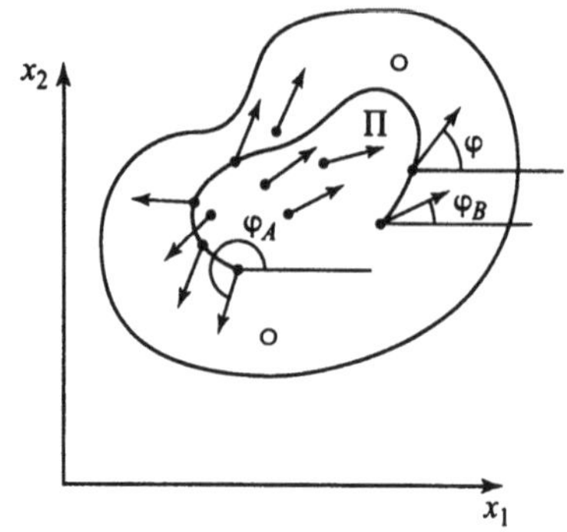

FIGURE 9.1.

vector field $\mathbf{\Pi}$ is defined on Σ. The points for which

$$\mathbf{\Pi}(x_1, x_2) = 0$$

are called critical points. Let us trace a smooth curve d in Σ (Figure 9.1) and let us examine the change in the angle φ between the vector $\mathbf{\Pi}$ and the x_1 axis as we move from the point A to the point B. For this we introduce the quantity

$$\gamma = \frac{1}{2\pi}(\varphi_B - \varphi_A),$$

which we call the winding number of the vector field Π on d. Here we are, of course, assuming that d contains no critical points of $\mathbf{\Pi}$. Clearly, if d is a closed curve, then γ is an integer (which can be positive, negative, or zero).

The winding number of a planar vector field was apparently first regarded as an important characteristic of boundary value problems in Riemann problems and in singular integral equations [86, 213]. In the case of a planar vector field, its winding number on a closed curve can be computed using the formula [106, 163]:

$$\gamma = \frac{1}{2\pi} \oint_d \frac{\Pi_2'\Pi_1 - \Pi_1'\Pi_2}{\Pi_1^2 + \Pi_2^2}\, d\sigma, \tag{9.3}$$

where $\Pi_i(\sigma)$ are the components of the vector $\mathbf{\Pi}$ expressed as functions of σ, the arc length of the curve d.

The formula (9.3) was suggested by Poincaré [162, 163, 245, 266]. We assume that $d \in C_d^1$.

Theorem 9.7 **([162, 163, 245, 266]).** *If the winding number γ of a planar vector field on a closed curve d is nonzero, then inside d there is at least one critical point.*

Theorem 9.7 can be used to prove solvability of equations in the plane. Indeed, suppose we need to find whether there are roots of the system

$$x_i = f_i(x_1, x_2), \quad i = 1, 2,$$

inside d. We introduce the vector field

$$\Pi_i(x_1, x_2) = f_i(x_1, x_2) - x_i$$

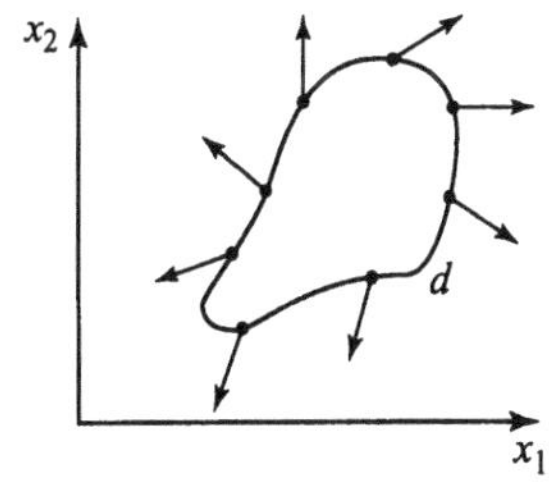

FIGURE 9.2.

and compute the winding number of this vector field on d using the Poincaré formula [245] (see Figure 9.2).

Let us note that the use of (9.3) is considerably simplified by the fact that γ is an integer. Let us also note that the concept of a planar vector field is a generalization of the concept of an analytic function, while (9.3) generalizes the well-known complex analysis formula

$$\mathcal{N} = \frac{1}{2\pi} \oint_d \frac{f'(z)}{f(z)} \, dz,$$

which is valid if $f(z)$ has no poles inside or on d [192].

The material presented above undoubtedly indicates the usefulness of the winding number of a vector field as a tool in proving solvability of systems in the plane. To use the latent capabilities of this method, we need a far-reaching generalization of the concept of winding number of a vector field to the case of operator equations and fields in infinite-dimensional spaces. The initial crucial step here was the generalization of winding number of a planar vector field to fields in n-dimensional Euclidean space. This was effected by Brouwer and Hopf [3, 39, 121] using a topological invariant, namely, the degree of a mapping [3, 162, 163]. We present here the gist of their arguments. First, using as an example a planar vector field, we shall show a different way of defining the winding number for the field on a curve d. For this note that every continuous vector field on a closed curve generates a mapping $\mathcal{M}$ of the curve onto itself or onto a part of that curve. The mapping $\mathcal{M}$ can be constructed as follows: Let us choose inside d an arbitrary point o and let us draw from o a vector $\mathbf{k}^1$ parallel to a vector $\mathbf{k}$ of the vector field $\mathbf{\Pi}$ at the point k (see Figure 9.3). The vector $\mathbf{k}^1$ crosses the curve d at a point k^1. Thus, we have constructed a correspondence between the point k and the point k^1; this is the mapping $\mathcal{M}$. Let us observe that the vector $\mathbf{k}^1$ can cross d in more than one point k^1; all these points are considered to be the image of k. The mapping $\mathcal{M}$ can thus turn out not to be single-valued. Conversely, vectors $\mathbf{k}$ corresponding to different points k can be parallel, and then more than one point k can map to the same point(s) k^1. Thus $\mathcal{M}$ does not have to be one-to-one either.

To understand the situations that can arise here, let us consider, for example, the vector field of Figure 9.4. It is generated by unit vectors originating at each point of a circle d along rays emanating from a point o that lies outside the circle. The winding number of the vector field on d is easily computed to be zero. Indeed,

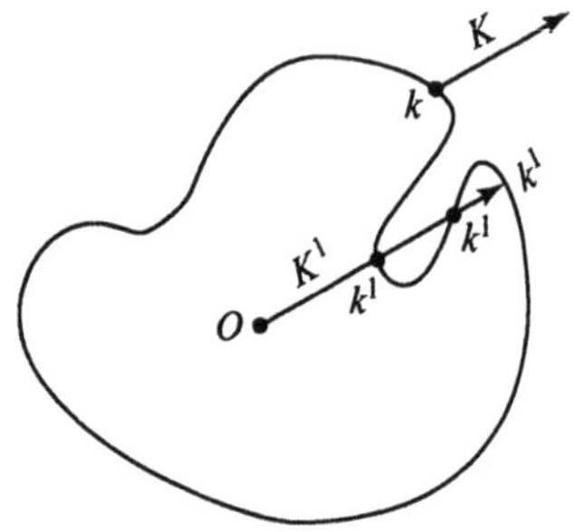

FIGURE 9.3.

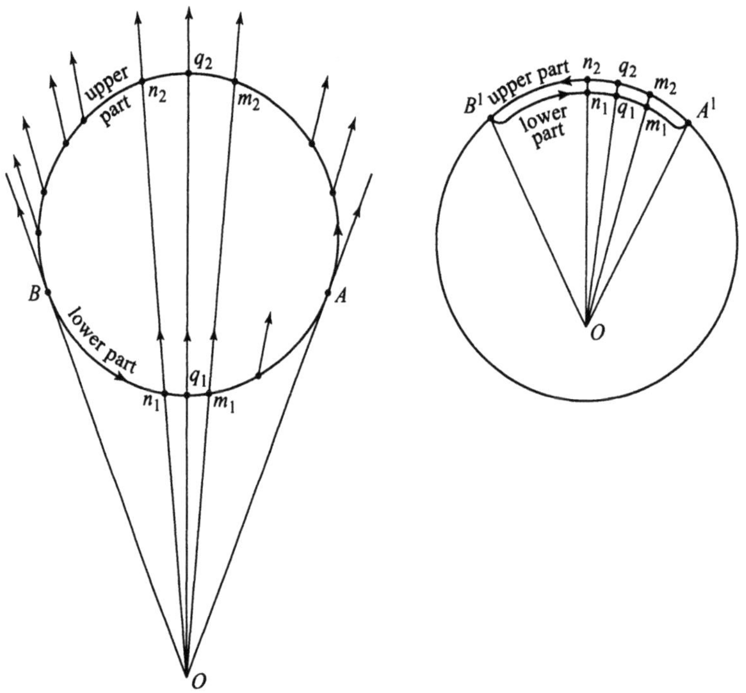

FIGURE 9.4.

as we go along the upper part of d from $\mathcal{A}$ to $\mathcal{B}$ so that the circle d is always to the left of us, vectors of the field **Π** rotate counterclockwise; as we traverse the lower part of d they rotate clockwise, and we return to $\mathcal{A}$ without having made a single complete rotation. The image of the circle d under the mapping $\mathcal{M}$ is in this case the two-sheeted arc $\mathcal{A}^1\mathcal{B}^1$. Indeed, to each point q^1 of this arc correspond two points q_1, q_2 of the circle d. However, we can observe a difference between q_1 and q_2 relative to the mapping $\mathcal{M}$. As we move from $\mathcal{A}^1$ to $\mathcal{B}^1$, we pass through these points in the order $m^1 \to q^1 \to n^1$. This is precisely the order the points are traversed along the upper arc $\mathcal{AB}$ of the circle d: $m_2 \to q_2 \to n_2$. As we go from $\mathcal{B}$ to $\mathcal{A}$, the order of encountering the points is changed: $n_1 \to q_1 \to m_1$. This

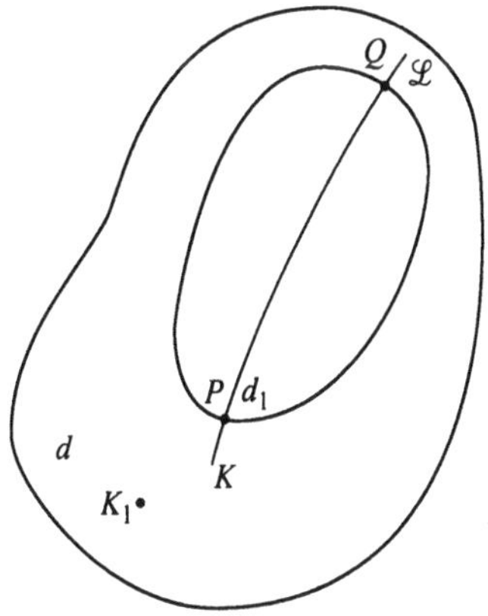

FIGURE 9.5.

qualitative difference allows us to assign different characteristics to the points q_1, q_2 that correspond to the same point on $\mathcal{A}^1\mathcal{B}^1$. If at a point q the order of traversal of the images under the mapping is preserved, we shall assign the point q characteristic $+1$, while in the opposite case it will be assigned the characteristic -1. At the points of degeneracy, $\mathcal{A}^1$, $\mathcal{B}^1$, we shall take the characteristic to be zero. Under this assignment of characteristic, we obtain a remarkable picture: Let us take any point q^1 of the image d^1 of the circle d under our mapping $\mathcal{M}$. If q^1 lies outside of $\mathcal{A}^1\mathcal{B}^1$, then to this point does not correspond any point q, and we let its index be zero (ind $= 0$). If q^1 is in the interval $\mathcal{A}^1\mathcal{B}^1$, then we set its index to be the sum of characteristics of all points q corresponding to it. Clearly, this again gives us zero. Thus, in this elementary example the index of q^1 is independent of its position on d and equals the winding number γ of the vector field $\mathbf{\Pi}$ on d. It is also easy to see that our construction is independent of the choice of the point o. This situation is not exceptional, but rather a reflection of a general result.

Theorem 9.8 ([121]). *Suppose we are given a continuous vector field* $\mathbf{\Pi}$ *with winding number* γ *on a planar curve* $d \in C^1_d$. *Let this vector field generate the mapping* $\mathcal{M} : d \to d^1$. *Then the sum of the characteristics of all the points d corresponding to a point on* d^1 *is an invariant that is independent of the position of the point on d; this invariant equals* γ.

This invariant is of importance in many topological arguments; it can be defined for arbitrary sufficiently smooth mappings, and is called the topological degree of a mapping. It is precisely this invariant that made it possible for Brouwer and Hopf to generalize the concept of winding number to vector fields in spaces of dimension higher than two.

Let us consider now a three-dimensional vector field $\mathbf{\Pi}$ on a closed surface d in three-dimensional space. In this case the field $\mathbf{\Pi}$ again generates a mapping of d into itself. Therefore, we shall study mappings of d and consider examples that clarify the situation in this case.

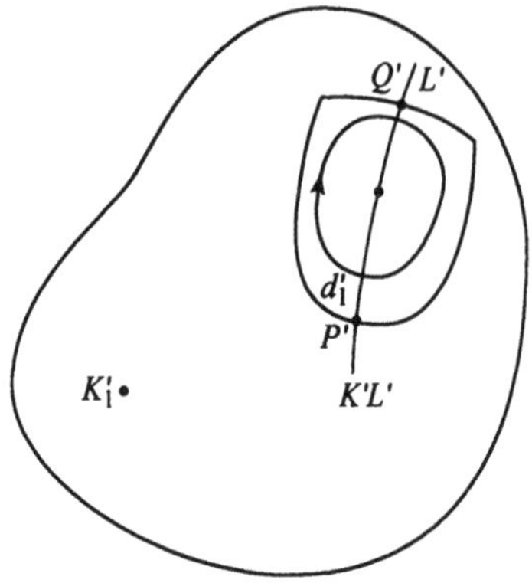

FIGURE 9.6.

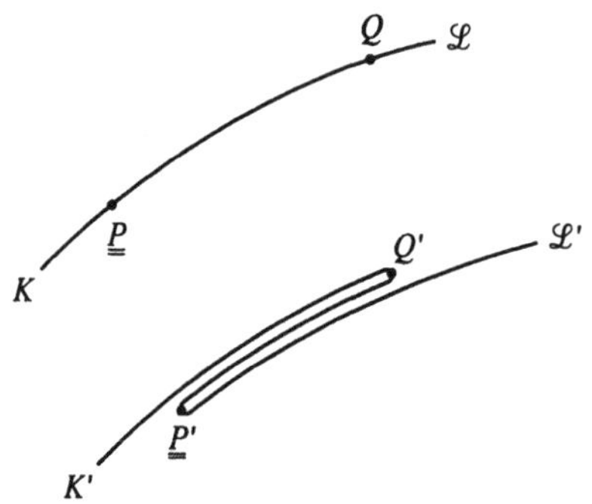

FIGURE 9.7.

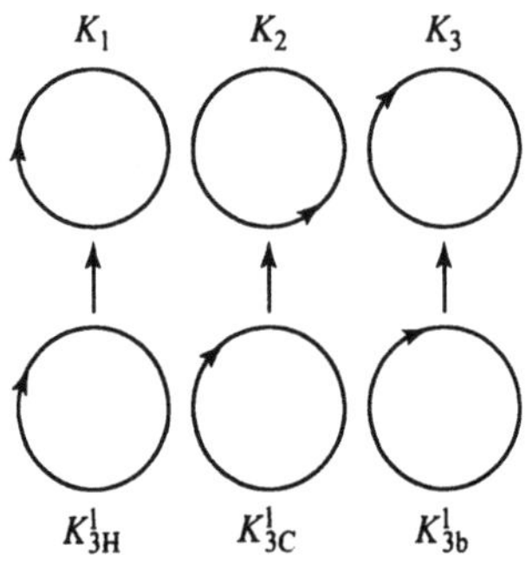

FIGURE 9.8.

For illustration, we shall represent the image of d as a physical deformation of this surface, and we shall denote the deformed surface by d'. In the general case this deformation may have folds (see Figure 9.6). Suppose that to the fold d_1' on the surface d corresponds a part d_1. We shall consider the simplest case when d_1' contains three sheets: the upper one, d_{1u}', the middle one, d_{1m}', and the lower one, d_{1l}'. Thus, to every point k_1 of the surface d_1 corresponds one point

k_1' if k_1 lies outside of d_1, two points k_2' if k_2 lies on the boundary of d_1, and three points k_3' if k_3 is in the interior of d_1. To illustrate the situation, let us take a section $\mathcal{KL}$ of the surface d; to this section corresponds the curve $\mathcal{K}^1\mathcal{L}^1$ (see Figure 9.7). Here $\mathcal{K}^1\mathcal{L}^1$ intersects the fold on a multisheeted interval $\mathcal{P}^1\mathcal{Q}^1$. We will show how to associate with each point of d a geometric characteristic, which, as before, we shall call its index (ind). To this we associate with each point of d' its characteristic. If k_1' is outside of d_1', so that to it corresponds one point k_1 of the surface d_1, we assign it the characteristic $+1$. The same characteristic is assigned to the point k_{2l}' on the lower sheet. To the coinciding points k_{2m}' and k_{2u}' we assign the characteristic 0. Let us consider the points $k_{3l}', k_{3m}', k_{3u}'$, which lie, respectively, on the lower, middle, and upper sheets. To determine their characteristics, let us consider on d_1 a circle of small radius with center at the point (Figure 9.8). Let us choose a positive orientation for it, so that the interior of the circle is on the left as the circle is traversed (see Figure 9.4). Then under deformation this circle is mapped into a closed curve on the lower sheet without changing the direction of traversal. The same will happen to its image on the upper sheet. It is only in the image on the middle sheet that the orientation will be reversed, as is easily seen. Therefore, it is natural to assign the characteristic $+1$ to the points k_{3l}' and k_{3u}', while k_{3m}' is assigned the characteristic -1. The index of the original point k_3 is naturally defined to be the sum of the characteristics of $k_{3l}', k_{3m}', k_{3u}'$. Here we again obtain the remarkable result that the index of any point of the surface d under the mapping turns out to be the same and is independent of its position on d. This is a general result: For any sufficiently smooth deformation of any sufficiently smooth surface, the index of any point is independent of its position on the surface. If the deformation of d is constructed using a vector field $\boldsymbol{\Pi}$ on d, then this index γ is the winding number of $\boldsymbol{\Pi}$.

The extension of our considerations to vector fields in the n-dimensional case can be done using analogous geometric considerations. However, we prefer to use here a different analytical method, which is simpler to formulate and is more constructive in the computation of ind k. Let us be given in the n-dimensional space R^n a sufficiently smooth vector field $\boldsymbol{\Pi}$, components of which are $\boldsymbol{\Pi}_i(x)$. Suppose there is a sufficiently smooth closed hypersurface d. The mapping $\mathcal{O}$ in this case is constructed as follows: We pick a point o in d and attach to it a vector k' parallel to a vector k of $\boldsymbol{\Pi}$ at a point k of d. This vector intersects d in one or several points that will correspond to k. Let one of these points be k' with coordinates x_i'. Let us consider the Jacobian $\mathcal{J}$ of the mapping at that point,

$$\mathcal{J} = \left\| \frac{\partial x_i^1}{\partial x_j} \right\|.$$

Now we can introduce the characteristic for each point k' of the deformed surface d': If k' is the only image of the point k, then we shall take its characteristic to be $+1$. If k' is not the unique image of k and in that point $\mathcal{J} \neq 0$, then the characteristic of k' is sign $\mathcal{J}$. Finally, if k' is not the unique image of k and in that point $\mathcal{J} = 0$, then we take the characteristic of k' to be zero. As before, we define the index of

k by

$$\operatorname{ind} k = \sum \text{characteristic } k'.$$

It turns out that under sufficient smoothness of $\mathbf{\Pi}$ and d, $\operatorname{ind} k$ is independent of the position of k in d. Thus it makes sense to call it the winding number of the vector field $\mathbf{\Pi}$ on d. These arguments are valid for any sufficiently smooth mapping, and so we have obtained an invariant, the topological degree of the mapping. In the case of the mapping $\mathcal{O}$ we have the winding number γ of the vector field $\mathbf{\Pi}$ on d.

9.5. In particular cases one can determine the winding number γ by a direct computation and summation of characteristics of all the points. However, for the winding number of finite-dimensional vector fields there is a formula derived by Kronecker [168]. Let us be given a vector field $\Pi_i(x_1, \ldots, x_n)$, $i = 1, \ldots, n$; we have to compute its winding number on a surface d, the equation of which is given by the relation

$$\mathcal{F}^0(x_1, \ldots, x_n) = 0.$$

Let us define the quantities

$$\mathcal{D} = \begin{vmatrix} \mathcal{F}^0 & \mathcal{F}^0_{x_1} & \cdots & \mathcal{F}^0_{x_n} \\ \Pi_1 & \Pi_{1x_1} & \cdots & \Pi_{1x_n} \\ \cdots & \cdots & \cdots & \cdots \\ \Pi_n & \Pi_{nx_1} & \cdots & \Pi_{nx_n} \end{vmatrix},$$

$$R = \|\Pi\|, \quad Q^2 = \sum_{i=1}^{n} \left| \frac{\partial \mathcal{F}^0}{\partial x_i} \right|^2.$$

Then

$$\gamma = -\mathcal{K}^n \int_V \frac{\mathcal{D}}{R^n Q}\, dv. \tag{9.4}$$

In (9.4) the integration is over an n-dimensional sphere; $\mathcal{K}^n$ is the volume of that sphere. For more details on the Kronecker formula and methods of calculation of winding number see [50, 168].

Theorem 9.9. *Let a vector field* $\mathbf{\Pi}$ *be defined on a domain* Σ_n *of* n*-dimensional space. If its winding number* γ *on a hypersurface* d *homeomorphic to a sphere in* R^n *is nonzero, then inside* D *there is at least one critical point of* $\mathbf{\Pi}$*, that is, a point such that*

$$\Pi_i(x_1, \ldots, x_n) = 0, \quad i = 1, \ldots, n.$$

If we apply this theorem to vector fields of the form

$$\Pi_i = x_i - f_i(x_1, \ldots, x_n), \quad i = 1, \ldots, n,$$

then frequently, using (9.4), we can obtain existence theorems for roots of systems of equations of the form

$$x_i = f_i(x_1, \ldots, x_n), \quad i = 1, \ldots, n.$$

Even though for large n computations using (9.4) are quite involved, they are still quite efficient, since γ is always either an integer or zero.

9.6. Generalization of the concept of winding number of a vector field to infinite-dimensional spaces was made by Leray and Schauder [178] on the basis of the pioneering work of Birkhoff and Kellogg [35]. Degrees of mappings were further extensively developed by Rotte [265], and in particular, in the works of Krasnosel'skii [162], where numerous applications may be found.

We shall explain the main ideas using as an example the computation of the winding number of a vector field defined on spheres of a separable Hilbert space H.

Suppose that at each point of a sphere of a Hilbert space H we have an element w such that

$$w - \mathbf{G}w \neq 0, \tag{9.5}$$

where $\mathbf{G}$ is a completely continuous operator on H, nonlinear in general. Assuming that the space is separable, we introduce in it an orthonormal basis φ_n and approximate (9.5) using a truncated Fourier series expansion in φ_n:

$$\begin{aligned} w &\approx \sum_{i=1}^{\mathcal{N}} x^i \varphi_i \\ \mathbf{G}_{\mathcal{N}}(w) &= \sum_{k=1}^{\mathcal{N}} \left(\mathbf{G}(\sum_{i=1}^{\mathcal{N}} x^i \varphi_i) \cdot \varphi_k \right)_H . \end{aligned} \tag{9.6}$$

In (9.6) and below we sum over indices repeated as sub- and superscripts. Then clearly, the field is approximated by a finite-dimensional field

$$x^k - (\mathbf{G}(x^i \varphi_i) \cdot \varphi_k)_H, \quad i, k = 1, \ldots, \mathcal{N}, \tag{9.7}$$

the winding number $\gamma_{\mathcal{N}}$ of which can be computed using (9.4).

Theorem 9.10. *For finite-dimensional vector fields of the form* (9.7) *the winding number $\gamma_{\mathcal{N}}$ is constant for all sufficiently large values of $\mathcal{N}$.*

This constant value is what we call the winding number of the vector field $w - \mathbf{G}w$.

The assumption that H is separable was used for simplicity of exposition. The winding number of vector fields of the form $w - \mathbf{G}w$ where $\mathbf{G}$ is a completely continuous operator can be defined for arbitrary Banach spaces [162, 165].

Theorem 9.11. *Suppose we are given a vector field $\mathbf{\Pi}$ in a Hilbert space H having the form*

$$\mathbf{\Pi} = w - \mathbf{G}(w).$$

Assume that the condition (9.5) *holds and that on a sphere* $\|w\| = R$ *the winding number* γ *of this vector field is nonzero. Then inside the sphere there is at least one critical point of the vector field, that is, a point such that*

$$w - \mathbf{G}w = 0.$$

This theorem offers a valuable tool in proving existence theorems for a large class of operator equations. Naturally, the hardest step in proofs of this sort is the computation of the winding number of the vector field γ, namely, the verification of (9.5).

9.7. As we already mentioned above, if **G** is a completely continuous operator, the computation of winding number of the vector field on spheres in a Hilbert space reduces to the computation of winding number of finite-dimensional vector fields of sufficiently high dimensionality, which could be done using a computer. Note that the winding number is always an integer, so that no high-precision computations are required. However, this method, even if effectively realizable, can be applied only in particular cases. The general method of computing winding numbers of vector fields is based on the homotopy theorem.

Definition 9.1. Let $\mathbf{\Pi}(w, t)$ be a field of the form

$$\mathbf{\Pi}(w, t) = w - \mathbf{G}(w, t),$$

where the operator $\mathbf{G}(w, t)$ is a completely continuous operator in H for all $t \in [0, 1]$. Furthermore, let the operator $\mathbf{G}(w, t)$ be uniformly continuous in t for all $w \in \Sigma = \{w \in H \mid \|w\|_H = R\}$. Let $\mathbf{G}(w, 0) = \mathbf{G}_0(w)$. Let $\mathbf{G}(w, 1) = \mathbf{G}_1(w)$, and $\mathbf{\Pi}(w, t) \neq 0$ if $w \in \Sigma$. Then we say that $\mathbf{G}_0$ and $\mathbf{G}_1$ are connected by a completely continuous homotopy.

Two vector fields $\mathbf{\Pi}_1$ and $\mathbf{\Pi}_2$ connected by a completely continuous homotopy are said to be homotopic.

Theorem 9.12 ([3, 162, 165]). *Two homotopic vector fields* $\mathbf{\Pi}_1$ *and* $\mathbf{\Pi}_2$ *have the same winding number on* Σ.

This theorem will be frequently used below.

9.8. Knowledge of the winding number of a vector field allows us sometimes to prove nonuniqueness of solutions of the corresponding operator equations and, in general, to estimate the number of solutions.

Theorem 9.13. *Let* w_0 *be an isolated solution* [162]*of the operator equation*

$$w_0 - \mathbf{G}w_0 = 0,$$

where **G** *is a completely continuous operator in a Hilbert space* H. *In this case the winding number of the vector field* $w - \mathbf{G}w$ *on spheres* $\Sigma(r, w_0)$ *with center at* w_0 *of sufficiently small radius* r *is independent of* r.

Definition 9.2. The winding number of $w - \mathbf{G}w$ for sufficiently small r is called the index of w_0 (ind w_0).

Theorem 9.14. *Assume that the winding number of a vector field on a sphere $\Sigma(R, 0)$ in a Hilbert space is γ, and that a finite number of critical points $w_1, \ldots, w_p$ are situated in the interior of $\Sigma(R, 0)$. Then*

$$\gamma = \sum_{k=1}^{p} \operatorname{ind} w_k.$$

10. General Mathematical Results (Continued)

10.1. On an arbitrary set π consider functions that are continuous together with their derivatives of orders up to and including k. By defining the norm

$$\|f\|_{C_\pi^k} = \sum_{i=0}^{k} \sum_{j=1}^{i} \max_{\alpha} \left| \frac{\partial^i f}{\partial \alpha_1^i \partial \alpha_2^{i-j}} \right|, \tag{10.1}$$

we make it into the Banach space C_π^k.

A vector $\mathbf{f} = (f_1, \ldots, f_n)$ is in C_π^k if $f_i \in C_\pi^k$ and

$$\|\mathbf{f}\|_{C_\pi^k} = \sum_{i=1}^{n} \|f_i\|_{C_\pi^k}.$$

Let us now introduce the space $H_\pi^{k,\lambda}$ of functions on π whose derivatives of order k are continuous, and such that each of these derivatives satisfies a Hölder condition with exponent λ. We have

$$\|f\|_{H_\pi^{k,\lambda}} = \|f\|_{C_\pi^k} + \sum_{|k|} \max_{\mathcal{P},\, \mathcal{Q} \in \bar{\pi}} \left| \frac{\partial^k f(\mathcal{P}) - \partial^k f(\mathcal{Q})}{r_{\mathcal{P},\mathcal{Q}}^{\lambda}} \right|, \quad r_{\mathcal{P},\mathcal{Q}} = |\mathcal{P} - \mathcal{Q}|. \tag{10.2}$$

The right-hand side of (10.2) contains the sum of Hölder differences of all derivatives of f of order k.

A vector $\mathbf{f} = (f_1, \ldots, f_n)$ is in $H_\pi^{k,\lambda}$ if $f_i \in H_\pi^{k,\lambda}$ and

$$\|\mathbf{f}\|_{C_\pi^k} = \sum_{i=1}^{n} \|f_i\|_{C_\pi^k}.$$

The space $L_{p\pi}$ is the space of all functions whose p-th power is integrable in π. We define

$$\|f\|_{L_{p\pi}} = \left(\int_\pi |f|^p d\pi \right)^{1/p}, \quad p \geq 1.$$

Also,

$$\|\mathbf{f}\|_{L_{p\pi}} = \sum_{i=1}^{n} \|f_i\|_{L_{p\pi}}.$$

As the set π we shall take either the domain Ω or a part or it, or, alternatively, the boundary curve Γ or a part of it. If π is Ω, then

$$\|f\|_{L_{p\Omega}} = \left(\int_{\Omega} |f|^p d\Omega\right)^{1/p}.$$

For $p = 2$ we have the Hilbert space H;

$$(f_1 \cdot f_2)_H = \int_{\Omega} f_1 f_2 d\Omega.$$

If $\mathbf{f}_1$, $\mathbf{f}_2$ are vectors, then

$$(\mathbf{f}_1 \cdot \mathbf{f}_2)_H = \int_{\Omega} \sum_{l=1}^{n} f_{1l} f_{2l} d\Omega.$$

10.2. Let the vector function $\rho(s)$ representing the closed curve Γ be such that for each point of Γ there is a neighborhood $\widetilde{\Gamma} \subset \Gamma$, such that $\rho \in C^k_{\widetilde{\Gamma}}$. In this case we shall say that $\Gamma \in C^k_{\Gamma}$. If Γ consists of a finite number of parts Γ_i, on each of which $\rho \in C^k_{\Gamma_i}$, then we say that Γ is a piecewise-smooth curve (PSC) belonging to C^k_{Γ}. Similarly we define curves $\Gamma \in H^{k,\lambda}_{\Gamma}$ and PSC $\Gamma \in H^{k,\lambda}_{\Gamma}$. The same definitions can be applied to the boundary of the domain Ω, the closed curve $\gamma(\sigma)$. Next, we introduce surfaces S of classes C^k_{Ω} and $H^{k,\lambda}_{\Omega}$. It is easy to see that if $S \in C^k_{\Omega}(H^{k,\lambda}_{\Omega})$ and PSC $\Gamma \in C^k_{\Gamma}(H^{k,\lambda}_{\Gamma})$, then PSC $\gamma \in C^k_{\gamma}(H^{k,\lambda}_{\gamma})$.

10.3. In this section we present results from the theory of Sobolev spaces [393, 31, 32, 130, 217, 407, 409, 285, 412] that will be necessary in the sequel. Let us consider a bounded domain Ω_n in the n-dimensional space of variables $\alpha^1, \ldots, \alpha^n$. Let $\overset{0}{C}{}^{\infty}_{\Omega_n}$ be the space of functions ψ with compact support in Ω_n, that is, the space of infinitely differentiable functions in Ω_n, each of which vanishes on a neighborhood of the boundary of Ω. Furthermore, assume that for a function $f(\alpha^1, \ldots, \alpha^n)$ there is a function $\varphi(\alpha^1, \ldots, \alpha^n)$ such that

$$\int_{\Omega_n} f \frac{\partial^k \psi}{\partial \alpha_1^{k_1} \cdots \partial \alpha_n^{k_n}} d\Omega_n = (-1)^k \int_{\Omega_n} \varphi \psi d\Omega_n$$

for any $\psi \in \overset{0}{C}{}^{\infty}_{\Omega_n}$. In this case we shall say that φ is the generalized Sobolev derivative of f. The following important theorem of Nikol'skii reveals the connections between Sobolev and classical derivatives.

Theorem 10.1 ([217]). *If the nonmixed derivative φ of f is integrable in Ω_n with an exponent $p > 1$, then f has a classical derivative of some order equal to φ almost everywhere.*

A mixed generalized derivative can exist even when a classical derivative does not.

To construct the Sobolev spaces $W^{(l)}_{p\Omega_n}$, consider the functions $f(\alpha^1, \ldots, \alpha^n) \in C^l_{\Omega_n}$. Let us equip $C^l_{\Omega_n}$ with a norm different from that of (10.1):

$$\|f\|_{W^{(l)}_{p\Omega_n}} = \left(\int_\Omega \sum_{k=0}^{l} \left| \frac{\partial^k f}{\partial\alpha_1^{k_1} \ldots \partial\alpha_n^{k_n}} \right|^p d\Omega \right)^{1/p}, \quad k_1+k_2+\cdots+k_n = k. \tag{10.3}$$

The closure of $C^l_{\Omega_n}$ in the norm (10.3) is called the Sobolev space $W^{(l)}_{p\Omega_n}$.

Thus $W^{(l)}_{p\Omega_n}$ contains all the functions in $C^l_{\Omega_n}$ and all the possible limits of sequences in $C^l_{\Omega_n}$ in the norm (10.3).

Instead of (10.3) we could take the norm

$$\|f\|_{W^{(l)}_{p\Omega_n}} = \sum_{k=0}^{l} \left(\int_\Omega \left| \frac{\partial^k f}{\partial\alpha_1^{k_1} \cdots \partial\alpha_n^{k_n}} \right|^p d\Omega \right)^{\frac{1}{p}} = \sum_{k=0}^{l} \left\| \frac{\partial^k f}{\partial\alpha_1^{k_1} \cdots \partial\alpha_n^{k_n}} \right\|_{L_{p\Omega_n}},$$

$$k_1 + k_2 + \cdots + k_n = k. \tag{10.4}$$

It is easy to see that the norms (10.3) and (10.4) are equivalent. If we have a vector **f** with components $f_1, \ldots, f_m$, then

$$\|\mathbf{f}\|_{W^{(l)}_{p\Omega_n}} = \sum_{k=0}^{m} \|f_k\|_{W^{(l)}_{p\Omega_n}}.$$

Theorem 10.2. *All generalized derivatives of a function $f \in W^{(l)}_{p\Omega_n}$ up to and including order l are integrable in Ω_n with exponent p.*

Functions in $W^{(l)}_{p\Omega_n}$ have a number of characteristic properties described by embedding theorems [285]. Initially, they were discovered by Sobolev and Kondrashev. Subsequently, they were developed in a number of publications; for more details see [393, 31, 32, 123, 124, 193, 194, 409, 314, 315, 85]. In embedding theorems, functions in $W^{(l)}_{p\Omega_n}$ are considered as elements of different function spaces, and properties of function spaces in which $W^{(l)}_{p\Omega_n}$ can be "embedded" are established. Since it is known that a ball in $W^{(l)}_{p\Omega_n}$ is weakly compact [393, 285, 409], then from Theorems 9.5 and 9.6 it follows that we can talk equally about complete and about strong continuity of the embedding operator. In applications it will be more convenient to use the strong continuity property of the embedding operator, and it will be noted in the statements of the theorems. To state these theorems let us introduce the number

$$\alpha = l - \frac{n}{p} = k + \lambda,$$

where k is the integer part of α, and λ is its fractional part.

Theorem 10.3. (1) *Let $n < lp$. Then $f \in W^{(l)}_{p\Omega_n}$ implies $f \in H^{k_1,\lambda_1}_{p\Omega_n}$ only if*

$$k_1 + \lambda_1 \leq \alpha. \tag{10.5}$$

Then the operator of embedding of $W^{(l)}_{p\Omega_n}$ *into* $H^{k_1,\lambda_1}_{\Omega_n}$ *is bounded, so that for any function* f *we have the inequality*

$$\|f\|_{H^{k_1,\lambda_1}_{\Omega_n}} \leq m\,\|f\|_{W^{(l)}_{p\Omega_n}},$$

with the same constant m *for all* $f \in W^{(l)}_{p\Omega_n}$. *If in* (10.5) *we have a strict inequality, then the embedding operator is completely continuous. That is, from the relation* $f_n \rightharpoonup f_0$ *in* $W^{(l)}_{p\Omega_n}$ *it follows that* $f_n \to f_0$ *in* $H^{k_1,\lambda_1}_{\Omega_n}$.

(2) *Let* $n < lp$ *and let*

$$\alpha = l - \frac{n}{p}$$

be an integer. Then $f \in H^{\alpha-1,\lambda^1}_{\Omega_n}$, *where* $\lambda^1 < 1$ *is a number arbitrarily close to one. Here the embedding is always completely continuous: From* $f_n \rightharpoonup f_0$ *in* $W^{(l)}_{p\Omega_n}$ *it follows that* $f_n \to f_0$ *in* $H^{\alpha-1,\lambda^1}_{\Omega_n}$, *and*

$$\|f\|_{H^{\alpha-1,\lambda^1}_{\Omega_n}} \leq m\,\|f\|_{W^{(l)}_{p\Omega_n}}.$$

(3) *Let* $n = lp$. *In this case* $f \in W^{(l)}_{p\Omega_n}$ *implies that* $f \in L_{q\Omega_n}$ *for any* $q \geq 1$, *and furthermore, the embedding operator is completely continuous. In other words, from* $f_n \rightharpoonup f_0$ *in* $W^{(l)}_{p\Omega_n}$ *it follows that* $f_n \to f_0$ *in* $L_{q\Omega_n}$. *There holds the inequality*

$$\|f\|_{L_{q\Omega_n}} \leq m\,\|f\|_{W^{(l)}_{p\Omega_n}}.$$

Next, suppose Ω_r *is a smooth set of dimension* $r < n$ *in* Ω_n. *Then* $f \in L_{q\Omega_r}$, *and the embedding operator is completely continuous: From* $f_n \rightharpoonup f_0$ *in* $W^{(l)}_{p\Omega_n}$ *it follows that* $f_n \to f_0$ *in* $L_{q\Omega_r}$; *in addition,*

$$\|f\|_{L_{q\Omega_r}} \leq m\,\|f\|_{W^{(l)}_{p\Omega_n}}.$$

(4) *Let* $n > lp$. *To characterize elements of* $W^{(l)}_{\Omega_n}$ *we define the number* q^* *by*

$$q^* = \frac{pr}{n - lp}.$$

In this case $f \in W^{(l)}_{p\Omega_n}$ *implies that* $f \in L_{q\Omega_r}$ *for any* $1 \leq q \leq q^*$, *and*

$$\|f\|_{L_{q\Omega_r}} \leq m\,\|f\|_{W^{(l)}_{p\Omega_n}}.$$

For all $1 \leq q < q^*$ *the embedding operator is completely continuous:* $f_n \rightharpoonup f_0$ *in* $W^{(l)}_{p\Omega_n}$ *implies that* $f_n \to f_0$ *in* $L_{q\Omega_r}$.

Theorem 10.3 holds if Ω_n satisfies certain conditions. The first sufficient conditions were formulated by Sobolev in terms of a domain being star-shaped. A domain Ω_n is called star-shaped with respect to a point o if any ray coming out of o intersects the boundary Γ of Ω_n only once. A domain Ω_n is star-shaped with respect to a ball B if it is star-shaped with respect to all points of B. The Sobolev

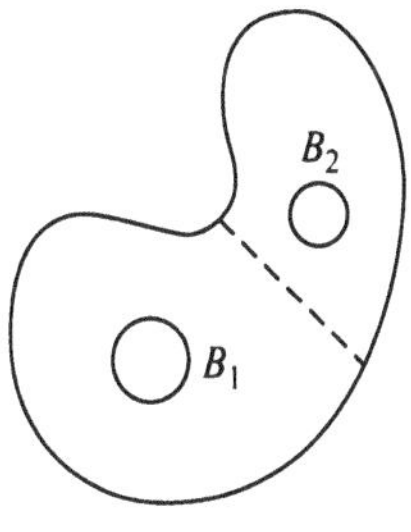

FIGURE 10.1.

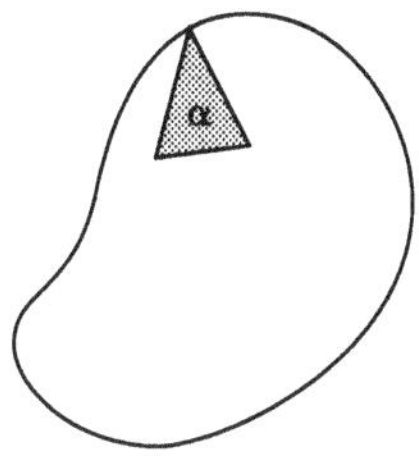

FIGURE 10.2.

condition [285] for the validity of Theorem 10.3 is to require Ω_n to be the union of a finite number of domains Ω_{n_i}, each of which is star-shaped with respect to a ball B_i in its interior (see Figure 10.1).

Subsequently, the cone condition was introduced: There exists a cone $\mathcal{K}$ of opening angle α such that its apex k can touch any point of Ω_n, in such a way that some fixed part of $\mathcal{K}$ adjacent to k is contained in Ω_n (see Figure 10.2). In the case of a planar domain Ω_n the cone condition is definitely satisfied if γ is a PSC of class C^1_Γ without interior zero angles. Below, the domains Ω_n for which the embedding Theorem 10.3 holds for $W^{(l)}_{p\Omega_n}$ will be called Sobolev domains of class (p, l, n). Thus, the property of being star-shaped, the cone condition, or of Γ being a PSC of class C^1_Γ are all sufficient conditions for Ω_n to be a Sobolev domain. More subtle conditions (some of which are necessary) that ensure that Ω_n is a Sobolev domain are presented in [123, 124].

10.4. We shall be most frequently using embedding theorems in the following cases:

A. The space $W^{(1)}_{2\Omega}$ with the norm

$$\|f\|_{W^{(1)}_{2\Omega}} = \int_\Omega (f^2 + f^2_{\alpha^1} + f^2_{\alpha^2}) d\alpha^1 d\alpha^2.$$

Sometimes we will use the equivalent norm

$$\|f\|^2_{W^{(1)}_{2\Omega}} = \int_\Omega (f^2 + f^2_{\alpha^1} + f^2_{\alpha^2})\, D d\alpha^1 d\alpha^2.$$

In this case $n = 2$, $p = 2$, $l = 1$, and $n = lp$, and we are in the situation of case (3) of the embedding Theorem 10.3, from which we deduce the following facts:

Theorem 10.4. (1) *If* $f \in W^{(1)}_{2\Omega}$, *then* $f \in L_{q\Omega}$ *and the embedding operator is completely continuous. That is, if*

$$f_n \rightharpoonup f_0 \text{ in } W^{(1)}_{2\Omega}, \text{ then } f_n \to f_0 \text{ in } L_{q\Omega}\ \forall q \geq 1,$$

and furthermore,

$$\|f\|_{L_{q\Omega}} \leq m\,\|f\|_{W^{(1)}_{2\Omega}}.$$

(2) *Let d be a PSC in* C^1_d *(which may partially coincide with* Γ*). Then* $f \in L_{qd}$ *for all* $q \geq 1$. *The embedding operator is completely continuous, and therefore from* $f_n \rightharpoonup f_0$ *in* $W^{(1)}_{2\Omega}$ *it follows that* $f_n \to f_0$ *in any* L_{qd}, $q \geq 1$; *in addition,*

$$\|f\|_{L_{qd}} \leq m\,\|f\|_{W^{(1)}_{2\Omega}}. \tag{10.6}$$

B. The space $W^{(2)}_{2\Omega}$ with the norm

$$\|f\|^2_{W^{(2)}_{2\Omega}} = \int_\Omega (f^2 + f^2_{\alpha^1} + f^2_{\alpha^2} + \sum_{i,j=1}^{2} f^2_{\alpha^i\alpha^j})\,d\alpha^1 d\alpha^2.$$

Sometimes we will use the equivalent norm

$$\|f\|^2_{W^{(2)}_{2\Omega}} = \int_\Omega (f^2 + f^2_{\alpha^1} + f^2_{\alpha^2} + \sum_{i,j=1}^{2} f^2_{\alpha^i\alpha^j})\,D\,d\alpha^1 d\alpha^2.$$

In this case $n = 2$, $l = 2$, $p = 2$, and $n < lp$, and

$$\alpha = 2 - \frac{2}{2} = 1,$$

so that we are in case (2) of the embedding Theorem 10.3, from which we deduce the following result.

Theorem 10.5. (1) *If* $f \in W^{(2)}_{2\Omega}$, *then* $f \in H^{0,\lambda^1}_\Omega$, *that is, it is Hölder continuous with the exponent arbitrarily close to 1; the embedding operator is completely continuous, so that from* $f_n \rightharpoonup f_0$ *in* $W^{(2)}_{2\Omega}$ *it follows that* $f_n \to f_0$ *in* H^{0,λ^1}_Ω *and moreover,*

$$\|f\|_{H^{0,\lambda^1}_\Omega} \leq m\,\|f\|_{W^{(2)}_{2\Omega}}. \tag{10.7}$$

(2) *If* $f \in W^{(2)}_{2\Omega}$, *then* $f_{\alpha^i} \in L_{q\Omega}$ *for any* $q \geq 1$. *Furthermore, the embedding operator is completely continuous, and* $f_n \rightharpoonup f_0$ *in* $W^{(2)}_{2\Omega}$ *implies* $f_{n\alpha^i} \to f_{0\alpha^i}$ *in* $L_{q\Omega}\ \forall q \geq 1$; *moreover,*

$$\|f_{\alpha^i}\|_{L_{q\Omega}} \leq m\,\|f\|_{W^{(2)}_{2\Omega}}. \tag{10.8}$$

(3) *Let d be a PSC in* C^1_d *(which may partially coincide with* Γ*). Then if* $f \in W^{(2)}_\Omega$, *then* $f_{\alpha^i} \in L_{qd}$ *for all* $q \geq 1$. *The embedding operator is completely continuous,*

and therefore from $f_n \rightharpoonup f_0$ *in* $W_{2\Omega}^{(2)}$ *it follows that* $f_{n\alpha^i} \to f_{0\alpha^i}$ *in* L_{qd}, $\forall q \geq 1$; *in addition,*

$$\|f_{\alpha^i}\|_{L_{qd}} \leq m(q)\,\|f\|_{W_{2\Omega}^{(2)}}, \quad i = 1,\, 2. \tag{10.9}$$

10.5. Below we shall need to know under what conditions a function $u(s)$ defined on Γ can be continued into Ω as a $W_{2\Omega}^{(1)}$ function. To state them, let us introduce the space $W_{2\Gamma}^{(\frac{1}{2})}$ defined as the closure of the set $u \in C_\Gamma^1$ in the norm

$$\|u\|_{W_{2\Gamma}^{(\frac{1}{2})}} = \|u\|_{L_{2\Gamma}} + \left(\int_\Gamma \int_\Gamma \frac{|u(\mathcal{P}) - u(\mathcal{Q})|^2}{r_{\mathcal{PQ}}^2}\, d\mathcal{P}\, d\mathcal{Q}\right)^{\frac{1}{2}}.$$

For non-integer l the spaces $W_{p\Gamma}^{(l)}$ were introduced in the work of Slobodetskii [282]; for more details see [31, 32, 22, 409, 314, 315].

Theorem 10.6. *Let* $\Gamma \in C_\Gamma^1$. *Then* $u \in W_{2\Omega}^{(1)}$ *if and only if* $u \in W_{2\Gamma}^{(\frac{1}{2})}$. *Furthermore, we have the inequality*

$$m\,\|u\|_{W_{2\Gamma}^{(\frac{1}{2})}} \leq \|u\|_{W_{2\Omega}^{(1)}} \leq M\,\|u\|_{W_{2\Gamma}^{(\frac{1}{2})}}.$$

Next, let us introduce the space $W_{2\Gamma}^{(\frac{3}{2})}$ defined as the closure of the set $u \in C_\Gamma^2$ in the norm

$$\|u\|_{W_{2\Gamma}^{(\frac{3}{2})}} = \|u\|_{L_{2\Gamma}} + \sum_{i=1}^{2} \|u_{\alpha^i}\|_{L_{2\Gamma}} + \sum_{i=1}^{2} \left(\int_\Gamma \int_\Gamma \frac{|u_{\alpha^i}(\mathcal{P}) - u_{\alpha^i}(\mathcal{Q})|^2}{r_{\mathcal{PQ}}^2}\, d\mathcal{P}\, d\mathcal{Q}\right)^{\frac{1}{2}}.$$

Theorem 10.7. *Let* $\Gamma \in C_\Gamma^2$. *Then* $u \in W_{2\Omega}^{(2)}$ *if and only if* $u \in W_{2\Gamma}^{(\frac{3}{2})}$. *Furthermore, we have the inequality*

$$m\,\|u\|_{W_{2\Gamma}^{(\frac{3}{2})}} \leq \|u\|_{W_{2\Omega}^{(2)}} \leq M\,\|u\|_{W_{2\Gamma}^{(\frac{3}{2})}}.$$

Moreover,

$$u_n \rightharpoonup u_0 \text{ in } W_{2\Gamma}^{(\frac{1}{2})} \text{ implies } u_n \to u_0 \text{ in } L_{q\Gamma}\ \forall q \geq 1;$$

$$u_n \rightharpoonup u_0 \text{ in } W_{2\Gamma}^{(\frac{3}{2})} \text{ implies } u_{n\alpha^i} \to u_{0\alpha^i} \text{ in } L_{q\Gamma}\ \forall q \geq 1,$$

and

$$u_n \to u_0 \text{ in } H_\Gamma^{0,\lambda}.$$

Sufficiency in Theorems 10.6, 10.7 is to be understood in the sense that if $u \in W_{2\Gamma}^{(\frac{1}{2})}$ (respectively, $W_{2\Gamma}^{(\frac{3}{2})}$), then there exists an extension of u into the interior of Ω such that $u \in W_{2\Omega}^{(1)}$ ($W_{2\Omega}^{(2)}$).

10.6.

Theorem 10.8. *Assume that a domain Ω in the plane is a Sobolev domain of class* $(2, l, 2)$. *Let us be given a quadratic functional on $W^{(l)}_{2\Omega}$ having the following structure:*

$$\mathcal{R}(\mathbf{w}) = \int_\Omega \left[\mathcal{P}_2 \left(\frac{\partial^l \mathbf{w}}{\partial \alpha_1^{k_1} \partial \alpha_2^{k_2}} \right) + \mathcal{P}_1 \left(\frac{\partial^l \mathbf{w}}{\partial \alpha_1^{k_1} \partial \alpha_2^{k_2}} \right) \right] d\Omega + \mathcal{P}_0(\mathbf{w}),$$

where $\mathbf{w}$ *is a two-dimensional vector and* $\mathcal{P}_2$ *is a quadratic form in its variables with coefficients that are continuous in* Ω. *Further, let* $\mathcal{P}_1$ *be a linear form in* l*th-order derivatives with coefficients that are linear forms in derivatives of lower order than* l, *so that* $\mathcal{P}_1$ *is a quadratic functional in* $W^{(l)}_{2\Omega}$. *Finally, let us assume that* $\mathcal{P}_0(\mathbf{w})$ *is a weakly continuous quadratic functional in* $W^{(l)}_{2\Omega}$. *Suppose furthermore that the following conditions are satisfied:*

1) $\mathcal{R}(\mathbf{w}) \geq 0$ *and* $\mathcal{R}(\mathbf{w}) = 0$ *implies* $\mathbf{w} \equiv 0$ *in* Ω; (10.10)

2) *if* $\mathbf{w}_n \rightharpoonup 0$ *in* $W^{(l)}_{2\Omega}$ *and* $\mathcal{R}(\mathbf{w}_n) \to 0$, *then* $\left\| \frac{\partial^l \mathbf{w}}{\partial \alpha_1^{k_1} \partial \alpha_2^{k_2}} \right\|_{L_{2\Omega}} \to 0$ (10.11)

for any $k_1, k_2, k_1 + k_2 = l$. *Then*

$$\mathcal{R}(\mathbf{w}) \geq m \, \|\mathbf{w}\|^2_{W^{(l)}_{2\Omega}} . \tag{10.12}$$

To prove the theorem, let us assume that (10.12) does not hold, so that there exists a sequence $\mathbf{w}_n$ that violates (10.12) such that

$$\mathcal{R}(\mathbf{w}_n) \to 0, \tag{10.13}$$

$$\|\mathbf{w}_n\|_{W^{(l)}_{2\Omega}} = 1. \tag{10.14}$$

Since we can consider $W^{(l)}_{2\Omega}$ to be a Hilbert space, we can assume that $\mathbf{w}_n$ converges weakly in $W^{(l)}_{2\Omega}$, and thus

$$\mathbf{w}_n \rightharpoonup \mathbf{w}_0 \text{ in } W^{(l)}_{2\Omega}; \; \mathbf{w}_n = \mathbf{w}_0 + \mathbf{v}_n; \; \mathbf{v}_n \rightharpoonup 0 \text{ in } W^{(l)}_{2\Omega}. \tag{10.15}$$

Next, by the structure of the functional $\mathcal{R}(\mathbf{w})$ we have

$$\begin{aligned} \mathcal{R}(\mathbf{w}_0 + \mathbf{v}) &= \mathcal{R}(\mathbf{w}_0) + \mathcal{R}(\mathbf{v}) + 2\mathcal{P}_2(\mathbf{w}_0, \mathbf{v}) \\ &+ \mathcal{P}_1(\mathbf{w}_0, \mathbf{v}) + \mathcal{P}_1(\mathbf{v}, \mathbf{w}_0) + \mathcal{P}_0(\mathbf{w}_0 + \mathbf{v}) - \mathcal{P}_0(\mathbf{w}_0) - \mathcal{P}_0(\mathbf{v}). \end{aligned} \tag{10.16}$$

Note that

$$\mathcal{P}_0(\mathbf{w}_0 + \mathbf{v}_k) - \mathcal{P}_0(\mathbf{w}_0) - \mathcal{P}_0(\mathbf{v}_k) \to 0.$$

The functionals $\mathcal{P}_2(\mathbf{w}_0, \mathbf{v})$, $\mathcal{P}_1(\mathbf{w}_0, \mathbf{v})$, $\mathcal{P}_1(\mathbf{v}, \mathbf{w}_0)$ are completely determined by the structure of $\mathcal{P}(w)$. The important fact for us is that $\mathcal{P}_2(\mathbf{w}_0, \mathbf{v})$, $\mathcal{P}_1(\mathbf{w}_0, \mathbf{v})$, $\mathcal{P}_1(\mathbf{v}, \mathbf{w}_0)$ are bounded linear functionals in $\mathbf{v}$ in $W^{(l)}_{2\Omega}$ for a fixed $\mathbf{w}_0 \in W^{(l)}_{2\Omega}$.

Indeed, it is easy to see that $\mathcal{P}_2(\mathbf{w}_0, \mathbf{v})$ contains terms of the form

$$\int_\Omega Q(\alpha^1, \alpha^2)\frac{\partial^l w_{0i}}{\partial\alpha_1^{k_1}\partial\alpha_2^{k_2}}\frac{\partial^l v_j}{\partial\alpha_1^{k_3}\partial\alpha_2^{k_4}}d\Omega, \; k_1+k_2=k_3+k_4=l,$$

where w_{0i}, v_j are the components of the vectors $\mathbf{w}_0$, $\mathbf{v}$, $i, j = 1, 2$, $Q \in C_\Omega$. Linearity of this functional with respect to $\mathbf{v}$ is obvious. By the Cauchy–Bunyakovsky–Schwarz inequality we have

$$\begin{aligned}\left|\int_\Omega Q(\alpha^1, \alpha^2)\frac{\partial^l w_{0i}}{\partial\alpha_1^{k_1}\partial\alpha_2^{k_2}}\frac{\partial^l v_j}{\partial\alpha_1^{k_3}\partial\alpha_2^{k_4}}d\Omega\right|& \\ \le m\left\|\frac{\partial^l w_{0i}}{\partial\alpha_1^{k_1}\partial\alpha_2^{k_2}}\right\|_{L_{2\Omega}}\left\|\frac{\partial^l v_j}{\partial\alpha_1^{k_3}\partial\alpha_2^{k_4}}\right\|_{L_{2\Omega}} &\le m_1\|\mathbf{v}\|_{W_{2\Omega}^{(l)}}.\end{aligned} \tag{10.17}$$

The last inequality in (10.17) follows because $\mathbf{w}_0 \in W_{2\Omega}^{(l)}$. Boundedness of $\mathcal{P}_2(\mathbf{w}_0, \mathbf{v})$ as a functional in $\mathbf{v}$ has been established. The functional $\mathcal{P}_1(\mathbf{v}, \mathbf{w}_0)$ is a linear combination of terms of the form

$$\int_\Omega Q(\alpha^1, \alpha^2)\frac{\partial^l v_j}{\partial\alpha_1^{k_1}\partial\alpha_2^{k_2}}\frac{\partial^t w_{0i}}{\partial\alpha_1^{k_3}\partial\alpha_2^{k_4}}d\Omega, \; k_1+k_2=l, \; k_3+k_4=t<l.$$

By the embedding Theorem 10.3, $\dfrac{\partial^t w_{0i}}{\partial\alpha_1^{k_3}\partial\alpha_2^{k_4}} \in L_{q\Omega}$ for every $q \ge 1$; $t < l$. If we have $t \le l-2$, then $\dfrac{\partial^t w_{0i}}{\partial\alpha_1^{k_3}\partial\alpha_2^{k_4}} \in C_\Omega$. In any case, we have

$$\begin{aligned}\left|\int_\Omega Q(\alpha^1, \alpha^2)\frac{\partial^l v_j}{\partial\alpha_1^{k_1}\partial\alpha_2^{k_2}}\frac{\partial^t w_{0i}}{\partial\alpha_1^{k_3}\partial\alpha_2^{k_4}}d\Omega\right|& \\ \le m\left\|\frac{\partial^l v_j}{\partial\alpha_1^{k_1}\partial\alpha_2^{k_2}}\right\|_{L_{2\Omega}}\left\|\frac{\partial^t w_{0i}}{\partial\alpha_1^{k_3}\partial\alpha_2^{k_4}}\right\|_{L_{2\Omega}} &\le m_2\|\mathbf{v}\|_{W_{2\Omega}^{(l)}},\end{aligned}$$

and we have shown that $\mathcal{P}_1(\mathbf{v}, \mathbf{w}_0)$ is bounded as a functional with respect to $\mathbf{v}$. Its linearity is obvious. The same arguments can be used to prove that $\mathcal{P}_1(\mathbf{w}_0, \mathbf{v})$ is a bounded linear functional in $\mathbf{v} \in W_{2\Omega}^{(l)}$. Since $\mathcal{P}_0(\mathbf{w})$ is weakly continuous in $W_{2\Omega}^{(l)}$, it follows from (10.13), (10.15), (10.16) that

$$\begin{aligned}\mathcal{R}(\mathbf{w}_0)+\mathcal{R}(\mathbf{v}_n)+\mathcal{P}_2(\mathbf{w}_0, \mathbf{v}_n)+\mathcal{P}_2(\mathbf{v}_n, \mathbf{w}_0)& \\ +\mathcal{P}_1(\mathbf{w}_0, \mathbf{v}_n)+\mathcal{P}_1(\mathbf{v}_n, \mathbf{w}_0) &\to 0.\end{aligned} \tag{10.18}$$

Since $\mathcal{P}_2(\mathbf{w}, \mathbf{v}_n) + \mathcal{P}_2(\mathbf{v}_n, \mathbf{w}) + \mathcal{P}_1(\mathbf{w}, \mathbf{v}_n) + \mathcal{P}_1(\mathbf{v}_n, \mathbf{w})$ is a continuous linear functional in $\mathbf{v}_n$, this tends to zero, and therefore from (10.18) we have

$$\mathcal{R}(\mathbf{w}_0)+\mathcal{R}(\mathbf{v}_n)\to 0.$$

By condition (10.10) it follows that

$$\mathcal{R}(\mathbf{w}_0)=0; \; \mathbf{w}_0 \equiv 0 \text{ in } \Omega; \; \mathcal{R}(\mathbf{v}_n)\to 0.$$

Therefore, $\mathbf{w}_n = \mathbf{v}_n$, and so

$$\mathbf{w}_n \rightharpoonup 0 \text{ in } W_{2\Omega}^{(l)}. \tag{10.19}$$

But then by the embedding Theorem 10.3,

$$\|\mathbf{w}_n\|_{W_{2\Omega}^{(l-1)}} \to 0. \tag{10.20}$$

By condition 2 of Theorem 10.8 it follows from (10.19) that

$$\left\| \frac{\partial^l \mathbf{w}_n}{\partial \alpha_1^{k_1} \partial \alpha_2^{k_2}} \right\|_{L_{2\Omega}} \to 0. \tag{10.21}$$

Therefore, we obtain from (10.20), (10.21)

$$\|\mathbf{w}_n\|_{W_{2\Omega}^{(l)}} \to 0,$$

which contradicts (10.14). Theorem 10.8 is proved..

The inequality (10.12) establishes the so-called coercivity property of the form $\mathcal{R}(\mathbf{w})$. The method of proof of coercivity used here can be used to produce a number of generalizations of the above theorem. In particular, the dimension of Ω is of no importance; some of the functionals can be nonlinear. Details on other coercivity inequalities can be found in [31, 32, 315].

11. The Function Spaces H_t, $t = 5, 6, 7, 8$. Properties of Their Elements

11.1. In this and the next section we shall introduce and study the properties of certain function spaces the norm in which is directly connected with the potential energy of the shell. The first family of spaces, denoted by H_t, is constructed by using the norm that originates from the part of the energy that relates to the tangential stresses in the shell. The square of the norm in these spaces contains the quadratic part of the energy of the shell obtained as the result of stretching/compression of the middle surface.

In all that follows we shall assume that Γ is a PSC of class C_Γ^2 and Ω is a Sobolev domain of class (2, 1, 2) in the plane, the middle surface $S \in C_\Omega^2$. Below we shall distinguish the case of essentially elastic tangential supports, for which

$$k_s^{\tau\tau} \geq M > 0, \tag{11.1}$$

$$k_s^{mm} \geq M > 0, \tag{11.2}$$

$$k_s^{ij} w_i w_j \geq M > 0, \;\; w_1^2 + w_2^2 = 1. \tag{11.3}$$

Recall that all $k_s^{\beta\gamma}$ are piecewise continuous. If the supports are not essentially elastic, then in (11.1)–(11.3) the constant M is zero. Let us introduce the set $\overset{0}{C}{}_\Omega^1$ of functions $\omega(w_1, w_2) \in C^1_\Omega$ that satisfy all the homogeneous tangential geometrical conditions (6.8), (6.9), (6.12), that is, when the right-hand side in those equations is zero.

Let us define on $\overset{0}{C}{}^1_\Omega$ the scalar product

$$(\omega_1 \cdot \omega_2)_{H_t} = \int_\Omega D_{\mathrm{s}}^{ijkl}(\alpha^1, \alpha^2)\overset{\widetilde{0}}{\gamma}_{ij}(\omega_1)\overset{\widetilde{0}}{\gamma}_{kl}(\omega_2)d\Omega + \int_{\Gamma_6} k_{\mathrm{s}}^{\tau\tau} w_{\tau 1} w_{\tau 2} ds$$
$$+ \int_{\Gamma_7} k_{\mathrm{s}}^{mm} w_{m1} w_{m2} ds + \int_{\Gamma_8} k_{\mathrm{s}}^{ij} w_{i1}\ w_{j2}\big|_{i,j=1,2}\ ds, \quad t = 5, 6, 7, \tag{11.4}$$

to which there corresponds the norm

$$\|\omega\|^2_{H_t} = \int_\Omega D_{\mathrm{s}}^{ijkl}(\alpha^1, \alpha^2)\overset{\widetilde{0}}{\gamma}_{ij}(\omega)\overset{\widetilde{0}}{\gamma}_{kl}(\omega)d\Omega + \int_{\Gamma_6} k_{\mathrm{s}}^{\tau\tau} w_\tau^2 ds$$
$$+ \int_{\Gamma_7} k_{\mathrm{s}}^{mm} w_m^2 ds + \int_{\Gamma_8} k_{\mathrm{s}}^{ij} w_i w_j\, ds, \quad t = 5, 6, 7. \tag{11.5}$$

Recall that in (11.4), (11.5) there is no summation over m and τ,

$$\overset{\widetilde{0}}{\gamma}_{11} = w_{1\alpha^1} - G^k_{11} w_k, \quad \overset{\widetilde{0}}{\gamma}_{12} = w_{1\alpha^2} + w_{2\alpha^1} - G^k_{12} w_k \ (1 \rightleftharpoons 2).$$

The space H_5 is introduced in the case when $\Gamma_5 > 0$. Here Γ_6, Γ_7, Γ_8 may be absent. If these are present, the supports there do not have to be essentially elastic, so that the constant M in (11.1)–(11.3) may be zero. Recall that Γ_i were introduced in Section 6. Let us introduce the set $\overset{0}{C}{}^1_{5\Omega}$ of vector functions that are differentiable in Ω and satisfy the homogeneous geometric conditions (6.8), (6.9), (6.12) under the requirement that $\Gamma_5 > 0$. Then H_5 is the closure of $\overset{0}{C}{}^1_{5\Omega}$ in the norm (11.5)..

The space H_6 is introduced if $\Gamma_6 > 0$. In this case the other parts of the boundary, Γ_5, Γ_7, Γ_8, may be absent. If these are present, the supports there do not have to be essentially elastic. However, the support on Γ_6 must be essentially elastic.

Furthermore, let $\overset{0}{C}{}^1_{6\Omega}$ be the set of continuously differentiable vector functions $\omega = (w_1, w_2)$ in Ω satisfying all the homogeneous geometric conditions (6.8), (6.9), (6.12) under the requirement that Γ_6 be present. Then H_6 is the closure of $\overset{0}{C}{}^1_{6\Omega}$ in the norm (11.5). .

The space H_7 is defined if $\Gamma_7 > 0$. In this case the support on Γ_7 has to be essentially elastic. The other parts, Γ_5, Γ_6, Γ_8, may be absent, while if they are present, the supports there do not have to be essentially elastic. If we introduce the set $\overset{0}{C}{}^1_{7\Omega}$ of continuously differentiable vector functions $\omega = (w_1, w_2)$ in Ω satisfying all the homogeneous geometric conditions (6.8), (6.9), (6.12) and take its closure in the norm (11.5), we obtain H_7. .

Finally, the space H_8 is introduced if $\Gamma_8 > 0$ by taking the closure of $\overset{0}{C}{}^1_{8\Omega}$ in the norm (11.5). In this case the support on Γ_8 is assumed to be essentially elastic, that is, (11.3) holds. The remaining parts of the boundary, $\Gamma_5, \Gamma_6, \Gamma_7$, may be absent, while if they are present, the supports there do not have to be essentially elastic.

The scalar product formula (11.4) satisfies all the requirements of a scalar product. It only remains to prove that

$$\|\omega\|_{H_t} = 0 \text{ implies } \omega \equiv 0. \tag{11.6}$$

We start by verifying this fact.

11.2. Let us introduce the two auxiliary spaces for vector functions $\omega = (w_1, w_2)$ with scalar products and norms:

$$(\omega_1 \cdot \omega_2)_{\overset{0}{W}{}^{(1)}_{2t\Omega}} = \int_\Omega (\omega_{1\alpha^1} \cdot \omega_{2\alpha^1} + \omega_{1\alpha^2} \cdot \omega_{2\alpha^2} + \omega_1 \cdot \omega_2)\, d\Omega, \tag{11.7}$$

$$\|\omega\|^2_{\overset{0}{W}{}^{(1)}_{2t\Omega}} = \int_\Omega \left(\omega^2_{\alpha^1} + \omega^2_{\alpha^2} + \omega^2\right) d\Omega, \tag{11.8}$$

$$(\omega_1 \cdot \omega_2)_{\overset{\tilde{0}}{W}{}^{(1)}_{2t\Omega}} = \int_\Omega (\omega_{1\alpha^1} \cdot \omega_{2\alpha^1} + \omega_{1\alpha^2} \cdot \omega_{2\alpha^2})\, d\Omega + \int_{\Gamma_6} k_s^{\tau\tau} w_{\tau 1} w_{\tau 2} ds$$
$$+ \int_{\Gamma_7} k_s^{mm} w_{m1} w_{m2} ds + \int_{\Gamma_8} k_s^{ij} w_{i1} w_{j2} ds, \tag{11.9}$$

$$\|\omega\|^2_{\overset{\tilde{0}}{W}{}^{(1)}_{2t\Omega}} = \int_\Omega \left(\omega^2_{\alpha^1} + \omega^2_{\alpha^2}\right) d\Omega + \int_{\Gamma_6} k_s^{\tau\tau} w_\tau^2 ds + \int_{\Gamma_7} k_s^{mm} w_m^2 ds$$
$$+ \int_{\Gamma_8} k_s^{ij} w_i w_j \, ds|_{i,j=1,2}\,. \tag{11.10}$$

The spaces $\overset{\tilde{0}}{W}{}^{(1)}_{2t\Omega}$ and $\overset{0}{W}{}^{(1)}_{2t\Omega}$ are formed by taking the closure of $\overset{0}{C}{}^1_{t\Omega}$ under the same conditions on elasticity coefficients as in the spaces H_t. The difference is just in the norm in which the closure is taken. We also remind the reader that in (11.9), (11.9) there is no summation over the indices "τ" and "m".

Lemma 11.1. *If Ω is a Sobolev domain of class* (2, 1, 2), *then for any Vector function $\omega \in \overset{0}{W}{}^{(1)}_{2t\Omega}$ we have the inequality*

$$\|\omega\|_{\overset{\tilde{0}}{W}{}^{(1)}_{2t\Omega}} \geq m\, \|\omega\|_{\overset{0}{W}{}^{(1)}_{2t\Omega}}\,. \tag{11.11}$$

To prove the lemma, we note that $\|\omega\|^2_{\overset{\tilde{0}}{W}{}^{(1)}_{2t\Omega}}$ has precisely the same structure as $\mathcal{R}(\omega)$, which we considered in Theorem 10.8. In this case $l = 1$ and

$$\mathcal{P}_2(\omega) = \int_\Omega \left(\omega^2_{\alpha^1} + \omega^2_{\alpha^2}\right) d\Omega, \; \mathcal{P}_1(\omega) \equiv 0,$$

$$\mathcal{P}_0(\omega) = \int_{\Gamma_6} k_s^{\tau\tau} w_\tau^2 ds + \int_{\Gamma_7} k_s^{mm} w_m^2 ds + \int_{\Gamma_8} k_s^{ij} w_i w_j ds. \tag{11.12}$$

Weak continuity of $\mathcal{P}_0(\omega)$ in $\overset{0}{W}{}^{(1)}_{2t\Omega}$ follows from the embedding Theorem 10.4. Furthermore, in this case it is easy to see that condition 1 of Theorem 10.8 (10.10) holds for all $\omega \in \overset{0}{W}{}^{(1)}_{2t\Omega}$. Now let $\omega_n \rightharpoonup 0$ in $\overset{0}{W}{}^{(1)}_{2t\Omega}$ and $\mathcal{R}(\omega_n) \to 0$.

By weak continuity of $\mathcal{P}_0(\omega)$ it follows from $\mathcal{R}(\omega_n) \to 0$ that

$$\mathcal{P}_2(\omega_n) \to 0, \ \|\omega_{n\alpha^i}\|_{L_{2\Omega}} \to 0,$$

so that condition 2 of Theorem 10.8 is satisfied as well. As all the conditions of Theorem 10.8 are satisfied, Lemma 11.1 is proved.

Theorem 11.1. *The spaces $\overset{\tilde{0}}{W}{}^{(1)}_{2t\Omega}$ and $\overset{0}{W}{}^{(1)}_{2t\Omega}$ contain the same elements, and their norms are equivalent.*

From (11.11) it follows that $\overset{\tilde{0}}{W}{}^{(1)}_{2t\Omega} \subset \overset{0}{W}{}^{(1)}_{2t\Omega}$. Furthermore, from (11.9) and the embedding Theorem 10.4 (see the relation (10.6)) we have

$$\|\omega\|_{\overset{\tilde{0}}{W}{}^{(1)}_{2t\Omega}} \leq M \, \|\omega\|_{\overset{0}{W}{}^{(1)}_{2t\Omega}} . \tag{11.13}$$

From (11.11) and (11.13) it follows that the norms of $\overset{\tilde{0}}{W}{}^{(1)}_{2t\Omega}$ and $\overset{0}{W}{}^{(1)}_{2t\Omega}$ are equivalent, and since these spaces are obtained by taking the closure of the same set $\overset{0}{C}{}^{1}_{t\Omega}$, they contain the same elements. Theorem 11.1 is proved.

Remark 11.1. In the definition of $\|\omega\|_{\overset{\tilde{0}}{W}{}^{(1)}_{2t\Omega}}$ on the right-hand side of (11.9) we can neglect some of the contour integrals. However, in doing so one has to take care that the homogeneous geometric boundary conditions (6.8), (6.9), (6.12), as well as the condition that the equality to zero of the contour integrals neglected in (11.9), would lead to the relation (11.6).

Let us introduce the notation

$$D(\omega, \mathbf{v}) = \int_\Omega \sum_{i,j=1}^{2} \left(w_{i\alpha^j} + w_{j\alpha^i}\right)\left(v_{i\alpha^j} + v_{j\alpha^i}\right) d\Omega + \int_{\Gamma_6} k_s^{\tau\tau} w_\tau v_\tau ds$$

$$+ \int_{\Gamma_7} k_s^{mm} w_m v_m ds + \int_{\Gamma_8} k_s^{ij} w_i v_j ds, \ \omega = (w_1, w_2), \ \mathbf{v} = (v_1, v_2).$$

Lemma 11.2. *Let the condition*

$$D(\omega, \omega) = 0$$

together with the geometric homogeneous boundary conditions (6.8), (6.9), (6.12) *imply that*

$$\omega \equiv 0 \ in \ \Omega.$$

Then there holds the inequality

$$D(\omega, \omega) \geq m \, \|\omega\|_{\overset{0}{W}{}^{(1)}_{2t\Omega}} . \tag{11.14}$$

To prove the lemma, observe that $D(\omega, \omega)$ has the same structure as $\mathcal{R}(\omega)$ of Theorem 10.8. Here $l = 1$, and

$$\mathcal{P}_2(\omega) = \int_\Omega \sum_{i,j=1}^{2} \left(w_{i\alpha^j} + w_{j\alpha^i}\right)\left(w_{i\alpha^j} + w_{j\alpha^i}\right) d\Omega, \ \mathcal{P}_1(\omega) \equiv 0;$$

$\mathcal{P}_0(\omega)$ is given by (11.12) and is a weakly continuous functional in $\overset{0}{W}{}^{(1)}_{2t\Omega}$ by Theorem 10.4. Condition 1 of Theorem 10.8 is guaranteed by the assumption of Lemma 11.2. Condition 2 of Theorem 10.8 is also satisfied, since if $\omega_n \rightharpoonup 0$ in $\overset{0}{W}{}^{(1)}_{2t\Omega}$ and furthermore $\mathcal{R}(\omega_n) \to 0$, then from the Korn inequality

$$\int_\Omega \sum_{i,j=1}^{2} (w_{i\alpha^i} + w_{j\alpha^i})^2 \, d\Omega \geq m \int_\Omega \sum_{i,j=1}^{2} w_{i\alpha^j}^2 d\Omega, \tag{11.15}$$

which is obviously true in our case [196, 207], it follows that

$$\|\omega_{\alpha^i}\|_{L_{2\Omega}} \to 0.$$

Thus all the condition of Theorem 10.8 are satisfied, and the inequality (11.7) has been proved.

Lemma 11.3. *Let the curve $\Gamma \in C_\Gamma$ contain a segment $\widetilde{\Gamma}$ on which $\omega \in W^{(1)}_{2\Omega}$ satisfies*

$$\omega|_{\widetilde{\Gamma}} = 0 \tag{11.16}$$

and assume that the following conditions hold on Ω:

$$w_{i\alpha^i} = G^k_{ii} w_k, \ \textit{(no summation in i)}. \tag{11.17}$$

Then Ω contains a neighborhood of $\widetilde{\Gamma}$, where

$$\omega = (w_1, w_2) = 0. \tag{11.18}$$

To prove the claim, let us first assume that $\widetilde{\Gamma}$ is not a segment of a straight line parallel to any of the axes α^i and that $\widetilde{\Gamma}$ is defined by one of the two equations

$$\alpha^1 = \alpha^1(\alpha^2), \ \alpha^2 = \alpha^2(\alpha^1).$$

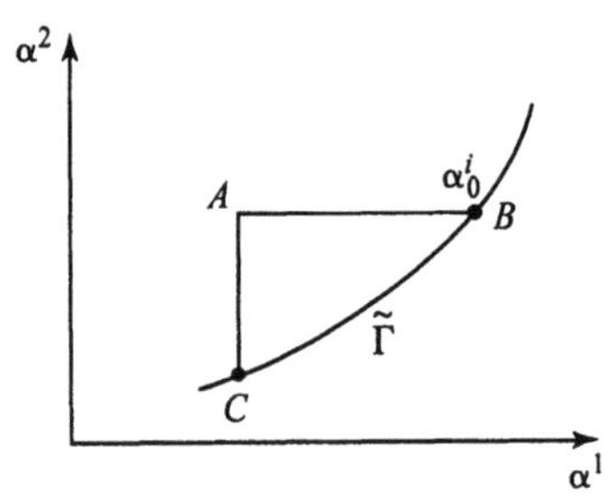

FIGURE 11.1.

By our assumption, there exists a point α_0^i on $\widetilde{\Gamma}$ where one of the following relations holds:

$$\frac{d\alpha^2}{d\alpha^1} \neq 0, \quad \frac{d\alpha^1}{d\alpha^2} \neq 0. \tag{11.19}$$

For definiteness, let us assume that the first of the relations (11.19) holds, and that $d\alpha^2/d\alpha^1 > 0$ at α_0^i (Figure 11.1). From α_0^i (the point $\mathcal{B}$) we trace a straight line parallel to the α^1-axis inside the domain. We follow it to some point $\mathcal{A}$ and trace out the straight line $\mathcal{AC}$ parallel to the α^2-axis. Thus we have formed the triangle $\mathcal{ABC}$. Clearly, if the segment $\mathcal{BA}$ is taken to be sufficiently small, then the whole triangle $\mathcal{ABC}$ can be inscribed in a circle of arbitrarily small diameter. Furthermore, if $\omega = (w_1, w_2) \in W_{2\Omega}^{(1)}$ and w_i satisfies (11.17), then it easily follows from Theorem 10.4 (part 2) that $w_i \in L_{q\Omega}$. Then, in view of (11.19), we have from (11.17),

$$w_1 = \int_{\mathcal{B}}^{\mathcal{A}} G_{11}^k w_k ds_1, \quad w_2 = \int_{\mathcal{C}}^{\mathcal{A}} G_{22}^k w_k ds_2, \tag{11.20}$$

which we rewrite in the form

$$w_1 = \int_{\alpha^1(\alpha^2)}^{\alpha^1} G_{11}^k w_k ds_1, \quad w_2 = \int_{\alpha^2(\alpha^1)}^{\alpha^2} G_{22}^k w_k ds_2. \tag{11.21}$$

Using the Cauchy–Bunyakovsky–Schwarz inequality, we have from (11.21)

$$\begin{aligned} w_1^2 &\le 2 \int_{\alpha^1(\alpha^2)}^{\alpha^1} |G_{11}^k|^2 ds_1 \int_{\alpha^1(\alpha^2)}^{\alpha^1} |w_k|^2 ds_1, \\ w_2^2 &\le 2 \int_{\alpha^2(\alpha^1)}^{\alpha^2} |G_{22}^k|^2 ds_2 \int_{\alpha^2(\alpha^1)}^{\alpha^2} |w_k|^2 ds_2. \end{aligned} \tag{11.22}$$

Furthermore, let

$$m = 2 \max \left\{ \int_{\alpha^1(\alpha^2)}^{\alpha^1} |G_{11}^k|^2 ds_1, \int_{\alpha^2(\alpha^1)}^{\alpha^2} |G_{22}^k|^2 ds_2 \right\}. \tag{11.23}$$

In (11.23) the maximum is taken over all α^1, $\alpha^2 \in \overline{\Omega}, k = 1, 2$. Relations (11.20)–(11.23) make sense for all $\omega \in W_{2\Omega}^{(1)}$, since by the embedding Theorem 10.4 $\omega \in L_{qd}$ for any $q \ge 1$ and any curve d, in particular, on the intervals of integration in the integrals (11.20)–(11.23). From (11.23) follows that

$$w_1^2 \le m \int_{\alpha^1(\alpha^2)}^{\alpha^1} (w_1^2 + w_2^2) ds_1, \quad w_2^2 \le m \int_{\alpha^2(\alpha^1)}^{\alpha^2} (w_1^2 + w_2^2) ds_2. \tag{11.24}$$

Integrating (11.24) over the triangle $\mathcal{ABC}$, we have

$$\begin{aligned} \int_{\mathcal{ABC}} w_1^2 d\alpha^1 d\alpha^2 &\le m \int_{\mathcal{ABC}} \int_{\alpha^1(\alpha^2)}^{\alpha^1} (w_1^2 + w_2^2) ds_1 \, d\alpha^1 d\alpha^2 \\ &= m \int_{\mathcal{C}}^{\mathcal{A}} \int_{\alpha^1(\mathcal{M}'')}^{\alpha^1(\mathcal{A})} \int_{\alpha^1(\alpha^2)}^{\alpha^1} (w_1^2 + w_2^2) ds_1 \, d\alpha^1 d\alpha^2 \end{aligned} \tag{11.25}$$

$$= m \int_C^A \int_{\alpha^1(\mathcal{M}'')}^{\alpha^1(\mathcal{A})} (\alpha^1(\mathcal{A}) - s_1)(w_1^2 + w_2^2)\, ds_1\, d\alpha^2.$$

From (11.25) it follows that

$$\int_{\mathcal{ABC}} w_1^2 d\alpha^1 d\alpha^2 \le m\delta_1 \int_{\mathcal{ABC}} (w_1^2 + w_2^2) d\alpha^1\, d\alpha^2, \qquad (11.26)$$

where $\delta_1 = \max|\alpha^1(\mathcal{P}) - \alpha^1(\mathcal{Q})|$, $\mathcal{P}$, $\mathcal{Q}$ being arbitrary points of $\mathcal{ABC}$. Exactly in the same way we obtain

$$\int_{\mathcal{ABC}} w_2^2 d\alpha^1 d\alpha^2 \le m\delta_2 \int_{\mathcal{ABC}} (w_1^2 + w_2^2) d\alpha^1\, d\alpha^2, \qquad (11.27)$$

where $\delta_2 = \max|\alpha^2(\mathcal{P}) - \alpha^2(\mathcal{Q})|$, $\mathcal{P}$, $\mathcal{Q}$ being arbitrary points of $\mathcal{ABC}$. From (11.26), (11.27) we obtain

$$\int_{\mathcal{ABC}} (w_1^2 + w_2^2) d\alpha^1 d\alpha^2 \le m(\delta_1 + \delta_2) \int_{\mathcal{ABC}} (w_1^2 + w_2^2) d\alpha^1\, d\alpha^2.$$

The triangle $\mathcal{ABC}$ can be chosen so small that

$$m(\delta_1 + \delta_2) < 1, \qquad (11.28)$$

and then from (11.28) we have that in this triangle

$$w_1 \equiv w_2 \equiv 0, \qquad (11.29)$$

and in this case the lemma is proved.

If $\widetilde{\Gamma}$ is an interval parallel to the α^1-axis (Figure 11.2), then we can pass to a system of coordinates ξ_i, rotated by 45°. The corresponding $W_{2\Omega}^{(1)}$ norm is invariant under the rotation of the axes. In these axes we can construct an isosceles right triangle the hypotenuse of which lies on $\widetilde{\Gamma}$, for which all the arguments above can be repeated. As a result we have that for a sufficiently small isosceles right triangle (11.29) will hold, so that Lemma 11.3 will have been proved in this case as well. The same considerations hold in the case when $\widetilde{\Gamma}$ is parallel to the α^2-axis (Figure 11.3). Lemma 11.3 has been completely proved.

It is important to note that in all the arguments used in the proof of Lemma 11.2 the sizes of all isosceles right triangles can be chosen independently of their location in Ω, so that (11.28) is satisfied.

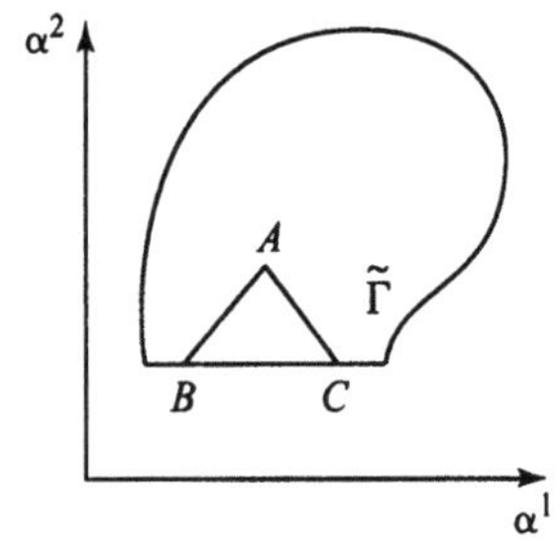

FIGURE 11.2.

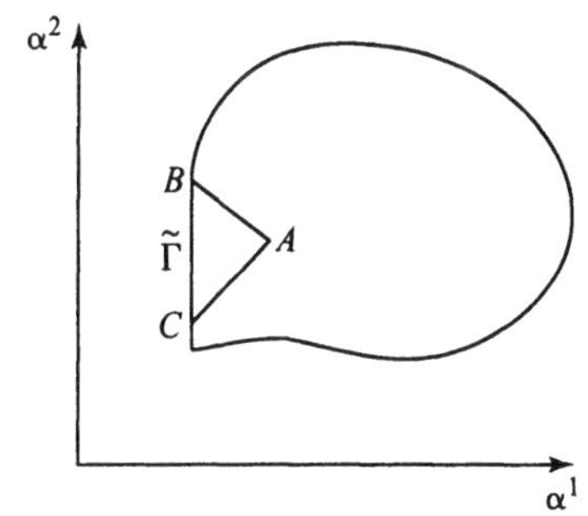

FIGURE 11.3.

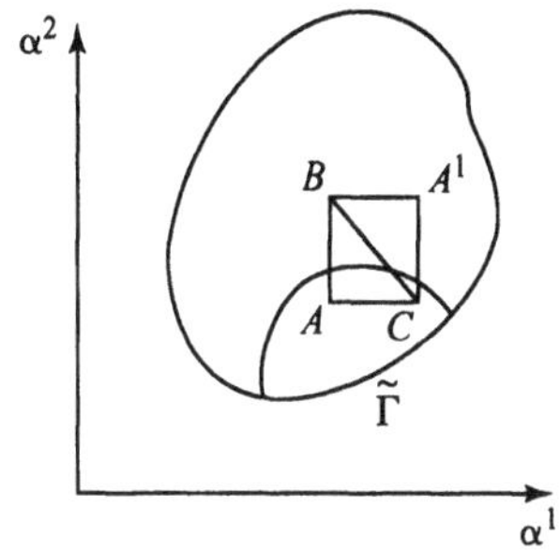

FIGURE 11.4.

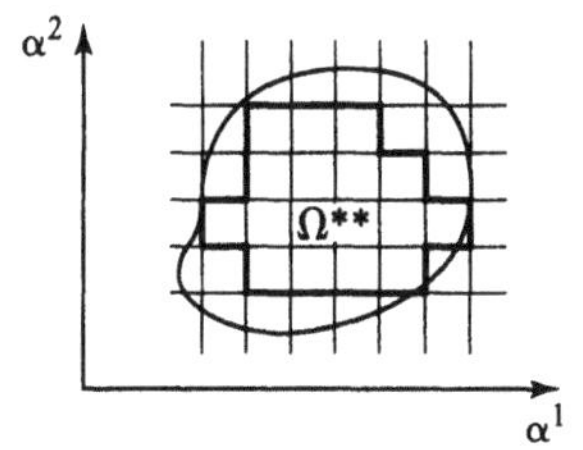

FIGURE 11.5.

Theorem 11.2. *Let all the conditions of Lemma 11.3 be satisfied. Then* (11.18) *holds in the entire domain* Ω.

To prove the claim, let us observe that by Lemma 11.3, (11.18) will hold in a neighborhood of $\widetilde{\Gamma}$, which we denote by Ω^*. But then one can extend it using isosceles right triangles (Figure 11.4). Indeed, from Ω^* relations (11.18) can be extended to the triangle $\mathcal{ABC}$, a side $\mathcal{AB}$ of which lies entirely in Ω^*. Then by Lemma 11.3, (11.18) can be extended to the triangle $\mathcal{BCA}'$, that is, to the whole square $\mathcal{ABA}'\mathcal{C}$. Obviously, in this way, (11.18) is established in any subdomain Ω^{**} of Ω^* composed of squares (Figure 11.5). Finally, since in our arguments the sides of the squares $\mathcal{ABA}'\mathcal{C}$ can be taken to be arbitrarily small, and the domain Ω is measurable, we can construct a subdomain Ω'' that will contain any interior point of Ω.

Theorem 11.2 has thus been proved.

Lemma 11.4. *Let* $\omega = (w_1, w_2) \in W^{(1)}_{2\Omega}$ *and* $\Gamma_t \in C^1_{\Gamma_t}$ *(*$t = 5, 6, 7, 8$*). Then the relation*

$$\|\omega\|_{H_t} = 0 \tag{11.30}$$

implies

$$\omega \equiv 0 \text{ in } \Omega. \tag{11.31}$$

We shall consider the situation separately for each of the spaces H_t.

From (11.30) and (11.5) by positive definiteness of the tensor $D_s^{ijkl}(\alpha^1, \alpha^2)$ and conditions on the supports, we obtain (11.17). Next, in the case of H_5 we necessarily have $\Gamma_5 > 0$, and then, taking $\widetilde{\Gamma} = \Gamma_5$, we are in the conditions of Theorem 11.2, from which we have (11.31). In the case of H_6, $\Gamma_6 > 0$ and (11.1) holds. Therefore, in addition to (11.17), it follows from (11.30) that

$$w_\tau|_{\Gamma_6} = 0.$$

Taking into account the fact that in the construction of H_6 we also assumed that the homogeneous condition (6.9) holds, we have

$$\omega|_{\Gamma_6} = 0.$$

Therefore, we are again under the conditions of Theorem 11.2 if we take $\widetilde{\Gamma} = \Gamma_6$.

For H_7, $\Gamma_7 > 0$ and (11.2) holds. Therefore, in addition to (11.17), it follows from (11.30) that

$$w_m|_{\Gamma_7} = 0.$$

Since in the construction of H_7 we assumed that the homogeneous condition (6.12) holds, we have

$$\omega|_{\Gamma_7} = 0.$$

and, taking $\widetilde{\Gamma} = \Gamma_7$, we obtain (11.31) from Lemma 11.3.

Finally, in the case of H_8, $\Gamma_8 > 0$, and k_s^{ij} is an essentially elastic support. From (11.30) we therefore obtain (11.29) and

$$\omega|_{\Gamma_8} = 0; \tag{11.32}$$

in this case $\widetilde{\Gamma} = \Gamma_8$, and we again have (11.31), so that Lemma 11.4 is proved.

This lemma justifies the introduction of the norms (11.5).

Lemma 11.5. *There holds the inequality*

$$\|\omega\|_{H_t} \geq m \, \|\omega\|_{\overset{0}{W}{}^{(1)}_{2t\Omega}}. \tag{11.33}$$

To prove it, let us consider $R(\omega) = \|\omega\|^2_{H_t}$. We have

$$R(\omega) = \int_\Omega [\mathcal{P}_2(\omega) + \mathcal{P}_1(\omega)] d\Omega + \mathcal{P}_0(\omega),$$

where

$$\mathcal{P}_2(\omega) = \frac{1}{4} D_{\mathrm{s}}^{klst} (w_{k\alpha^l} + w_{l\alpha^k})(w_{s\alpha^t} + w_{t\alpha^s}),$$

$$\mathcal{P}_1(\omega) = \frac{1}{2} D_{\mathrm{s}}^{klst} G_{st}^{\gamma} w_{\gamma} (w_{k\alpha^l} + w_{l\alpha^k}),$$

$$\mathcal{P}_0(\omega) = \frac{1}{4} \int_{\Omega} D_{\mathrm{s}}^{klst} G_{kl}^{\gamma} G_{st}^{\delta} w_{\gamma} w_{\delta} d\Omega + \int_{\Gamma_6} k^{\tau\tau} w_{\tau}^2 ds + \int_{\Gamma_7} k^{mm} w_m^2 ds + \int_{\Gamma_8} k_{\mathrm{s}}^{ij} w_i w_j ds.$$

We intend to apply Theorem 10.8, and must verify its conditions. The structure of $\mathcal{P}_2(\omega), \mathcal{P}_1(\omega)$ clearly satisfies the requirements of that theorem for $l = 1$. From the embedding Theorem 10.4, it also immediately follows that the functional $\mathcal{P}_0(\omega)$ is weakly continuous in $\overset{0}{W}{}^{(1)}_{2\Omega}$. Condition 1 of Theorem 10.8 is satisfied automatically by Lemma 11.4. It remains to verify that condition 2 of Theorem 10.8 holds. Let $\omega_n \rightharpoonup 0$ in $\overset{0}{W}{}^{(1)}_{2\Omega}$. Let us prove that $\mathcal{P}_1(\omega_n) \to 0$. This follows from an elementary result: If f_n, φ_n are two sequences in $L_{2\Omega}$ and

$$f_n \to f_0 \text{ and } \varphi_n \rightharpoonup \varphi_0 \text{ in } L_{2\Omega},$$

then

$$\int_{\Omega} f_n \varphi_n d\Omega \to \int_{\Omega} f_0 \varphi_0 d\Omega.$$

In our case

$$\varphi_n = w_{nk\alpha^l} + w_{nl\alpha^k} \rightharpoonup 0 \text{ in } L_{2\Omega}, \quad f_n = w_{ni} \to 0 \text{ in } L_{2\Omega}.$$

Thus, from $\mathcal{P}_1(\omega_n) \to 0$ in $\overset{0}{W}{}^{(1)}_{2\Omega}$ and $\mathcal{R}(\omega_n) \to 0$, we obtain

$$\int_{\Omega} [\mathcal{P}_2(\omega_n) + \mathcal{P}_0(\omega_n)] d\Omega \to 0. \tag{11.34}$$

Next, by the embedding Theorem 10.4 we have

$$\int_{\Omega} D_{\mathrm{s}}^{klst} G_{kl}^{\gamma} G_{st}^{\delta} w_{n\gamma} w_{n\delta} d\Omega \to 0 \tag{11.35}$$

and from (11.34), (11.35) we obtain

$$\int_{\Omega} \mathcal{P}_2(\omega_n) d\Omega + \int_{\Gamma_6} k^{\tau\tau} w_{n\tau}^2 ds + \int_{\Gamma_7} k^{mm} w_{nm}^2 ds + \int_{\Gamma_8} k_{\mathrm{s}}^{ij} w_{ni} w_{nj} ds \to 0 \; (i, \; j = 1, \; 2). \tag{11.36}$$

There is no summation over τ and m. Since $\mathcal{P}_2(\omega)$ is a positive definite quadratic form of $w_{k\alpha^l} + w_{l\alpha^k}$, it follows from (11.36) that

$$\int_\Omega \sum_{k,l=1}^{2} (w_{nk\alpha^l} + w_{nl\alpha^k})^2 + \int_{\Gamma_6} k^{\tau\tau} w_{n\tau}^2 ds + \int_{\Gamma_7} k^{mm} w_{nm}^2 ds$$
$$+ \int_{\Gamma_8} k_s^{ij} w_{ni} w_{nj} ds \to 0,$$

and from the inequality (11.15) we obtain

$$\|\omega_n\|_{\overset{0}{W}{}^{(1)}_{2t\Omega}} \to 0,$$

and therefore

$$\|w_{nl\alpha^k}\|_{L_{2\Omega}} \to 0.$$

Thus, condition 2 of Theorem 10.8 is satisfied, and inequality (11.33) has been proved.

Theorem 11.3. *The spaces H_t, $t = 5, 6, 7, 8$, are equivalent to the corresponding spaces $\overset{0}{W}{}^{(1)}_{2t\Omega}$ obtained by the closure of the corresponding sets $\overset{0}{C}{}^1_{t\Omega}$.*

To prove the theorem, we note that from (11.33) it follows that $H_t \subset \overset{0}{W}{}^{(1)}_{2t\Omega}$. Next, we have the obvious inequality

$$\|\omega\|_{H_t} \le M \|\omega\|_{\overset{0}{W}{}^{(1)}_{2t\Omega}}. \tag{11.37}$$

Thus we have

$$m \|\omega\|_{\overset{0}{W}{}^{(1)}_{2t\Omega}} \le \|\omega\|_{H_t} \le M \|\omega\|_{\overset{0}{W}{}^{(1)}_{2t\Omega}},$$

which shows that the norms in H_t and $\overset{0}{W}{}^{(1)}_{2t\Omega}$ are equivalent. If they are constructed from the same set $\overset{0}{C}{}^1_{t\Omega}$, they contain the same elements. Theorem 11.3 is proved.

From Theorem 11.3 we obtain a very important corollary: The embedding Theorem 10.4 holds for the spaces H_t with $l = 1$, $p = 2$, $n = 2$. Since we shall be constantly using this conclusion, we shall write out all its consequences.

Theorem 11.4. *Let $\omega = (w_1, w_2) \in H_t$, $t = 5, 6, 7, 8$. Then*

(1) $w_{i\alpha^j} \in L_{2\Omega}$, $\nabla_i w_j \in L_{2\Omega}$, *and* $\|w_{i\alpha^j}\|_{L_{2\Omega}}, \left\|\nabla_i w_j\right\|_{L_{2\Omega}} \le m \|\omega\|_{H_t}$.

(2) $w_i \in L_{q\Omega}$, L_{qd} *for any* $q \ge 1$. (11.38)

Here d is any C_d^1 PSC. There hold the inequalities

$$\|w_i\|_{L_{q\Omega}}, \|w_i\|_{L_{qd}} \le m \|\omega\|_{H_t}. \tag{11.39}$$

(3) *The embedding operator of H_t into $L_{q\Omega}$ and L_{qd} is completely continuous, that is,*

$$\begin{aligned}&\omega_n \rightharpoonup \omega_0 \text{ in } H_t \text{ implies } \omega_n \to \omega_0 \text{ in any } L_{q\Omega},\\ &\omega_n \to \omega_0 \text{ in any } L_{qd} \text{ for all } q \geq 1.\end{aligned} \tag{11.40}$$

Let us note that the curve d can intersect or be contained in Γ.

Lemma 11.6. *All elements $\omega \in \overset{0}{W}{}^{(1)}_{2t\Omega} \sim \overset{\tilde{0}}{W}{}^{(1)}_{2t\Omega} \sim H_t$ satisfy the homogeneous boundary conditions* (6.8), (6.9), (6.12) *almost everywhere.*

Indeed, if $\omega \in \overset{0}{W}{}^{(1)}_{2t\Omega}$, there is a sequence $\omega_n \in C^1_\Omega$ for which these conditions are satisfied pointwise, and moreover,

$$\|\omega - \omega_n\|_{\overset{0}{W}{}^{(1)}_{2t\Omega}} \to 0. \tag{11.41}$$

By the embedding Theorem 10.4 and the inequalities (11.39), we have

$$\|\omega\|_{L_{2\Gamma_5}} = \|w_m\|_{L_{2\Gamma_6}} = \|w_\tau\|_{L_{2\Gamma_7}} = 0. \tag{11.42}$$

11.3.

Lemma 11.7. *Let a part $\tilde{\Gamma}$ of the boundary curve Γ be in $C^{\tilde{\Gamma}}$, while on a part $\tilde{\tilde{\Gamma}}$, completely contained in $\tilde{\Gamma}$, we have $w_i \in W^{\frac{1}{2}}_{2\tilde{\tilde{\Gamma}}}$. Then w_i can be extended from $\tilde{\tilde{\Gamma}}$ to $\tilde{\Gamma}$ in such a way that outside of $\tilde{\Gamma}$, w_i vanishes, $w_i \in W^{\frac{1}{2}}_{2\tilde{\Gamma}}$, and furthermore,*

$$\|w_i\|_{W^{\frac{1}{2}}_{2\tilde{\Gamma}}} \leq m \, \|w_i\|_{W^{\frac{1}{2}}_{2\tilde{\tilde{\Gamma}}}}, \tag{11.43}$$

where m does not depend on w_i.

To prove the lemma, we establish first that it is in principle possible to extend $w_i \in W^{\frac{1}{2}}_{2\tilde{\tilde{\Gamma}}}$ to $W^{\frac{1}{2}}_{2\tilde{\Gamma}}$ so that (11.43) holds (that is, without requiring it be zero outside of $\tilde{\Gamma}$). In Figure 11.6 the endpoints of $\tilde{\tilde{\Gamma}}$ are $\mathcal{C}$ and $\mathcal{D}$, while the endpoints of $\tilde{\Gamma}$ are $\mathcal{A}$ and $\mathcal{B}$.

Next, it is known that if $w_i \in W^{\frac{1}{2}}_{2\tilde{\tilde{\Gamma}}}$, then there exists a function $f_i \in L_{2\tilde{\tilde{\Gamma}}}$ such that

$$w_i(s) = \int_{\mathcal{C}}^{s} \frac{f_i}{\sqrt{\sigma - s}} d\sigma, \quad \mathcal{C} \leq s \leq \mathcal{D}. \tag{11.44}$$

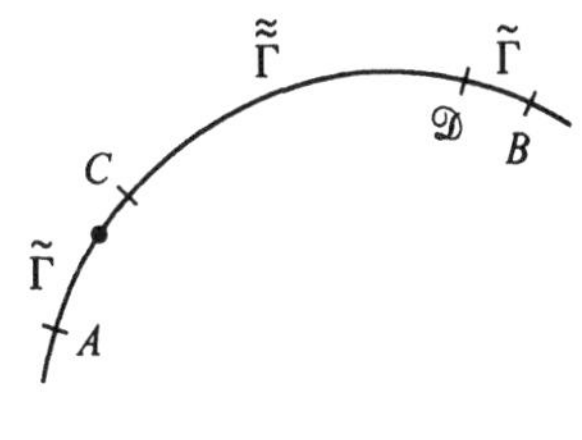

FIGURE 11.6.

Usually, the representation (11.44) is used when Γ is an interval of straight line. However, it still holds when Γ is a C^1_Γ curve. The function $f_i(\sigma)$ can be extended to the segment $\mathcal{DB}$ as well as to the segment $\mathcal{CA}$ in such a way that the resulting function belongs to $L_{2\widetilde{\Gamma}}$; moreover, it can be done so that the inequality

$$\| f_i \|_{L_{2AB}} \leq m \, \| f_i \|_{L_{2CD}}$$

is satisfied.

Therefore, the function

$$w_i(s) = \int_C^s \frac{f_i d\sigma}{\sqrt{\sigma - s}} \in W^{(\frac{1}{2})}_{2AB},$$

but at this stage it is not zero outside of $\widetilde{\Gamma}$. This can be accomplished by using cutoff functions of the type seen in Figure 11.7. Such a cutoff function $\chi_1(s)$ can be introduced on $\mathcal{DB}$, and then $w_i(s)\chi_1(s)$ will be zero for $s \geq \mathcal{B}$. A similar cutoff function $\chi_2(s)$ can be defined on $\mathcal{CA}$, and then $\chi_2(s)w_1(s)\chi_1(s)$ satisfies the conditions of Lemma 11.7.

Lemma 11.8. *Let* $\Gamma \in C^1_\Gamma$, *and assume that there is a segment* $\widetilde{\Gamma} \in C^{\frac{3}{2}}_{\widetilde{\Gamma}}$ *(which can be disconnected) containing* $\Gamma_5, \Gamma_6, \Gamma_7$. *Then for the existence of a vector function* $\overset{0}{\omega} \in W^{(1)}_{2\Omega}$ *such that*

$$\left. \overset{0}{w}_i \right|_{\Gamma_5} = \widetilde{w}_i \ (i = 1, 2) \ \left. \overset{0}{w}_m \right|_{\Gamma_6} = \widetilde{w}_m \ \left. \overset{0}{w}_\tau \right|_{\Gamma_7} = \widetilde{w}_\tau \ \Gamma_i \cap \Gamma_j = \emptyset,$$

it is necessary and sufficient that

$$\widetilde{w}_i |_{\Gamma_5} \in W^{(\frac{1}{2})}_{2\Gamma_5}, \ \widetilde{w}_m |_{\Gamma_6} \in W^{(\frac{1}{2})}_{2\Gamma_6}, \ \widetilde{w}_\tau |_{\Gamma_7} \in W^{(\frac{1}{2})}_{2\Gamma_7}. \tag{11.45}$$

Let us sketch the proof. By assumption, $\Gamma_5, \Gamma_6, \Gamma_7$ have no points in common. Let as take $\Gamma_5 = \widetilde{\widetilde{\Gamma}}$, and we have $w_i \in W^{(\frac{1}{2})}_{2\Gamma_5}$. By Lemma 11.7 we can assume that there is a segment of $\widetilde{\Gamma}$ containing $\Gamma_5 \equiv \widetilde{\widetilde{\Gamma}}$ such that $w_i \in W^{(\frac{1}{2})}_{2\widetilde{\Gamma}}$ and w_i vanishes outside of $\widetilde{\Gamma}$. Therefore, we can assume that $w_i \in W^{(\frac{1}{2})}_{2\Gamma}$ and that

$$w_i |_{\Gamma_6} \equiv 0, \ \ w_i |_{\Gamma_7} \equiv 0.$$

Then there must exist a vector function $\omega_1 \in W^{(1)}_{2\Omega}$, and

$$\|\omega_1\|_{W^{(1)}_{2\Omega}} \leq m \, \|\widetilde{\omega}\|_{W^{(\frac{1}{2})}_{2\Gamma_5}}. \tag{11.46}$$

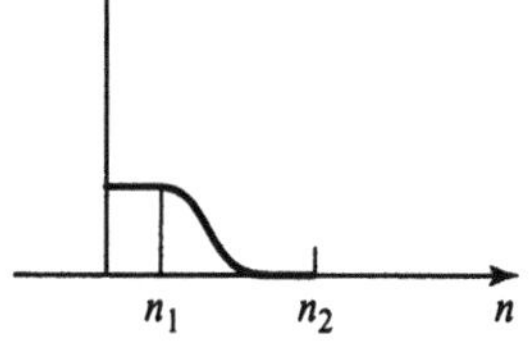

FIGURE 11.7.

Similarly, we extend $\widetilde{w}_m$ from Γ_6 to $\widetilde{\Gamma}_6$, so that $\widetilde{w}_m \in W_{2\widetilde{\Gamma}_6}^{(\frac{1}{2})}$, and outside of $\widetilde{\Gamma}_6$, $\widetilde{w}_m = 0$; in particular,

$$\widetilde{w}_m|_{\Gamma_5} \equiv \widetilde{w}_m|_{\Gamma_7} \equiv 0.$$

Then there will exist a function $\widetilde{w}_m \in W_{2\Gamma}^{(\frac{1}{2})}$, and to it will correspond a vector function $\omega_2 \in W_{2\Omega}^{(1)}$, such that $\omega_{2m}|_{\Gamma_6} = \widetilde{\omega}_m$, $\omega_{2\tau}|_{\Gamma_6} \equiv 0$, and

$$\|\omega_2\|_{W_{2\Omega}^{(1)}} \leq m\, \|w_m\|_{W_{2\Gamma_6}^{(\frac{1}{2})}}. \tag{11.47}$$

Similarly, we extend $\widetilde{w}_\tau$ from Γ_7 to $\widetilde{\Gamma}_7$, so that $\widetilde{w}_\tau \in W_{2\widetilde{\Gamma}_7}^{(1)}$, and outside of $\widetilde{\Gamma}_7$, $\widetilde{w}_\tau = 0$; in particular,

$$\widetilde{w}_\tau|_{\Gamma_5} \equiv \widetilde{w}_\tau|_{\Gamma_6} \equiv 0.$$

Then there will exist a function $\widetilde{w}_3 \in W_{2\Omega}^{(1)}$. Clearly, the vector function

$$\overset{0}{\omega} = \omega_1 + \omega_2 + \omega_3$$

satisfies the inequality

$$\left\|\overset{0}{\omega}\right\|_{W_{2\Omega}^{(1)}} \leq m\left(\|\widetilde{\omega}\|_{W_{2\Gamma_5}^{(\frac{1}{2})}} + \|\widetilde{w}_m\|_{W_{2\Gamma_6}^{(\frac{1}{2})}} + \|\widetilde{w}_\tau\|_{W_{2\Gamma_7}^{(\frac{1}{2})}}\right).$$

The boundary values of $\widetilde{\omega}|_{\Gamma_5}$, $\widetilde{w}_m|_{\Gamma_6}$, $\widetilde{w}_\tau|_{\Gamma_7}$ that satisfy (11.45) will be called admissible.

Necessity follows from [282]; see also [31, 32, 181, 314, 315].

11.4. In this section we introduce function spaces, which will be used to characterize lateral stresses acting on the shell. Let us consider the quantity

$$\int_\Omega R^i w_i d\Omega + \int_{\Gamma_6+\Gamma_8} \widetilde{T}^\tau w_\tau ds + \int_{\Gamma_7+\Gamma_8} \widetilde{T}^m w_m ds \tag{11.48}$$

and the set $\overline{H}_t$ of such lateral loads $[R^s, \widetilde{T}^\tau, \widetilde{T}^m]$, for which (11.48) defines a bounded functional in H_t. It is a bounded linear functional m if and only if there exists a constat m such that

$$\sup \frac{\left|\int_\Omega R^i w_i d\Omega + \int_{\Gamma_6+\Gamma_8} \widetilde{T}^\tau w_\tau ds + \int_{\Gamma_7+\Gamma_8} \widetilde{T}^m w_m ds\right|}{\|\omega\|_{H_t}} \leq m \leq \infty \tag{11.49}$$

for all $\omega \in H_t$. In (11.48), (11.49) there is no summation on τ, m. Next, if (11.49) holds, by the Riesz theorem there exists an element ω_{p} such that

$$\int_\Omega R^i w_i d\Omega + \int_{\Gamma_6+\Gamma_8} \widetilde{T}^\tau w_\tau ds + \int_{\Gamma_7+\Gamma_8} \widetilde{T}^m w_m ds = (\omega_{\mathrm{p}} \cdot \omega)_{H_t}. \tag{11.50}$$

The relation (11.50) allows us to make $\overline{H}_t$ into a Hilbert space. For that let us assume that $[R_1^s, \widetilde{T}_1^\tau, \widetilde{T}_1^m]$ and $[R_2^s, \widetilde{T}_2^\tau, \widetilde{T}_2^m]$ are two elements of $\overline{H}_t$, and let us take

$$([R_1^s, \widetilde{T}_1^\tau, \widetilde{T}_1^m] \cdot [R_2^s, \widetilde{T}_2^\tau, \widetilde{T}_2^m])_{\overline{H}_t} = (\omega_{\mathrm{p}1} \cdot \omega_{\mathrm{p}2})_{H_t}, \tag{11.51}$$

where ω_{pi} is the element of H_t corresponding to $[R_i^s, \widetilde{T}_i^\tau, \widetilde{T}_i^m]_i$. From (11.51) we have

$$\left\|[R^s, \widetilde{T}^\tau, \widetilde{T}^m]\right\|_{\overline{H}_t} = \left\|\omega_p\right\|_{H_t}. \tag{11.52}$$

At this point in (11.49) instead of sup we can put max and instead of m on the right-hand side, $\left\|\omega_p\right\|$. From this it follows that if $\omega = 0$ and $\left\|[R^s, \widetilde{T}^\tau, \widetilde{T}^m]\right\|_{\overline{H}_t} = 0$, then $R^s \equiv \widetilde{T}^\tau \equiv \widetilde{T}^m \equiv 0$.

12. The Function Spaces H_κ, $\kappa = 1, 2, 3, 4$. Properties of Their Elements

12.1. In all that follows we shall assume that Γ is a PSC of class C_Γ^3; Ω is a Sobolev domain of class (2, 2, 2), the material of the shell is regular, and $S \in C_\Omega^2$.

Definition 12.1. We shall say that the supports are essentially elastic if

$$k_f^{44} \geq m_0 > 0, \tag{12.1}$$

$$k_f^{33} \geq m_0 > 0, \tag{12.2}$$

$$k_f^{ij} w_i w_j \,|_{i,j=3,4} \geq m_0 > 0, \text{ for } w_3^2 + w_4^2 = 1. \tag{12.3}$$

If the supports are not essentially elastic, then in (12.1)–(12.3) the constant m_0 is zero.

Let us consider the set of functions $w \in \overset{0}{C}{}_\Omega^2$ that satisfy the homogeneous conditions (6.1)–(6.3), on which we define the scalar product

$$\begin{aligned}(w_1 \cdot w_2)_{H_\kappa} = &\int_\Omega D_f^{ijkl} \overset{1}{\gamma}_{ij}(w_1) \overset{1}{\gamma}_{kl}(w_2) d\Omega + \int_{\Gamma_2} k_f^{44} w_{14} w_{24} ds \\ &+ \int_{\Gamma_3} k_f^{33} w_1 w_2 ds + \int_{\Gamma_4} k_f^{ij} w_{1i}\, w_{2j}\Big|_{i,j=3,4} ds,\end{aligned} \tag{12.4}$$

$$w_{i3} = w_i, \quad w_{i4} = \frac{\partial w_i}{\partial m}.$$

To (12.4) corresponds the norm

$$\begin{aligned}\|w\|_{H_\kappa}^2 = &\int_\Omega D_f^{ijkl} \overset{1}{\gamma}_{ij}(w) \overset{1}{\gamma}_{kl}(w) d\Omega + \int_{\Gamma_2} k_f^{44} w_4^2 ds \\ &+ \int_{\Gamma_3} k_f^{33} w^2 ds + \int_{\Gamma_4} k_f^{ij} w_i\, w_j\Big|_{i,j=3,4} ds.\end{aligned} \tag{12.5}$$

The formula (12.4) satisfies all the requirements of a scalar product in a Hilbert space. We need to prove only that

$$\|w\|_{H_\kappa} = 0, \ \kappa = 1, 2, 3, 4,$$

implies

$$w \equiv 0 \text{ in } \overline{\Omega},$$

and this question will be resolved in Lemma 12.4.

To construct the space H_1, we assume that $\Gamma_1 > 0$. Here the notation $\Gamma_i > 0$ is taken to mean that Γ_i contains a connected segment of positive length. Let us now form the closure of $C^2_{1\Omega}$ in the norm (12.5). Here $C^2_{1\Omega}$ is $\overset{0}{C}{}^2_{\Omega}$, but under the condition $\Gamma_1 > 0$. Here $\Gamma_2, \Gamma_3, \Gamma_4$ may be absent, while if these are present, the supports there do not have to be essentially elastic. Thus (12.1)-(12.3) may hold with $m_0 = 0$.

In the case of the space H_2 we assume that $\Gamma_2 > 0$, and in this case there are two further subvariants. In case of subvariant (a) we assume that for at least one point of Γ_2 we have

$$\kappa_2 \neq 0, \tag{12.6}$$

where κ_2 is the geodesic curvature of Γ_2. We call the closure of $\overset{0}{C}{}^2_{2\Omega}$ in the norm (12.5) H_{2a}. Here $\overset{0}{C}{}^2_{2\Omega}$ is $\overset{0}{C}{}^2_{\Omega}$ but with the condition $\Gamma_2 > 0$ necessarily holding. In this subvariant the supports do not have to be essentially elastic. It is sufficient that (12.1)–(12.3) hold with $m_0 = 0$. In subvariant (b), (12.6) does not have to hold, but the support on Γ_2 has to be essentially elastic, that is, (12.1) has to be satisfied. The other supports do not have to be essentially elastic. The closure of $\overset{0}{C}{}^2_{2\Omega}$ in the norm (12.5) with the condition $\Gamma_2 > 0$ necessarily holding and with (12.1) will be called H_{2b}. In both cases (a) and (b) the segments $\Gamma_1, \Gamma_3, \Gamma_4$ may be completely absent.

The space H_3 is defined if $\Gamma_3 > 0$ and the support on Γ_3 is essentially elastic. The closure of $\overset{0}{C}{}^2_{3\Omega}$ in the norm (12.5) will be called H_3. Here $\Gamma_1, \Gamma_2, \Gamma_4$ may be absent, while if they are present, the supports there do not have to be essentially elastic.

Finally, the space H_4 is introduced if $\Gamma_4 > 0$ and in the case when the matrix k_{f}^{ij} is positive definite, that is, if the support on Γ_4 is essentially elastic. The remaining parts of the boundary, $\Gamma_5, \Gamma_6, \Gamma_7$, may be absent, while if they are present, the supports there do not have to be essentially elastic. Let $\overset{0}{C}{}^2_{4\Omega}$ be $\overset{0}{C}{}^2_{\Omega}$ under the condition that Γ_4 is present. H_4 is the closure of $\overset{0}{C}{}^2_{4\Omega}$ in the norm (12.5).

Let us introduce two auxiliary spaces with the scalar products:

$$(w_1 \cdot w_2)_{\overset{0}{W}{}^{(2)}_{2\kappa\Omega}} = \int_\Omega \left(\sum_{i,j=1}^{2} w_{1\alpha^i\alpha^j} w_{2\alpha^i\alpha^j} + \sum_{i=1}^{2} w_{1\alpha^i} w_{2\alpha^i} + w_1 w_2 \right) d\Omega,$$

$$(w_1 \cdot w_2)_{\tilde{\overset{0}{W}}{}^{(2)}_{2\kappa\Omega}} = \int_\Omega (\sum_{i,j=1}^{2} w_{1\alpha^i\alpha^j} w_{2\alpha^i\alpha^j}) d\Omega + \int_{\Gamma_2} k_{\mathrm{f}}^{44} w_{14} w_{24} ds$$
$$+ \int_{\Gamma_3} k_{\mathrm{f}}^{33} w_1 w_2 ds + \int_{\Gamma_4} k_{\mathrm{f}}^{ij} w_{i1} w_{j2}\, ds|_{i,j=3,4},$$

$$w_{i3} \equiv w_i, \ w_{i4} = \frac{\partial w_i}{\partial m}.$$

To these spaces correspond the norms

$$\|w\|^2_{\overset{0}{W}{}^{(2)}_{2\kappa\Omega}} = \int_\Omega \left(\sum_{i,j=1}^{2} w^2_{\alpha^i\alpha^j} + \sum_{i=1}^{2} w^2_{\alpha^i} + w^2 \right) d\Omega, \tag{12.7}$$

$$\|w\|^2_{\overset{\tilde{0}}{W}{}^{(2)}_{2\kappa\Omega}} = \int_\Omega \sum_{i,j=1}^{2} w^2_{\alpha^i\alpha^j}\, d\Omega + \int_{\Gamma_2} k_{\mathrm{f}}^{44} w_4^2 ds + \int_{\Gamma_3} k_{\mathrm{f}}^{33} w^2 ds$$

$$+ \int_{\Gamma_4} k_{\mathrm{f}}^{ij} w_i \ w_j \big|_{i,j=3,4} ds. \tag{12.8}$$

Here it is assumed that the combination of the bending boundary conditions is such that

$$\|w\|_{\overset{\tilde{0}}{W}{}^{(2)}{2\kappa\Omega}} = 0$$

for each κ implies that

$$w \equiv 0 \text{ in } \Omega.$$

The main aim of the present section is a study of the properties of H_κ, $\kappa = 1$, 2, 3, 4.

12.2.

Lemma 12.1. *Any function* $w \in \overset{0}{W}{}^{(2)}_{2\kappa\Omega}$ *satisfies the inequality*

$$\|w\|_{\overset{\tilde{0}}{W}{}^{(2)}_{2\kappa\Omega}} \geq m \, \|w\|_{\overset{0}{W}{}^{(2)}_{2\kappa\Omega}}. \tag{12.9}$$

To prove the lemma, we use Theorem 10.8; here $l = 2$. Let us introduce $\mathcal{R}(w) = \|w\|^2_{\overset{\tilde{0}}{W}{}^{(2)}_{2\kappa\Omega}}$, and by (12.8) we have

$$\mathcal{P}_2(w) = \int_\Omega \sum_{i,j=1}^{2} w^2_{\alpha^i\alpha^j}\, d\Omega, \ \ \mathcal{P}_1(w) \equiv 0,$$

$$\mathcal{P}_0(w) = \int_{\Gamma_2} k_{\mathrm{f}}^{44} w_4^2 ds + \int_{\Gamma_3} k_{\mathrm{f}}^{33} w_3^2 ds + \int_{\Gamma_4} k_{\mathrm{f}}^{ij} w_i \ w_j \big|_{i,j=3,4} ds.$$

The structure of $\mathcal{P}_1$, $\mathcal{P}_2$ obviously satisfies the conditions of Theorem 10.8. The functional $\mathcal{P}_0(w)$ is weakly continuous in $\overset{0}{W}{}^{(2)}_{2\kappa\Omega}$, which follows from Theorem 10.5. Furthermore, let us note that from the relation

$$\mathcal{R}(w) = \|w\|^2_{\overset{\tilde{0}}{W}{}^{(2)}_{2\kappa\Omega}},$$

in view of the boundary conditions used in the construction of H_κ, we obtain

$$w \equiv 0 \text{ in } \Omega, \tag{12.10}$$

and thus condition 1 of Theorem 10.8 holds for all $w \in \overset{0}{W}{}^{(2)}_{2\kappa\Omega}$. Now let $w_n \rightharpoonup 0$ in $\overset{0}{W}{}^{(2)}_{2\kappa\Omega}$. Then by weak continuity of $\mathcal{P}_0(w)$ in $\overset{0}{W}{}^{(2)}_{2\kappa\Omega}$ we have

$$\mathcal{P}_0(w_n) \to 0,$$

and from $\mathcal{R}(w_n) \to 0$ we have that

$$\mathcal{P}_2(w_n) \to 0,$$

whence

$$\|w_{n\alpha^i\alpha^j}\|_{L_{2\Omega}} \to 0.$$

Therefore, condition 2 of Theorem 10.8, (10.11), is also satisfied, and Lemma 12.1 is proved.

Theorem 12.1. *The spaces* $\overset{\tilde{0}}{W}{}^{(2)}_{2\kappa\Omega}$ *and* $\overset{0}{W}{}^{(2)}_{2\kappa\Omega}$ *contain the same elements, and their norms are equivalent.*

From (12.9) it follows that $\overset{\tilde{0}}{W}{}^{(2)}_{2\kappa\Omega} \subset \overset{0}{W}{}^{(2)}_{2\kappa\Omega}$. Furthermore, from (12.9), (12.7), and the embedding Theorem 10.5 (see the relations (10.7), (10.8), (10.9)) we obtain the obvious inequality

$$\|w\|_{\overset{\tilde{0}}{W}{}^{(2)}_{2\kappa\Omega}} \leq M \|w\|_{\overset{0}{W}{}^{(2)}_{2\kappa\Omega}}. \tag{12.11}$$

From (12.9) and (12.11) it follows that the norms of $\overset{\tilde{0}}{W}{}^{(2)}_{2\kappa\Omega}$ and $\overset{0}{W}{}^{(2)}_{2\kappa\Omega}$ are equivalent. Since these spaces are obtained by taking the closure of the same set $\overset{0}{C}{}^{2}_{\kappa\Omega}$, they are the same.

Remark 12.1. In the definition of $\|w\|_{\overset{\tilde{0}}{W}{}^{(2)}_{2\kappa\Omega}}$ on the right-hand side of (12.7), depending on κ, some of the integrals can be ignored. However, in doing so one has to take care that the geometric boundary conditions (6.1)–(6.3), (6.5), as well as the condition that all the contour integrals neglected in (11.9) vanish, together with

$$\|w\|_{\overset{0}{W}{}^{(2)}_{2\kappa\Omega}} = 0$$

lead to the relation (12.10).

Lemma 12.2. *Let* $w \in \overset{0}{W}{}^{(2)}_{2\kappa\Omega}$, *the PSC* $\Gamma \in C^1_\Gamma$, *and almost everywhere in* Ω *the following condition be satisfied:*

$$-\overset{1}{e}_{ij}(w) = \nabla_{\alpha^i\alpha^j} w = w_{\alpha^i\alpha^j} - G^k_{ij} w_{\alpha^k} = 0, \; i, \; j = 1, 2, \tag{12.12}$$

and moreover, assume that there exists a part $\tilde{\Gamma}$ *of the contour* Γ *such that*

$$w_{\alpha^i}|_{\tilde{\Gamma}} = 0, \; i = 1, 2. \tag{12.13}$$

Then

$$w = \text{const in } \Omega. \tag{12.14}$$

To prove the lemma, we set

$$w_i = w_{\alpha^i}.$$

Then (12.12) takes the form

$$w_{i\alpha^j} + w_{j\alpha^i} = G_{ij}^k w_k,$$

and in essence we have obtained the system of equations (11.17). Condition (12.13) corresponds to (11.16), and we are entirely in the conditions of Lemma 11.3, from which we deduce (12.14).

Lemma 12.3. *Assume that the PSC* $\Gamma \in C_\Gamma^1$ *and* $\Gamma_1 > 0$. *Then the relation*

$$\|w\|_{H_1} = 0 \tag{12.15}$$

implies

$$w \equiv 0. \tag{12.16}$$

In the case of H_1, from (12.15) we obtain (12.12). Since $\Gamma_1 > 0$, we can take $\widetilde{\Gamma} = \Gamma_1$. Thus we are in the conditions of Lemma 12.2, from which (12.16) follows. Lemma 12.3 is proved.

Lemma 12.4. *Assume that the PSC* $\Gamma \in C_\Gamma^1$, $\Gamma_2 > 0$, *and* Γ_2 *contains a segment* $\widetilde{\Gamma}_2 \in C_{\widetilde{\Gamma}_2}^2$ *on which there is a point where condition* (12.6) *is satisfied. Then*

$$\|w\|_{H_2} = 0$$

implies (12.16).

To prove this lemma, we note that $\|w\|_{H_2} = 0$ implies (12.12) almost everywhere. Naturally, we can again start with the fact that $w \in \overset{0}{W}_{2\kappa\Omega}$. But then by the embedding Theorem 10.5 (see (10.8)), $w_{\alpha^i} \in L_{q\Omega}$, $i = 1, 2$, for all $q \geq 1$. Then from (12.12) we have that $w_{\alpha^i\alpha^j} \in L_{q\Omega}$ for all $q \geq 1$. Using now the embedding Theorem 10.3 for $n = 2$, $q \geq 2$, and $l = 2$ (case 1), we have that $w_{\alpha^i} \in H_\Omega^{1-2/q}$. From this fact and (12.12) it follows that $w \in C_\Omega^2$ if (12.12) holds for a function $w \in \overset{0}{W}_{2\kappa\Omega}$. But then from (6.3) we have on Γ_2

$$\frac{dw}{ds} = w_{\alpha^i}\frac{d\alpha^i}{ds} = 0. \tag{12.17}$$

$$\left.\frac{d^2w}{ds^2}\right|_{\Gamma_2} = w_{\alpha^i\alpha^j}\frac{d\alpha^i}{ds}\frac{d\alpha^j}{ds} + w_{\alpha^k}\frac{d^2\alpha^k}{ds^2} = 0. \tag{12.18}$$

Substituting $w_{\alpha^i\alpha^j}$ from (12.12) in (12.18), we obtain

$$\left.\frac{d^2w}{ds^2}\right|_{\widetilde{\Gamma}_2} \equiv w_{\alpha^k}\left(\frac{d^2\alpha^k}{ds^2} + G_{ij}^k\frac{d\alpha^i}{ds}\frac{d\alpha^j}{ds}\right) = 0. \tag{12.19}$$

Equations (12.17), (12.19) can be considered as a system of equations with respect to w_{α^k}. We easily see that the determinant of this system is κ_2, in complete agreement with (1.16). By the assumption of Lemma 12.4, $\widetilde{\Gamma}_2$ contains a point where $\kappa_2 \neq 0$, and therefore, due to the fact that $\widetilde{\Gamma}_2 \in C^2_{\widetilde{\Gamma}_2}$, there is a whole segment $\widetilde{\widetilde{\Gamma}}_2 \subset \widetilde{\Gamma}_2 \subset \Gamma_2$ where $\kappa_2 \neq 0$. Then on $\widetilde{\Gamma}_2$, (12.13) follows from (12.19), and again the conditions of Lemma 12.2 apply, from which Lemma 12.4 follows.

Lemma 12.5. *Assume that* $\Gamma_2 > 0$, *the PSC* $\Gamma \in C_\Gamma$, *and* Γ_2 *contains a segment* $\widetilde{\Gamma}_2$ *where the support is essentially elastic. Then* $\|w\|_{H_2} = 0$ *implies* (12.16).

To prove the claim, we note that since on $\widetilde{\Gamma}_2$ the support is essentially elastic, then $\|w\|_{H_2} = 0$ implies (12.12) and

$$\left.\frac{\partial w}{\partial m}\right|_{\widetilde{\Gamma}_2} \equiv w_4|_{\widetilde{\Gamma}_2} = 0.$$

Moreover, (6.3) is satisfied, from which (12.13) follows. We are again under the conditions of Lemma 12.2. Lemma 12.5 is proved.

Lemma 12.6. *Assume that the PSC* $\Gamma \in C^1_\Gamma$, $\Gamma_3 > 0$, *and there is a segment* $\widetilde{\Gamma}_3$ *where the support is essentially elastic. Then*

$$\|w\|_{H_3} = 0 \tag{12.20}$$

implies (12.16).

By (12.20) we have (12.12), and moreover,

$$w|_{\widetilde{\Gamma}_3} = 0. \tag{12.21}$$

Furthermore, note that for all $w \in H_{3\Omega}$, (6.5) is necessarily satisfied, so that (12.13) follows from (12.21). We are again in the conditions of Lemma 12.2. Lemma 12.6 is proved.

Lemma 12.7. *Assume the PSC* $\Gamma \in C^1_\Gamma$, $\Gamma_4 > 0$, *and there is a segment* $\widetilde{\Gamma}_4$ *where the support is essentially elastic. Then*

$$\|w\|_{H_4} = 0 \tag{12.22}$$

implies (12.16).

The relation (12.12) follows from (12.22) and

$$w|_{\widetilde{\Gamma}_4} = \left.\frac{\partial w}{\partial m}\right|_{\widetilde{\Gamma}_4} \equiv w_4|_{\widetilde{\Gamma}_4} = 0,$$

from which (12.13) follows. Again we are in the conditions of Lemma 12.2. Thus (12.16) follows, and Lemma 12.7 is proved..

Lemma 12.8. *All elements* $w \in H_\kappa$ *satisfy the inequality*

$$\|w\|_{H_\kappa} \geq m\,\|w\|_{\overset{0}{W}{}^{(2)}_{2\kappa\Omega}}. \tag{12.23}$$

The proof of Lemma 12.8 uses Theorem 10.8. Let us introduce $\mathcal{R}(w) = \|w\|^2_{H_\kappa \Omega}$. By (12.5) we have

$$R(w) = \int_\Omega [\mathcal{P}_2(w_{\alpha^k\alpha^l}) + \mathcal{P}_1(w_{\alpha^k\alpha^l})]d\Omega + \mathcal{P}_0 w,$$

where

$$\mathcal{P}_2(w_{\alpha^k\alpha^l}) = \frac{1}{4} D_{\mathrm{f}}^{ijkl} w_{\alpha^i\alpha^j} w_{\alpha^k\alpha^l},$$

$$\mathcal{P}_1(w_{\alpha^k\alpha^l}) = -D_{\mathrm{f}}^{ijkl} \left(w_{\alpha^i\alpha^j} G^t_{kl} + w_{\alpha^k\alpha^l} G^t_{ij}\right) w_{\alpha^t},$$

$$\mathcal{P}_0(w) = \int_\Omega D_{\mathrm{f}}^{ijkl} G^s_{kl} w_{\alpha^t} w_{\alpha^s} d\Omega + \int_{\Gamma_2} k_{\mathrm{f}}^{44} w_4^2 ds$$
$$+ \int_{\Gamma_3} k_{\mathrm{f}}^{33}(s) w^2 ds + \int_{\Gamma_4} k_{\mathrm{f}}^{ij} w_i\, w_j \big|_{i,j=3,4}\, ds.$$

Thus in this case $l = 2$. The structure of $\mathcal{P}_2$, $\mathcal{P}_1$ corresponds to Theorem 10.8. The functional $\mathcal{P}_0(w)$ is weakly continuous by Theorem 10.5. Let us verify the remaining conditions of that theorem. Condition 1 of Theorem 10.8 is an immediate consequence of Lemma 12.3. To verify Condition 2, let $w_n \rightharpoonup 0$ in $\overset{0}{W}{}^{(2)}_{2\kappa\Omega}$. Then, by weak continuity of $\mathcal{P}_0(w)$,

$$\mathcal{P}_0(w_n) \to 0.$$

Next, since $w_{n\alpha^i\alpha^j} \rightharpoonup 0$ in $L_{2\Omega}$, while $w_{n\alpha^t} \to 0$ in $L_{2\Omega}$,

$$\mathcal{P}_1(w_n) \to 0,$$

and from $\mathcal{P}_1(w_n) \to 0$ it follows that $\mathcal{P}_2(w_{n\alpha^i\alpha^j}) \to 0$; by positive definiteness of D_{f}^{ijkl} we have

$$\|w_{n\alpha^l\alpha^k}\|_{L_{2\Omega}} \to 0.$$

Thus, condition 2 of Theorem 10.8, (10.11), is satisfied, and inequality (12.23) has been proved.

Theorem 12.2. *The spaces H_κ and the spaces $\overset{0}{W}{}^{(2)}_{2\kappa\Omega}$ contain the same elements, and their norms are equivalent.*

To prove the theorem, we note that from (12.23) it follows that $H_\kappa \subset \overset{0}{W}{}^{(2)}_{2\kappa\Omega}$. Next, we have the obvious inequality

$$\|w\|_{H_\kappa} \le M\, \|w\|_{\overset{0}{W}{}^{(2)}_{2\kappa\Omega}}. \tag{12.24}$$

The inequalities (12.23) and (12.24) show that the norms $\|w\|_{H_\kappa}$ and $\|w\|_{\overset{0}{W}{}^{(2)}_{2\kappa\Omega}}$ are equivalent. As these spaces are constructed as the closure of the same set $\overset{0}{C}{}^2_{\kappa\Omega}$ in equivalent norms, they are the same. Theorem 12.2 is proved.

Theorem 12.2 establishes a fact that will be quite important below: The embedding Theorem 10.7 holds for all elements $w \in H_\kappa$.

Theorem 12.3. *Let $w \in H_\kappa$. Then*

(1) $w_{\alpha^i\alpha^j} \in L_{2\Omega},\ \nabla_{ij} w \in L_{2\Omega};\ \|w_{\alpha^i\alpha^j}\|_{L_{2\Omega}}$
and $\left\|\nabla_{ij} w\right\|_{L_{2\Omega}} \le m\, \|w\|_{H_\kappa}$. (12.25)

(2) $w_{\alpha^i} \in L_{q\Omega}$ *for any* $q \ge 1$, *and* $\|w_{\alpha^i}\|_{L_{q\Omega}} \le m\, \|w\|_{H_\kappa}$. (12.26)

(3) $w_{\alpha^j} \in L_{qd}$ *for any* $q \ge 1$, *and* $\|w_{\alpha^j}\|_{L_{qd}} \le m\, \|w\|_{H_\kappa}$. (12.27)

(4) $w \in H_\Omega^{0,\lambda^1}$ *and* $\|w_{\alpha^j}\|_{H_\Omega^{0,\lambda^1}} \le m\, \|w\|_{H_\kappa}$. (12.28)

We remind the reader that here d is a PSC of class C_d^1 that can intersect the boundary Γ.

Thus H_κ is embedded in $W_{2\Omega}^{(2)}, W_{2\Omega}^{(1)}, L_{qd}, H_\Omega^{0,\lambda^1}$. Here the operator of embedding H_κ into $W_{2\Omega}^{(1)}, L_{qd}, H_\Omega^{0,\lambda^1}$ is completely continuous, so that

$$
\begin{aligned}
&f_n \rightharpoonup f_0 \text{ in } H_\kappa \text{ implies } f_n \to f_0 \text{ in } W_{q\Omega}^{(1)},\ f_n \to f_0 \text{ in } L_{qd},\\
&f_n \to f_0 \text{ in } H_\Omega^{0,\lambda^1} \text{ for all } q \ge 1.
\end{aligned}
\tag{12.29}
$$

12.4. Now let us introduce function spaces for Airy stress functions. Let $\overset{00}{C}{}_\Omega^2$ be the set of functions Ψ that satisfy the boundary conditions

$$
\Psi|_\Gamma = \left.\frac{\partial \Psi}{\partial m}\right|_\Gamma = 0. \tag{12.30}
$$

Let us define a scalar product on $\overset{00}{C}{}_\Omega^2$ by

$$
(\Psi_1 \cdot \Psi_2)_{H_9} = \int_\Omega C_{ijkl,\mathbf{s}} C^{i\lambda} C^{j\mu} C^{ks} C^{lt} \nabla_{\lambda\mu} \Psi_1 \cdot \nabla_{st} \Psi_2 \, d\Omega, \tag{12.31}
$$

to which corresponds the norm

$$
\|\Psi\|_{H_9}^2 = \int_\Omega C_{ijkl,\mathbf{s}} C^{i\lambda} C^{j\mu} C^{ks} C^{lt} \nabla_{\lambda\mu} \Psi \cdot \nabla_{st} \Psi \, d\Omega, \tag{12.32}
$$

where $C_{ijkl,\mathbf{s}}$ are the elastic and geometric constants defined by (4.16), and $C^{\gamma\delta}$ are the components of the discriminant tensor of the surface S given by (1.12).

It is easily seen that by condition (4.18) of regularity of the material, the quadratic form in $\nabla_{\lambda\mu}\Psi$ of the integrand of (12.32) is positive definite. Indeed, by (7.1) we have

$$
\|\Psi\|_{H_9}^2 = \int_\Omega C_{ijkl,\mathbf{s}} T^{ij} T^{kl} d\Omega,
$$

and the tensor $C_{ijkl,\mathbf{s}}$ is positive definite, while (7.1) is trivially invertible. The closure of $\overset{00}{C}{}_\Omega^2$ in the norm (12.32) will be called $H_{9\Omega}$. Clearly, we have the inequality

$$
m \int_\Omega \sum_{i,j=1}^{2} (\nabla_{ij}\Psi)^2 d\Omega \le \|\Psi\|_{H_9}^2 \le M \int_\Omega \sum_{i,j=1}^{2} (\nabla_{ij}\Psi)^2 d\Omega.
$$

Under the boundary conditions (12.30) for $\Psi = w$, in (12.4), (12.5) the line integrals vanish, and we have

$$\|w\|^2_{H_1} = \int_\Omega D_{\mathrm{f}}^{ijkl}(\alpha^1, \alpha^2)\overset{1}{\gamma}_{ij}(w)\overset{1}{\gamma}_{kl}(w)d\Omega. \tag{12.33}$$

By positive definiteness of D_{f}^{ijkl} we have

$$m\int_\Omega \sum_{i,j=1}^{2}(\nabla_{ij}w)^2 d\Omega \le \|w\|^2_{H_1} \le M\int_\Omega \sum_{i,j=1}^{2}(\nabla_{ij}w)^2 d\Omega. \tag{12.34}$$

From (12.33), (12.34) we have that H_1 and H_9 are equivalent in the case when $\Gamma_1 = \Gamma$. In view of this equivalence, if $\Psi \in H_9$, then (12.30) holds: For Ψ it is satisfied pointwise and for $\frac{\partial\Psi}{\partial m}$ almost everywhere on Γ; for all $\Psi \in H_9$ the embedding Theorem 10.5 holds.

12.5. For the analysis of boundary value problems in displacements (6.34), (6.35), we introduce spaces of vector functions $\mathbf{a} = (w_1, w_2, w)$, where $w \in H_\kappa$, $\kappa = 1, 2, 3, 4$, and the vector $\boldsymbol{\omega} = (w_1, w_2) \in H_t$, $t = 5, 6, 7, 8$. A scalar product in these spaces is defined as follows. Let $\mathbf{a}_1, \mathbf{a}_2$ have the components $\mathbf{a}_1 = (\boldsymbol{\omega}_1, w_1) = (w_{11}, w_{12}\, w_1)$, $\mathbf{a}_2 = (\boldsymbol{\omega}_2, w_2) = (w_{21}, w_{22}, w_2)$; then

$$(\mathbf{a}_1 \cdot \mathbf{a}_2)_{H_{\kappa t}} = (w_1 \cdot w_2)_{H_\kappa} + (\boldsymbol{\omega}_1 \cdot \boldsymbol{\omega}_2)_{H_t}.$$

For the analysis of problems with an Airy stress function (7.51), (7.60) and of the corresponding boundary conditions we shall need spaces of pairs of functions (w, Ψ), $w \in H_\kappa$ $(\kappa = 1, 2, 3, 4)$, $\Psi \in H_9$, with the scalar product

$$((w_1, \Psi_1)\cdot(w_2, \Psi_2))_{H_{\kappa 9}} = (w_1 \cdot w_2)_{H_\kappa} + (\Psi_1 \cdot \Psi_2)_{H_9}.$$

Properties of the elements of $H_{\kappa t}$, $H_{\kappa 9}$ $(\kappa = 1, 2, 3, 4;\ t = 5, 6, 7, 8)$ are determined directly by the properties of elements of H_κ, H_t, H_9.

12.6. Below we shall need certain facts concerning the possibility of extending a function w from the boundary Γ into the interior of Ω.

Lemma 12.9. *Let a part $\tilde{\Gamma}$ of the boundary curve Γ be in $C^2_{\tilde{\Gamma}}$, while on a part $\approx\Gamma$, completely contained in $\tilde{\Gamma}$, we have $w \in W^{(\frac{3}{2})}_{2\approx\Gamma}$. Then w can be extended from $\approx\Gamma$ to $\tilde{\Gamma}$ in such a way that w vanishes outside of $\tilde{\Gamma}$, $w \in W^{(\frac{3}{2})}_{2\tilde{\Gamma}}$, and furthermore,*

$$\|w\|_{W^{(\frac{3}{2})}_{2\tilde{\Gamma}}} \le m\,\|w\|_{W^{(\frac{3}{2})}_{2\approx\Gamma}}.$$

The proof of Lemma 12.9 is similar to that of Lemma 11.7, with the difference that instead of (11.44) we use the relation

$$w = \frac{2}{\sqrt{\pi}}\int_0^s f(\sigma)\sqrt{s-\sigma}\,d\sigma,\ \ f(\sigma) \in L_{2\Gamma}.$$

Lemma 12.10. *Let $\Gamma \in C^1_\Gamma$, and assume that there is a segment $\tilde{\Gamma} \in C^4_{\tilde{\Gamma}}$ (which can be disconnected) containing $\Gamma_1, \Gamma_2, \Gamma_3$. Then for the existence of a function*

$\overset{0}{w} \in W_{2\Omega}^{(2)}$ *such that*

$$\overset{0}{w}\Big|_{\Gamma_1+\Gamma_2} = \widetilde{w}, \quad \overset{0}{w}_4\Big|_{\Gamma_1+\Gamma_2} \equiv \frac{\partial \overset{0}{w}}{\partial m}\Bigg|_{\Gamma_1+\Gamma_3} = \widetilde{w}_4, \tag{12.35}$$

it is necessary and sufficient that

$$\widetilde{w}|_{\Gamma_1} \in W_{2\Gamma_1}^{(\frac{3}{2})}, \quad \widetilde{w}_4|_{\Gamma_1} \in W_{2\Gamma_1}^{(\frac{1}{2})}, \quad \widetilde{w}|_{\Gamma_2} \in W_{2\Gamma_2}^{(\frac{3}{2})}, \quad \widetilde{w}_4|_{\Gamma_3} \in W_{2\Gamma_3}^{(\frac{1}{2})}. \tag{12.36}$$

When Γ_1 and Γ_2 (or Γ_1 and Γ_3) have a boundary point in common, we must have, respectively,

$$\widetilde{w}|_{\Gamma_1+\Gamma_2} \in W_{2,\Gamma_1+\Gamma_2}^{(\frac{3}{2})}, \quad (\widetilde{w}_4|_{\Gamma_1+\Gamma_3} \in W_{2,\Gamma_1+\Gamma_3}^{(\frac{1}{2})}). \tag{12.37}$$

Here $\overset{0}{w}$ can be constructed always in such a way that the inequality

$$\left\| \overset{0}{w} \right\|_{W_{2\Omega}^{(2)}} \le m \left(\|\widetilde{w}\|_{W_{2,\Gamma_1+\Gamma_2}^{(\frac{3}{2})}} + \|\widetilde{w}_4\|_{W_{2,\Gamma_1+\Gamma_3}^{(\frac{1}{2})}} \right)$$

holds for some constant m.

We do not present a proof. Below we shall say that boundary values $\widetilde{w}|_{\Gamma_1+\Gamma_2}$, $\widetilde{w}_4|_{\Gamma_1+\Gamma_3}$ that satisfy conditions (12.36), (12.37) of Lemma 12.10 are admissible.

12.7. In this section we introduce function spaces that will be used to characterize transverse loads R^3, $\widetilde{M}^m$, $\widetilde{Q}$ acting on the shell. Let us consider the quantity

$$\int_\Omega R^3 w \, d\Omega + \int_{\Gamma_2+\Gamma_4} \widetilde{M}^m w_4 \, ds + \int_{\Gamma_3+\Gamma_4} \widetilde{Q} w \, ds \tag{12.38}$$

and the set $\overline{H}_\kappa$ of such triples $[R^3, \widetilde{M}^m, \widetilde{Q}]$, for which (12.38) defines a bounded functional in H_κ. Clearly, the existence of a constant m such that

$$\sup \frac{\left| \int_\Omega R^3 w \, d\Omega + \int_{\Gamma_2+\Gamma_4} \widetilde{M}^m w_4 \, ds + \int_{\Gamma_3+\Gamma_4} \widetilde{Q} w \, ds \right|}{\|w\|_{H_\kappa}} = m^0 \le \infty \tag{12.39}$$

is necessary and sufficient for this functional to be bounded for all $w \in H_\kappa$. From (12.39) it follows that there exists an element $w_{\mathrm{p}} \in H_\kappa$ such that

$$\int_\Omega R^3 w \, d\Omega + \int_{\Gamma_2+\Gamma_4} \widetilde{M}^m w_4 \, ds + \int_{\Gamma_3+\Gamma_4} \widetilde{Q} w \, ds = (w_{\mathrm{p}} \cdot w)_{H_\kappa}. \tag{12.40}$$

The relation (12.40) allows us to make $\overline{H}_\kappa$ into a Hilbert space. For that let us assume that $[R_i^3, \widetilde{M}_i^m, \widetilde{Q}_i]$, $i = 1, 2$, are two elements of $\overline{H}_\kappa$ and let us define

$$\left([R_1^3, \widetilde{M}_1^m, \widetilde{Q}_1] \cdot [R_2^3, \widetilde{M}_2^m, \widetilde{Q}_2] \right)_{\overline{H}_\kappa} = (w_{\mathrm{p}1} \cdot w_{\mathrm{p}2})_{H_\kappa}, \tag{12.41}$$

where $w_{\mathrm{p}i}$ is the element of H_κ corresponding to $[R_i^3, \widetilde{M}_i^m, \widetilde{Q}_i]$. From (12.41) we have

$$\|[R^3, \widetilde{M}^m, \widetilde{Q}]\|_{\overline{H}_\kappa} = \|w_{\mathrm{p}}\|_{H_\kappa}.$$

Now, on the left-hand side of (12.39) instead of sup we can put max and take $m^0 = \|w_p\|_{H_\kappa}$. From this it follows that if $w_p = 0$, then $\|[R^3, \widetilde{M}^m, \widetilde{Q}]\|_{\overline{H}_\kappa} = 0$, that is, $R^3 \equiv \widetilde{M}^m \equiv \widetilde{Q} \equiv 0$.

12.8. The spaces $\overline{H}_t$ and $\overline{H}_\kappa$ allow us to introduce a Hilbert space for the entire load complex R^s, $\widetilde{T}^\tau$, $\widetilde{T}^m$, R^3, $\widetilde{M}^m$, $\widetilde{Q}$ by using the relations

$$([R_1^s, \widetilde{T}_1^\tau, \widetilde{T}_1^m, R_1^3, \widetilde{M}_1^m, \widetilde{Q}_1] \cdot [R_2^s, \widetilde{T}_2^\tau, \widetilde{T}_2^m, R_2^3, \widetilde{M}_2^m, \widetilde{Q}_2])_{\overline{H}_{t\kappa}}$$
$$= (w_{p1} \cdot w_{p2})_{H_\kappa} + (\omega_{p1} \cdot \omega_{p2})_{H_t}.$$

Loads belonging to $\overline{H}_{t\kappa}$ will be called admissible.

12.9. Let us introduce the load complex $[\widetilde{T}_p^{ij}, R^3, \widetilde{M}^m]$ and let us consider

$$\int_\Omega (\widetilde{T}_p^{ij} B_{ij} + R^3)\varphi d\Omega + \int_{\Gamma_2} \widetilde{M}^m \varphi_4 ds, \quad \varphi_4 = \frac{\partial \varphi}{\partial m}. \tag{12.42}$$

Let the complex $[\widetilde{T}_p^{ij}, R^3, \widetilde{M}^m]$ be such that (12.42) defines a bounded functional in H_9. Then

$$\int_\Omega \left(\widetilde{T}_p^{ij} B_{ij} + R^3\right) \varphi d\Omega + \int_{\Gamma_2} \widetilde{M}^m \varphi_4 ds = (\widetilde{w}_p \cdot \varphi)_{H_\kappa}. \tag{12.43}$$

On complexes $[\widetilde{T}_p^{ij}, R^3, \widetilde{M}^m]$ we can define the scalar product

$$\left([\widetilde{T}_{p1}^{ij}, R_1^3, \widetilde{M}_1^m] \cdot [\widetilde{T}_{p2}^{ij}, R_2^3, \widetilde{M}_2^m]\right)_{\overline{\overline{H}}_\kappa} = (\widetilde{w}_{p1} \cdot \widetilde{w}_{p2})_{H_\kappa},$$

thus turning it into the Hilbert space $\overline{\overline{H}}_\kappa$.

12.10. Using (11.51), (12.5), (12.40), we can transform the expression (6.38) for $\mathcal{I}_{t\kappa}$, which now assumes the form

$$\mathcal{I}_{t\kappa} = \frac{1}{2}\Big(\|w\|_{H_\kappa}^2 + \int_\Omega D_s^{ijkl} \overset{0}{\epsilon}_{ij}(\omega) \overset{0}{\epsilon}_{kl}(\omega)\, d\Omega + \int_{\Gamma_6} k_s^{\tau\tau} w_\tau^2 ds$$
$$+ \int_{\Gamma_7} k_s^{mm} w_m^2 ds + \int_{\Gamma_8} k_s^{ij} w_i\, w_j\big|_{i,j=1,2} ds\Big) - (w_p \cdot w)_{H_\kappa} - (\omega_p \cdot \omega)_{H_t}. \tag{12.44}$$

CHAPTER III

Topological Methods Applied to Solvability of the Main Boundary Value Problems of the Nonlinear Theory of Shallow Shells in Displacements

13. The Generalized Formulation of Boundary Value Problems in Displacements. Reduction to Operator Equations. The Physical Meaning of Generalized Solutions

13.1. In this chapter we prove the solvability of the main boundary value problems of shallow shell theory in displacements. The scheme of the proof is as follows. First of all we introduce the concept of a generalized solution using the principle of virtual displacements. Then we show how to pass from the generalized formulation to a nonlinear operator equation in an energy space and establish properties of the operator such as its complete continuity and its representation as a sum of homogeneous operators. Finally, we apply to the operator equation the topological method of proving solvability. For that we introduce a real parameter into the operator equation and prove that that the corresponding vector field is homotopic to the identity vector field on spheres of sufficiently large radius in the energy space, to which end we prove that the corresponding family of operator equations has no solutions on these spheres. This last assertion is the most substantial part of the proof (Theorem 16.1).

We shall assume that the following ***Conditions*** hold:

(1) $S \in C_{\Omega}^{2}$ is a regular surface.

(2) Ω is an admissible domain, that is, it is a Sobolev domain that belongs in both classes (2, 1, 2) and (2, 2, 2). Sufficient conditions for this were presented in Section 10.

(3) Γ is a piecewise-continuous curve belonging to C_{Γ}^{1}.

(4) The material properties of the shell are regular, that is, conditions (4.4) hold and D_s^{ijkl} and D_f^{ijkl} are piecewise continuous in Ω.

(5) Tangential boundary conditions on Γ and the elastic coefficients of the supports must be compatible in such a way that one of the spaces H_t can be defined. In other words, the conditions of Lemma 11.4 must be satisfied. The elastic coefficients of the supports $k_s^{\tau\tau}$, k_s^{mn}, k_s^{ij} are assumed to be piecewise continuous on Γ_6, Γ_7, Γ_8, respectively.

(6) The bending boundary conditions on Γ and the elastic coefficients of the supports must be compatible in such a way that one of the spaces H_κ can be defined. In other words, the conditions of Lemmas 12.3–12.7 must be satisfied. The elastic coefficients of the supports k_f^{44}, k_f^{33}, k_f^{ij} are assumed to be piecewise continuous on Γ_2, Γ_3, Γ_4, respectively; in addition, $\Gamma_1+\Gamma_2>0$.

(7) The loads R^s, $\widetilde{T}^\tau$, $\widetilde{T}^m$, R^3, $\widetilde{M}^m$, $\widetilde{Q}$ are admissible, which means that

$$[R^s, \widetilde{T}^\tau, \widetilde{T}^m] \in \overline{H}_t, \quad [R^3, \widetilde{M}^m, \widetilde{Q}] \in \overline{H}_\kappa, \quad [R^s, \widetilde{T}^\tau, \widetilde{T}^m, R^3, \widetilde{M}^m, \widetilde{Q}] \in \overline{H}_{t\kappa}.$$

The spaces $\overline{H}_t$, $\overline{H}_\kappa$, $\overline{H}_{t\kappa}$ were introduced in Sections 11, 12.

(8) The given displacements $\widetilde{w} \equiv \widetilde{w}_3, \widetilde{w}_4, \widetilde{w}_1, \widetilde{w}_2, \widetilde{w}_m, \widetilde{w}_\tau$ are admissible, which means that there exists a function $\overset{0}{\mathbf{a}} = (\overset{0}{w}_1, \overset{0}{w}_2, \overset{0}{w}_3) \in W_{2\Omega}^{(1)} \times W_{2\Omega}^{(1)} \times W_{2\Omega}^{(2)}$ that satisfies the boundary conditions

$$\overset{0}{w}\,|_{\Gamma_1+\Gamma_2} = \widetilde{w}_3, \tag{13.1}$$

$$\frac{\partial \overset{0}{w}}{\partial m}\,|_{\Gamma_1+\Gamma_3} \equiv w_4\,|_{\Gamma_1+\Gamma_3} = \widetilde{w}_4, \tag{13.2}$$

$$\overset{0}{w}_m\,|_{\Gamma_5+\Gamma_6} = \widetilde{w}_m, \tag{13.3}$$

$$\overset{0}{w}_\tau\,|_{\Gamma_5+\Gamma_7} = \widetilde{w}_\tau. \tag{13.4}$$

13.2. The classical formulation of the boundary value problems (6.34), (6.35) and (6.1)–(6.3), (6.5), (6.8), (6.9), (6.12), (6.20)–(6.23), (6.28)–(6.31) assumes that the solution $\mathbf{a} = (w_1, w_2, w)$ has the number of continuous derivatives determined by the order of the system. In our case, the vector function $\omega = (w_1, w_2)$ has to have continuous derivatives of up to second order, while w must have continuous derivatives of up to fourth order. Like the equations themselves, the boundary conditions have to be satisfied pointwise. At the boundary, the only exceptions can be the points where the type of the boundary condition changes. It is known that even for infinitely smooth initial data there will be loss of regularity of one sort or another at the points of change of type of boundary conditions.

We shall return to a more detailed consideration of conditions of existence of classical solutions in Section 20. Here we just observe that the classical formulation entails strict requirements on the smoothness of the data of the problem. Thus, in this case we have to demand that $S \in C_\Omega^3$, $D_s^{ijkl} \in C_\Omega^1$, $D_f^{ijkl} \in C_\Omega^2$. External loads have to be at least continuous. The restrictions on the boundary curve Γ have to be quite substantial as well.

It is easily seen that by themselves the main relations of elasticity (4.14) contain only first derivatives of w_1, w_2 and second derivatives of w. It can be observed that the doubling of the order of derivatives is due to the fact that the equilibrium equations of an elements are written using the methods of geometrical statics [97, 218].

The outstanding papers of Sobolev introduced into mathematical physics the concept of a generalized formulation of boundary value problems and generalized solutions. Having arisen from purely mathematical considerations, having to do with the desire to make the formulation of problems more general, this concept turns out to have a substantial physical content as well. We introduce a generalized solution of problem $t\kappa$ from a purely mathematical viewpoint initially, and then give it a physical interpretation.

Definition 13.1. By a *generalized solution of the boundary value Problem $t\kappa$ in displacements* we understand a vector function $\mathbf{a} = (w_1, w_2, w)$ for which $\omega = (w_1, w_2) \in W^{(1)}_{2\Omega}$, $w \in W^{(2)}_{2\Omega}$ that satisfies the boundary conditions (6.1)–(6.3), (6.5), (6.8), (6.9), (6.11) and satisfies the integral relation

$$(\mathbf{a}\cdot\mathbf{b})_{H_{t\kappa}} = \int_\Omega \Big[\Big(B_{kl} w - \frac{1}{2} w_{\alpha^k} w_{\alpha^l} \Big) D_{\rm s}^{ijkl} \nabla_j \varphi_i + R^i \varphi_i + R^3 \varphi + T^{ij} \left(B_{ij}\varphi - w_{\alpha^i}\varphi_{\alpha^j}\right) \Big] d\Omega + \int_{\Gamma_2+\Gamma_4} \widetilde{M}^m \frac{\partial\varphi}{\partial m}\, ds + \int_{\Gamma_3+\Gamma_4} \widetilde{Q}\varphi\, ds + \int_{\Gamma_6+\Gamma_8} \widetilde{T}^\tau \varphi_\tau\, ds + \int_{\Gamma_7+\Gamma_8} \widetilde{T}^m \varphi_m\, ds \tag{13.5}$$

for any vector function $\mathbf{b} = (\varphi_1, \varphi_2, \varphi) \in H_{t\kappa}$.

The equation (13.5) is just the equality (6.16) written in terms of scalar products we had introduced above. It could have been obtained by more conventional methods, namely, by multiplying the equilibrium equations by appropriate test functions and then integrating by parts to transfer derivatives to the test functions and taking into consideration boundary conditions. In a sense, these computations can be considered as being the "inverse" of the computations of Section 6, where by "direct" computations we obtained the equilibrium equations and the natural boundary conditions.

The relation (13.5) can be replaced by two equivalent integral identities. Set in (13.5) $\varphi \equiv 0$, which can be done as $\varphi \equiv 0 \in H_\kappa$. As a result, for $\omega = (w_1, w_2)$ and $\chi = (\varphi_1, \varphi_2)$ we have

$$(\omega\cdot\chi)_{H_t} = \int_\Omega \Big[\Big(B_{kl} w - \frac{1}{2} w_{\alpha^k} w_{\alpha^l} \Big) D_{\rm s}^{ijkl} \nabla_i \varphi_j + R^i \varphi_i \Big] d\Omega + \int_{\Gamma_6+\Gamma_8} \widetilde{T}^\tau \varphi_\tau\, ds + \int_{\Gamma_7+\Gamma_8} \widetilde{T}^m \varphi_m\, ds. \tag{13.6}$$

Next, set in (13.5) $\chi = (\varphi_1, \varphi_2) \equiv 0$. As a result we have

$$(w \cdot \varphi)_{H_\kappa} = \int_\Omega \left[T^{ij}(B_{ij}\varphi - w_{\alpha^i}\varphi_{\alpha^j}) + R^3\varphi\right]d\Omega + \int_{\Gamma_2+\Gamma_4} \widetilde{M}^m \frac{\partial\varphi}{\partial m}\, ds + \int_{\Gamma_3+\Gamma_4} \widetilde{Q}\varphi\, ds. \tag{13.7}$$

13.3. Let us introduce a vector $\omega_{\mathrm{p}} = (w_{1\mathrm{p}}, w_{2\mathrm{p}}) \in H_t$, defined by the relation

$$(\omega_{\mathrm{p}} \cdot \chi)_{H_t} = \int_\Omega R^i\varphi_i\, d\Omega + \int_{\Gamma_6+\Gamma_8} \widetilde{T}^\tau\varphi_\tau\, ds + \int_{\Gamma_7+\Gamma_8} \widetilde{T}^m\varphi_m\, ds. \tag{13.8}$$

Indeed, by condition 5 of Section 13.1, on the right-hand side of (13.8) we have a linear functional in χ, and by the Riesz theorem there exists an element ω_{p} that is defined by this functional. Next, let $\overset{0}{\omega} = (\overset{0}{w}_1, \overset{0}{w}_2)$ be defined by condition 6 of section 13.1, so that (13.3), (13.4) also hold, and that furthermore $\overset{0}{\omega} \in W^{(1)}_{2\Omega}$. Finally, let us introduce the vector function

$$\omega^* = \omega_{\mathrm{p}} + \overset{0}{\omega} \tag{13.9}$$

and make the change of variable

$$\omega \mapsto \omega^* + \omega, \tag{13.10}$$

using the same notation ω for the new unknown vector function; now, however, on the right-hand side of (13.10) ω is already contained in H_t. Let us now consider a function $w_{\mathrm{p}} \in H_\kappa$ defined by the relation

$$(w_{\mathrm{p}} \cdot \varphi)_{H_\kappa} = \int_\Omega R^3\varphi\, d\Omega + \int_{\Gamma_2+\Gamma_4} \widetilde{M}^m \frac{\partial\varphi}{\partial m}\, ds + \int_{\Gamma_3+\Gamma_4} \widetilde{Q}\varphi\, ds.$$

Existence of w_{p} follows from the Riesz theorem and from conditions 5 of Section 13.1. Next, let $w \in W^{(2)}_{2\Omega}$ be given by condition 6 in Section 13.1, so that in addition, (13.1), (13.2) are satisfied. Finally, let us introduce the function

$$w^* = w_{\mathrm{p}} + \overset{0}{w} \tag{13.11}$$

and make the change of variable

$$w \mapsto w^* + w, \tag{13.12}$$

preserving the old notation for the new variable w.

Substituting (13.10), (13.11) into (13.5)–(13.7), we have

$$\begin{aligned}(\mathbf{a} \cdot \mathbf{b})_{H_{t\kappa}} &= (\overset{0}{\mathbf{a}} \cdot \mathbf{b})_{H_{t\kappa}} \\ &\quad + \int_\Omega \left[B_{kl}(w + w^*) - \frac{1}{2}\left(w_{\alpha^k} + w^*_{\alpha^k}\right)\left(w_{\alpha^l} + w^*_{\alpha^l}\right)\right] D^{ijkl}_{\mathrm{s}} \nabla_i\varphi_j\, d\Omega \\ &\quad + \int_\Omega T^{ij}(\mathbf{a} + \mathbf{a}^*)[B_{ij}\varphi - (w + w^*)_{\alpha^i}\varphi_{\alpha^j}]\, d\Omega,\end{aligned} \tag{13.13}$$

where the notation $T^{ij}(\mathbf{a}+\mathbf{a}^*)$ means that in the expression of T^{ij} in terms of displacements we substitute the vector function $\mathbf{a}+\mathbf{a}^*$. Setting $\overset{0}{\omega} = (\overset{0}{w}_1, \overset{0}{w}_2)$, we have

$$(\omega \cdot \chi)_{H_t} = -(\overset{0}{\omega} \cdot \chi)_{H_t} + \int_\Omega \left[B_{kl}(w+w^*) - \frac{1}{2}(w+w^*)_{\alpha^k}(w+w^*)_{\alpha^l}\right] \times D_s^{ijkl} \nabla_i \varphi_j \, d\Omega, \quad \overset{0}{\omega} = (\overset{0}{\omega}_1, \overset{0}{\omega}_2), \ \chi = \varphi_1, \varphi_2), \tag{13.14}$$

$$(w \cdot \varphi)_{H_\kappa} = -(\overset{0}{w} \cdot \varphi)_{H_\kappa} + \int_\Omega T^{ij}(\mathbf{a}+\mathbf{a}^*)[B_{ij}\varphi - (w+w^*)_{\alpha^i}\varphi_{\alpha^j}]\, d\Omega \tag{13.15}$$

Thus a generalized solution of the problem $t\kappa$ in our understanding is given by (13.10), (13.12), where $\mathbf{a} = (w_1, w_2, w) \equiv (\omega, w)$ is in $H_{t\kappa}$ and satisfies the integral identities (13.13)–(13.15) for any vector function $\mathbf{b} = (\varphi_1, \varphi_2, \varphi) \in H_{t\kappa}$.

13.4. We start the analysis of the above definition of a generalized solution by showing that it is well-defined.

Lemma 13.1. *Assume that all the conditions 1–8 of Section 13.1 are satisfied. Then if* $\mathbf{a} = (w_1, w_2, w)$, $\mathbf{b} = (\varphi_1, \varphi_2, \varphi) \in H_{t\kappa}$, *then all the terms on the right-hand side of* (13.5)–(13.7), (13.13)–(13.15) *make sense. In fact, the right-hand sides of* (13.5)–(13.7), (13.13)–(13.15) *define linear functionals in* H_t, H_κ, $H_{t\kappa}$, *respectively, with respect to* $\mathbf{b} = (\chi, \varphi) = (\varphi_1, \varphi_2, \varphi)$ *for a fixed vector function* $\mathbf{a} = (w_1, w_2, w) \in H_{t\kappa}$.

To prove Lemma 13.1, we note that by conditions 7–8 of section 13.1 and the embedding Theorems 10.4, 11.3, we have

$$w^*_{i\alpha^j}, \quad \nabla_j w^*_i \in L_{2\Omega}; \quad w^*_i \in L_{q\Omega}, \quad \forall q \geq 1,$$
$$\left\|w^*_{i\alpha^j}\right\|_{L_{2\Omega}}, \quad \left\|\nabla_j w^*_i\right\|_{L_{2\Omega}}, \quad \left\|w^*_i\right\|_{L_{2\Omega}} \leq m \left\|\omega^*\right\|_{W^{(1)}_{2\Omega}}. \tag{13.16}$$

Moreover,

$$w^*_{\alpha^i\alpha^j}, \quad \nabla_{ij} w^* \in L_{2\Omega}; \quad w^*_{\alpha^i} \in L_{q\Omega}, \quad \forall q \geq 1; \quad w^* \in H^{0,\lambda^0}, \quad \lambda^0 < 1, \tag{13.17}$$

and

$$\left\|w^*_{\alpha^i\alpha^j}\right\|_{L_{2\Omega}}, \quad \left\|w^*_{\alpha^i}\right\|_{L_{2\Omega}}, \quad \left\|w^*\right\|_{H^{0,\lambda^0}} \leq m \left\|w^*\right\|_{W^{(2)}_{2\Omega}}, \tag{13.18}$$

and furthermore,

$$w_{i\alpha^j}, \ \varphi_{i\alpha^j}; \ \nabla_j w_i, \ \nabla_j \varphi_i \in L_{2\Omega}; \ w_i, \ \varphi_i \in L_{q\Omega}, \ \forall q \geq 1. \tag{13.19}$$

in addition,

$$\max\{\|w_{i\alpha^j}\|_{L_{2\Omega}}, \ \left\|\nabla_j w_i\right\|_{L_{2\Omega}}, \ \|w_i\|_{L_{q\Omega}}\} \leq m\,\|\omega\|_{H_t} \leq m\,\|\mathbf{a}\|_{H_{t\kappa}},$$
$$\max\{\|\varphi_{i\alpha^j}\|_{L_{2\Omega}}, \ \left\|\nabla_j \varphi_i\right\|_{L_{2\Omega}}, \ \|\varphi_i\|_{L_{q\Omega}}\} \leq m\,\|\chi\|_{H_t} \leq m\,\|\mathbf{b}\|_{H_{t\kappa}}, \tag{13.20}$$

$$w_{\alpha^i\alpha^j}, \ \nabla_{ij} w, \ \varphi_{\alpha^i\alpha^j}, \ \nabla_{ij}\varphi \in L_{2\Omega}, \ w_{\alpha^i}, \ \varphi_{\alpha^i} \in L_{q\Omega}, \ \forall q \geq 1;$$
$$w, \ \varphi \in H^{0,\lambda^0}_\Omega; \ \lambda^0 < 1, \ \lambda^0 \text{ arbitrarily close to } 1. \tag{13.21}$$

Moreover,

$$\max\{\|w_{\alpha^i\alpha^j}\|_{L_{2\Omega}},\ \left\|\nabla_{ij}w\right\|_{L_{2\Omega}},\ \|w_{\alpha^i}\|_{L_{q\Omega}},\ \|w\|_{H_\Omega^{0,\lambda 0}}\} \le m\ \|w\|_{H_K} \le m\ \|\mathbf{a}\|_{H_{tK}}, \tag{13.22}$$

$$\max\{\|\varphi_{\alpha^i\alpha^j}\|_{L_{2\Omega}},\ \left\|\nabla_{ij}\varphi\right\|_{L_{2\Omega}},\ \|\varphi_{\alpha^i}\|_{L_{q\Omega}},\ \|\varphi\|_{H_\Omega^{0,\lambda 0}}\} \le m\ \|\varphi\|_{H_\kappa} \le m\ \|\mathbf{b}\|_{H_{tK}}. \tag{13.23}$$

Let us first establish the claim of Lemma 13.1 with respect to (13.14), and let us consider, in order, the two terms on the right-hand side of that expression. From (13.16), (13.22) we have

$$\begin{aligned}
&\left\|\left[B_{kl}(w+w^*) - \frac{1}{2}(w+w^*)_{\alpha^k}(w+w^*)_{\alpha^l}\right]D_{\mathrm{s}}^{ijkl}\right\|_{L_{2\Omega}} \\
&\quad\le \left\|B_{kl}D_{\mathrm{s}}^{ijkl}\right\|_{L_{2\Omega}} \left\|w+w^*\right\|_{L_{2\Omega}} \\
&\qquad + \frac{1}{2}\left\|D_{\mathrm{s}}^{ijkl}\right\|_{C_\Omega} \left\|(w+w^*)_{\alpha^k}\right\|_{L_{4\Omega}} \left\|(w+w^*)_{\alpha^l}\right\|_{L_{4\Omega}} \\
&\quad\le m\ \left\|w+w^*\right\|_{W_{2\Omega}^{(2)}} \left(1+\left\|w+w^*\right\|_{W_{2\Omega}^{(2)}}\right).
\end{aligned} \tag{13.24}$$

By assumption the D_{s}^{ijkl} and D_{f}^{ijkl} are piecewise continuous; everywhere below, $\left\|D_{\mathrm{s}}^{ijkl}\right\|_{C_\Omega}$ and $\left\|D_{\mathrm{f}}^{ijkl}\right\|_{C_\Omega}$ means $\max\left|D_{\mathrm{s}}^{ijkl}\right|$ and $\max\left|D_{\mathrm{f}}^{ijkl}\right|$ respectively. Furthermore, from (13.20) it follows that

$$\begin{aligned}
\left\|\nabla_i\varphi_j\right\|_{L_{2\Omega}} = \left\|\varphi_{i\alpha^j} - G_{ij}^{\lambda}\varphi_\lambda\right\|_{L_{2\Omega}} &\le \|\varphi_{i\alpha^j}\|_{L_{2\Omega}} + \left\|G_{ij}^{\lambda}\right\|_{C_\Omega} \|\varphi_\lambda\|_{L_{2\Omega}} \\
&\le m\ \|\chi\|_{H_t} \le m\ \|\mathbf{b}\|_{H_{tK}}.
\end{aligned} \tag{13.25}$$

From (13.24), (13.25) we have

$$\begin{aligned}
&\left|\int_\Omega \left[B_{kl}(w+w^*) - \frac{1}{2}(w+w^*)_{\alpha^k}(w+w^*)_{\alpha^l}\right]D_{\mathrm{s}}^{ijkl}\nabla_i\varphi_j\,d\Omega\right| \\
&\quad\le \left\|\left[B_{kl}(w+w^*) - \frac{1}{2}(w+w^*)_{\alpha^k}(w+w^*)_{\alpha^l}\right]D_{\mathrm{s}}^{ijkl}\right\|_{L_{2\Omega}} \left\|\nabla_i\varphi_j\right\|_{L_{2\Omega}} \\
&\quad\le m\ \left\|w+w^*\right\|_{W_{2\Omega}^{(2)}} \left(1+\left\|w+w^*\right\|_{W_{2\Omega}^{(2)}}\right) \|\chi\|_{H_t} \\
&\quad\le m\ \left\|w+w^*\right\|_{W_{2\Omega}^{(2)}} \left(1+\left\|w+w^*\right\|_{W_{2\Omega}^{(2)}}\right) \|\mathbf{b}\|_{H_{tK}}.
\end{aligned} \tag{13.26}$$

From (13.26) it follows that

$$\begin{aligned}
&|(\overset{0}{\omega}\cdot\chi)_{H_t}| + \left|\int_\Omega \left[B_{kl}(w+w^*) - \frac{1}{2}(w+w^*)_{\alpha^k}(w+w^*)_{\alpha^l}\right]D_{\mathrm{s}}^{ijkl}\nabla_i\varphi_j\,d\Omega\right| \\
&\quad\le m\left[\left\|\overset{0}{\omega}\right\|_{H_t} + \left\|w+w^*\right\|_{W_{2\Omega}^{(2)}} \left(1+\left\|w+w^*\right\|_{W_{2\Omega}^{(2)}}\right)\right] \|\chi\|_{H_t} \\
&\quad\le m\left[\left\|\overset{0}{\omega}\right\|_{H_t} + \left\|w+w^*\right\|_{W_{2\Omega}^{(2)}} \left(1+\left\|w+w^*\right\|_{W_{2\Omega}^{(2)}}\right]\right. \|\mathbf{b}\|_{H_{tK}},
\end{aligned}$$

from which we deduce that Lemma 13.1 holds for the right-hand side of (13.14).

Moreover, if $\mathbf{a}$, $\mathbf{a}^* \in H_{t\kappa}$, then as can be seen from (3.16), $\overset{0}{\epsilon}_{ij}(\mathbf{a}+\mathbf{a}^*) \in L_{2\Omega}$; from (4.14) it then follows that $T^{ij}(\mathbf{a}+\mathbf{a}^*) \in L_{2\Omega}$. Finally, from (13.18)–(13.23) we have

$$\begin{aligned}\left\| B_{ij}\varphi - (w+w^*)_{\alpha^i}\varphi_{\alpha^j} \right\|_{L_{2\Omega}} &\le m \left\| B_{ij} \right\|_{C_\Omega} \|\varphi\|_{L_{2\Omega}} + \left\| (w+w^*)_{\alpha^i}\varphi_{\alpha^j} \right\|_{L_{2\Omega}} \\ &\le m \left\| B_{ij} \right\|_{C_\Omega} \|\varphi\|_{L_{2\Omega}} + \left\| w+w^* \right\|_{L_{4\Omega}} \|\varphi_{\alpha^j}\|_{L_{4\Omega}} \\ &\le m\left(1 + \left\| w+w^* \right\|_{W_{2\Omega}^{(2)}}\right) \|\varphi\|_{H_\kappa} \\ &\le m\left(1 + \left\| w+w^* \right\|_{W_{2\Omega}^{(2)}}\right) \|\mathbf{b}\|_{H_{t\kappa}} .\end{aligned} \tag{13.27}$$

From (13.27), we obtain for the right-hand side of (13.15)

$$\left| -(\overset{0}{w}\cdot\varphi)_{H_\kappa} \right| \le \left\| \overset{0}{w} \right\|_{W_{2\Omega}^{(2)}} \|\varphi\|_{H_\kappa} , \tag{13.28}$$

while for the entire right-hand side it follows from (13.27), (13.28) that

$$\begin{aligned}&\left| -(\overset{0}{w}\cdot\varphi)_{H_\kappa} + \int_\Omega T^{ij}(\mathbf{a}+\mathbf{a}^*)\left[B_{ij}\varphi - (w+w^*)_{\alpha^i}\varphi_{\alpha^j}\right] d\Omega \right| \\ &\le \left\| \overset{0}{w} \right\|_{W_{2\Omega}^{(2)}} \|\varphi\|_{H_\kappa} + \left\| T^{ij}(\mathbf{a}+\mathbf{a}^*) \right\|_{L_{2\Omega}} \left\| B_{ij}\varphi - (w+w^*)_{\alpha^i}\varphi_{\alpha^j} \right\|_{L_{2\Omega}} \\ &\le m\left[\left\| \overset{0}{w} \right\|_{W_{2\Omega}^{2}} + \left\| T^{ij}(\mathbf{a}+\mathbf{a}^*) \right\|_{L_{2\Omega}} \left(1 + \left\| w+w^* \right\|_{W_{2\Omega}^{(2)}}\right)\right] \|\varphi\|_{H_\kappa} \\ &\le m\left[\left\| \overset{0}{w} \right\|_{W_{2\Omega}^{(2)}} + \left\| T^{ij}(\mathbf{a}+\mathbf{a}^*) \right\|_{L_{2\Omega}} \left(1 + \left\| w+w^* \right\|_{W_{2\Omega}^{(2)}}\right)\right] \|\mathbf{b}\|_{H_{t\kappa}} .\end{aligned} \tag{13.29}$$

Relation (13.29) proves Lemma 13.1 for the right-hand side of (13.15) as well. Thus Lemma 13.1 has been proved also for the right-hand side of (13.13), that is, completely.

13.5. In this section we reduce boundary value Problems $t\kappa$ to certain operator equations in H_t, H_κ, $H_{t\kappa}$. To this end, let us consider the integral identity (13.13). By Lemma 13.14 the right-hand side of (13.13) is a bounded linear functional in $H_{t\kappa}$ with respect to $\mathbf{b} = (\varphi_1, \varphi_2, \varphi)$. By the Riesz theorem there must exist an element $\mathbf{G}_{t\kappa} \in H_{t\kappa}$ that represents this functional as a scalar product,

$$\begin{aligned}-(\overset{0}{\mathbf{a}}\cdot\mathbf{b})_{H_{t\kappa}} + \int_\Omega \Big\{ \Big[B_{kl}(w+w^*) - \frac{1}{2}(w+w^*)_{\alpha^k}(w+w^*)_{\alpha^l} \Big] D_{\text{s}}^{ijkl}\nabla_i\varphi_j \\ + T^{ij}(\mathbf{a}+\mathbf{a}^*)[B_{ij}\varphi - (w+w^*)_{\alpha^i}\varphi_{\alpha^j}] \Big\} d\Omega = (\mathbf{G}_{t\kappa}\cdot\mathbf{b})_{H_{t\kappa}} .\end{aligned} \tag{13.30}$$

The element $\mathbf{G}_{t\kappa}$ is obviously uniquely determined by $\mathbf{a}$, $\mathbf{a}^*$, $\overset{0}{\mathbf{a}}$ and thus is an operator that will be denoted by $G_{t\kappa}(\mathbf{a}, \mathbf{a}^*, \overset{0}{\mathbf{a}})$. Comparing (13.30) and (13.13), we obtain the relation

$$\mathbf{a} = G_{t\kappa}(\mathbf{a},\ \mathbf{a}^*,\ \overset{0}{\mathbf{a}}), \tag{13.31}$$

which is the required nonlinear operator equation (NOE). The NOE (13.31) is equivalent to three integro-differential equations, which can also be written out using the corresponding Green's functions. Obviously, we have the following:

Lemma 13.2. *A vector function* $\mathbf{a} \in H_{t\kappa}$ *will be a generalized solution of problem* $t\kappa$ *if and only if it satisfies* (13.31).

By equations (13.13)–(13.15), the equation (13.31) constitutes two operator equations. Furthermore, since the right-hand side of (13.14) is a bounded linear functional in $H_{t\kappa}$ with respect to χ, by the theorem of Riesz there is an element $\mathbf{K}_{t\kappa} \in H_t$ that represents this functional as a scalar product,

$$\begin{aligned} &-(\overset{0}{\omega}\cdot\chi)_{H_t} + \int_\Omega \left[B_{kl}(w+w^*) - \frac{1}{2}(w+w^*)_{\alpha^k}(w+w^*)_{\alpha^l}\right] \\ &\quad \times D_{\mathrm{s}}^{ijkl}\nabla_i\varphi_j\, d\Omega = (\mathbf{K}_{t\kappa}\cdot\chi)_{H_t}. \end{aligned} \tag{13.32}$$

The element $\mathbf{K}_{t\kappa}$ will obviously depend on w, w^*, $\overset{0}{\omega}$, and so it defines an operator in those variables, for which we use the notation $\mathbf{K}_{t\kappa}(w,\ w^*,\ \overset{0}{\omega})$. Comparing (13.14) and (13.30), we have

$$\omega = \mathbf{K}_{t\kappa}(w,\ w^*,\ \overset{0}{\omega}). \tag{13.33}$$

Finally, since by Lemma 13.1 the right-hand side of (13.15) is also a linear functional in H_κ with respect to φ, by Riesz's theorem there is an element $\mathbf{G}_{\kappa\kappa} \in H_\kappa$ such that

$$-(\overset{0}{w}\cdot\varphi)_{H_\kappa} + \int_\Omega T^{ij}(\mathbf{a}+\mathbf{a}^*)[B_{ij}\varphi - (w+w^*)_{\alpha^i}\varphi_{\alpha^j}]\, d\Omega = (\mathbf{G}_{\kappa\kappa}\cdot\varphi)_{H_\kappa}. \tag{13.34}$$

From relation (13.34), which defines $\mathbf{G}_{\kappa\kappa}$, it can be seen that this element depends on $\mathbf{a}$, $\mathbf{a}^*$, $\overset{0}{w}$ or on w, ω, $\mathbf{a}^*$, $\overset{0}{w}$, and so the operator $\mathbf{G}_{\kappa\kappa}(w,\ \omega,\ \mathbf{a}^*,\ \overset{0}{w})$ is defined. Comparing (13.15) and (13.34), we obtain

$$w = \mathbf{G}_{\kappa\kappa}(w,\ \omega,\ \mathbf{a}^*,\ \overset{0}{w}). \tag{13.35}$$

The nonlinear operator equations (13.33) and (13.35) comprise a system of operator equations that is equivalent to (13.31).

Lemma 13.3. *A vector function* $\mathbf{a} = (\omega,\ w)$ *is a generalized solution of problem* $t\kappa$ *if and only if* w *and* ω *satisfy the nonlinear operator equation* (13.33), (13.35).

This result can be interpreted to mean that the generalized solution $\mathbf{a} = (\omega,\ w)$ of problem $t\kappa$ lies on a hypersurface HS1 of the space $H_{t\kappa}$, which is defined by (13.33). Obviously, if we substitute (13.33) into (13.35), we obtain an operator equation with respect to w,

$$w = \mathbf{G}_{\kappa\kappa}(w,\ \mathbf{K}_{t\kappa}(w,\ w^*,\ \overset{0}{\omega}),\ \mathbf{a}^*,\ \overset{0}{w}) = \mathbf{G}_{\kappa}(w,\ \mathbf{a}^*,\ \overset{0}{\mathbf{a}}). \tag{13.36}$$

Let us state the main results of this section.

By equations (13.13)–(13.15), equation (13.31) decomposes into two operator equations. Furthermore, the following result would holds.

Theorem 13.1. *A vector function* $\mathbf{a} = (w_1,\ w_2,\ w) \in H_{t\kappa}$ *is a generalized solution of problem* $t\kappa$ *if and only if* w *is determined by the nonlinear operator equation* (13.35), *while* ω *is determined by the relation* (13.33).

13.6. We define a generalized solution of problem $t\kappa$ in displacements for a properly shallow shell to be a vector function $\mathbf{a} = (w_1,\ w_2,\ w)$ in which $\omega = (w_1,\ w_2) \in W^{(1)}_{2\Omega}$, $w \in W^{(2)}_{2\Omega}$, satisfying the boundary conditions (6.1)–(6.3), (6.5), (6.8), (6.9), (6.11) and the integral equation

$$\begin{aligned}(\mathbf{a},\ \mathbf{b})_{H_{t\kappa}} = &\int_\Omega \Big[\Big(f_{\alpha^k\alpha^l} - \frac{1}{2} w_{\alpha^k} w_{\alpha^l}\Big) D_{\mathrm{s}}^{ijkl} \varphi_{i\alpha^j} + R^i \varphi_i + R^3 \varphi \\ &+ T^{kl}\Big(f_{\alpha^k\alpha^l}\varphi - \frac{1}{2} w_{\alpha^k}\varphi_{\alpha^l}\Big)\Big]\, d\alpha^1 d\alpha^2 + \int_{\Gamma_2+\Gamma_4} \widetilde{M}^m \varphi_m\, ds \\ &+ \int_{\Gamma_3+\Gamma_4} \widetilde{Q}\varphi\, ds + \int_{\Gamma_6+\Gamma_8} \widetilde{T}^\tau \varphi_\tau\, ds + \int_{\Gamma_7+\Gamma_8} \widetilde{T}^m \varphi_m\, ds\end{aligned} \tag{13.37}$$

for any vector function $\mathbf{b} = (\varphi_1,\ \varphi_2,\ \varphi) \in H_{t\kappa}$. Here the $H_{t\kappa}$ remain the same, but the expressions for the norms are changed in the way prescribed by the theory of properly shallow shells.

13.7. Let us go back to analyze the method of introducing generalized solutions (13.5)–(13.7) proposed here; these are well-defined by Lemma 13.1. The definition of generalized solutions presents certain difficulties in the case of nonlinear boundary value problems. The point is that the main criterion of validity of the generalized concept of solution is the preservation of the main properties: unique solvability, Fredholm and Noether properties. However, when dealing with nonlinear problems it is not possible to preserve the Fredholm and Noether properties. The criterion of preserving unique solvability is also not applicable, since for the type of question we are considering here, unique solvability does not necessarily always obtain by the nature of the phenomena being described. More than that, the most interesting situations in nonlinear shell theory have to do exactly with nonuniqueness of solutions. Therefore, the most reasonable method of introducing generalized solutions would appear to be one that follows from some mechanical principle. It is easily seen that (13.5)–(13.7) express the virtual displacements principle due to Lagrange (or, which is the same, the virtual work principle) for a shell under external forces. Here $\mathbf{b} = (\varphi_1,\ \varphi_1,\ \varphi)$ are the virtual (admissible) displacements for a system that satisfies by the construction itself all the homogeneous geometrical conditions of the problem. An important fact is that if we impose additional smoothness conditions on the initial data of the problem, every generalized (in the above sense) solution will also be a classical one. At the same time, the reader should notice that Definition 13.1 itself requires certain minimal conditions on the smoothness of the data of the problem to be satisfied.

14. Some Properties of the Operators $\mathbf{K}_{t\kappa}$, $\mathbf{G}_{\kappa\kappa}$

14.1. The nonlinearity of the integro-differential equations derived above is of a polynomial nature. This fact will be of importance in our arguments below.

Let us make a definition.

Definition. A operator $A(x)$ in a Banach space is said to be *homogeneous of order k in x* if for every real number a we have $A(ax) = a^k A(x)$.

Everywhere below we indicate explicitly only the dependence on w. Thus, $\mathbf{K}_{t\kappa\mu}(w,\ w^*,\ \omega) = \mathbf{K}_{t\kappa\mu}(w)$, $\mathbf{G}_{\kappa\kappa\mu}(w,\ \omega,\ \mathbf{a}^*,\ \overset{0}{\mathbf{w}}) = \mathbf{G}_{\kappa\kappa\mu}(w)$.

Lemma 14.1. *The operators $\mathbf{K}_{t\kappa}$, $\mathbf{G}_{\kappa\kappa}$ admit the representation*

$$\mathbf{K}_{t\kappa}(w) = \mathbf{K}_{t\kappa 0} + \mathbf{K}_{t\kappa 1}(w) + \mathbf{K}_{t\kappa 2}(w), \tag{14.1}$$

$$\mathbf{G}_{\kappa\kappa}(w) = \mathbf{G}_{\kappa\kappa 0} + \mathbf{G}_{\kappa\kappa 1}(w) + \mathbf{G}_{\kappa\kappa 2}(w) + \mathbf{G}_{\kappa\kappa 3}(w), \tag{14.2}$$

where $\mathbf{K}_{t\kappa\mu}$, $\mathbf{G}_{\kappa\kappa\mu}$ are homogeneous operators of order μ in w. In addition, we have the estimates

$$\left\|\mathbf{K}_{t\kappa\mu}(w)\right\|_{H_t} \le m\, \|w\|^{\mu}_{H_\kappa}\,,\ \ \mu = 0,\ 1,\ 2, \tag{14.3}$$

$$\left\|\mathbf{G}_{\kappa\kappa\mu}(w)\right\|_{H_\kappa} \le m\, \|w\|^{\mu}_{H_\kappa}\,,\ \ \mu = 0,\ 1,\ 2,\ 3. \tag{14.4}$$

To prove (14.1) let us consider (13.32), which defines $K_{t\kappa}$. From (13.32) it follows that the $K_{t\kappa\mu}$ are defined by the relations

$$(\mathbf{K}_{t\kappa 0}\cdot\chi)_{H_t} = -(\omega^*\cdot\chi)_{H_t} + \int_\Omega \left(B_{kl}w^* - \frac{1}{2}w^*_{\alpha^k}w^*_{\alpha^l}\right)D_{\mathrm{s}}^{ijkl}\nabla_i\varphi_j\, d\Omega, \tag{14.5}$$

$$(\mathbf{K}_{t\kappa 1}\cdot\chi)_{H_t} = \int_\Omega \left[B_{kl}w - \frac{1}{2}\left(w_{\alpha^k}w^*_{\alpha^l} + w_{\alpha^l}w^*_{\alpha^k}\right)\right]D_{\mathrm{s}}^{ijkl}\nabla_i\varphi_j\, d\Omega, \tag{14.6}$$

$$(\mathbf{K}_{t\kappa 2}\cdot\chi)_{H_t} = -\frac{1}{2}\int_\Omega w_{\alpha^k}w_{\alpha^l}D_{\mathrm{s}}^{ijkl}\nabla_i\varphi_j\, d\Omega. \tag{14.7}$$

The estimates (14.3) follow directly from (13.18), (13.20). Let us consider, for example, the case $\mu = 1$. From (14.6), (13.27) we have

$$
\begin{aligned}
\left|(\mathbf{K}_{\iota\kappa 1}(w)\cdot\chi)_{H_\iota}\right| &\leq \left|\int_\Omega D_{\mathrm{s}}^{stkl}\Big[B_{kl}w - \frac{1}{2}\big(w_{\alpha^k}w_{\alpha^l}^* + w_{\alpha^l}w_{\alpha^k}^*\big)\Big]\nabla_s\varphi_t \, d\Omega\right| \\
&\leq \left\| D_{\mathrm{s}}^{stkl}\Big[B_{kl}w - \frac{1}{2}\big(w_{\alpha^k}w_{\alpha^l}^* + w_{\alpha^l}w_{\alpha^k}^*\big)\Big]\right\|_{L_{2\Omega}} \|\nabla_s\varphi_t\|_{L_{2\Omega}} \\
&\leq \sum_{s,t=1}^{2}\Big(\left\|D_{\mathrm{s}}^{stkl}B_{kl}\right\|_{C_\Omega}\|w\|_{L_{2\Omega}} + \frac{1}{2}\left\|D_{\mathrm{s}}^{stkl}w_{\alpha^l}^*\right\|_{L_{4\Omega}}\|w_{\alpha^k}\|_{L_{4\Omega}} \\
&\qquad + \frac{1}{2}\left\|D_{\mathrm{s}}^{s\rho kl}w_{\alpha^k}^*\right\|_{L_{4\Omega}}\|w_{\alpha^l}\|_{L_{4\Omega}}\Big) m\,\|\chi\|_{H_\iota} \\
&\leq m\sum_{s,\rho,k,l=1}^{2}\Big(\left\|D_{\mathrm{s}}^{s\rho kl}B_{kl}\right\|_{C_\Omega} + \left\|D_{\mathrm{s}}^{s\rho kl}w_{\alpha^l}^*\right\|_{L_{4\Omega}}\Big)\|\chi\|_{H_\iota}\|w\|_{H_\kappa}.
\end{aligned}
\tag{14.8}
$$

Since $\chi \in H_\iota$ is arbitrary, we have from (14.8),

$$
\|\mathbf{K}_{\iota\kappa 1}(w)\|_{H_\iota} \leq m\sum_{s,\rho,k,l=1}^{2}\Big(\left\|D_{\mathrm{s}}^{s\rho kl}B_{kl}\right\|_{C_\Omega} + \left\|D_{\mathrm{s}}^{s\rho kl}w_{\alpha^l}^*\right\|_{L_{4\Omega}}\Big)\|w\|_{H_\kappa}.
$$

In a similar way, from (14.7), (13.27) it follows that

$$
\begin{aligned}
(\mathbf{K}_{\iota\kappa 2}(w)\cdot\chi)_{H_\iota} &\leq \frac{1}{2}\left|\int_\Omega D_{\mathrm{s}}^{s\rho kl}\nabla_s\varphi_\rho w_{\alpha^k}w_{\alpha^l}\, d\Omega\right| \\
&\leq \left\|D_{\mathrm{s}}^{s\rho kl}\right\|_{C_\Omega}\left\|\nabla_\rho\varphi_s\right\|_{L_{2\Omega}}\|w_{\alpha^k}w_{\alpha^l}\|_{L_{2\Omega}} \\
&\leq m\sum_{s,\rho,k,l=1}^{2}\left\|D_{\mathrm{s}}^{s\rho kl}\right\|_{C_\Omega}\|\chi\|_{H_\iota}\left\|w_\alpha^k\right\|_{L_{4\Omega}}\|w_{\alpha^l}\|_{L_{4\Omega}} \\
&\leq m\sum_{s,\rho,k,l=1}^{2}\left\|D_{\mathrm{s}}^{s\rho kl}\right\|_{C_\Omega}\|\chi\|_{H_\iota}\|w\|_{H_\kappa}^2.
\end{aligned}
\tag{14.9}
$$

Equation (14.2) for $\mu = 2$ follows from (14.9), since χ is arbitrary. For $\mu = 0$, (14.3) is established in precisely the same way. Thus, Lemma 14.1 has been established for the operators $K_{\iota\kappa\mu}$.

From (13.33), (14.1) we have

$$
\omega = \omega_0 + \omega_1 + \omega_2, \quad \omega_\mu = \mathbf{K}_{\iota\kappa\mu}(w), \quad \mu = 0,\ 1,\ 2, \tag{14.10}
$$

so that

$$
w_k = w_{k0} + w_{k1}(w) + w_{k2}(w), \quad k = 1,\ 2. \tag{14.11}
$$

In (14.10), (14.11), ω_μ, $w_{k\mu}$ are homogeneous operators of order μ in w. Finally, from (14.10)–(14.11), (3.20), (3.22), (4.14) we have

$$\overset{0}{\epsilon}_{ij}(\mathbf{a}) \mapsto \overset{0}{\epsilon}_{ij}(w) = \overset{0}{\epsilon}_{ij0} + \overset{0}{\epsilon}_{ij1} + \overset{0}{\epsilon}_{ij2}, \tag{14.12}$$

$$T^{ij}(\mathbf{a}) \mapsto T_{ij}(w) = T_0^{ij} + T_1^{ij} + T_2^{ij}, \tag{14.13}$$

$$T^{ij}(\mathbf{a}+\mathbf{a}^*) = T^{ij}(\mathbf{a}) + T^{ij}(\mathbf{a}^*) + \frac{1}{2} D_{\mathrm{s}}^{ijkl}\left(w_{\alpha^k} w^*_{\alpha^l} + w_{\alpha^l} w^*_{\alpha^k}\right), \tag{14.14}$$

where $\overset{0}{\epsilon}_{ij\mu}$, T_μ^{ij} are given by the relations

$$\begin{aligned}
2\overset{0}{\epsilon}_{ij0} &= w_{i0\alpha^j} + w_{j0\alpha^i} - 2G_{ij}^k w_{k0} - 2B_{ij} w^* + w^*_{\alpha^i} w^*_{\alpha^j},\\
2\overset{0}{\epsilon}_{ij1} &= w_{i1\alpha^j} + w_{j1\alpha^i} - 2G_{ij}^k w_{k1} - 2B_{ij} w + \left(w^*_{\alpha^i} w_{\alpha^j} + w^*_{\alpha^j} w_{\alpha^i}\right),\\
2\overset{0}{\epsilon}_{ij2} &= w_{i2\alpha^j} + w_{j2\alpha^i} - 2G_{ij}^k w_{k2} + w_{\alpha^i} w_{\alpha^j}.
\end{aligned} \tag{14.15}$$

From (4.14), we have for T_μ^{ij},

$$T_\mu^{ij} = D_{\mathrm{s}}^{ijkl} \overset{0}{\epsilon}_{kl\mu}. \tag{14.16}$$

Clearly, $\overset{0}{\epsilon}_{ij\mu}$, T_μ^{ij} are homogeneous operators in w of order $\mu = 0,\ 1,\ 2$ that map H_κ into H_t. Let us note the inequalities

$$\max\left\{\left\|\overset{0}{\epsilon}_{ij\mu}\right\|_{L_{2\Omega}},\ \left\|T_\mu^{ij}\right\|_{L_{2\Omega}}\right\} \le m\,\|w\|_{H_\kappa}^{\mu}. \tag{14.17}$$

Let us now consider the operators $\mathbf{G}_{\kappa\kappa\mu}$. To prove (14.2), let us take into account the relations that defined these operators. From (14.13), (14.16), we see that we should take

$$(\mathbf{G}_{\kappa\kappa0}(w)\cdot\varphi)_{H_\kappa} = -(w^*\cdot\varphi)_{H_\kappa} + \int_\Omega [T_0^{ij} + T^{ij}(\mathbf{a}^*)]\left(B_{ij}\varphi - w^*_{\alpha^i}\varphi_{\alpha^j}\right) d\Omega,$$

$$\begin{aligned}
(\mathbf{G}_{\kappa\kappa1}(w)\cdot\varphi)_{H_\kappa} = \int_\Omega \Big\{&\Big[T_1^{ij} + \frac{1}{2} D_{\mathrm{s}}^{ijkl}\left(w_{\alpha^k} w^*_{\alpha^l} + w_{\alpha^l} w^*_{\alpha^k}\right)\Big]\left(B_{ij}\varphi - w^*_{\alpha^i}\varphi\right)\\
&-[T_0^{ij} + T^{ij}(\mathbf{a}^*)] w_{\alpha^i}\varphi_{\alpha^j}\Big\}\, d\Omega,
\end{aligned} \tag{14.18}$$

$$\begin{aligned}
(\mathbf{G}_{\kappa\kappa2}(w)\cdot\varphi)_{H_\kappa} = \int_\Omega \Big\{&\Big[T_2^{ij}\left(B_{ij}\varphi - w^*_{\alpha^i}\varphi_{\alpha^j}\right)\\
&+\Big[T_1^{ij} + \frac{1}{2} D_{\mathrm{s}}^{ijkl}\left(w_{\alpha^k} w^*_{\alpha^l} + w_{\alpha^l} w^*_{\alpha^k}\right)\Big] w_{\alpha^i}\varphi_{\alpha^j}\Big]\Big\}\, d\Omega
\end{aligned} \tag{14.19}$$

$$(\mathbf{G}_{\kappa\kappa3}(w)\cdot\varphi)_{H_\kappa} = -\int_\Omega T_2^{ij} w_{\alpha^i}\varphi_{\alpha^j}\, d\Omega. \tag{14.20}$$

It is easy to see that the right-hand sides in (14.18)–(14.20) are linear functionals with respect to φ in H_κ, and therefore the $\mathbf{G}_{\kappa\kappa\mu}$ are defined by Riesz's theorem. It is also easy to see that

$$\mathbf{G}_{\kappa\kappa} = \sum_{\mu=0}^{3} \mathbf{G}_{\kappa\kappa\mu},$$

and it remains to prove (14.4). Let us present some auxiliary inequalities. From (13.17), (13.18), (13.21)–(13.23) we have

$$\begin{aligned} \left\| w_{\alpha^k} w^*_{\alpha^l} \right\|_{L_{2\Omega}} &\leq \| w_{\alpha^k} \|_{L_{4\Omega}} \left\| w^*_{\alpha^l} \right\|_{L_{4\Omega}} \leq m \, \| w \|_{H_\kappa} , \\ \| w_{\alpha^k} \varphi_{\alpha^l} \|_{L_{2\Omega}} &\leq \| w_{\alpha^k} \|_{L_{4\Omega}} \, \| \varphi_{\alpha^l} \|_{L_{4\Omega}} \leq m \, \| w \|_{H_\kappa} \, \| \varphi \|_{H_\kappa} . \end{aligned} \tag{14.21}$$

Let us give, for example, the estimate for $\mathbf{G}_{\kappa\kappa 1}$. We have

$$\begin{aligned} &\left\| T_1^{ij} + \frac{1}{2} D_{\mathrm{s}}^{ijkl} \left(w_{\alpha^k} w^*_{\alpha^l} + w_{\alpha^l} w^*_{\alpha^k} \right) \right\|_{L_{2\Omega}} \\ &\leq \left\| T_1^{ij} \right\|_{L_{2\Omega}} + \frac{1}{2} \left\| D_{\mathrm{s}}^{ijkl} \right\|_{C_\Omega} \left(\left\| w_{\alpha^k} w^*_{\alpha^l} \right\|_{L_{2\Omega}} + \left\| w_{\alpha^l} w^*_{\alpha^k} \right\|_{L_{2\Omega}} \right) \\ &\leq m \, \| w \|_{H_\kappa} . \end{aligned} \tag{14.22}$$

Inequality (14.22) follows from (14.17), (14.21). From (14.18) we obtain

$$\begin{aligned} \left| (\mathbf{G}_{\kappa\kappa 1}(w) \cdot \varphi)_{H_\kappa} \right| \leq{} & \left\| T_1^{ij} + \frac{1}{2} D_{\mathrm{s}}^{ijkl} \left(w_{\alpha^k} w^*_{\alpha^l} + w_{\alpha^l} w^*_{\alpha^k} \right) \right\|_{L_{2\Omega}} \\ & \times \left\| B_{ij} \varphi - w^*_{\alpha^i} \varphi_{\alpha^j} \right\|_{L_{2\Omega}} \\ & + \left\| T_0^{ij} + T^{ij}(\mathbf{a}^*) \right\|_{L_{2\Omega}} \| w_{\alpha^i} \varphi_{\alpha^j} \|_{L_{2\Omega}} , \end{aligned} \tag{14.23}$$

and using (14.21), (14.22) in (13.26) for $w^* \equiv 0$, from (14.23) we have

$$\left| (\mathbf{G}_{\kappa\kappa 1}(w) \cdot \varphi)_{H_\kappa} \right| \leq m \, \| w \|_{H_\kappa} \, \| \varphi \|_{H_\kappa} . \tag{14.24}$$

Since φ is arbitrary, from (14.24) we have (14.2) for $\mu = 1$.

For $\mathbf{G}_{\kappa\kappa 2}$, it follows from (14.19) that

$$\begin{aligned} \left| (\mathbf{G}_{\kappa\kappa 2}(w) \cdot \varphi)_{H_\kappa} \right| \leq{} & \left\| T_2^{ij} \right\|_{L_{2\Omega}} \left\| B_{ij} \varphi - w^*_{\alpha^i} \varphi_{\alpha^j} \right\|_{L_{2\Omega}} \\ & + \left\| T_1^{ij} + \frac{1}{2} D_{\mathrm{s}}^{ijkl} (w_{\alpha^k} w^*_{\alpha^l} + w_{\alpha^l} w^*_{\alpha^k}) \right\|_{L_{2\Omega}} \| w_{\alpha^i} \varphi_{\alpha^j} \|_{L_{2\Omega}} . \end{aligned} \tag{14.25}$$

From (14.17) for $\mu = 2$, (13.26) for $w \equiv 0$, (14.21), (14.22), (14.25), we obtain

$$\left| (\mathbf{G}_{\kappa\kappa 2}(w) \cdot \varphi)_{H_\kappa} \right| \leq m \, \| \varphi \|_{H_\kappa} \, \| w \|^2_{H_\kappa} , \tag{14.26}$$

and (14.4) has been proved for $\mu = 2$.

Finally, for $\mathbf{G}_{\kappa\kappa 3}$ from (14.20) it follows that

$$\left| (\mathbf{G}_{\kappa\kappa 3}(w) \cdot \varphi)_{H_\kappa} \right| \leq \left\| T_2^{ij} \right\|_{L_{2\Omega}} \| w_{\alpha^i} \|_{L_{4\Omega}} \, \| \varphi_{\alpha^j} \|_{L_{4\Omega}} .$$

From (14.17) for $\mu = 3$ and from (13.21)–(13.23) we have

$$\left| (\mathbf{G}_{\kappa\kappa 3}(w) \cdot \varphi)_{H_\kappa} \right| \leq m \, \| w \|^3_{H_\kappa} \, \| \varphi \|_{H_\kappa} . \tag{14.27}$$

Since φ is arbitrary, from (14.27) follows (14.4) for $\mu = 3$. Lemma 14.1 is completely proved.

14.2.

Theorem 14.1. *Each of the operators* $\mathbf{K}_{t\kappa}$, $\mathbf{K}_{t\kappa\mu}$ *is a completely continuous operator from* H_κ *into* H_t.

To prove this theorem, we note that by complete continuity of the operator of embedding H_κ into C_Ω and $W^{(1)}_{q\Omega}$ (Theorem 12.3, relation (12.52)) from $w_m \rightharpoonup w_0$ in H_κ, we have the relations

$$w_n \to w_0 \text{ in } L_{2\Omega}, \quad w_{n\alpha^k} \to w_{0\alpha^k} \text{ in } L_{q\Omega} \tag{14.28}$$

for any $q \geq 1$. From (14.6) it is easy to show that then

$$((\mathbf{K}_{t\kappa 1}(w_0) - \mathbf{K}_{t\kappa 1}(w_n)) \cdot \chi)_{H_t} = \int_\Omega \Big[B_{kl}(w_0 - w_n) - \frac{1}{2} w^*_{\alpha^l}(w_{0\alpha^k} - w_{n\alpha^k})$$
$$- \frac{1}{2} w^*_{\alpha^k}(w_{0\alpha^l} - w_{n\alpha^l})\Big] D^{ijkl}_{\mathrm{s}} \nabla_i \varphi_j \, d\Omega,$$

from which

$$\begin{aligned} &|((\mathbf{K}_{t\kappa 1}(w_0) - \mathbf{K}_{t\kappa 1}(w_n)) \cdot \chi)_{H_t}| \\ &\quad \leq m \Big\| B_{kl}(w_0 - w_n) - \frac{1}{2} w^*_{\alpha^l}(w_{0\alpha^k} - w_{n\alpha^k}) \\ &\qquad - \frac{1}{2} w^*_{\alpha^k}(w_{0\alpha^l} - w_{n\alpha^l})\Big\|_{L_{2\Omega}} \sum_{i,j=1}^{2} \|D^{ijkl}_{\mathrm{s}}\|_{C_\Omega} \|\chi\|_{H_t} . \end{aligned} \tag{14.29}$$

In the derivation of (14.29) we have used the inequality (13.25).

From (14.28), (14.29) it follows that

$$|((\mathbf{K}_{t\kappa 1}(w_0) - \mathbf{K}_{t\kappa 1}(w_n)) \cdot \chi)_{H_t}| \leq \epsilon_n \|\chi\|_{H_t} ,$$

and hence, since χ is arbitrary, putting

$$\chi = \mathbf{K}_{t\kappa 1}(w_0) - \mathbf{K}_{t\kappa 1}(w_n),$$

we have

$$\|\mathbf{K}_{t\kappa 1}(w_0) - \mathbf{K}_{t\kappa 1}(w_n)\|_{H_t} \leq \epsilon_n, \ \epsilon_n \to 0 \text{ as } n \to \infty.$$

Complete continuity of $\mathbf{K}_{t\kappa 1}(w)$ has been established.

For $\mathbf{K}_{t\kappa 2}$ we have from (14.7),

$$\begin{aligned} |((\mathbf{K}_{t\kappa 2}(w_0) - \mathbf{K}_{t\kappa 2}(w_n)) \cdot \chi)_{H_t}| &\leq \frac{1}{2} \Big| \int_\Omega (w_{0\alpha^k} w_{0\alpha^l} - w_{n\alpha^k} w_{n\alpha^l}) D^{ijkl}_{\mathrm{s}} \nabla_i \varphi_j \, d\Omega \Big| \\ &\leq m \, \|w_{0\alpha^k} w_{0\alpha^l} - w_{n\alpha^k} w_{n\alpha^l}\|_{L_{2\Omega}} \|\chi\|_{H_t} . \end{aligned} \tag{14.30}$$

By (14.28) and the arbitrariness of χ we have from (14.30),

$$\|\mathbf{K}_{t\kappa 2}(w_0) - \mathbf{K}_{t\kappa 2}(w_n)\|_{H_t} \leq \epsilon^1_n, \ \epsilon^1_n \to 0 \text{ as } n \to \infty,$$

and the complete continuity of $\mathbf{K}_{\iota\kappa 2}(w)$ is established, and so is the complete continuity of $\mathbf{K}_{\iota\kappa}(w)$. Theorem 14.1 is proved.

Let us consider now the properties of operators $\mathbf{G}_{\kappa\kappa\mu}(w)$; for that, we consider T_{μ}^{ij}, defined by (14.12)–(14.16). Let us show that from $w_n \rightharpoonup w_0$ it follows that

$$T_{\mu}^{ij}(w_n) \to T_{\mu}^{ij}(w_0) \text{ in } L_{2\Omega}. \tag{14.31}$$

To prove this claim, we note that by Theorem 14.1, we have

$$\mathbf{K}_{\iota\kappa\mu}(w_n) \to \mathbf{K}_{\iota\kappa\mu}(w_0) \text{ in } H_{\iota}.$$

If we also take into account (14.27), (14.24), (14.26), then from (14.15) we have

$$\overset{0}{\epsilon}_{ij\mu}(w_n) \to \overset{0}{\epsilon}_{ij\mu}(w_0) \text{ in } L_{2\Omega}, \tag{14.32}$$

and (14.31) follows from (14.16), (14.32).

14.3. We have the following theorem.

Theorem 14.2. *Each of the operators $\mathbf{G}_{\kappa\kappa\mu}(w)$is completely continuous.*

To prove this, we note that from (14.18),

$$\begin{aligned}
&|((\mathbf{G}_{\kappa\kappa 1}(w_0) - (\mathbf{G}_{\kappa\kappa 1}(w_n)) \cdot \varphi)_{H_\kappa}| \\
&\quad \le \left\| T_1^{ij}(w_0) - T_1^{ij}(w_n) + \frac{1}{2} D_{\mathrm{s}}^{ijkl} \left[w_{\alpha^l}^{*}(w_{0\alpha^k} - w_{n\alpha^k}) \right.\right. \\
&\qquad \left.\left. + w_{\alpha^k}^{*}(w_{0\alpha^l} - w_{n\alpha^l}) \right] \right\|_{L_{2\Omega}} \left\| B_{ij}\varphi - w_{\alpha^i}^{*}\varphi_{\alpha^j} \right\|_{L_{2\Omega}} \\
&\quad + \left\| T_0^{ij} + T^{ij}(\mathbf{a}^{*}) \right\|_{L_{2\Omega}} \|\varphi_{\alpha^j}\|_{L_{4\Omega}} \|w_{0\alpha^i} - w_{n\alpha^i}\|_{L_{4\Omega}}.
\end{aligned} \tag{14.33}$$

From (14.28), (14.31), (14.33), and (13.26) for $w = 0$ it follows that if $w_m \rightharpoonup w_0$ in H_κ, then

$$\left|((\mathbf{G}_{\kappa\kappa 1}(w_0) - \mathbf{G}_{\kappa\kappa 1}(w_n)) \cdot \varphi)_{H_\kappa}\right| \le \epsilon_n \|\varphi\|_{H_\kappa},$$

from which, taking $\varphi = \mathbf{G}_{\kappa\kappa 1}(w_0) - \mathbf{G}_{\kappa\kappa 1}(w_n)$, which is certainly allowed, we have that

$$\mathbf{G}_{\kappa\kappa 1}(w_n) \to \mathbf{G}_{\kappa\kappa 1}(w_0) \text{ in } H_\kappa,$$

and we have proved complete continuity of $\mathbf{G}_{\kappa\kappa 1}(w)$.

To study $\mathbf{G}_{\kappa\kappa 2}(w)$, let us consider the first term on the right-hand side of (14.19). We have

$$\begin{aligned}
&\left| \int_{\Omega} [T_2^{ij}(w_0) - T_2^{ij}(w_n)](B_{ij}\varphi - w_{\alpha^i}^{*}\varphi_{\alpha^j})\, d\Omega \right| \\
&\quad \le \sum_{i,j=1}^{2} \left\| T_2^{ij}(w_0) - T_2^{ij}(w_n) \right\|_{L_{2\Omega}} m \|\varphi\|_{H_\kappa} \le \epsilon_n \|\varphi\|_{H_\kappa}.
\end{aligned} \tag{14.34}$$

In the derivation of (14.34) we used (14.31). Next, for the second term on the right-hand side of (14.19) we have

$$\Big|\int_\Omega \Big\{\Big[T_1^{ij}(w_0)+\frac{1}{2}D_s^{ijkl}\big(w_{0\alpha^k}w^*_{\alpha^l}+w_{0\alpha^l}w^*_{\alpha^k}\big)\Big]w_{0\alpha^i}$$

$$-\Big[T_1^{ij}(w_n)+\frac{1}{2}D_s^{ijkl}\big(w_{n\alpha^k}w^*_{\alpha^l}+w_{n\alpha^l}w^*_{\alpha^k}\big)\Big]w_{n\alpha^i}\Big\}\varphi_{\alpha^j}\,d\Omega\Big|$$

$$\le\Big|\int_\Omega \Big[T_1^{ij}(w_0)-T_1^{ij}(w_n)+\frac{1}{2}D_s^{ijkl}\big(w_{0\alpha^k}-w_{n\alpha^k}\big)w^*_{\alpha^l}$$

$$+\frac{1}{2}D_s^{ijkl}\big(w_{0\alpha^l}-w_{n\alpha^l}\big)w^*_{\alpha^k}\Big]w_{0\alpha^i}\varphi_{\alpha^j}\,d\Omega\Big|$$

$$+\Big|\int_\Omega \Big[T_1^{ij}(w_n)+\frac{1}{2}D_s^{ijkl}\big(w_{n\alpha^k}w^*_{\alpha^l}+w_{n\alpha^l}w^*_{\alpha^k}\big)\Big]\big(w_{0\alpha^i}-w_{n\alpha^i}\big)\varphi_{\alpha^j}\,d\Omega\Big|$$

$$\le\Big\|T_1^{ij}(w_0)-T_1^{ij}(w_n)+\frac{1}{2}D_s^{ijkl}\big(w_{0\alpha^k}-w_{n\alpha^k}\big)w^*_{\alpha^l}$$

$$+\frac{1}{2}D_s^{ijkl}\big(w_{0\alpha^l}-w_{n\alpha^l}\big)w^*_{\alpha^k}\Big\|_{L_{2\Omega}}\|w_{0\alpha^i}\|_{L_{4\Omega}}\|\varphi_{\alpha^j}\|_{L_{4\Omega}}$$

$$+\Big\|T_1^{ij}(w_n)+\frac{1}{2}D_s^{ijkl}\big(w_{n\alpha^k}w^*_{\alpha^l}+w_{n\alpha^l}w^*_{\alpha^k}\big)\Big\|_{L_{2\Omega}}$$

$$\times\big\|w_{0\alpha^i}-w_{n\alpha^i}\big\|_{L_{4\Omega}}\|\varphi_{\alpha^j}\|_{L_{4\Omega}}$$

$$\le m\sum_{j=1}^{2}\Big(\Big\|T_1^{ij}(w_0)-T_1^{ij}(w_n)+\frac{1}{2}D_s^{ijkl}\big(w_{0\alpha^k}-w_{n\alpha^k}\big)w^*_{\alpha^l}$$

$$+\frac{1}{2}D_s^{ijkl}\big(w_{0\alpha^l}-w_{n\alpha^l}\big)w^*_{\alpha^k}\Big\|_{L_{2\Omega}}\|w_{0\alpha^i}\|_{L_{4\Omega}}$$

$$+\Big\|T_1^{ij}(w_n)+\frac{1}{2}D_s^{ijkl}\big(w_{n\alpha^k}w^*_{\alpha^l}+w_{n\alpha^l}w^*_{\alpha^k}\big)\Big\|_{L_{2\Omega}}$$

$$\times\big\|w_{0\alpha^i}-w_{n\alpha^i}\big\|_{L_{4\Omega}}\Big)\|\varphi\|_{H_\kappa}\,. \tag{14.35}$$

From (14.19), (14.34), (14.35), taking into account (14.31) and complete continuity of the embedding operator of H_κ into $W_{2\Omega}^{(1)}$ (see (14.28)), we obtain

$$\big|((\mathbf{G}_{\kappa\kappa2}(w_0)-\mathbf{G}_{\kappa\kappa2}(w_n))\cdot\varphi)_{H_\kappa}\big|\le\epsilon_n\|\varphi\|_{H_\kappa},\quad \epsilon_n\to 0 \text{ as } n\to\infty,$$

whence

$$\|\mathbf{G}_{\kappa\kappa2}(w_0)-\mathbf{G}_{\kappa\kappa2}(w_n)\|_{H_\kappa}\le\epsilon_n,$$

and complete continuity of $\mathbf{G}_{\kappa\kappa2}(w)$ is established.

For $\mathbf{G}_{\kappa\kappa3}(w)$ we obtain from (14.20)

$$\big|((\mathbf{G}_{\kappa\kappa3}(w_0)-\mathbf{G}_{\kappa\kappa3}(w_n))\cdot\varphi)_{H_\kappa}\big|\le\Big|\int_\Omega\Big[T_2^{ij}(w_0)w_{0\alpha^i}-T_2^{ij}(w_n)w_{n\alpha^i}\Big]\varphi_{\alpha^j}\,d\Omega\Big|$$

$$\le \left| \int_\Omega [T_2^{ij}(w_0) - T_2^{ij}(w_n)] w_{0\alpha^i} \varphi_{\alpha^j}\, d\Omega \right| (14.36)$$
$$+ \left| \int_\Omega T_2^{ij}(w_n)\big(w_{0\alpha^i} - w_{n\alpha^i}\big)\varphi_{\alpha^j}\, d\Omega \right|.$$

From (14.28), (14.31), and (14.36) we obtain complete continuity of $\mathbf{G}_{\kappa\kappa 3}(w)$. From complete continuity of $\mathbf{G}_{\kappa\kappa\mu}(w)$, $\mu = 1, 2, 3$, we also have complete continuity of $\mathbf{G}_{\kappa\kappa}(w)$. Theorem 14.2 is finally proved.

15. Computation of the Winding Number of the Vector Field $w - \mathbf{G}_{\kappa\kappa}(w)$ on Spheres of Large Radius in H_κ: Preliminary Lemmas

15.1. In this section we obtain results dealing with uniqueness of solutions of certain auxiliary nonlinear boundary value problems. They play a crucial part in obtaining a priori bounds for generalized solutions.

Lemma 15.1. *Assume that the middle surface belongs to* $H_\Omega^{2,\lambda}$, $\mathbf{a} = (w_1, w_2, w) \in H_{t\kappa}$, *and that the following conditions hold:* $\Gamma_1 + \Gamma_2 > 0$,

$$2\overset{0}{\epsilon}_{ij}(w) = \nabla_i w_j + \nabla_j w_i + w_{\alpha^i} w_{\alpha^j} \equiv 0 \textit{ in } \Omega, \tag{15.1}$$
$$\omega|_\Gamma = 0; \quad w|_{\Gamma_1+\Gamma_2} = 0. \tag{15.2}$$

Then

$$\omega \equiv 0, \ w \equiv 0 \textit{ in } \Omega. \tag{15.3}$$

To prove the lemma, we note that under the conditions of Lemma 15.1 we can pass to the isothermal coordinates (1.23), in which relations (15.1) will assume the form

$$w_{1\alpha^1} + \frac{\lambda_{\alpha^1}}{2\lambda} w_1 - \frac{\lambda_{\alpha^2}}{2\lambda} w_2 + \frac{1}{2} w_{\alpha^1}^2 = 0, \tag{15.4}$$
$$w_{2\alpha^2} - \frac{\lambda_{\alpha^1}}{2\lambda} w_1 + \frac{\lambda_{\alpha^2}}{2\lambda} w_2 + \frac{1}{2} w_{\alpha^2}^2 = 0, \tag{15.5}$$
$$w_{1\alpha^2} + w_{2\alpha^1} + \frac{\lambda_{\alpha^2}}{2\lambda} w_1 + \frac{\lambda_{\alpha^1}}{2\lambda} w_1 + w_{\alpha^1} w_{\alpha^2} = 0, \tag{15.6}$$

from which we have

$$w_{1\alpha^1} + w_{2\alpha^2} + \frac{1}{2}\left(w_{\alpha^1}^2 + w_{\alpha^2}^2\right) = 0.$$

Taking into account (15.2), from (15.4), (15.5) we obtain

$$\int_\Omega (w_{1\alpha^1} + w_{2\alpha^2})\, d\alpha^1 d\alpha^2 + \frac{1}{2}\int_\Omega \left(w_{\alpha^1}^2 + w_{\alpha^2}^2\right) d\alpha^1 d\alpha^2$$
$$= \frac{1}{2}\int_\Omega \left(w_{\alpha^1}^2 + w_{\alpha^2}^2\right) d\alpha^1 d\alpha^2 = 0,$$

whence

$$w_{\alpha^i} \equiv 0, \text{ in } \Omega, \; i = 1, 2, \; w = \text{const}. \tag{15.7}$$

Substituting (15.7) into (15.4)–(15.6), we arrive at (11.17), and then using Lemmas 11.3 and 11.4, we deduce (15.3). Lemma 15.1 is proved.

Lemma 15.2. *Assume that Conditions* (15.1), (15.2) *hold in a developable shell. If* $\Gamma_1 + \Gamma_2 > 0$, *relations* (15.7) *are satisfied.*

To prove the lemma, we note that in a developable shell the metric of the middle surface is Euclidean. Therefore, the parametrization α^1, α^2 can be chosen such that

$$G_{ij}^k \equiv 0, \; i, \; j, \; k = 1, 2,$$

and relations (15.1) become

$$w_{i\alpha^i} + \frac{1}{2} w_{\alpha^i}^2 = 0,$$

from which by (15.2) we have

$$\int_\Omega w_{i\alpha^i} \, d\alpha^1 d\alpha^2 + \frac{1}{2} \int_\Omega w_{\alpha^i}^2 \, d\alpha^1 d\alpha^2 = \frac{1}{2} \int_\Omega w_{\alpha^i}^2 \, d\alpha^1 d\alpha^2 = 0,$$

whence

$$w_{\alpha^i} \equiv 0, \; w_{i\alpha^i} \equiv 0;$$

and now, arguing as in Lemma 15.1, we obtain (15.3). Lemma 15.2 is proved.

Let us stress the difference in the conditions of Lemmas 15.1 and 15.2. In Lemma 15.1 the shell might also be physically shallow; that is, it is not necessary for the metric A_{ij} to be Euclidean, but here it is assumed that $S \in H_\Omega^{2,\lambda}$. This requirement is not imposed in Lemma 15.2, but the metric of S is assumed to be Euclidean, since the shell is developable.

15.2.

Lemma 15.3. *Let* $w \in W_{2\Omega}^{(2)}$ *and assume that in addition, the following conditions are satisfied:*

$$w|_\Gamma = 0, \tag{15.8}$$

and almost everywhere

$$w_{\alpha^1\alpha^1} w_{\alpha^2\alpha^2} - w_{\alpha^1\alpha^2}^2 = 0. \tag{15.9}$$

Then

$$w \equiv 0. \tag{15.10}$$

If we had $w \in C_\Omega^2$, the proof of the lemma would have been obvious. From (15.9) it would follow that $w = w(\alpha^1, \alpha^2)$ is a smooth developable surface that contains the planar closed curve Γ, and then (15.10) has to hold.

In the case $w \in W^{(2)}_{2\Omega}$, the situation is more complex, and additional arguments are required. First of all, if $w \in W^{(2)}_{2\Omega}$ and the condition (15.8) holds, there is a sequence $w_n \in C^2_\Omega$ that satisfies (15.8) and such that

$$\|w - w_n\|_{W^{(2)}_{2\Omega}} \to 0 \text{ as } n \to \infty. \tag{15.11}$$

This means that

$$\begin{aligned} \|w_{\alpha^i\alpha^j} - w_{n\alpha^i\alpha^j}\|_{L_{2\Omega}} &\to 0. \\ \|w_{\alpha^i} - w_{n\alpha^i}\|_{L_{p\Omega}} &\to 0, \\ \|w - w_n\|_{C_\Omega} &\to 0. \end{aligned} \tag{15.12}$$

The last of these relations means that we have uniform convergence of w_n to w in $\overline{\Omega}$.

Let us now introduce

$$|K_n| = \frac{\left|w_{n\alpha^1\alpha^1}w_{n\alpha^2\alpha^2} - w^2_{n\alpha^1\alpha^2}\right|}{\left(1 + w^2_{n\alpha^1} + w^2_{n\alpha^2}\right)^2},$$

the absolute value of the Gaussian curvature of the surface $w = w_n(\alpha^1, \alpha^2)$ [254]. Let us prove that for our surface $w = w(\alpha^1, \alpha^2)$, the corresponding sequence w_n satisfying (15.11) also satisfies

$$\int_\Omega |K_n|\, d\Omega \to 0 \text{ as } n \to \infty. \tag{15.13}$$

Indeed, let us set

$$w_n(\alpha^1, \alpha^2) = w(\alpha^1, \alpha^2) + \epsilon_n(\alpha^1, \alpha^2).$$

From (15.11) and (15.12) it follows that

$$\|\epsilon_n\|_{W^{(2)}_{2\Omega}} \to 0, \quad \|\epsilon_{n\alpha^i\alpha^j}\|_{L_{2\Omega}} \to 0, \quad \text{as } n \to \infty. \tag{15.14}$$

Furthermore, we have

$$\begin{aligned} \int_\Omega |K_n|\, d\Omega &\le \int_\Omega \left|w_{n\alpha^1\alpha^1}w_{n\alpha^2\alpha^2} - w^2_{n\alpha^1\alpha^2}\right| d\Omega \\ &= \int_\Omega \Big|\left(\epsilon_{n\alpha^1\alpha^1} + w_{\alpha^1\alpha^1}\right)\left(\epsilon_{n\alpha^2\alpha^2} + w_{\alpha^2\alpha^2}\right) \\ &\quad - \left(\epsilon_{n\alpha^1\alpha^2} + w_{\alpha^1\alpha^2}\right)^2\Big|\, d\Omega. \end{aligned}$$

From this, by (15.9), we obtain

$$\begin{aligned} \int_\Omega |K_n|\, d\Omega \le& \int_\Omega \left|w_{\alpha^1\alpha^1}\epsilon_{n\alpha^2\alpha^2} + w_{\alpha^2\alpha^2}\epsilon_{n\alpha^1\alpha^1} - 2w_{\alpha^1\alpha^2}\epsilon_{n\alpha^1\alpha^2}\right| d\Omega \\ &+ \int_\Omega \left|\epsilon_{n\alpha^1\alpha^1}\epsilon_{n\alpha^2\alpha^2} - \epsilon^2_{n\alpha^1\alpha^2}\right| d\Omega. \end{aligned}$$

Then by (15.14) we obtain (15.13).

Our subsequent arguments are based on the concept of the spherical image [254]. Let us be given a surface $z = z(\alpha^1, \alpha^2) \in C^1_{\Omega}$, that is, the function $z(\alpha^1, \alpha^2)$ has continuous first derivatives in $\overline{\Omega}$. At each point M on the surface $z(\alpha^1, \alpha^2)$ we can define a unit normal vector. Let us translate the point of origin of all the normal vectors to some point O. Then the set of endpoints of all these vectors lies on a sphere of radius one. A point on the sphere corresponding to a point of the surface $z(\alpha^1, \alpha^2)$ is called the *spherical image* of that point, and the set of spherical images of all the points of the surface is called the *spherical image* of that surface. We use the fact [254] that the element of area of the original surface, $d\sigma$, is related to the element of area, $d\sigma'$, of its spherical image by the relation

$$\frac{d\sigma'}{d\sigma} = |K|,$$

where $K = K(\alpha^1, \alpha^2)$ is the Gaussian curvature of the surface at a point. Thus we have the following estimate of the area S of the spherical image of the surface (taking into account the fact that a point of the image can have several preimages):

$$S \leq \int_{\Omega} |K(\alpha^1, \alpha^2)| \, d\Omega. \tag{15.15}$$

Let us inscribe a sphere of radius 1 in a right circular cone with base radius $r > 1$ and height h, as shown in Figure 15.1. The circle of tangencies of the sphere and the cone, the radius of which is $h/\sqrt{r^2 + h^2}$, divides the sphere into two parts. Let us denote the "upper" cap of the sphere by Σ and let us call it the *cap corresponding to the cone*. It is clear that the area of Σ, S_{Σ}, admits the estimate

$$S_{\Sigma} > \frac{\pi h^2}{r^2 + h^2}. \tag{15.16}$$

Such a cap corresponding to a cone can be constructed for a cone of any radius r by extending it to a sufficiently large similar cone. This does not change the formula (15.16).

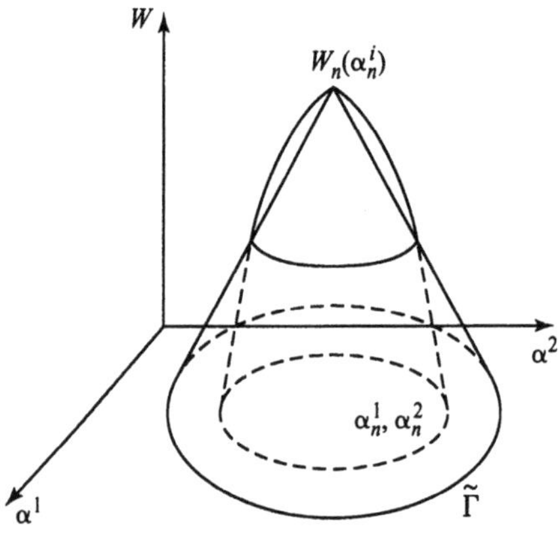

FIGURE 15.1.

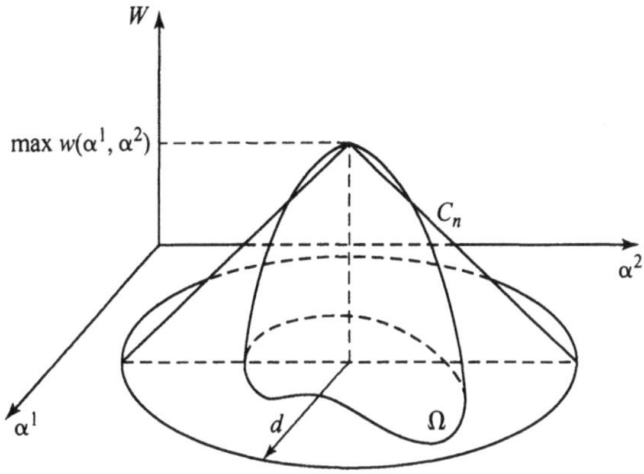

FIGURE 15.2.

Let us return now to the original problem dealing with the sequence $w_n(\alpha^1, \alpha^2)$. Recall that $w_n(\alpha^1, \alpha^2) \in C^2_{\Omega}$, so that all the considerations concerning spherical images for it still hold. Let F_n be the spherical image of the surface $w_n(\alpha^1, \alpha^2)$ and let $w_n(\alpha^1, \alpha^2)$ take its maximal value at $w_{n\max}$, which without loss of generality we take to be positive. Let us construct a right cone C_n, "joined" to the surface $w_n(\alpha^1, \alpha^2)$, as follows (see Figure 15.2): The axis of the cone is perpendicular to the plane of the variables (α^1, α^2); its apex coincides with the point where $w_n(\alpha^1, \alpha^2)$ assumes its maximal value, and the radius of the base of the cone is taken to be d, where d is the diameter of the domain Ω (that is, the maximum of lengths of segments connecting pairs of points in Ω). Above we introduced Σ, the cap of the unit sphere corresponding to a cone. Let us denote by Σ_n the cap corresponding to the nth cone, and place F_n on the same sphere.

By (15.16), for the area S_{Σ_n} we have

$$\frac{\pi w^2_{n\max}}{w^2_{n\max} + d^2} < S_{\Sigma_n}. \tag{15.17}$$

Let us prove that

$$\Sigma_n \subseteq F_n, \tag{15.18}$$

or, in other words, that the domain Σ_n is completely contained in the domain F_n.

Let us make an additional construction. Let us pass a plane through a tangent to the circular base of the cone (see Figure 15.2). Consider the set of all such planes that intersect the cone C_n. It is clear that to prove (15.18), it suffices to show that for any such plane there is a point on the surface $w = w_n(\alpha^1, \alpha^2)$ such that the normal at that point is perpendicular to the plane. But this is almost obvious. Indeed, from geometrical considerations, this plane divides the surface into two parts, one of which, lying above the plane, is supported by a closed curve lying in the plane. By continuity, there exists a point that lies furthest away from the plane. Since this is not a boundary point, the normal $\mathbf{n}'$ at that point is perpendicular to the plane, which proves (15.18).

From (15.18) and (15.17) we deduce the inequality

$$\frac{\pi w_{n\max}^2}{w_{n\max}^2 + d^2} < S_{F_n},$$

where S_{F_n} is the area of F_n. From (15.15) we have that

$$S_{F_n} \leq \int_\Omega \left|K_n(\alpha^1, \alpha^2)\right| d\Omega = \delta_n,$$

so that

$$\frac{\pi w_{n\max}^2}{w_{n\max}^2 + d^2} \leq \delta_n.$$

Solving this elementary inequality, we have that

$$w_{n\max} \leq \sqrt{\frac{\delta_n d}{\pi - \delta_n}}.$$

By (15.13), $\delta_n \to 0$ as $n \to \infty$, so that $w_{n\max} \to 0$ as $n \to \infty$, which completes the proof of Lemma 15.3.

Lemma 15.4. *Assume that for a developable shell the vector of displacements* $\mathbf{a} = (w_1, w_2, w)$ *is in* H_{tK}, *that the relations* (15.1) *are satisfied and that, in addition, condition* (15.1) *holds. Then* (15.10) *holds as well.*

To prove the lemma, let us write (15.1) in Euclidean coordinates, which always exist for the middle surface S, a developable surface,

$$w_{i\alpha^i} + \frac{1}{2} w_{\alpha^i}^2 = w_{1\alpha^2} + w_{2\alpha^1} + w_{\alpha^1} w_{\alpha^2} \equiv 0. \tag{15.19}$$

Were it the case that $w \in C_\Omega^4$, $w_i \in C_\Omega^3$, we would have had from (15.19) that

$$\frac{1}{2}(w_{1\alpha^1})_{\alpha^2\alpha^2} + \frac{1}{2}(w_{2\alpha^2})_{\alpha^1\alpha^1} - \left(w_{1\alpha^2} + w_{2\alpha^1}\right)_{\alpha^1\alpha^2}$$
$$\equiv 0 \equiv w_{\alpha^1\alpha^2}^2 - w_{\alpha^1\alpha^1} w_{\alpha^2\alpha^2}.$$

Therefore $w = w(\alpha^1, \alpha^2)$ is a developable surface, and we immediately obtain (15.10) from (15.8). But in our case $w_{i\alpha^j} \in L_{2\Omega}$ and $w_{\alpha^i} \in L_{q\Omega}$, $\forall q \geq 1$ only, and therefore we need some additional considerations. Let $\psi(\alpha^1, \alpha^2)$ be a twice continuously differentiable function that vanishes in a band around Γ. From (15.19) we have

$$\begin{aligned}\int_\Omega \Big[\Big(w_{1\alpha^1} + \frac{1}{2} w_{\alpha^1}^2\Big)\psi_{\alpha^2\alpha^2} + \Big(w_{2\alpha^2} + \frac{1}{2} w_{\alpha^2}^2\Big)\psi_{\alpha^1\alpha^1} \\ - \left(w_{1\alpha^2} + w_{2\alpha^1} + w_{\alpha^1} w_{\alpha^2}\psi_{\alpha^1\alpha^2}\right)\Big] d\alpha^1 d\alpha^2 = 0.\end{aligned} \tag{15.20}$$

Let us show that

$$\int_\Omega [w_{1\alpha^1}\psi_{\alpha^2\alpha^2} + w_{2\alpha^2}\psi_{\alpha^1\alpha^1} - (w_{2\alpha^1} + w_{1\alpha^2})\psi_{\alpha^1\alpha^2}]\, d\alpha^1\, d\alpha^2 = 0 \tag{15.21}$$

for any function ψ. Indeed, since $\omega = (w_1, w_2) \in H_t$, there is a sequence $\omega_n = (w_{n1}, w_{n2}) \in C^1_\Omega$ such that

$$\|\omega - \omega_n\|_{H_t} \to 0,$$

and therefore

$$\|w_{i\alpha^j} - w_{ni\alpha^j}\|_{L_{2\Omega}} \to 0, \quad i = 1, 2.$$

Next, since now ω_n is a vector in C^1_Ω, we can construct a sequence of vectors $\omega_{n_k} \in C^3_\Omega$ such that

$$\left\| w_{ni\alpha^j} - w_{n_k i\alpha^j} \right\|_{L_{2\Omega}} \to 0,$$

so that

$$\left\| w_{i\alpha^j} - w_{n_k i\alpha^j} \right\|_{L_{2\Omega}} \to 0. \tag{15.22}$$

However, since $w_{n_k i} \in C^3_\Omega$, but transferring the derivatives on the right-hand side of (15.21) from ψ to $w_{n_k i}$, we easily see that the relation

$$\int_\Omega [w_{n_k 1\alpha^1}\psi_{\alpha^2\alpha^2} + w_{n_k 2\alpha^2}\psi_{\alpha^1\alpha^1} - (w_{n_k 2\alpha^1} + w_{n_k 1\alpha^2})\psi_{\alpha^1\alpha^2}]\, d\alpha^1\, d\alpha^2 = 0 \tag{15.23}$$

holds. But from (15.22), (15.23) we easily obtain (15.21). Then from (15.20) it follows that

$$\int_\Omega \left[w^2_{\alpha^1}\psi_{\alpha^2\alpha^2} + w^2_{\alpha^2}\psi_{\alpha^1\alpha^1} - 2w_{\alpha^1}w_{\alpha^2}\psi_{\alpha^1\alpha^2}\right] d\alpha^1\, d\alpha^2 = 0. \tag{15.24}$$

Since $w \in H_\kappa$, there is a sequence $w_n \in C^2_\Omega$, such that

$$\|w - w_n\|_{H_\kappa} \to 0,\ n \to \infty, \tag{15.25}$$

and clearly there is a sequence $w_{n_k} \in C^3_\Omega$ such that

$$\left\|w_n - w_{n_k}\right\|_{H_\kappa} \to 0,\ k \to \infty. \tag{15.26}$$

From (15.25), (15.26) it follows that

$$\left\|w - w_{n_k}\right\|_{H_\kappa} \to 0, \tag{15.27}$$

and furthermore,

$$\left\|w_{\alpha^i} - w_{n_k\alpha^i}\right\|_{L_{q\Omega}} \to 0 \text{ for } q \geq 1. \tag{15.28}$$

Finally, we note that by smoothness of w_{n_k} we have

$$\begin{aligned}\int_\Omega \left(w^2_{n_k\alpha^1}\psi_{\alpha^2\alpha^2} + w^2_{n_k\alpha^2}\psi_{\alpha^1\alpha^1} - 2w_{n_k\alpha^1}w_{n_k\alpha^2}\psi_{\alpha^1\alpha^2}\right) d\alpha^1\, d\alpha^2 \\ = 2\int_\Omega \left(w^2_{n_k\alpha^1\alpha^2} - w_{n_k\alpha^1\alpha^1}w_{n_k\alpha^2\alpha^2}\right)\psi\, d\alpha^1\, d\alpha^2.\end{aligned} \tag{15.29}$$

Relation (15.29) is proved by transferring the derivatives from ψ onto the terms next to it. Passing to the limit as $n \to \infty$, we obtain from (15.26)–(15.29)

$$\int_\Omega \left(w_{\alpha^1}^2 \psi_{\alpha^2\alpha^2} + w_{\alpha^2}^2 \psi_{\alpha^1\alpha^1} - 2w_{\alpha^1} w_{\alpha^2} \psi_{\alpha^1\alpha^2}\right) d\alpha^1\, d\alpha^2 = 2\int_\Omega \left(w_{\alpha^1\alpha^2}^2 - w_{\alpha^1\alpha^1} w_{\alpha^2\alpha^2}\right)\psi\, d\alpha^1\, d\alpha^2, \tag{15.30}$$

and from (15.24), (15.30) we have that for any ψ,

$$\int_\Omega \left(w_{\alpha^1\alpha^2}^2 - w_{\alpha^1\alpha^1} w_{\alpha^2\alpha^2}\right)\psi\, d\alpha^1\, d\alpha^2 = 0. \tag{15.31}$$

Since $w_{\alpha^1\alpha^2}^2 - w_{\alpha^1\alpha^1} w_{\alpha^2\alpha^2} \in L_{1\Omega}$, from (15.31) follows (15.9). Using Lemma 15.3, we have (15.10) from (15.8), (15.9). Lemma 15.4 is proved.

16. Computation of the Winding Number of the Vector Field $\mathbf{w} - \mathbf{G}_{\kappa\kappa}(\mathbf{w})$ on Spheres of Large Radius in H_κ. Solvability of the Main Boundary Value Problems in Displacements

16.1. In this section we establish the solvability of the main boundary value problems in displacements (Theorem 16.5). The proof is based on a computation of the winding number of the corresponding vector field, which turns out to be equal to $+1$. The main, and technically the most demanding, point of the proof is obtaining a priori estimates for solutions of the operator equation $w - t\mathbf{G}_{\kappa\kappa}(w) = 0$ for all $0 \le t \le 1$.

Below we shall need a decomposition of the sphere $\Sigma_{H_\kappa}(R,\ 0)$ of radius R with center at the origin in the Hilbert space H_κ. To construct it, let us consider a sphere of radius 1, $\Sigma_{H_\kappa}(1,\ 0)$, and define on it the set $\Sigma'_{H_\kappa}(1,\ 0)$, the elements v of which satisfy the inequality

$$\|v\|_{H_\kappa}^2 - \frac{1}{\epsilon}\sum_{i,j=1}^{2} \left\|B_{ij}v - v_{\alpha^i} w_{\alpha^j}^*\right\|_{L_{2\Omega}}^2 - c\,\|\omega_2(v)\|_{L_{2\Omega}} < \frac{1}{2}, \tag{16.1}$$

where ϵ, c are some fixed positive constants, $\|v\|_{H_\kappa} = 1$, and the vector function $\omega_2(v)$ is given by the decomposition (14.1), (14.10). The set $\Sigma'_{H_\kappa}(R,\ 0)$ is the radial projection of $\Sigma'_{H_\kappa}(1,\ 0)$ onto the sphere $\Sigma_{H_\kappa}(R,\ 0)$. Next, let $\Sigma''_{H_\kappa}(1,\ 0)$ be the set of elements of $v \in \Sigma_{H_\kappa}(1,\ 0)$ for which

$$\|v\|_{H_\kappa}^2 - \frac{1}{\epsilon}\sum_{i,j=1}^{2} \left\|B_{ij}v - v_{\alpha^i} w_{\alpha^j}^*\right\|_{L_{2\Omega}}^2 - c\,\|\omega_2(v)\|_{L_{2\Omega}} \ge \frac{1}{2}. \tag{16.2}$$

Clearly, $\Sigma''_{H_\kappa}(1,\ 0)$ is the complement of $\Sigma'_{H_\kappa}(1,\ 0)$ in $\Sigma_{H_\kappa}(1,\ 0)$. Next, let $\Sigma''_{H_\kappa}(R,\ 0)$ be the radial projection of $\Sigma''_{H_\kappa}(1,\ 0)$ onto $\Sigma_{H_\kappa}(R,\ 0)$. Obviously,

$\Sigma''_{H_\kappa}(R,\,0)$ is the complement of $\Sigma'_{H_\kappa}(R,\,0)$ in the sphere $\Sigma_{H_\kappa}(R,\,0)$, so that

$$\Sigma_{H_\kappa}(R,\,0)=\Sigma'_{H_\kappa}(R,\,0)\cup\Sigma''_{H_\kappa}(R,\,0).$$

In general, the set $\Sigma''_{H_\kappa}(1,\,0)$ (and thus $\Sigma''_{H_\kappa}(R,\,0)$) can turn out to be empty; this will serve to simplify our considerations below.

Finally, let us introduce $\overline{\Sigma}'_{H_\kappa}(R,\,0)$, the weak closure of $\Sigma'_{H_\kappa}(R,\,0)$. Thus, the set $\overline{\Sigma}'_{H_\kappa}(R,\,0)$ contains $\Sigma'_{H_\kappa}(R,\,0)$ and all its weak limits, which do not have to belong to $\Sigma'_{H_\kappa}(R,\,0)$. Below, unless the opposite is explicitly stated, we assume that conditions 1–9 of Section 13 hold.

Lemma 16.1. *The set $\overline{\Sigma}'_{H_\kappa}(1,\,0)$ does not contain zero.*

To prove this, we note that $\Sigma'_{H_\kappa}(1,\,0)$ does not contain zero. Next, let $v_n\in\Sigma'_{H_\kappa}(1,\,0)$ and $v_n\rightharpoonup 0$ in H_κ. By complete continuity and homogeneity of the operator $\omega_2=\mathbf{K}_{t\kappa 2}(w)$ (Theorem 14.1), we have that

$$\|\mathbf{K}_{t\kappa 2}(v_n)\|_{H_t}\to 0.$$

By complete continuity of the operator of embedding H_κ in $L_{2\Omega}$ and $W^{(1)}_{2\Omega}$ (Theorem 11.3, relation (11.42)), we shall have

$$\frac{1}{\epsilon}\sum_{i,j=1}^{2}\left\|B_{ij}v_n-v_{n\alpha^i}w^*_{\alpha^j}\right\|^2_{L_{2\Omega}}-c\,\|\omega_2(v_n)\|_{L_{2\Omega}}\to 0, \tag{16.3}$$

and (16.1) becomes impossible. This contradiction proves Lemma 16.1.

Let us consider a functional $\Phi(w,\,t)$ of the form

$$\Phi(w,\,t)=((w-t\mathbf{G}_{\kappa\kappa}(w))\cdot w)_{H_\kappa}=\|w\|^2_{H_\kappa}-t(\mathbf{G}_{\kappa\kappa}(w)\cdot w)_{H_\kappa}, \tag{16.4}$$

defined on $H_\kappa\times[0,\,1]$.

By (13.34), we have from (16.4),

$$\begin{aligned}\Phi(w,\,t)=\|w\|^2_{H_\kappa}+t(\overset{0}{w}\cdot w)_{H_\kappa}-t\int_\Omega T^{ij}(\mathbf{a}+\mathbf{a}^*)\\ \times[B_{ij}w-(w+w^*)_{\alpha^i}w_{\alpha^j}]\,d\Omega.\end{aligned} \tag{16.5}$$

Lemma 16.2. *The functional $\Phi(w,\,t)$ can be written in the following form:*

$$\begin{aligned}\Phi(w,\,t)=\|w\|^2_{H_\kappa}&+t(\overset{0}{w}\cdot w)_{H_\kappa}+2t\int_\Omega Q_{\mathrm{s}}\,d\Omega\\ &+t\int_\Omega T^{ij}(\mathbf{a}+\mathbf{a}^*)[B_{ij}(w+2w^*)-(w+w^*)_{\alpha^i}w^*_{\alpha^j}\\ &\qquad-\nabla_i w^*_j-\nabla_j w^*_i]\,d\Omega\\ &-2t(\omega_{\mathrm{p}}\cdot\omega)_{H_t}+2t\Big(\int_{\Gamma_6}k^{\tau\tau}_{\mathrm{s}}(w_\tau+w^*_\tau)w_\tau\,ds\\ &\qquad+\int_{\Gamma_7}k^{mm}_{\mathrm{s}}(w_m+w^*_m)w_m\,ds+\int_{\Gamma_8}k^{ij}_{\mathrm{s}}(w_i+w^*_i)w_j\,ds\Big).\end{aligned} \tag{16.6}$$

In (16.6), Q_s *is defined by* (4.13).

To derive (16.6), let us use some identities. From (16.5) we have

$$\begin{aligned}\Phi(w, t) = {}& \|w\|_{H_\kappa}^2 + t(\overset{0}{w} \cdot w)_{H_\kappa} \\ & - t\int_\Omega T^{ij}(\mathbf{a}+\mathbf{a}^*)[B_{ij}(w+w^*) - (w+w^*)_{\alpha^i}(w+w^*)_{\alpha^j}]\, d\Omega \\ & + t\int_\Omega T^{ij}(\mathbf{a}+\mathbf{a}^*)[B_{ij}w^* - (w+w^*)_{\alpha^i} w^*_{\alpha^j}]\, d\Omega \\ = {}& \|w\|_{H_\kappa}^2 + t(\overset{0}{w} \cdot w)_{H_\kappa} \\ & - 2t\int_\Omega T^{ij}(\mathbf{a}+\mathbf{a}^*)\Big[B_{ij}(w+w^*) - \frac{1}{2}(w+w^*)_{\alpha^i}(w+w^*)_{\alpha^j}\Big]\, d\Omega \\ & + \int_\Omega T^{ij}(\mathbf{a}+\mathbf{a}^*)[B_{ij}(w+2w^*) - (w+w^*)_{\alpha^i} w^*_{\alpha^j}]\, d\Omega.\end{aligned} \tag{16.7}$$

Next, let us observe that if the second term on the right-hand side of (13.14) is transferred to the left-hand side, we shall have for $\chi = \omega$,

$$\begin{aligned}\int_\Omega T^{ij}(\mathbf{a}+\mathbf{a}^*)\nabla_i w_j\, d\Omega = {}& (\omega_{\mathrm{p}} \cdot \omega)_{H_t} - \int_{\Gamma_6} k_{\mathrm{s}}^{\tau\tau}(w_\tau + w_\tau^*) w_\tau\, ds \\ & - \int_{\Gamma_7} k_{\mathrm{s}}^{mm}(w_m + w_m^*) w_m\, ds \\ & - \int_{\Gamma_8} k_{\mathrm{s}}^{ij}(w_i + w_i^*) w_j\, ds.\end{aligned} \tag{16.8}$$

From (16.8) we have

$$\begin{aligned}\int_\Omega T^{ij}(\mathbf{a}+\mathbf{a}^*)\nabla_j(w_i + w_i^*)\, d\Omega = {}& (\omega_{\mathrm{p}} \cdot \omega)_{H_t} - \int_{\Gamma_6} k_{\mathrm{s}}^{\tau\tau}(w_\tau + w_\tau^*) w_\tau\, ds \\ & - \int_{\Gamma_7} k_{\mathrm{s}}^{mm}(w_m + w_m^*) w_m\, ds \\ & - \int_{\Gamma_8} k_{\mathrm{s}}^{ij}(w_i + w_i^*) w_j\, ds \\ & + \int_\Omega T^{ij}(\mathbf{a}+\mathbf{a}^*)\nabla_j w_i\, d\Omega,\end{aligned} \tag{16.9}$$

and then from (16.7), (16.9) we obtain

$$\begin{aligned}
\Phi(w, t) = {} & \|w\|^2_{H_\kappa} + t(\overset{0}{w} \cdot w)_{H_\kappa} \\
& - 2t \int_\Omega T^{ij}(\mathbf{a}+\mathbf{a}^*)\Big[B_{ij}(w+w^*) - \frac{1}{2}(w+w^*)_{\alpha^i}(w+w^*)_{\alpha^j} \\
& \qquad - \nabla_j(w_i + w_i^*)\Big]\, d\Omega \\
& + t \int_\Omega T^{ij}(\mathbf{a}+\mathbf{a}^*)\Big[B_{ij}(w+2w^*)(w+w^*)_{\alpha^i} w_{\alpha^j} - 2\nabla_i w_j^*\Big]\, d\Omega \\
& - 2t(\boldsymbol{\omega}_{\mathrm{p}} \cdot \boldsymbol{\omega})_{H_t} \\
& + 2t\Big(\int_{\Gamma_6} k_{\mathrm{s}}^{\tau\tau}(w_\tau + w_\tau^*)w_\tau\, ds + \int_{\Gamma_7} k_{\mathrm{s}}^{mm}(w_m + w_m^*)w_m\, ds \\
& \qquad + \int_{\Gamma_8} k_{\mathrm{s}}^{ij}(w_i + w_i^*)w_j\, ds\Big).
\end{aligned} \tag{16.10}$$

Now we take into account the relations

$$\begin{aligned}
& B_{ij}(w+w^*) - \frac{1}{2}(w+w^*)_{\alpha^i}(w+w^*)_{\alpha^j} - \frac{1}{2}[\nabla_i(w_j+w_j^*) + \nabla_j(w_i+w_i^*)] \\
& \quad = -\overset{0}{\epsilon}_{ij}(\mathbf{a}+\mathbf{a}^*),
\end{aligned} \tag{16.11}$$

and substituting (16.11) into (16.10) we obtain (16.6), since

$$\begin{aligned}
Q_{\mathrm{s}} &= T^{ij}(\mathbf{a}+\mathbf{a}^*)\overset{0}{\gamma}_{ij}(\mathbf{a}+\mathbf{a}^*) = D_{\mathrm{s}}^{ijkl}\overset{0}{\gamma}_{ij}(\mathbf{a}+\mathbf{a}^*)\overset{0}{\gamma}_{kl}(\mathbf{a}+\mathbf{a}^*) \\
&= C_{\mathrm{s}}^{\lambda\mu qs} T_{\lambda\mu}(\mathbf{a}+\mathbf{a}^*)T_{qs}(\mathbf{a}+\mathbf{a}^*).
\end{aligned}$$

Here we used (4.17).

16.2. Let us now consider the functional $\Pi(w)$ defined by the relation

$$\begin{aligned}
\Pi(w) &= \int_\Omega Q_{\mathrm{s}}\, d\Omega + \int_{\Gamma_6} k_{\mathrm{s}}^{\tau\tau} w_\tau^2\, ds + \int_{\Gamma_7} k_{\mathrm{s}}^{mm} w_m^2\, ds + \int_{\Gamma_8} k_{\mathrm{s}}^{ij} w_i w_j\, ds \\
&= \int_\Omega D_{\mathrm{s}}^{ijkl}\overset{0}{\gamma}_{ij}(\mathrm{w})\overset{0}{\gamma}_{kl}(\mathrm{w})\, d\Omega + \int_{\Gamma_6} k_{\mathrm{s}}^{\tau\tau} w_\tau^2\, ds + \int_{\Gamma_7} k_{\mathrm{s}}^{mm} w_m^2\, ds \\
&\quad + \int_{\Gamma_8} k_{\mathrm{s}}^{ij} w_i w_j\, ds.
\end{aligned} \tag{16.12}$$

In (16.12) the vector function $\boldsymbol{\omega} = (w_1, w_2)$ has been expressed in terms of w through (13.33) and (14.1).

Lemma 16.3. *We have the representation*

$$\Pi(w) = \sum_{\mu=0}^{4} \Pi_\mu(w), \tag{16.13}$$

where the $\Pi_\mu(w)$ are homogeneous of order μ and weakly continuous functionals of w in H_κ, so that we have the inequalities

$$\left|\Pi_\mu(w)\right| \leq m \left\|w\right\|^{\mu}_{H_\kappa}. \tag{16.14}$$

To prove the lemma, we take into account (14.1), (14.10)–(14.12), (14.15), from which it follows that (the last index in $\overset{0}{\gamma}_{kli}$, w_{mi}, $w_{\tau i}$, w_{ji} shows the order of homogeneity in the variable w)

$$\begin{aligned}
\Pi_0 &= \int_\Omega D_s^{ijkl} \overset{0}{\gamma}_{kl0} \overset{0}{\gamma}_{ij0}\, d\Omega + \int_{\Gamma_6} k_s^{\tau\tau} w_{\tau 0}^2\, ds \\
&\quad + \int_{\Gamma_7} k_s^{mm} w_{m0}^2\, ds + \int_{\Gamma_8} k_s^{ij} w_{i0} w_{j0}\, ds,
\end{aligned} \tag{16.15}$$

$$\begin{aligned}
\Pi_1(w) &= 2\int_\Omega D_s^{ijkl} \overset{0}{\gamma}_{kl0} \overset{0}{\gamma}_{ij1}\, d\Omega + 2\int_{\Gamma_6} k_s^{\tau\tau} w_{\tau 0} w_{\tau 1}\, ds \\
&\quad + 2\int_{\Gamma_7} k_s^{mm} w_{m0} w_{m1}\, ds + 2\int_{\Gamma_8} k_s^{ij} w_{i1} w_{j0}\, ds, \\
\Pi_2(w) &= \int_\Omega D_s^{ijkl} \left[\overset{0}{\gamma}_{ij1}(w) \overset{0}{\gamma}_{kl1}(w) + 2\overset{0}{\gamma}_{ij0}(w) \overset{0}{\gamma}_{kl2}(w)\right] d\Omega \\
&\quad + \int_{\Gamma_6} k_s^{\tau\tau} (2w_{\tau 0} w_{\tau 2} + w_{\tau 1}^2)\, ds + \int_{\Gamma_7} k_s^{mm} (2w_{m0} w_{m2} + w_{m1}^2)\, ds \\
&\quad + \int_{\Gamma_8} k_s^{ij} (w_{i0} w_{j2} + w_{i1} w_{j1})\, ds, \\
\Pi_3(w) &= 2\left\{ \int_\Omega D_s^{ijkl} \overset{0}{\gamma}_{ij1}(w) \overset{0}{\gamma}_{kl2}(w)\, d\Omega + \int_{\Gamma_6} k_s^{\tau\tau} w_{\tau 1} w_{\tau 2}\, ds \right. \\
&\quad \left. + \int_{\Gamma_7} k_s^{mm} w_{m1} w_{m2}\, ds + \int_{\Gamma_8} k_s^{ij} w_{i1} w_{j2}\, ds \right\}, \\
\Pi_4(w) &= \int_\Omega D_s^{ijkl} \overset{0}{\gamma}_{ij2}(w) \overset{0}{\gamma}_{kl2}(w)\, d\Omega + \int_{\Gamma_6} k_s^{\tau\tau} w_{\tau 2}^2\, ds \\
&\quad + \int_{\Gamma_7} k_s^{mm} w_{m2}^2\, ds + \int_{\Gamma_8} k_s^{ij} w_{i2} w_{j2}\, ds.
\end{aligned} \tag{16.16}$$

Weak continuity of all the functionals $\Pi_\mu(w)$ in H_κ follows from Theorem 14.1 on complete continuity of the operators $\mathbf{K}_{t\kappa\mu}(w)$, acting from H_t into H_κ and the embedding Theorem 11.3 (relation (11.37)). The estimates (16.14) follow from (14.3), (14.17). Lemma 16.2 has been proved.

Lemma 16.4. *Assume that Conditions 2–6, 8 of Section 13 are satisfied, $S \in H_\Omega^{2,\lambda}$, and the shell is essentially elastically supported on Γ_6, Γ_7, Γ_8. Then on $\Sigma'_{H_\kappa}(R, 0)$*

we have the inequality

$$\Pi_4(w) \geq m_0 R^4, \quad m_0 > 0. \tag{16.17}$$

Since $\Pi_4(w)$ is homogeneous of order 4 in w, it suffices to show that on $\Sigma'_{H_\kappa}(1, 0)$ we have the inequality

$$\Pi_4(w) \geq m_0 > 0. \tag{16.18}$$

If (16.18) does not hold, then there would exist a sequence $w_n \in \overline{\Sigma}'_{H_\kappa}(1, 0)$ such that

$$\Pi_4(w_n) \to 0.$$

Obviously, this sequence can be taken to be weakly convergent, and we let $w_n \rightharpoonup \overline{w}$ in H_κ. Clearly, $\overline{w} \in \overline{\Sigma}'_{H_\kappa}(1, 0)$, and since $\Pi_4(w)$ is a weakly continuous functional,

$$\Pi_4(\overline{w}) = 0. \tag{16.19}$$

Let $\overset{2}{\overline{\omega}}$ correspond to $\overline{w}$. Then from (16.16), (16.19) we have

$$2\epsilon_{ij2}(\overline{w}) = \overline{w}_{i2\alpha^j} + \overline{w}_{j2\alpha^i} + \overline{w}_{\alpha^i}\overline{w}_{\alpha^j} = 0, \tag{16.20}$$

$$\overline{w}_{\tau 2}|_{\Gamma_6} = \overline{w}_{m2}|_{\Gamma_7} = \overline{w}_i|_{\Gamma_8} = 0. \tag{16.21}$$

Taking now into account the homogeneous boundary conditions (6.9), (6.12), which hold on H_t, we obtain from (16.21),

$$\overset{2}{\overline{\omega}}|_\Gamma = 0. \tag{16.22}$$

By Lemma 15.2, we have from (16.20), (16.22) that

$$\overline{w} \equiv 0, \tag{16.23}$$

which contradicts Lemma 16.1. Thus, (16.18) is established. We can easily pass now to (16.17) from considerations of homogeneity.

The same result also holds for a developable shell. We state it for two types of boundary conditions.

Lemma 16.5. *Assume that a developable shell satisfies Conditions 1–6 of Section 13, and that the shell is essentially elastically supported on* Γ_6, Γ_7, Γ_8. *Then inequality* (16.17) *holds on* $\Sigma'_{H_\kappa}(1, 0)$.

Lemma 16.6. *Assume that a developable shell satisfies Conditions 1–6 of Section 13, and that*

$$w|_\Gamma = 0. \tag{16.24}$$

Then inequality (16.17) *holds on* $\Sigma'_{H_\kappa}(1, 0)$.

Proofs of Lemmas 16.4, 16.5 essentially repeat the proof of Lemma 16.3. Here we obtain (16.20), (16.21) from (16.19), and the passage to (16.23) is then done using Lemmas 15.3, 15.5, respectively.

16.3. Let us return now to estimate $\Phi(w, t)$. We formulate this estimate as follows.

Theorem 16.1. *Assume that conditions 2–8 of Section 13 are satisfied and that in addition,*

$$S \in H_{\kappa}^{2,\lambda}.$$

Moreover, assume that the shell is essentially elastically supported on $\Gamma_6, \Gamma_7, \Gamma_8$. *Then we have the inequality*

$$\Phi(w, t) \geq \overset{0}{m} R^2, \ \overset{0}{m} > 0, \ 0 \leq t \leq 1, \tag{16.25}$$

on spheres of sufficiently large radius in H_{κ}.

Let us first estimate $\Phi(w, t)$ on $\Sigma'_{H_{\kappa}}(R, 0)$. From (16.6), (16.13) we have

$$\begin{aligned}
\Phi(w, t) \geq \|w\|^2_{H_{\kappa}} &+ t\Pi(w) + t\int_{\Omega} Q_{\mathrm{s}}\, d\Omega \\
&- t\left|\int_{\Omega} T^{ij}(\mathbf{a}+\mathbf{a}^*)[B_{ij}(w+2w^*) - (w+w^*)_{\alpha^i} w^*_{\alpha^j} - 2\nabla_j w^*_i]\, d\Omega\right| \\
&- 2t\left|(\omega_{\mathrm{p}} \cdot \omega)_{H_l}\right| - 2t\left|\int_{\Gamma_6} k_{\mathrm{s}}^{\tau\tau} w^*_{\tau} w_{\tau}\, ds + \int_{\Gamma_7} k_{\mathrm{s}}^{mm} w^*_m w_m\, ds \right.\\
&\left. + \int_{\Gamma_8} k_{\mathrm{s}}^{ij} w^*_i w_j\, ds\right| \\
&+ t\left(\int_{\Gamma_6} k_{\mathrm{s}}^{\tau\tau} w_{\tau}^2\, ds + \int_{\Gamma_7} k_{\mathrm{s}}^{mm} w_m^2\, ds + \int_{\Gamma_8} k_{\mathrm{s}}^{ij} w_i w_j\, ds\right) - t\left|(\overset{0}{w} \cdot w)_{H_{\kappa}}\right|.
\end{aligned} \tag{16.26}$$

Taking into account the fact that $k_{\mathrm{s}}^{\tau\tau}$, k_{s}^{mm}, and the form $K^{ij} w_i w_j$ are positive, we obtain

$$\begin{aligned}
\Phi(w, t) \geq \|w\|^2_{H_{\kappa}} &+ t\Pi(w) + t\int_{\Omega} Q_{\mathrm{s}}\, d\Omega \\
&- t\left|\int_{\Omega} T^{ij}(\mathbf{a}+\mathbf{a}^*)[B_{ij}(w+2w^*) - (w+w^*)_{\alpha^i} w^*_{\alpha^j} \right.\\
&\qquad \left. - \nabla_i w^*_j - \nabla_j w^*_i]\, d\Omega\right| \\
&- 2t\left|(\omega_{\mathrm{p}} \cdot \omega)_{H_l}\right| \\
&- 2t\left|\int_{\Gamma_6} k_{\mathrm{s}}^{\tau\tau} w^*_{\tau} w_{\tau}\, ds + \int_{\Gamma_7} k_{\mathrm{s}}^{mm} w^*_m w_m\, ds + \int_{\Gamma_8} k_{\mathrm{s}}^{ij} w^*_i w_j\, ds\right| \\
&- t\left|(\overset{0}{w} \cdot w)_{H_{\kappa}}\right|.
\end{aligned} \tag{16.27}$$

Let us now write down the obvious inequality, which clearly holds for any $\epsilon > 0$:

$$\Big| \int_\Omega T^{ij}(\mathbf{a}+\mathbf{a}^*)[B_{ij}(w+2w^*) - (w+w^*)_{\alpha^i} w^*_{\alpha^j} - \nabla_i w^*_j - \nabla_j w^*_i]\, d\Omega \Big|$$

$$\leq \sum_{i,j=1}^{2} \Big\{ \frac{\epsilon}{2} \big\| T^{ij}(\mathbf{a}+\mathbf{a}^*) \big\|^2_{L_{2\Omega}} + \frac{1}{2\epsilon} \| B_{ij}(w+2w^*) - (w+w^*)_{\alpha^i} w^*_{\alpha^j}$$

$$- \nabla_i w^*_j - \nabla_j w^*_i \|^2_{L_{2\Omega}} \Big\}$$

$$\leq \sum_{i,j=1}^{2} \Big\{ \frac{\epsilon}{2} \big\| T^{ij}(\mathbf{a}+\mathbf{a}^*) \big\|^2_{L_{2\Omega}} + \frac{1}{2\epsilon} \{ \big\| B_{ij} w - w_{\alpha^i} w^*_{\alpha^j} \big\|_{L_{2\Omega}} + \big\| B_{ij} 2w^*$$

$$- w^*_{\alpha^i} w^*_{\alpha^j} - \nabla_i w^*_j - \nabla_j w^*_i \big\|_{L_{2\Omega}} \}^2 \Big\}$$

$$\leq \sum_{i,j=1}^{2} \Big\{ \frac{\epsilon}{2} \big\| T^{ij}(\mathbf{a}+\mathbf{a}^*) \big\|^2_{L_{2\Omega}} + \frac{1}{\epsilon} \big\| B_{ij} w - w_{\alpha^i} w^*_{\alpha^j} \big\|^2_{L_{2\Omega}}$$

$$+ \frac{1}{\epsilon} \big\| B_{ij} 2w^* - w^*_{\alpha^i} w^*_{\alpha^j} - \nabla_i w^*_j - \nabla_j w^*_i \big\|^2_{L_{2\Omega}} \Big\}. \tag{16.28}$$

Next, let us observe that the representation (14.1) and (14.10), (14.11), which follow from it, give us

$$w_\tau = w_{\tau 0} + w_{\tau 1}(w) + w_{\tau 2}(w), \quad w_m = w_{m0} + w_{m1}(w) + w_{m2}(w), \tag{16.29}$$

and by (14.3) and the embedding Theorem 11.4 (relation 2 of (11.38)), we have

$$\max \Big\{ \|w_{\tau k}\|_{L_{qd}},\ \|w_{mk}\|_{L_{qd}},\ \|\omega_k\|_{L_{qd}} \Big\} \leq m \,\|\omega_k(w)\|_{H_t} \leq m_1 \,\|w\|^k_{H_\kappa}, \tag{16.30}$$

where d is any part of the boundary curve Γ. From (16.29), (16.30), and Condition 8 of Section 13, we obtain

$$\Big| \int_{\Gamma_6} k_s^{\tau\tau} w^*_\tau w_\tau \, ds + \int_{\Gamma_7} k_s^{mm} w^*_m w_m \, ds + \int_{\Gamma_8} k_s^{ij} w_i w^*_j \, ds \Big|$$

$$\leq \Big(\big\| k_s^{\tau\tau} w^*_\tau \big\|_{L_{2\Omega}} + \big\| k_s^{mm} w^*_m \big\|_{L_{2\Omega}} + \sum_{i=1}^{2} \big\| k_s^{ij} w^*_j \big\|_{L_{2\Omega}} \Big)$$

$$\times \Big(\|\omega_0\|_{L_{2\Omega}} + \|\omega_1\|_{L_{2\Omega}} + \|\omega_2\|_{L_{2\Omega}} \Big)$$

$$\leq m\Big(\|\omega_0\|_{L_{2\Omega}} + \|\omega_1\|_{L_{2\Omega}} + \|\omega_2\|_{L_{2\Omega}} \Big) \leq m_2(1 + \|w\|_{H_\kappa} + \|w\|^2_{H_\kappa}). \tag{16.31}$$

Finally, let us note the inequalities

$$\big|(\omega_p \cdot \omega)_{H_t}\big| \leq \big\|\omega_p\big\|_{H_t} m\Big(\|\omega_0\|_{H_t} + \|\omega_1\|_{H_t} + \|\omega_2\|_{H_t} \Big)$$

$$\leq m_3(1 + \|w\|_{H_\kappa} + \|w\|^2_{H_\kappa}), \tag{16.32}$$

$$\Big| (\overset{0}{w} \cdot w)_{H_\kappa} \Big| \leq \Big\| \overset{0}{w} \Big\|_{H_\kappa} \|w\|_{H_\kappa}. \tag{16.33}$$

From (16.27)–(16.33) we have

$$\Phi(w,\,t) \geq \|w\|^2_{H_\kappa} + t\Pi(w) + t\Big(\int_\Omega \Big[Q_s - \frac{\epsilon}{2}\sum_{i,j=1}^{2}\big(T^{ij}(\mathbf{a}+\mathbf{a}^*)\big)^2\Big]d\Omega\Big)$$

$$- \frac{t}{\epsilon}\sum_{i,j=1}^{2}\big\| B_{ij}w - w_{\alpha^i}w^*_{\alpha^j}\big\|^2_{L_{2\Omega}} - \frac{t}{\epsilon}\mathcal{A} - \left\|\begin{matrix}0\\ w\end{matrix}\right\|_{H_\kappa}\|w\|_{H_\kappa}$$

$$- 2tm\big(\|\omega_0\|_{L_{2\Omega}} + \|\omega_1\|_{L_{2\Omega}} + \|\omega_2\|_{L_{2\Omega}}\big), \tag{16.34}$$

where

$$\mathcal{A} = \sum_{i,j=1}^{2}\big\|2B_{ij}w^* - w^*_{\alpha^i}w^*_{\alpha^j} - \nabla_i w^*_j - \nabla_j w^*_i\big\|^2_{L_{2\Omega}}.$$

Next, since by Condition 1 of Section 13 (regularity of the material; condition (4.18)) Q_s is a positive definite form of the variables $T^{ij}(\mathbf{a}+\mathbf{a}^*)$, ϵ can be chosen so small such that the condition

$$\int_\Omega \Big[Q_s - \frac{\epsilon}{2}\sum_{i,j=1}^{2}\big(T^{ij}(\mathbf{a}+\mathbf{a}^*)\big)^2\Big]d\Omega > 0 \tag{16.35}$$

holds. Furthermore, from (12.26), (12.28) it follows that

$$\sum_{i,j=1}^{2}\big\| B_{ij}w - w_{\alpha^i}w^*_{\alpha^j}\big\|^2_{L_{2\Omega}} < m\,\|w\|^2_{H_\kappa}. \tag{16.36}$$

Finally, we take into account relations (6.9)–(6.11) and the main result of Lemma 16.3, the inequality (16.17). Lemma 16.3 is valid under the conditions of Theorem 16.1. Then on $\Sigma'_{H_\kappa}(R,\,0)$ we obtain

$$\Pi(w) \geq \Pi_4(w) - \sum_{\mu=0}^{3}\big|\Pi_\mu(w)\big| \geq m_0\Big(\|w\|^4_{H_\kappa} - \sum_{\mu=0}^{3}\|w\|^\mu_{H_\kappa}\Big). \tag{16.37}$$

From (16.34)–(16.37) we have

$$\Phi(w,\,t) \geq \|w\|^2_{H_\kappa} + tm_0\Big(\|w\|^4_{H_\kappa} - \sum_{\mu=0}^{3}\|w\|^\mu_{H_\kappa}\Big) - \frac{t}{\epsilon}\|w\|^2_{H_\kappa}$$
$$- t\left\|\begin{matrix}0\\ w\end{matrix}\right\|_{H_\kappa}\|w\|_{H_\kappa} - 2tm\big(1 + \|w\|^2_{H_\kappa}\big) - \frac{t}{\epsilon}\mathcal{A}. \tag{16.38}$$

Now we note that on the right-hand side of (16.38) ϵ is already fixed, and let us combine all the terms on the right-hand side involving t. We have a fourth-degree polynomial with a positive leading coefficient. Thus for $\|w\|_{H_\kappa}$ large enough, the t terms on the right-hand side of (16.38) will be positive, so that on $\Sigma'_{H_\kappa}(R,\,0)$ for sufficiently large R, we have

$$\Phi(w,\,t) \geq \|w\|^2_{H_\kappa} = R^2. \tag{16.39}$$

On $\Sigma''_{H_\kappa}(R, 0)$, which complements $\Sigma'_{H_\kappa}(R, 0)$ to the entire sphere $\Sigma_{H_\kappa}(R, 0)$, we obtain from (16.34),

$$\Phi(w, t) \geq \|w\|^2_{H_\kappa} - \left(\frac{1}{\epsilon}\sum_{i,j=1}^{2} \|B_{ij}w - w_{\alpha^i} w^*_{\alpha^j}\|^2_{L_{2\Omega}} + 2m\,\|w\|_{L_{2\Omega}}\right)$$
$$- \mathcal{A}\frac{1}{\epsilon} - \left\|\begin{matrix}0\\ w\end{matrix}\right\|_{H_\kappa} \|w\|_{H_\kappa} - 2m\big(1 + \|w\|_{H_\kappa}\big),$$

and by (16.2), setting on $c = 2m$, $\Sigma''_{H_\kappa}(R, 0)$, we obtain the inequality

$$\Phi(w, t) \geq \frac{1}{2}\|w\|^2_{H_\kappa} - \frac{\mathcal{A}}{\epsilon} - \left\|\begin{matrix}0\\ w\end{matrix}\right\|_{H_\kappa} \|w\|_{H_\kappa} - 2m\big(1 + \|w\|_{H_\kappa}\big). \tag{16.40}$$

From (16.40) it obviously follows that on $\Sigma''_{H_\kappa}(R, 0)$ for sufficiently large R we have the inequality

$$\Phi(w, t) \geq \frac{1}{3}\|w\|^2_{H_\kappa}. \tag{16.41}$$

From (16.39), (16.41) we see that Theorem 16.1 is completely proved.

For a developable shell, we formulate two modifications of the last theorem.

Theorem 16.2. *Assume that Conditions 1–8 of Section 13 are satisfied and that furthermore, the shell is essentially elastically supported on Γ_6, Γ_7, Γ_9 and is developable. Then on spheres of sufficiently large radius in H_κ we have the inequality* (16.25).

The proof of Theorem 16.2 uses the same arguments as in Theorem 16.1; furthermore, it is somewhat simplified, since on a developable shell we can introduce a Euclidean parametrization and use Lemma 16.4.

Theorem 16.3. *Assume that a developable shell satisfies Conditions 1–8 of Section 13 and, in addition, condition* (16.24). *Then we have the inequality* (16.25).

The proof of Theorem 16.3 follows the scheme of Theorem 16.1; however, instead of Lemma 16.3 we use Lemma 16.5.

Using Theorem 16.1, we obtain the following result.

Theorem 16.4. *Let all the conditions of Theorem 16.1 be satisfied. Then the vector field $w - \mathbf{G}_{\kappa\kappa}(w)$ is homotopic to the vector field w on spheres of sufficiently large radius R, and therefore its winding number on these spheres is $+1$.*

To prove this claim, let us construct the vector field $\mathbf{\Pi}(w, t) = w - t\mathbf{G}_{\kappa\kappa}(w)$. Clearly, $\mathbf{\Pi}(w, 0) = w$, $\mathbf{\Pi}(w, 1) = w - \mathbf{G}_{\kappa\kappa}(w)$. Let us prove now that

$$\mathbf{\Pi}(w, t) = w - t\mathbf{G}_{\kappa\kappa}(w) \neq 0 \text{ for } \|w\|_{H_\kappa} = R,\ 0 \leq t \leq 1,$$

if R is sufficiently large. Indeed, if for some $0 \leq t_0 \leq 1$ and w_0 we have the relation

$$\mathbf{\Pi}(w_0, t_0) = w_0 - t_0\mathbf{G}_{\kappa\kappa}(w_0) = 0,$$

then we have

$$(\mathbf{\Pi}(w_0,\, t_0)\cdot w_0)_{H_\kappa} = \|w_0\|^2_{H_\kappa} - t_0(\mathbf{G}_{\kappa\kappa}(w_0)\cdot w_0) = \Phi(w_0,\, t) = 0,$$

which contradicts Theorem 16.1. Theorem 16.4 is proved.

Now we can formulate the central result of this chapter.

Theorem 16.5. *Assume that the following conditions hold:* (1) $S \in H^{2,\lambda}_\Omega$; (2) *the shell is essentially elastically supported on* Γ_6, Γ_7, Γ_8, *and in addition, Conditions 2–6 of Section 13 hold. Then problem* $t\kappa$ *is solvable if and only if Conditions 7–8 of Section 13 hold. For a given solution* $\mathbf{a} = (w_1,\, w_2,\, w^3) \in W^{(1)}_{2\Omega} \times W^{(1)}_{2\Omega} \times W^{(2)}_{2\Omega}$, ω_p, w_p *(and thus the load complex* $[R^s,\, \widetilde{T},\, \widetilde{T}^m,\, \widetilde{M}^m,\, \widetilde{Q}]$*) are determined uniquely.*

The proof of sufficiency of conditions of Theorem 16.5 follows immediately from the fact that under these conditions, by Theorem 16.4, the winding number of the vector field $w - \mathbf{G}_{\kappa\kappa}(w)$ on spheres of large radius in H_κ is $+1$. Let us demonstrate the necessity of conditions 7, 8 of Section 13 if all the other conditions of Theorem 16.5 hold. Let us note that if there is a solution $\mathbf{a} = (w_1,\, w_2,\, w^3)$, this function is in $W^{(1)}_{2\Omega} \times W^{(1)}_{2\Omega} \times W^{(2)}_{2\Omega}$ (Definition (13.5)–(13.7)). Then conditions (6.1)–(6.3), (6.5), (6.8), (6.9), (6.12) are satisfied. But by well-known results (see, for example, [22, 32, 181]) the geometrical boundary assignments of $\widetilde{w}$, $\widetilde{w}_4$, $\widetilde{w}_m$, $\widetilde{w}_\tau$ necessarily have to be admissible, that is, (11.32), (12.36), (12.37) must hold. Necessity of Condition 8 of Section 13 has been established. Let us turn to prove the necessity of Condition 7. Assume that there exists a generalized solution $\mathbf{a} = (w_1,\, w_2,\, w)$ of Problem $t\kappa$ and $\omega = (w_1,\, w_2) \in H_t$, $w \in H_\kappa$. From (13.6), (13.7) we have

$$\begin{aligned}(\omega_p\cdot\chi)_{H_t} &\equiv \int_\Omega R^i\varphi_i\, d\Omega + \int_{\Gamma_6+\Gamma_8} \widetilde{T}^\tau\varphi_\tau\, ds + \int_{\Gamma_7+\Gamma_8} \widetilde{T}^m\varphi_m\, ds \\ &= (\omega\cdot\chi)_{H_t} - \int_\Omega \Big[\Big(B_{kl}w - \frac{1}{2}w_{\alpha^k}w_{\alpha^l}\Big)D^{ijkl}_{\mathrm{s}}\nabla_i\varphi_j\Big]\, d\Omega,\end{aligned} \tag{16.42}$$

$$\begin{aligned}(w_p\cdot\varphi)_{H_\kappa} &\equiv \int_\Omega R^3\varphi\, d\Omega + \int_{\Gamma_2+\Gamma_4} \widetilde{M}^m\varphi_4\, ds + \int_{\Gamma_3+\Gamma_8} \widetilde{Q}\varphi\, ds \\ &= (w\cdot\varphi)_{H_\kappa} - \int_\Omega T^{ij}\big(B_{ij}\varphi - w_{\alpha^i}\varphi_{\alpha^j}\, d\Omega.\end{aligned} \tag{16.43}$$

From (16.42), which must hold for any $\chi = (\varphi_1,\, \varphi_2) \in H_t$, we have

$$\begin{aligned}\big|(\omega_p\cdot\chi)_{H_t}\big| &= \Big|\int_\Omega R^i\varphi_i\, d\Omega + \int_{\Gamma_6+\Gamma_8} \widetilde{T}^\tau\varphi_\tau\, ds + \int_{\Gamma_7+\Gamma_8} \widetilde{T}^m\varphi_m\, ds\Big| \\ &\le \big|(\omega\cdot\chi)_{H_t}\big| + \int_\Omega \Big|B_{kl}w - \frac{1}{2}w_{\alpha^k}w_{\alpha^l}\Big|\Big|D^{ijkl}_{\mathrm{s}}\nabla_i\varphi_j\Big|\, d\Omega \\ &\le \|\omega\|_{H_t}\,\|\chi\|_{H_t} + m\Big(\sum_{k,l=1}^{2}|B_{kl}|\,\|w\|_{W^{(2)}_{2\Omega}} + \|w\|^2_{W^{(2)}_{2\Omega}}\Big)\,\|\chi\|_{H_t}\,.\end{aligned} \tag{16.44}$$

The second inequality in (16.44) is established using (13.24) for $w^* \equiv 0$ and (13.25). From (16.44) it is seen that the load complex $[R^s, \widetilde{T}^\tau, \widetilde{T}^m]$ is in $\overline{H}_t$ by definition (12.40), (12.41). Necessity of a part of Condition 7 in Section 13 is established.

To complete the proof of the necessity of Condition 7 of Section 13, we have from (16.43),

$$\begin{aligned} |(w_{\mathrm{p}} \cdot \varphi)_{H_\kappa}| &= \left| \int_\Omega R^3 \varphi \, d\Omega + \int_{\Gamma_2+\Gamma_4} \widetilde{M}^m \varphi_4 \, ds + \int_{\Gamma_3+\Gamma_4} \widetilde{Q} \varphi \, ds \right| \\ &\le \|w\|_{W^{(2)}_{2\Omega}} \|\varphi\|_{H_\kappa} + m\left(1 + \|w\|_{W^{(2)}_{2\Omega}}\right) \sum_{i,j=1}^{2} \left\|T^{ij}\right\|_{L_{2\Omega}} \|\varphi\|_{H_\kappa} \\ &\le m\,\|\varphi\|_{H_\kappa} \left[\|w\|_{W^{(2)}_{2\Omega}} + \left(1 + \|w\|_{W^{(2)}_{2\Omega}}\right) \sum_{i,j=1}^{2} \left\|T^{ij}\right\|_{L_{2\Omega}} \right]. \end{aligned} \tag{16.45}$$

The second inequality in (16.45) was proved using (13.29) for $w^* \equiv 0$. From (16.45) it is seen that the loading complex $[R^3, \widetilde{M}^m, \widetilde{Q}]$ is in $\overline{H}_\kappa$ by definition (12.40), (12.41). Necessity of Condition 7 in Section 13 is completely established. Unique definition of the entire loading complex $[R^s, \widetilde{T}^m, \widetilde{T}^\tau, R^3, \widetilde{M}^m, \widetilde{Q}]$ is obvious. The proof of Theorem 16.5 is complete.

As a direct corollary of Theorem 16.1, we obtain the following theorem.

Theorem 16.6. *Assume that all the conditions of Theorem 16.1 hold. Then all the solutions of the boundary value problem $t\kappa$ are contained in a sphere of some radius R_0 in H_κ.*

Similarly, the following theorem is a corollary of Theorem 16.2.

Theorem 16.7. *Assume that all the conditions of Theorem 16.2 hold. Then the vectorfield $w - \mathbf{G}_{\kappa\kappa}(w)$ is homotopic to the vectorfield w on spheres of a sufficiently large radius $R \ge R_0$, so that its winding number on these spheres is $+1$.*

Let us state now a solvability theorem under two types of conditions.

Theorem 16.8. *Assume that Conditions 1–6 of Section 13 are satisfied, that the supports on Γ_6, Γ_7, Γ_8 are essentially elastic, and that the shell is developable. Then a necessary and sufficient condition of solvability of the boundary value problem $t\kappa$ is that Conditions 7–8 of Section 13 hold. Here the loading complex $[R^s, \widetilde{T}^m, \widetilde{T}^\tau, R^3, \widetilde{M}^m, \widetilde{Q}]$ is uniquely determined for a given solution $\mathbf{a} = (w_1, w_2, w_3) \in W^{(1)}_{2\Omega} \times W^{(1)}_{2\Omega} \times W^{(2)}_{2\Omega}$.*

Theorem 16.8 is proved using the scheme of proof of Theorem 16.5, with the difference that sufficiency of its conditions follows from Theorem 16.2.

Theorem 16.9. *Assume that all the conditions of Theorem 16.2 hold. Then all the solutions of Problem $t\kappa$ are contained in a sphere of some radius R_0.*

Theorem 16.9 follows from Theorem 16.7.

Theorem 16.10. *Assume that all the conditions of Theorem 16.3 hold. Then the vector field $w - \mathbf{G}_{\kappa\kappa}(w)$ is homotopic to the vector field w on spheres of a sufficiently large radius $R \geq R_0$, so that its winding number on these spheres is $+1$.*

Theorem 16.11. *Assume that a developable shell satisfies Conditions 2–6 of Section 13 as well as condition* (16.24). *Then a necessary and sufficient condition of solvability of Problem $t\kappa$ is that Conditions 7–8 of Section 13 hold. A given solution $\mathbf{a} = (w_1, w_2, w_3) \in W^{(1)}_{2\Omega} \times W^{(1)}_{2\Omega} \times W^{(2)}_{2\Omega}$ uniquely determines the loading complex $[R^s, \widetilde{T}^m, \widetilde{T}^\tau, R^3, \widetilde{M}^m, \widetilde{Q}]$.*

Theorem 16.11 follows from Theorem 16.3.

Theorem 16.12. *Assume that a developable shell satisfies Conditions 2–6 of Section 13 as well as* (16.24). *Then all solutions of Problem $t\kappa$ lie inside a sphere of some radius R_0.*

Remark 16.1. Solvability Theorems 16.5, 16.6, 16.8 have been proved without any assumptions on the smallness of deformations or any other factors defining the problem. Such assumptions were used only in the derivation and formulation of the boundary value problems. The mathematical analysis of the obtained nonlinear boundary value problems is in no way based on them. Solvability theorems proved above cover a reasonably wide range of problems. Let us describe in more detail the boundary value conditions covered by the above theorem. In Theorem 16.5, a slightly more stringent condition, $S \in H^{2,\lambda}_{\Omega}$ is required; this is not unduly restrictive in practice. Here the clamping conditions with respect to the deflection w can be quite arbitrary, as long as the conditions allowing us to construct the spaces H_κ are met. For example, along the whole boundary curve Γ we could impose rigid clamping conditions 6.1, 6.2, that is, $\Gamma = \Gamma_1$. The shell could be hinged along the entire boundary curve, in which case $\Gamma = \Gamma_2$. Mixed clamping conditions can be satisfied, $\Gamma = \Gamma_1 + \Gamma_2$. Regions of elastic support Γ_3, Γ_4 are allowed. In these cases, if for example $\Gamma_1 > 0$ or $\Gamma_2 > 0$ so that conditions 6.1, 6.2 hold, the supports on Γ_3, Γ_4 do not have to be essentially elastic. In all these cases, with respect to lateral displacements, the supports on Γ_6, Γ_7, Γ_8 (if these segments of the boundary at all exist) are assumed to be essentially elastic. In Theorem 16.8 the condition $S \in H^{2,\lambda}_{\Omega}$ has been weakened, since it is required that $S \in C^2_{\Omega}$. It is, however, assumed that the shell is developable. To estimate the degree of generality of these results, we observe that the theories of shallow shells of Marguerre, Vlasov, Galimov, Mushtari, Chien deal with properly shallow shells, the middle surface of which is close to a plane. In other words, they cover a particular case of a developable shell.

In Theorem 16.11 we have removed the condition of essential elasticity of lateral supports but introduced condition (16.26), which covers now a new class of problems.

Remark 16.2. An important step in the proof is the demonstration of the inequality (16.17) in Theorems 16.1–16.3. The method of decomposing the sphere $\Sigma_{H_\kappa}(R, 0)$ into the two sets $\Sigma'_{H_\kappa}(R, 0)$ and $\Sigma''_{H_\kappa}(R, 0)$ proposed here has mechanical foundations. The point is that $\Sigma'_{H_\kappa}(R, 0)$ is "bounded away" from a weak neighborhood of zero. Roughly speaking, the higher-order derivatives are not much larger than the lower-order ones, and the potential energy of stretching is important in the deformation process. Taking it into account ensures that (16.26) holds on $\Sigma'_{H_\kappa}(R, 0)$ in Theorems 16.1–16.3. On $\Sigma''_{H_\kappa}(R, 0)$ validity of (16.18) is guaranteed due to the domination of the potential energy of bending.

Remark 16.3. Let us clarify the meaning of Condition 7 of Section 13. From (11.52), which defines $\overline{H}_t$, it follows that if $R^s \in L_{p\Omega}$ and $\widetilde{T}^\tau \in L_{p,\Gamma_6+\Gamma_8}$, $\widetilde{T}^m \in L_{p,\Gamma_7+\Gamma_8}$ for some $p > 1$, then $[R^s, \widetilde{T}^\tau, \widetilde{T}^m] \in \overline{H}_t$, which is simply shown using the Hölder inequality, since by Theorem 11.3 (see (11.41)), $w_i \in L_{q\Omega}$ for any $q \geq 1$. Here, clearly, the loads admit jumps of the form $r^{-2+\epsilon}$, $\epsilon > 0$, and ϵ can be arbitrarily small. We also easily obtain that $[R^s, \widetilde{T}^\tau, \widetilde{T}^m] \in \overline{H}_t$ if on Γ_6, Γ_7, Γ_8 there are discontinuities of the form $r^{-1+\epsilon}$, $\epsilon > 0$, and ϵ can be arbitrarily small. From (12.40) it follows that $[R^3, \widetilde{M}^m, \widetilde{Q}] \in \overline{H}_\kappa$ if R^3 admits δ-function components. Indeed, we have

$$\int_\Omega \delta(x_1 - \xi_1, x_2 - \xi_2) w(\xi_1, \xi_2)\, d\Omega_\xi = w(x_1, x_2),$$

while by the embedding Theorem 12.3 (see (12.28)) we conclude that on the right-hand side we have a linear functional in H_κ. It is also easy to see that $[R^3, \widetilde{M}^m, \widetilde{Q}] \in \overline{H}_\kappa$ if $\widetilde{M}^m \in L_{p,\Gamma_2+\Gamma_4}$ for some $p > 1$. Thus $\widetilde{M}^m$ can have discontinuities of the form $r^{-1+\epsilon}$, $\epsilon > 0$, and ϵ can be arbitrarily small. The shear force $\widetilde{Q}$ on $\Gamma_3 + \Gamma_4$ can have discontinuities of δ-function type.

Remark 16.4. By Condition 8 of Section 13, the geometrical boundary conditions must be admissible. As far as the lateral geometrical data is concerned, the meaning of this is explained in Lemma 11.8. Here we just observe that relations (11.46), (11.47) will hold if the segment $\widetilde{\Gamma}$ containing Γ_5, Γ_6, Γ_7 is in $C^2_{\widetilde{\Gamma}}$, while $\widetilde{w}_i \in C^1_{\Gamma_5}$, $\widetilde{w}_m \in C^1_{\Gamma_6}$, $\widetilde{w}_\tau \in C^1_{\Gamma_7}$. For transverse geometrical data, admissibility is described in Lemma 12.4. We observe as well that relations (12.36), (12.37) will be satisfied if the segment $\widetilde{\Gamma}$ containing Γ_1, Γ_2, Γ_3 is in $C^3_{\widetilde{\Gamma}}$, while $\widetilde{w} \in C^2_{\Gamma_1+\Gamma_2}$, $\widetilde{w}_4 \in C^1_{\Gamma_1+\Gamma_3}$.

Remark 16.5. Some minor modifications in our constructions lead to a solvability theorem also in the case when the middle surface S has a more complex topological structure. If, for example, S is homeomorphic to a piece of a closed cylinder (Figure 1.2), then in the construction of the space $H_{t\kappa}$ we have to introduce the set of vector functions $\mathbf{a} = (w_1, w_2, w_3)$ that satisfy, in addition to the geometrical boundary conditions on $\mathcal{BD}$ and $\mathcal{AC}$, the following compatibility conditions on the section $\mathcal{AB} = \mathcal{CD}$:

$$(w_i, w)|_{\mathcal{AB}} \equiv (w_i, w)|_{\mathcal{CD}}; \quad (w_{i\alpha^j}, w_{\alpha^i})|_{\mathcal{AB}} \equiv (w_{i\alpha^j}, w_{\alpha^i})|_{\mathcal{CD}};$$

$$w_{\alpha^i\alpha^j}|_{\mathcal{AB}} \equiv w_{\alpha^i\alpha^j}|_{\mathcal{CD}}.$$

The remainder of the scheme of proof of solvability theorems can then be extended to the case of a complex topological structure of S.

16.4.

Theorem 16.13. *Assume that a properly shallow shell satisfies Conditions 1–6 of Section 13 and that the supports on* Γ_6, Γ_7, Γ_8 *are essentially elastic. Then a boundary value problem* $t\kappa$ *will be solvable if and only if Conditions 7–8 of Section 13 hold. Then for a given solution* $\mathbf{u}(w_1, w_2, w) \in \widetilde{W}^{(1)}_{2\Omega} \times \widetilde{W}^{(1)}_{2\Omega} \times \widetilde{W}^{(2)}_{2\Omega}$, *the components* ω_p, w_p *(and therefore the load complex* $[R^s, \widetilde{T}^m, \widetilde{T}^\tau, R^3, \widetilde{M}^m, \widetilde{Q}])$ *are uniquely determined.*

In this section we shall consider the question of existence of solutions of the NOE (nonlinear operator equations) (13.14), (13.6) of the theory of shallow shells in various classes of function. It frequently happens that the shape of the shell, boundary data, and external loads have various types of symmetry. In such a case it is of interest to prove existence of solutions having the same type of symmetry. Thus, in [205], under certain additional conditions, Morozov shows existence (and even uniqueness) of an axisymmetric equilibrium state for a circular plate with an axisymmetric load and axisymmetric boundary data. In [204] he considers shells having r axes of symmetry. In this section we shall state a general result proved by the methods developed in this chapter. Let us first clarify some issues. Let us assume that all the initial data of the nonlinear operator equations (13.14), (13.15), the shape of the shell and of the boundary curve Γ, boundary conditions, and external load are such that there exist transformations

$$\mathbf{R} = \mathbf{R}(\rho)$$

with respect to which all the equations and loads are invariant. Below we shall call such transformations automorphisms of the boundary value problem of shallow shell theory, or automorphisms $\mathcal{X}$ of the NOE (13.14), (13.15). Clearly, the automorphisms form a group [173]. Therefore, in spaces $H_{t\kappa}$ we can choose subsets $H_{t\kappa s}$ of vector functions $\mathbf{a} = (w_1, w_2, w)$ that are invariant under $\mathcal{X}$. It is easily seen that the $H_{t\kappa s}$ are subspaces of $H_{t\kappa}$. Indeed, $H_{t\kappa s}$ clearly is a linear space, and furthermore, limits in the norm of $H_{t\kappa}$ of sequences in $H_{t\kappa s}$ are in $H_{t\kappa s}$. Similarly, we can define the subspaces H_{ts} and $H_{\kappa s}$. The approach developed in this chapter for studying the NOE (13.14), (13.15) leads to the following results.

Theorem 16.14. *Assume that the NOE* (13.14), (13.15) *have a group of automorphisms* $\mathcal{X}$. *Assume, furthermore, that all the conditions of Theorem 16.5 (respectively, 16.8, 16.11) are satisfied. Then the operators* $\mathbf{G}_{t\kappa}$, $\mathbf{G}_{\kappa\kappa}$, $\mathbf{K}_{t\kappa}$ *are completely continuous on* $H_{t\kappa s}$, $H_{\kappa s}$, *and from* H_{ts} *into* $H_{\kappa s}$. *The vector field* $w - \mathbf{G}_{\kappa\kappa}(w)$ *has winding number* $+1$ *on spheres of large radius in* $H_{\kappa s}$. *The NOE* (13.14), (13.15) *have at least one solution in* $H_{t\kappa s}$. *All the solutions of the NOE* (13.14), (13.15) *are contained in a sphere of some radius* R_0 *in* $H_{t\kappa s}$, *or, respectively, in* $H_{\kappa s}$.

Applications of Theorem 16.13 can be found in [378].

CHAPTER IV

The Topological Method in the Problem of Solvability of the Main Boundary Value Problems in the Nonlinear Theory of Shallow Shells with an Airy Stress Function

17. The Generalized Formulation of the Boundary Value Problems of Shallow Shells with an Airy Stress Function. Reduction to Operator Equations. Physical Interpretation of Generalized Solutions

In the present chapter (Sections 17, 18, 19) we shall establish solvability of the main nonlinear boundary value problems for equations (7.51), (7.60) and the boundary conditions (6.2), (6.20), (7.9), (7.24). Let us sketch the general direction of our arguments. In Section 17 we shall introduce generalized solutions of the above boundary value problems; these can be given a well-defined physical interpretation. The search for these generalized solutions will be completely reduced to a solution of the nonlinear operator equation (17.19), in which the operator $\mathbf{G}_\kappa$ turns out to be completely continuous in H_κ. Therefore, we can compute the winding number of the vector field $w - \mathbf{G}_\kappa(w)$ on spheres of sufficiently large radius in H_κ. This computation is based on the homotopy of the field $w - \mathbf{G}_\kappa(w)$ to the field w on spheres of large radius in H_κ. The homotopy is established in Section 19, using inequality (19.17); this is Theorem 19.1. Thus, the crucial step in the proof is this inequality. The inequality (19.17) comes from the properties of the functional $\Pi_4(w)$, which is introduced in Section 19. The functional $\Pi_4(w)$ turns out to be a weakly continuous functional, and on a part of the sphere $\Sigma_{H_\kappa}(R, 0)$ we have the inequality (19.15), from which follows (19.17) on the same part of the sphere $\Sigma_{H_\kappa}(R, 0)$. On the remainder of $\Sigma_{H_\kappa}(R, 0)$, (19.15) is established using different considerations (see Theorem 19.1). This is the main argument in the proof of the solvability theorem.

17.1. First of all, let us state the ***Conditions*** that are to hold in the subsequent analysis.

(1) S is a regular surface of class C^2_Ω.
(2) Ω is a Sobolev domain of class (2, 2, 2).
(3) Γ is a piecewise-smooth curve of class C^1_Γ.
(4) The material of the shell is regular, so that condition (4.18) holds; furthermore, D^{ijkl}_{f}, $C^{\lambda\mu st}_*$ are piecewise smooth in $\overline{\Omega}$.
(5) A combination of boundary conditions on Γ and the elasticity coefficients k^{44}_{f} are such that one of the spaces H_κ, $\kappa = 1, 2$, can be constructed.
Therefore, by condition 5, our theory covers the cases when $\Gamma_1 > 0$, and if $\Gamma_1 = 0$, then on Γ_2 there must be points where $\kappa_2 \neq 0$; otherwise, the support on Γ_2 must be essentially elastic, that is, $k^{44}_{\text{f}}(s) \geq m_0 > 0$. In other words, in the situation we are considering here, conditions of Lemmas 12.3–12.5 are satisfied, and $\kappa = 1, 2$. Condition (7.8) always holds.
(6) Furthermore, there is a function $\overset{0}{w} \in W^{(2)}_{2\Omega}$ such that

$$\overset{0}{w}\Big|_{\Gamma} = \widetilde{w}; \qquad \frac{\partial \overset{0}{w}}{\partial m}\Bigg|_{\Gamma_1} = \widetilde{w}_4. \tag{17.1}$$

In other words, $\widetilde{w}$, $\widetilde{w}_4$ are taken to be admissible. Sufficient conditions on Γ and both necessary and sufficient conditions on $\widetilde{w}$ and $\widetilde{w}_4$ are given in Lemma 12.4 (see relation (12.36)).
(7) The loading complex $[\widetilde{T}^{ij}_{\text{p}}, R^3, \widetilde{M}^m]$ is in $\overline{H}_\kappa$; $T^{ij}_{\text{p}} \in L_{2\Omega}$.
(8) The shell is developable.

17.2. In the derivation of equations (7.51), (7.60) and the static boundary conditions (6.20), it was assumed that both the solution w, Ψ and the initial data of the problem are sufficiently smooth. Namely, it was assumed that $w, \Psi \in C^4_\Omega$ and that this solution satisfies (7.51), (7.60) at each interior point of Ω and the boundary conditions (6.2), (6.20), (7.9), (7.24) at each point of Γ, apart from the points at which boundary conditions change type; then D^{ijkl}_{f}, $C^{\lambda\mu st}_* \in C^2_\Omega$; T^{ij}_{p}; R^3, $R^s \in C_\Omega, k^{44}_{\text{f}}, \widetilde{M}^m \in C_{\Gamma_2}$. Such strict requirements on initial data compel us in this case as well to turn to a formulation of the problems in generalized solutions.

Definition 17.1. A generalized solution of Problem 9κ is taken to be a pair of functions $w \in W^{(2)}_{2\Omega}$, $\Psi \in H_9$ that satisfy the integral identities

$$\begin{aligned}(w \cdot \varphi)_{H_\kappa} &= \int_\Omega C^{ik}C^{jl}\left(B_{ij}\varphi - \varphi_{\alpha^i} w_{\alpha^j}\right)\nabla_{kl}\Psi d\Omega - \int_\Omega \widetilde{T}^{ij}_{\text{p}} w_{\alpha^i}\varphi_{\alpha^j} d\Omega \\ &\quad + \int_\Omega \left(\widetilde{T}^{ij}_{\text{p}} B_{ij} + R^3\right)\varphi d\Omega + \int_{\Gamma_2} \widetilde{M}^m \varphi_4 ds, \quad \varphi_4 = \frac{\partial \varphi}{\partial m},\end{aligned} \tag{17.2}$$

$$\begin{aligned}(\Psi \cdot \theta)_{H_9} &= -\int_\Omega C^{ik}C^{jl}\left(wB_{ij} - \frac{1}{2}w_{\alpha^i}w_{\alpha^j}\right)\nabla_{kl}\theta d\Omega \\ &\quad - \int_\Omega C^{\lambda\mu}_{*ij}\widetilde{T}^{ij}_{\text{p}}\nabla_{\lambda\mu}\theta d\Omega.\end{aligned} \tag{17.3}$$

If we compare (17.2), (17.3) and (7.51), (7.60), we see that they are identical if we set $\delta w = \varphi$ and $\delta\psi = \theta$. Therefore, the physical meaning of the generalized solution by Definition 17.1 is that φ and Ψ satisfy the Alumyae mixed variational principle, which, as is well known, includes both equilibrium and compatibility equations.

Lemma 17.1. *Let Conditions 1-5 and 7 of Section 17.1 hold. In this case Definition 17.1 of the generalized solution* (17.2), (17.3) *is well-defined in the sense that each term on the right-hand sides of* (17.2)–(17.3) *is defined if* $w \in H_\kappa$, $\Psi \in H_9$. *Then the right-hand sides define linear bounded functionals in* H_κ, $H_{9\Omega}$, *respectively, in* φ, θ.

To prove Lemma 17.1, we take into account the fact that by Theorem 12.3, if φ, $w \in H_\kappa$, then $\varphi,\ w \in C_\Omega$; $\varphi_{\alpha^i},\ w_{\alpha^j} \in L_{q\Omega}$ for any $q \geq 1$; and $\nabla_{ij}\varphi,\ \nabla_{ij}w \in L_{2\Omega}$. Furthermore, if $\Psi \in H_9$, then $\nabla_{kl}\Psi \in L_{2\Omega}$. Therefore, we have

$$\begin{aligned}
&\left|\int_\Omega C^{ik}C^{jl}\left(B_{ij}\varphi - \varphi_{\alpha^i} w_{\alpha^j}\right)\nabla_{kl}\Psi d\Omega\right| \\
&\leq \left\|C^{ik}C^{jl}B_{ij}\right\|_{C_\Omega} \|\varphi\|_{L_{2\Omega}} \|\nabla_{kl}\Psi\|_{L_{2\Omega}} + \left\|C^{ik}C^{jl}\varphi_{\alpha^i} w_{\alpha^j}\right\|_{L_{2\Omega}} \|\nabla_{kl}\Psi\|_{L_{2\Omega}} \\
&\leq m\sum_{k,l=1}^{2} \left\|C^{ik}C^{jl}B_{ij}\right\|_{C_\Omega} \|\varphi\|_{H_\kappa} \|\Psi\|_{H_9} \\
&\quad + m\sum_{i,j,k,l=1}^{2} \left\|C^{ik}C^{jl}\right\|_{C_\Omega} \|\varphi\|_{L_{4\Omega}} \|w\|_{L_{4\Omega}} \|\Psi\|_{H_9} \\
&\leq m\, \|\varphi\|_{H_\kappa} \|\Psi\|_{H_9} \left(1 + \|w\|_{H_\kappa}\right).
\end{aligned} \tag{17.4}$$

Next, by condition 7 and the formulae (12.4) we have

$$\left|\int_\Omega \widetilde{T}^{ij}_{\mathrm{p}} w_{\alpha^i}\varphi_{\alpha^j} d\Omega\right| \leq \sum_{i,j=1}^{2} \left\|\widetilde{T}^{ij}_{\mathrm{p}}\right\|_{L_{2\Omega}} \|w\|_{L_{4\Omega}} \|\varphi\|_{L_{4\Omega}}, \tag{17.5}$$

whence by Theorem 12.3 we obtain

$$\left|\int_\Omega \widetilde{T}^{ij}_{\mathrm{p}} w_{\alpha^i}\varphi_{\alpha^j} d\Omega\right| \leq m\sum_{i,j=1}^{2} \left\|\widetilde{T}^{ij}_{\mathrm{p}}\right\|_{L_{2\Omega}} \|w\|_{H_\kappa} \|\varphi\|_{H_\kappa}. \tag{17.6}$$

Finally, by the same condition 7 we have

$$\left|\int_\Omega (\widetilde{T}^{ij}_{\mathrm{p}} B_{ij} + R^3)\varphi d\Omega + \int_{\Gamma_2} \widetilde{M}^m \varphi_4 ds\right| = \left|(w^*_{\mathrm{p}}\cdot\varphi)_{H_k}\right| \leq \left\|w^*_{\mathrm{p}}\right\|_{H_\kappa} \|\varphi\|_{H_\kappa}. \tag{17.7}$$

Therefore, all the terms on the right-hand side of (17.2) are defined, and moreover,

$$\Big|\int_\Omega C^{ik}C^{jl}(B_{ij}-\varphi_{\alpha^i}w_{\alpha^j})\nabla_{kl}\Psi d\Omega+\int_\Omega \widetilde{T}_{\mathrm{p}}^{ij}w_{\alpha^i}\varphi_{\alpha^j}d\Omega$$

$$+\int_\Omega(\widetilde{T}_{\mathrm{p}}^{ij}B_{ij}+R^3)\varphi d\Omega+\int_{\Gamma_2}\widetilde{M}^m\varphi_4 ds\Big|$$

$$\le m\,\|\varphi\|_{H_\kappa}\Big[\,\|\Psi\|_{H_9}\big(1+\|w\|_{H_\kappa}\big)+\sum_{i,j=1}^{2}\Big\|\widetilde{T}_{\mathrm{p}}^{ij}\Big\|_{L_{2\Omega}}\|w\|_{H_\kappa}+\big\|w^*\big\|_{H_\kappa}\Big]. \tag{17.8}$$

The part of Lemma 17.1 that relates to the right-hand side of (17.2) is proved.

By similar considerations, for the first term of the right-hand side of (17.3) we have

$$\Big|\int_\Omega C^{ik}C^{jl}\big(wB_{ij}-\frac{1}{2}w_{\alpha^i}w_{\alpha^j}\big)\nabla_{kl}\theta d\Omega\Big|$$

$$\le \big\|C^{ik}C^{jl}B_{ij}\big\|_{C_\Omega}\|w\|_{L_{2\Omega}}\|\nabla_{kl}\theta\|_{L_{2\Omega}}+\frac{1}{2}\big\|C^{ik}C^{jl}w_{\alpha^i}w_{\alpha^j}\big\|_{L_{2\Omega}}\|\nabla_{kl}\theta\|_{L_{2\Omega}}$$

$$\le m\,\|\theta\|_{H_9}\Big(\sum_{k,l=1}^{2}\big\|C^{ik}C^{jl}B_{ij}\big\|_{C_\Omega}\|w\|_{H_\kappa}+\sum_{i,j=1}^{2}\|w_{\alpha^i}\|_{L_{4\Omega}}\|w_{\alpha^j}\|_{L_{4\Omega}}\Big)$$

$$\le m\,\|\theta\|_{H_9}\|w\|_{H_\kappa}\big(1+\|w\|_{H_\kappa}\big). \tag{17.9}$$

In addition, by condition 7, the last term on the right-hand side of (17.3) is estimated as follows (see relation (12.43)):

$$\Big|\int_\Omega C_{*ij}^{\lambda\mu}\widetilde{T}_{\mathrm{p}}^{ij}\nabla_{\lambda\mu}d\Omega\Big|\le m\sum_{i,j=1}^{2}\Big\|\widetilde{T}_{\mathrm{p}}^{ij}\Big\|_{L_{2\Omega}}\|\theta\|_{H_9}. \tag{17.10}$$

From (17.9), (17.10) it follows that

$$\Big|-\int_\Omega C^{ik}C^{jl}\big(wB_{ij}-\frac{1}{2}w_{\alpha^i}w_{\alpha^j}\big)\nabla_{kl}\theta d\Omega-\int_\Omega C_{*ij}^{\lambda\mu}\widetilde{T}_{\mathrm{p}}^{ij}\nabla_{\lambda\mu}\theta d\Omega\Big|$$

$$\le m\,\|\theta\|_{H_9}\Big[\,\|w\|_{H_\kappa}\big(1+\|w\|_{H_\kappa}\big)+\sum_{i,j=1}^{2}\Big\|\widetilde{T}_{\mathrm{p}}^{ij}\Big\|_{L_{2\Omega}}\Big]. \tag{17.11}$$

Thus the part of Lemma 17.1 that relates to the right-hand side of (17.3) is established, and its proof is concluded.

For our considerations below it is natural to introduce the change of variables

$$w\sim w^*+w, \tag{17.12}$$

$$w^*=\overset{0}{w}+w_{\mathrm{p}},\ \Psi\sim\Psi, \tag{17.13}$$

where $\overset{0}{w}$ is given by Condition 6, Section 17 (relation (17.1)), while w_{p} is given by condition 7 (relation (12.43)). Thus, having introduced a new function to measure the deflection of the shell, we still use the old notation, but now $w \in H_\kappa$. Then the definition of the generalized solution (17.2), (17.3) takes the following form:

$$(w \cdot \varphi)_{H_\kappa} = -(\overset{0}{w} \cdot \varphi)_{H_\kappa} + \int_\Omega C^{ik} C^{jl} \big[B_{ij}\varphi - \varphi_{\alpha^i}(w^* + w)_{\alpha^j}\big] \nabla_{kl} \Psi \, d\Omega$$
$$- \int_\Omega \widetilde{T}^{ij}_{\text{p}} (w + w^*)_{\alpha^i} \varphi_{\alpha^j} \, d\Omega, \tag{17.14}$$

$$(\Psi \cdot \theta)_{H_9} = -\int_\Omega C^{ik} C^{jl} \Big[(w + w^*) B_{ij} - \frac{1}{2}(w + w^*)_{\alpha^i} (w + w^*)_{\alpha^j}\Big]$$
$$\times \nabla_{kl} \theta \, d\Omega - \int_\Omega C^{\lambda\mu}_{*ij} \widetilde{T}^{ij}_{\text{p}} \nabla_{\lambda\mu} \theta \, d\Omega. \tag{17.15}$$

17.3. To reduce the boundary value Problem 9κ to an operator equation, we note that the right-hand side of (17.3) and thus of (17.15), is a linear functional in H_9 with respect to θ. Therefore, by the Riesz theorem we have

$$\int_\Omega C^{ik} C^{jl} \Big[\frac{1}{2}(w + w^*)_{\alpha^i} (w + w^*)_{\alpha^j} - B_{ij}(w + w^*)\Big] \nabla_{kl} \theta \, d\Omega$$
$$- \int_\Omega C^{\lambda\mu}_{*ij} \widetilde{T}^{ij}_{\text{p}} \nabla_{\lambda\mu} \theta \, d\Omega = \big(\mathbf{K}_{9\kappa}(w, w^*, \widetilde{T}^{ij}_{\text{p}}) \cdot \theta\big)_{H_9}, \tag{17.16}$$

where $\mathbf{K}_{9\kappa}(w, w^*, \widetilde{T}^{ij}_{\text{p}})$ depends on $w^* + w$. Since θ is arbitrary, from (17.15), (17.16) we have

$$\Psi = \mathbf{K}_{9\kappa}\big(w, w^*, \widetilde{T}^{ij}_{\text{p}}\big). \tag{17.17}$$

Substituting (17.17) into the right-hand side of (17.14), we again obtain a linear functional, though now in φ, which by the Riesz theorem can be represented in the form

$$-(\overset{0}{w} \cdot \varphi)_{H_\kappa} + \int_\Omega C^{ik} C^{jl} \big[B_{ij}\varphi - \varphi_{\alpha^i}(w^* + w)_{\alpha^j}\big] \nabla_{kl} \Psi \, d\Omega$$
$$- \int_\Omega \widetilde{T}^{ij}_{\text{p}} (w + w^*)_{\alpha^i} \varphi \alpha^j \, d\Omega = \big(\mathbf{G}_\kappa(w) \cdot \varphi\big)_{H_\kappa}, \tag{17.18}$$

where $\mathbf{G}_\kappa(w)$ is an element of H_κ that supplies the required representation. Clearly, $\mathbf{G}_\kappa(w)$ depends on w^*, $\overset{0}{w}$. Since we are most interested in the dependence of $\mathbf{G}_\kappa$ on w, while w^* and $\overset{0}{w}$ are fixed, we reflect this fact in the notation we use. From (17.14), (17.18), we have

$$w = \mathbf{G}_\kappa(w). \tag{17.19}$$

Therefore, the search for a generalized solution of Problem 9κ has been reduced to the determination of a fixed point of the mapping given by the operator $\mathbf{G}_\kappa$ and the consequent determination of Ψ from (17.15), (17.17).

Sometimes it will be convenient to write the system (17.9), (17.17) as a single equation,

$$\mathbf{c} = \mathbf{G}_{9\kappa}(\mathbf{c}), \tag{17.20}$$

where $\mathbf{c} \in H_{9\kappa}$ has components Ψ and w. The operator $G_{9\kappa}$ has components $\mathbf{K}_{9\kappa}$, $\mathbf{G}_{\kappa}$.

Using the above, we conclude that the following theorem holds:

Theorem 17.1. *A pair of functions w, $\Psi \in H_{9\kappa}$ will be a generalized solution of problem 9κ if and only if it satisfies* (17.19) *and* (17.17) *or, which is the same,* (17.20).

In conclusion, let us note that (17.17) can be interpreted as a surface in $H_{9\kappa}$ (a "hyperparaboloid"), which we shall call HP2. Thus we deduce that all the generalized solutions of problem 9κ lie in $H_{9\kappa}$ on HP2.

17.4. Let us show that the definition of a generalized solution (17.2) can be recast in a different, also quite useful, form. For this we have to obtain some auxiliary relations.

Lemma 17.2. *Let $\widetilde{T}_{\mathrm{p}}^{ij} \in L_{2\Omega}$. In this case the sum $R^s w_{\alpha^s}$ generates a linear functional in H_κ. In other words, the integral*

$$\int_\Omega R^i w_{\alpha^i} \varphi d\Omega \tag{17.21}$$

converges, and

$$\left|\int_\Omega R^i w_{\alpha^i} \varphi d\Omega\right| \le m \, \|\varphi\|_{H_\kappa} \, \|w\|_{H_\kappa} \, . \tag{17.22}$$

To prove the lemma, we observe that $\widetilde{T}_{\mathrm{p}}^{ij}$ is the solution of the inhomogeneous system of equations (6.17), and therefore

$$\begin{aligned}\int_\Omega \widetilde{T}_{\mathrm{p}}^{ij} w_{\alpha^j} \varphi_{\alpha^i} d\Omega &= \int_\Omega \widetilde{T}_{\mathrm{p}}^{ij} w_{\alpha^j} D\varphi_{\alpha^i} d\alpha^1 d\alpha^2 = -\int_\Omega \left(\widetilde{T}_{\mathrm{p}}^{ij} w_{\alpha^j} D\right)_{\alpha^i} \varphi d\alpha^1 d\alpha^2 \\ &= -\int_\Omega \left[\left(\widetilde{T}_{\mathrm{p}}^{ij} D\right)_{\alpha^i} w_{\alpha^j} + \widetilde{T}_{\mathrm{p}}^{ij} D w_{\alpha^i \alpha^j}\right] \varphi d\alpha^1 d\alpha^2 .\end{aligned} \tag{17.23}$$

Taking (6.17) into account, we have

$$\left(\widetilde{T}_{\mathrm{p}}^{ij} D\right)_{\alpha^i} = -G_{st}^j \widetilde{T}_{\mathrm{p}}^{st} - DR^j, \tag{17.24}$$

whence

$$\int_\Omega \widetilde{T}_{\mathrm{p}}^{ij} w_{\alpha^j} \varphi_{\alpha^i} d\Omega = -\int_\Omega \widetilde{T}_{\mathrm{p}}^{ij} \nabla_{ij} w \varphi d\Omega + \int_\Omega R^i w_{\alpha^i} d\Omega, \tag{17.25}$$

and therefore

$$\begin{aligned}\left|\int_\Omega R^i w_{\alpha^i} \varphi d\Omega\right| &= \left|\int_\Omega \widetilde{T}_{\mathrm{p}}^{ij} \nabla_{ij} w \varphi d\Omega\right| + \left|\int_\Omega \widetilde{T}_{\mathrm{p}}^{ij} w_{\alpha^j} \varphi_{\alpha^i} d\Omega\right| \\ &\le \left\|\widetilde{T}_{\mathrm{p}}^{ij}\right\|_{L_{2\Omega}} \left(\left\|\nabla_{ij} w\right\|_{L_{2\Omega}} \|\varphi\|_{C_\Omega} + \|w_{\alpha^i}\|_{L_{4\Omega}} \|\varphi_{\alpha^j}\|_{L_{4\Omega}}\right).\end{aligned} \tag{17.26}$$

Taking into account the inequalities of Theorem 12.3 (relations (12.25), (12.26)), we obtain from (17.26),

$$\left|\int_\Omega R^i w_{\alpha^i}\varphi d\Omega\right| \leq m \sum_{i,j=1}^{2} \left\|\widetilde{T}^{ij}_{\mathrm{p}}\right\|_{L_{2\Omega}} \|w\|_{H_\kappa} \|\varphi\|_{H_\kappa}, \tag{17.27}$$

which proves Lemma 17.2.

Lemma 17.3. *Let*

$$\Psi \in H_9, \ w \in W^{(2)}_{2\Omega}, \ \varphi \in H_\kappa, \tag{17.28}$$

and assume that the shell is developable. Then the following relation holds:

$$\int_\Omega C^{ik}C^{jl}\nabla_{kl}\Psi w_{\alpha^i}\varphi_{\alpha^j} d\Omega = -\int_\Omega C^{ik}C^{jl}\nabla_{kl}\Psi\nabla_{ij}w\varphi d\Omega. \tag{17.29}$$

To prove it, note that for a developable shell ($K \equiv 0$), (17.29) is already proved in Section 7 (relation (7.46)), where one puts $\delta w = \varphi$. However, there we assumed that

$$\Psi \in C^3_\Omega; \quad \varphi, \ w \in C^2_\Omega.$$

Under our assumptions we choose sequences

$$\Psi_n \in C^4_\Omega; \quad \varphi_n, \ w_n \in C^2_\Omega,$$

such that

$$\begin{aligned} &w_n \to w \text{ in } W^{(2)}_{2\Omega}; \\ &\Psi_n \to \Psi \text{ in } H_9; \\ &\varphi_n \to \varphi \text{ in } H_\kappa. \end{aligned} \tag{17.30}$$

Furthermore, we have

$$\begin{aligned} &\int_\Omega C^{ik}C^{jl}\left(\varphi\nabla_{ij}w + w_{\alpha^i}\varphi_{\alpha^j}\right)\nabla_{kl}\Psi d\Omega \\ &\quad - \int_\Omega C^{ik}C^{jl}\left(\varphi_n\nabla_{ij}w_n + w_{n\alpha^i}\varphi_{n\alpha^j}\right)\nabla_{kl}\Psi_n d\Omega \\ &= \int_\Omega C^{ik}C^{jl}\left[\left(\nabla_{ij}w - \nabla_{ij}w_n\right)\varphi + \nabla_{ij}w_n\left(\varphi - \varphi_n\right)\right]\nabla_{kl}\Psi d\Omega \\ &\quad + \int_\Omega C^{ik}C^{jl}\left[\left(\varphi - \varphi_n\right)\nabla_{ij}w_n + \left(\varphi_{\alpha^j} - \varphi_{n\alpha^j}\right)w_{n\alpha^i}\right]\nabla_{kl}\Psi d\Omega \\ &\quad + \int_\Omega C^{ik}C^{jl}\left[\left(\nabla_{kl}\Psi - \nabla_{kl}\Psi_n\right)\left(\varphi_n\nabla_{ij}w_n + w_{n\alpha^i}\varphi_{n\alpha^j}\right)d\Omega.\right. \end{aligned} \tag{17.31}$$

Here by Theorem 10.5 for w, inequalities (10.7), (10.8), and the embedding Theorem 12.3 for φ, Ψ (formulae (12.25), (12.26), (12.28)), we have from (17.30),

$$\begin{aligned} &\nabla_{ij}w_n \to \nabla_{ij}w, \ \nabla_{kl}\Psi_n \to \nabla_{kl}\Psi \text{ in } L_{2\Omega}, \ \varphi_n \to \varphi \text{ in } C_\Omega, \\ &\varphi_{n\alpha^i} \to \varphi_{\alpha^i} \text{ in } L_{q\Omega} \text{ for any } q \geq 1. \end{aligned} \tag{17.32}$$

From (17.31), (17.32) it is easily concluded that the right-hand side of (17.31) vanishes as $n \to \infty$. For example, for the first term of the right-hand side of (17.31) we have

$$\begin{aligned}
&\left|\int_\Omega C^{ik}C^{jl}\left[(\nabla_{ij}w - \nabla_{ij}w_n)\varphi + \nabla_{ij}w_n(\varphi - \varphi_n)\right]\nabla_{kl}\Psi d\Omega\right| \\
&\quad \le \left|\int_\Omega C^{ik}C^{jl}(\nabla_{ij}w - \nabla_{ij}w_n)\varphi\nabla_{kl}\Psi d\Omega\right| \\
&\quad + \left|\int_\Omega C^{ik}C^{jl}\nabla_{ij}w_n(\varphi - \varphi_n)\nabla_{kl}\Psi d\Omega\right| \qquad (17.33)\\
&\quad \le \left\|C^{ik}C^{jl}\varphi\right\|_{C_\Omega}\left\|\nabla_{ij}w - \nabla_{ij}w_n\right\|_{L_{2\Omega}}\|\nabla_{kl}\Psi\|_{L_{2\Omega}} \\
&\quad + \left\|C^{ik}C^{jl}\right\|_{C_\Omega}\|\varphi - \varphi_n\|_{C_\Omega}\left\|\nabla_{ij}w_n\right\|_{L_{2\Omega}}\|\nabla_{kl}\Psi\|_{L_{2\Omega}} \to 0.
\end{aligned}$$

Similarly, we establish that the other terms on the right-hand side of (17.31) also vanish. Now we recall that the second term of the left-hand side of (17.31) is zero by (7.46).

Lemma 17.3 is proved.

Lemma 17.4. *Let $w \in W^{(2)}_{2\Omega}$, $\theta \in H_9$. Then*

$$\int_\Omega C^{ik}C^{jl}\nabla_{kl}wB_{ij}\theta\, d\Omega = \int_\Omega C^{ik}C^{jl}wB_{ij}\nabla_{kl}\theta d\Omega, \tag{17.34}$$

$$\int_\Omega C^{ik}C^{jl}\theta\nabla_{kl}w\nabla_{ij}w\, d\Omega = -\int_\Omega C^{ik}C^{jl}w_{\alpha^i}w_{\alpha^j}\nabla_{kl}\theta d\Omega. \tag{17.35}$$

Relation (17.34) was proved in Section 7 (formulae (7.52)), where one should put $\delta\Psi \equiv \theta$. However, there we assumed that $w,\ \theta \in C^2_\Omega$, $S \in C^3_\Omega$.

Under the conditions of Lemma 17.4 this formula is proved by passing to a limit, as in relation (17.29). Then (17.35) is obtained in the same way as (17.34). Initially, (17.35) is proved for smooth w, θ, and a passage to a limit is effected.

Lemma 17.5. *Let $w + w^* \in W^{(2)}_{2\Omega}$, $\varphi \in H_\kappa$, $\Psi \in H_9$. Then we have the relation*

$$\begin{aligned}
&\int_\Omega C^{ik}C^{jl}\left[B_{ij}\varphi - \varphi_{\alpha^i}(w + w^*)_{\alpha^j}\right]\nabla_{kl}\Psi d\Omega - \int_\Omega \widetilde{T}^{ij}_{\mathrm{p}}(w + w^*)_{\alpha^i}\varphi_{\alpha^j}d\Omega \\
&\quad = \int_\Omega \left\{C^{ik}C^{jl}\left[B_{ij} + \nabla_{ij}(w + w^*)\right]\nabla_{kl}\Psi\right. \qquad (17.36)\\
&\quad \left. + \widetilde{T}^{ij}_{\mathrm{p}}\nabla_{ij}(w + w^*) - R^i w_{\alpha^i}\right\}\varphi d\Omega.
\end{aligned}$$

In the proof of (17.36) we use (17.29), (17.25) under the assumption $w \sim w+w^*$.

Lemma 17.6. *Assume that the conditions of Lemma 17.4 are satisfied. Then we have the relation*

$$\begin{aligned}
&\int_\Omega \left[C^{ik}C^{jl}\nabla_{kl}(w + w^*)\Big(B_{ij} + \frac{1}{2}\nabla_{ij}(w + w^*)\Big)\theta + C^{\lambda\mu}_{*ij}\widetilde{T}^{ij}_{\mathrm{p}}\nabla_{\lambda\mu}\theta\right]d\Omega \qquad (17.37)\\
&= \int_\Omega \left\{C^{ik}C^{jl}\left[(w + w^*)B_{ij} - \frac{1}{2}(w + w^*)_{\alpha^i}(w + w^*)_{\alpha^j}\right] + C^{kl}_{*ij}\widetilde{T}^{ij}_{\mathrm{p}}\right\}\nabla_{kl}\theta d\Omega.
\end{aligned}$$

In the derivation of (17.37) we used (17.34), (17.35), where we set $w \sim w+w^*$.

Theorem 17.2. *A pair of functions $w \in H_\kappa$ and $\Psi \in H_9$ will be a generalized solution if and only if the following integral equation holds,*

$$(w \cdot \varphi)_{H_\kappa} = -(\overset{0}{w} \cdot \varphi)_{H_\kappa} + \int_\Omega \Big\{ C^{ik}C^{jl}\big[B_{ij} + \nabla_{ij}(w+w^*)\big]\nabla_{kl}\Psi + \widetilde{T}^{ij}_{\mathrm{p}}\nabla_{ij}(w+w^*) - R^i w_{\alpha^i} \Big\}\varphi d\Omega, \tag{17.38}$$

in addition to the integral equation

$$(\Psi \cdot \theta)_{H_9} = -\int_\Omega \Big[C^{ik}C^{jl}\nabla_{kl}(w+w^*)\Big(B_{ij} - \frac{1}{2}\nabla_{ij}(w+w^*)\Big)\theta + C^{\lambda\mu}_{*ij}\widetilde{T}^{ij}_{\mathrm{p}}\nabla_{\lambda\mu}\theta \Big] d\Omega. \tag{17.39}$$

17.5. A generalized solution of Problem 9κ is taken to be a pair of functions $w \in W^{(2)}_{2\Omega}$, $\Psi \in H_9$ that satisfy the integral identities

$$\begin{aligned}(w \cdot \varphi)_{H_\kappa} &= \int_\Omega C^{ik}C^{jl}\left(f_{\alpha^i\alpha^j}\varphi - \frac{1}{2}\varphi_{\alpha^i}w_{\alpha^j}\right)\Psi_{\alpha^k\alpha^l}\,d\alpha^1 d\alpha^2 \\ &\quad - \int_\Omega \widetilde{T}^{ij}_{\mathrm{p}} w_{\alpha^i}\varphi_{\alpha^j}\,d\alpha^1 d\alpha^2 + \int_\Omega (\widetilde{T}^{ij}_{\mathrm{p}} f_{\alpha^i\alpha^j} + R^3)\varphi\,d\alpha^1 d\alpha^2 \\ &\quad + \int_{\Gamma_2} \widetilde{M}\frac{\partial\varphi}{\partial m}\,ds,\end{aligned} \tag{17.40}$$

$$\begin{aligned}(\Psi \cdot \theta)_{H_9} &= -\int_\Omega C^{ik}C^{jl}\left(f_{\alpha^i\alpha^j}w - \frac{1}{2}w_{\alpha^i}w_{\alpha^j}\right)\theta_{\alpha^k\alpha^l}\,d\alpha^1 d\alpha^2 \\ &\quad - \int_\Omega C^{kl}_{*ij}\widetilde{T}^{ij}_{\mathrm{p}}\,\theta_{\alpha^k\alpha^l}\,d\alpha^1 d\alpha^2.\end{aligned} \tag{17.41}$$

18. Main Properties of the Operators $\mathbf{K}_{9\kappa}(w)$, $\mathbf{G}_\kappa(w)$

Let us now study the properties of the operators introduced above. They are similar to the properties of operators of the last section and are obtained by the same methods.

Lemma 18.1. *We have the representations*

$$\mathbf{K}_{9\kappa}(w) = \mathbf{K}_{9\kappa 0} + \mathbf{K}_{9\kappa 1}(w) + \mathbf{K}_{9\kappa 2}(w), \tag{18.1}$$

$$\mathbf{G}_\kappa(w) = \mathbf{G}_{\kappa 0} + \mathbf{G}_{\kappa 1}(w) + \mathbf{G}_{\kappa 2}(w) + \mathbf{G}_{\kappa 3}(w), \tag{18.2}$$

where $\mathbf{K}_{9\kappa\mu}$, $\mathbf{G}_{\kappa\mu}$ are homogeneous operators of order μ. We also have the estimates

$$\left\|\mathbf{K}_{9\kappa\mu}(w)\right\|_{H_9} \le m\,\|w\|^\mu_{H_\kappa}, \qquad \mu = 0, 1, 2, \tag{18.3}$$

$$\left\|\mathbf{G}_{\kappa\mu}(w)\right\|_{H_\kappa} \le m\,\|w\|^\mu_{H_\kappa}, \qquad \mu = 0, 1, 2, 3. \tag{18.4}$$

To prove the lemma, let us consider the relation (17.16) which defines $\mathbf{K}_{9\kappa}(w)$. From it we immediately have

$$\int_\Omega \Big[C^{i\lambda}C^{j\mu}\Big(\frac{1}{2}w^*_{\alpha^i}w^*_{\alpha^j} - w^* B_{ij}\Big) - C^{\lambda\mu}_{*ij}\widetilde{T}^{ij}_{\mathrm{p}}\Big]\nabla_{\lambda\mu}\theta d\Omega = \big(\mathbf{K}_{9\kappa 0}\cdot\theta\big)_{H_9}. \quad (18.5)$$

Equation (18.5) defines $\mathbf{K}_{9\kappa 0}$ as an element of the space H_9.

Next, from (17.15) we also have

$$\big(\mathbf{K}_{9\kappa 1}\cdot\theta\big)_{H_9} = -\int_\Omega C^{ik}C^{jl}\big(-wB_{ij} + w^*_{\alpha^i}w_{\alpha^j}\big)\nabla_{kl}\theta d\Omega. \quad (18.6)$$

From (18.6) it follows that

$$\|\mathbf{K}_{9\kappa 1}\|_{H_9} \le m\Big(\big\|C^{ik}C^{jl}B_{ij}\big\|_{C_\Omega} + \sum_{j,k,l=1}^{2}\big\|C^{ik}C^{jl}\big\|_{C_\Omega}\big\|w^*_{\alpha^i}\big\|_{L_4}\Big)\|w\|_{H_{\kappa 1}}.$$

Finally, we have

$$\big(\mathbf{K}_{9\kappa 2}\cdot\theta\big)_{H_9} = -\frac{1}{2}\int_\Omega C^{ik}C^{jl}w_{\alpha^i}w_{\alpha^j}\nabla_{kl}\theta d\Omega, \quad (18.7)$$

whence

$$\|\mathbf{K}_{9\kappa 2}\|_{H_9} \le m\sum_{i,j,k,l=1}^{2}\big\|C^{ik}C^{jl}\big\|_{C_\Omega}\|w\|^2_{H_\kappa}.$$

The inequalities (18.3) are proved. Since the elements $\mathbf{K}_{9\kappa 1}$ and $\mathbf{K}_{9\kappa 2}$ depend on w, they define the operators $\mathbf{K}_{9\kappa 1}(w)$ and $\mathbf{K}_{9\kappa 2}(w)$, respectively, and so the relation (18.1) is established.

Let us consider now (17.18), which defines $\mathbf{G}_\kappa$. We have

$$\begin{aligned}\big(\mathbf{G}_{\kappa 0}\cdot\varphi\big)_{H_\kappa} &= -(\overset{0}{w}\cdot\varphi)_{H_\kappa} + \int_\Omega C^{ik}C^{jl}\big(B_{ij}\varphi - \varphi_{\alpha^i}w^*_{\alpha^j}\big)\nabla_{kl}\mathbf{K}_{9\kappa 0}d\Omega \\ &+ \int_\Omega \widetilde{T}^{ij}_{\mathrm{p}}w^*_{\alpha^i}\varphi_{\alpha^j}d\Omega.\end{aligned} \quad (18.8)$$

Relation (18.8) defines $\mathbf{G}_{\kappa 0}$ as an element of H_κ.

Taking into account (18.1), by the Riesz theorem we have from (17.18),

$$\begin{aligned}\big(\mathbf{G}_{\kappa 1}(w)\cdot\varphi\big)_{H_\kappa} = \int_\Omega \Big[&C^{ik}C^{jl}\nabla_{kl}\mathbf{K}_{9\kappa 1}(w)\big(B_{ij}\varphi - \varphi_{\alpha^i}w^*_{\alpha^j}\big) \\ &- C^{ik}C^{jl}\nabla_{kl}\mathbf{K}_{9\kappa 0}\varphi_{\alpha^i}w_{\alpha^j} + \widetilde{T}^{ij}_{\mathrm{p}}w_{\alpha^i}\varphi_{\alpha^j}\Big]d\Omega,\end{aligned} \quad (18.9)$$

from which it follows that

$$
\begin{aligned}
\left|\left(\mathbf{G}_{\kappa 1}(w)\cdot\varphi\right)_{H_\kappa}\right| \le{}& \left\|C^{ik}C^{jl}B_{ij}\right\|_{C\Omega} \|\varphi\|_{L_{2\Omega}} \|\nabla_{kl}\mathbf{K}_{9\kappa 1}(w)\|_{L_{2\Omega}} \\
&+ \left\|C^{ik}C^{jl}\right\|_{C\Omega} \Big(\|\nabla_{kl}\mathbf{K}_{9\kappa 1}(w)\|_{L_{2\Omega}} \|\varphi_{\alpha^i}\|_{L_{4\Omega}} \left\|w^*_{\alpha^j}\right\|_{L_{4\Omega}} \\
&\qquad + \|\nabla_{kl}\mathbf{K}_{9\kappa 0}\|_{L_{2\Omega}} \|\varphi_{\alpha^i}\|_{L_{4\Omega}} \left\|w^*_{\alpha^j}\right\|_{L_{4\Omega}} \\
&\qquad + \|\nabla_{kl}\mathbf{K}_{9\kappa 0}\|_{L_{2\Omega}} \|\varphi_{\alpha^i}\|_{L_{4\Omega}} \|w\|_{L_{4\Omega}} \Big) \\
&+ \left\|\widetilde{T}^{ij}_{\mathrm{p}}\right\|_{L_{2\Omega}} \|w_{\alpha^i}\|_{L_{4\Omega}} \|\varphi_{\alpha^j}\|_{L_{4\Omega}} \\
\le{}& m\,\|\varphi\|_{H_\kappa} \|w\|_{H_\kappa} \Big(\sum_{k,l=1}^{2} \left\|C^{ik}C^{jl}B_{ij}\right\|_{C\Omega} \\
&\qquad + \sum_{i,k,l=1}^{2} \left\|C^{ik}C^{jl}\right\|_{C\Omega} \left\|w^*_{\alpha^j}\right\|_{L_{4\Omega}} \\
&\qquad + \sum_{i,j=1}^{2} \left\|C^{ik}C^{jl}\right\|_{C\Omega} \|\nabla_{kl}\mathbf{K}_{9\kappa 0}\|_{L_{2\Omega}} + \left\|\widetilde{T}^{ij}_{\mathrm{p}}\right\|_{L_{2\Omega}} \Big).
\end{aligned}
\tag{18.10}
$$

From (18.10) we deduce

$$
\begin{aligned}
\|\mathbf{G}_{\kappa 1}(w)\|_{H_\kappa} \le m\Big(& \sum_{k,l=1}^{2} \left\|C^{ik}C^{jl}B_{ij}\right\|_{C\Omega} + \sum_{i,k,l=1}^{2} \left\|C^{ik}C^{jl}\right\|_{C\Omega} \left\|w^*_{\alpha^j}\right\|_{L_{4\Omega}} \\
&+ \sum_{i,j=1}^{2} \left\|C^{ik}C^{jl}\right\|_{C\Omega} \|\nabla_{kl}\mathbf{K}_{9\kappa 0}\|_{L_{2\Omega}} \\
&+ \sum_{i,j=1}^{2} \left\|\widetilde{T}^{ij}_{\mathrm{p}}\right\|_{L_{2\Omega}} \Big) \|w\|_{H_\kappa}.
\end{aligned}
$$

Thus (18.4) is established for $\mu = 1$.

For $\mu = 2$ it follows from (17.18), (18.1) that

$$
\begin{aligned}
\left(\mathbf{G}_{\kappa 2}(w)\cdot\varphi\right)_{H_\kappa} = \int_\Omega \Big[& C^{ik}C^{jl}\nabla_{kl}\mathbf{K}_{9\kappa 2}(w)\big(B_{ij}\varphi - \varphi_{\alpha^i} w^*_{\alpha^j}\big) \\
&- C^{ik}C^{jl}\nabla_{kl}\mathbf{K}_{9\kappa 1}(w)\varphi_{\alpha^i} w^*_{\alpha^j}\Big] d\Omega,
\end{aligned}
\tag{18.11}
$$

and from (18.4) we obtain

$$\|\mathbf{G}_{\kappa 2}(w)\|_{H_\kappa} \le m\Big(\big\|C^{ik}C^{jl}B_{ij}\big\|_{C\Omega} \|\nabla_{kl}\mathbf{K}_{9\kappa 2}(w)\|_{L_2\Omega} + \sum_{i=1}^{2} \big\|C^{ik}C^{jl}w^*_{\alpha^j}\big\|_{L_2\Omega} \|\nabla_{kl}\mathbf{K}_{9\kappa 2}(w)\|_{L_2\Omega} + \sum_{i=1}^{2} \big\|C^{ik}C^{jl}\big\|_{C\Omega} \|\mathbf{K}_{9\kappa 1}(w)\|_{L_2\Omega} \|w_{\alpha^j}\|_{L_4\Omega}\Big)$$

$$\le m \sum_{k,l=1}^{2} \Big(\big\|C^{ik}C^{jl}B_{ij}\big\|_{C\Omega} + \sum_{i=1}^{2} \big\|C^{ik}C^{jl}w^*_{\alpha^j}\big\|_{L_4\Omega} + \sum_{i,j=1}^{2} \big\|C^{ik}C^{jl}\big\|_{C\Omega}\Big) \|w\|^2_{H_\kappa},$$

so that (18.4) has been established for $\mu = 2$. Finally, for $\mu = 3$ it follows from (17.18), (18.1) that

$$\big(\mathbf{G}_{\kappa 3}(w)\cdot\varphi\big)_{H_\kappa} = -\int_\Omega C^{ik}C^{jl}\nabla_{kl}\mathbf{K}_{9\kappa 2}(w)\varphi_{\alpha^i} w_{\alpha^j}\, d\Omega, \tag{18.12}$$

and therefore

$$\|\mathbf{G}_{\kappa 3}(w)\|_{H_\kappa} \le m \sum_{i,j=1}^{2} \big\|C^{ik}C^{jl}\big\|_{C\Omega} \|\nabla_{kl}\mathbf{K}_{9\kappa 2}(w)\|_{L_2\Omega} \|w\|_{L_4\Omega} \le m \sum_{i,j,k,l=1}^{2} \big\|C^{ik}C^{jl}\big\|_{C\Omega} \|w\|^3_{H_\kappa}.$$

Lemma 18.1 is completely proved.

Theorem 18.1. *Each of the operators $\mathbf{K}_{9\kappa}(w)$, $\mathbf{K}_{9\kappa\mu}(w)$ is a completely continuous operator from H_κ into H_9.*

Let

$$w_n \rightharpoonup w \text{ in } H_\kappa.$$

From (18.6) we have

$$\big((\mathbf{K}_{9\kappa 1}(w) - \mathbf{K}_{9\kappa 1}(w_n))\cdot\theta\big)_{H_9} = \int_\Omega C^{ik}C^{jl}B_{ij}(w - w_n)\nabla_{kl}\theta\, d\Omega + \int_\Omega C^{ik}C^{jl}w^*_{\alpha^j}\nabla_{kl}\theta(w_{n\alpha^j} - w_{\alpha^j})\, d\Omega,$$

whence

$$\begin{aligned}
&\left|\left((\mathbf{K}_{9\kappa 1}(w) - \mathbf{K}_{9\kappa 1}(w_n)) \cdot \theta\right)_{H_9}\right| \\
&\quad \le \left| \int_\Omega C^{ik} C^{jl} B_{ij}(w - w_n) \nabla_{kl} \theta d\Omega \right| \\
&\quad + \left| \int_\Omega C^{ik} C^{jl} w^*_{\alpha^j}(w_{n\alpha^j} - w_{\alpha^j}) \nabla_{kl} \theta d\Omega \right| \qquad (18.13)\\
&\quad \le m \sum_{k,l=1}^{2} \left(\left\| C^{ik} C^{jl} B_{ij} \right\|_{C_\Omega} \times \| w - w_n \|_{L_{2\Omega}} \right. \\
&\qquad \left. + \left\| C^{ik} C^{jl} \right\|_{C_\Omega} \left\| w^*_{\alpha^j} \right\|_{L_{4\Omega}} \| w_{n\alpha^j} - w_{\alpha^j} \|_{L_{4\Omega}} \right) \| \theta \|_{H_9} .
\end{aligned}$$

By complete continuity of the embedding operator (Theorem 12.3, see relation (12.29)), we conclude that as $n \to \infty$ the coefficient at $\|\theta\|_{H_9}$ on the right-hand side of (18.13) vanishes. Therefore,

$$\|\mathbf{K}_{9\kappa 1}(w) - \mathbf{K}_{9\kappa 1}(w_n)\|_{H_9} \to 0. \qquad (18.14)$$

Thus we have proved the assertion of Theorem 18.1 as it relates to $\mathbf{K}_{9\kappa 1}$.

For $\mathbf{K}_{9\kappa 2}(w)$ it follows from (18.7) that

$$\left((\mathbf{K}_{9\kappa 2}(w) - \mathbf{K}_{9\kappa 2}(w_n)) \cdot \theta\right)_{H_9} = \frac{1}{2} \int_\Omega C^{ik} C^{jl} (w_{n\alpha^i} w_{n\alpha^j} - w_{\alpha^i} w_{\alpha^j}) \nabla_{kl} \theta d\Omega,$$

whence

$$\|\mathbf{K}_{9\kappa 2}(w) - \mathbf{K}_{9\kappa 2}(w_n)\|_{H_9} \le m \sum_{k,l=1}^{2} \left\| C^{ik} C^{jl} \right\|_{C_\Omega} \| w_{n\alpha^i} w_{n\alpha^j} - w_{\alpha^i} w_{\alpha^j} \|_{L_{2\Omega}} .$$

Next, by the same embedding Theorem 12.3 (relations (12.29)) we have

$$w_{n\alpha^i} \to w_{\alpha^i} \text{ in any } L_{q\Omega},\ q \ge 1,$$

from which we conclude that

$$\|\mathbf{K}_{9\kappa 2}(w) - \mathbf{K}_{9\kappa 2}(w_n)\|_{H_9} \to 0, \qquad (18.15)$$

which concludes the proof of Theorem 18.1 for $\mathbf{K}_{9\kappa 2}(w)$.

Theorem 18.1 is proved.

Theorem 18.2. *All the operators* $\mathbf{G}_{\kappa\mu}$, $\mathbf{G}_\kappa$, $\mu = 1, 2, 3$, *are completely continuous operators on* H_κ.

To prove the theorem, we consider the relation (18.9), which defines the operator $\mathbf{G}_{\kappa 1}(w)$. We have

$$\begin{aligned}
&\left((\mathbf{G}_{\kappa 1}(w) - \mathbf{G}_{\kappa 1}(w_n)) \cdot \varphi\right)_{H_\kappa} \\
&= \int_\Omega \Big[C^{ik} C^{jl} \nabla_{kl} \big(\mathbf{K}_{9\kappa 1}(w) - \mathbf{K}_{9\kappa 1}(w_n)\big)\Big(B_{ij}\varphi - \varphi_{\alpha^i} w^*_{\alpha^j}\Big) \qquad (18.16) \\
&\quad - \big(C^{ik}C^{jl}\nabla_{kl}\mathbf{K}_{9\kappa 0} - \widetilde{T}^{ij}_{\mathrm{p}}\big)\varphi_{\alpha^i}\big(w_{\alpha^j} - w_{n\alpha^j}\big)\Big] d\Omega,
\end{aligned}$$

whence

$$\begin{aligned}
&\|\mathbf{G}_{\kappa 1}(w) - \mathbf{G}_{\kappa 1}(w_n)\|_{H_\kappa} \\
&\le m\Big[\Big\| \nabla_{kl}\big(\mathbf{K}_{9\kappa 1}(w) - \mathbf{K}_{9\kappa 1}(w_n)\big)\Big\|_{L_{2\Omega}} \\
&\quad \times \Big(\big\|C^{ik}C^{jl}B_{ij}\big\|_{C_\Omega} + \sum_{i=1}^{2} \|C^{ik}C^{jl} w^*_{\alpha^j}\|_{L_{4\Omega}} \Big) \\
&\quad + \Big\| C^{ik}C^{jl}\nabla_{kl}\mathbf{K}_{9\kappa 0} + \widetilde{T}^{ij}_{\mathrm{p}} \Big\|_{L_{2\Omega}} \cdot \sum_{i=1}^{2} \|w_{\alpha^j} - w_{n\alpha^j}\|_{L_{4\Omega}} \Big].
\end{aligned}$$

By the embedding Theorem 12.3 we have complete continuity of the embedding operator (see (12.29)), whence we conclude that the right-hand side of (18.16) vanishes as $n \to \infty$, and strong continuity of $\mathbf{G}_{\kappa 1}(w)$ is proved.

For our arguments below, let us consider (18.11), which defines $\mathbf{G}_{\kappa 2}(w)$. We have

$$\begin{aligned}
&\left((\mathbf{G}_{\kappa 2}(w) - \mathbf{G}_{\kappa 2}(w_n)) \cdot \varphi\right)_{H_\kappa} \\
&= \int_\Omega C^{ik} C^{jl} \Big[\nabla_{kl}\big(\mathbf{K}_{9\kappa 2}(w) - \mathbf{K}_{9\kappa 2}(w_n)\big)\Big(B_{ij}\varphi - \varphi_{\alpha^i} w^*_{\alpha^j}\Big) \\
&\quad - \nabla_{kl}\big(\mathbf{K}_{9\kappa 1}(w) w_{\alpha^j} - \mathbf{K}_{9\kappa 1}(w_n) w_{n\alpha^j}\big)\varphi_{\alpha^i}\Big] d\Omega,
\end{aligned}$$

from which

$$\begin{aligned}
&\|\mathbf{G}_{\kappa 2}(w) - \mathbf{G}_{\kappa 2}(w_n)\|_{H_\kappa} \\
&\le m\Big[\Big\| \nabla_{kl}\big(\mathbf{K}_{9\kappa 2}(w) - \mathbf{K}_{9\kappa 2}(w_n)\big)\Big\|_{L_{2\Omega}} \\
&\quad \times \Big(\big\|C^{ik}C^{jl}B_{ij}\big\|_{C_\Omega} + \sum_{i=1}^{2} \|C^{ik}C^{jl} w^*_{\alpha^j}\|_{L_{4\Omega}} \Big) \\
&\quad + \sum_{i,k,l=1}^{2} \|\mathbf{K}_{9\kappa 1}(w) w_{\alpha^j} - \mathbf{K}_{9\kappa 1}(w_n) w_{n\alpha^j}\|_{L_{2\Omega}} \big\|C^{ik}C^{jl}\big\|_{C_\Omega} \Big].
\end{aligned} \tag{18.17}$$

Since (18.14), (18.15) hold, and by the embedding Theorem 12.3, we have $w_{n\alpha^j} \to w_{\alpha^j}$ in any space $L_{q\Omega}$, $q \ge 1$, the right-hand side of (18.17) vanishes as

$n \to \infty$, so that

$$\|\mathbf{G}_{\kappa 2}(w) - \mathbf{G}_{\kappa 2}(w_n)\|_{H_\kappa} \to 0, \text{ as } n \to \infty.$$

Finally, from (18.12) we obtain

$$\begin{aligned}&\left((\mathbf{G}_{\kappa 3}(w) - \mathbf{G}_{\kappa 3}(w_n)) \cdot \varphi\right)_{H_\kappa}\\ &= \int_\Omega C^{ik} C^{jl} \left[\nabla_{kl} \mathbf{K}_{9\kappa 2}(w) w_{\alpha^j} - \nabla_{kl} \mathbf{K}_{9\kappa 2}(w_n) w_{n\alpha^j}\right] \varphi_{\alpha^i} \, d\Omega,\end{aligned}$$

from which we have

$$\begin{aligned}&\|\mathbf{G}_{\kappa 3}(w) - \mathbf{G}_{\kappa 3}(w_n)\|_{H_\kappa}\\ &\leq m \sum_{i=1}^{2} \|C^{ik} C^{jl}\|_{C_\Omega} \|\nabla_{kl} \mathbf{K}_{9\kappa 2}(w) w_{\alpha^j} - \nabla_{kl} \mathbf{K}_{9\kappa 2}(w_n) w_{n\alpha^j}\|_{L_{2\Omega}} .\end{aligned} \tag{18.18}$$

Taking into account (18.15) and the embedding Theorem 12.3, we conclude that the right-hand side of (18.18) vanishes as $n \to \infty$, and therefore

$$\|\mathbf{G}_{\kappa 3}(w) - \mathbf{G}_{\kappa 3}(w_n)\|_{H_\kappa} \to 0, \text{ as } n \to \infty.$$

Theorem 18.2 is proved.

19. Computation of the Winding Number of the Vector Field $w - \mathbf{G}_\kappa(w)$ on Spheres of Large Radius in H_κ. Solvability of the Main Boundary Value Problems of the Theory of Shallow Shells with an Airy Stress Function

19.1. We shall follow the plan proposed in the introduction of Section 17: First we construct the functional Φ (19.1), analyze some of its properties, and show in particular that it is increasing. This will allow us to claim that on spheres of sufficiently large radius in the energy space the equation $w - t\mathbf{G}_\kappa(w) = 0$ has no solutions for all $0 \leq t \leq 1$. This fact allows us to claim that the winding number of the vector field $w - \mathbf{G}_\kappa(w)$ of spheres of sufficiently large radius is $+1$ and that there is a solution of the equation $w = \mathbf{G}_\kappa(w)$ of bounded norm.

Let us consider the functional $\Phi(w, t)$ defined in $H_\kappa \times [0, 1]$ by the relation

$$\Phi(w, t) = ((w - t\mathbf{G}_\kappa(w)) \cdot w)_{H_\kappa} = \|w\|^2_{H_\kappa} - t(\mathbf{G}_\kappa(w) \cdot w)_{H_\kappa}. \tag{19.1}$$

Lemma 19.1. *The functional $\Phi(w,\ t)$ can be represented in the form*

$$\Phi(w,t) = \|w\|^2_{H_\kappa} + t\langle 2\,\|\Psi\|^2_{H_9} + (\overset{0}{w}\cdot w)_{H_\kappa}$$
$$+ \int_\Omega \{C^{ik}C^{jl}[-B_{ij}(w+2w^*) + w^*_{\alpha^i}w^*_{\alpha^j} + w^*_{\alpha^i}w_{\alpha^j} + w_{\alpha^i}w^*_{\alpha^j}]$$
$$+ 2C^{kl}_{*ij}\widetilde{T}^{ij}_{\mathrm{p}}\}\nabla_{kl}\Psi d\Omega - \int_\Omega \widetilde{T}^{ij}_{\mathrm{p}}(w+w^*)_{\alpha^i}w_{\alpha^j}d\Omega\rangle. \tag{19.2}$$

To prove Lemma 19.1, we set $\varphi = w$ in equation (17.18), which defines $\mathbf{G}_\kappa$. Then we obtain

$$\Phi(w,t) = \|w\|^2_{H_\kappa} - t\Big\{ -(\overset{0}{w}\cdot w)_{H_\kappa}$$
$$+ \int_\Omega C^{ik}C^{jl}[B_{ij}w - w_{\alpha^i}(w+w^*)_{\alpha^j}]\nabla_{kl}\Psi d\Omega \tag{19.3}$$
$$- \int_\Omega \widetilde{T}^{ij}_{\mathrm{p}}(w+w^*)_{\alpha^i}w_{\alpha^j}d\Omega\Big\}.$$

Next, in equation (17.15), which defines Ψ, we set $\theta = \Psi$:

$$\|\Psi\|^2_{H_9} = \int_\Omega C^{ik}C^{jl}\Big[\frac{1}{2}(w+w^*)_{\alpha^i}(w+w^*)_{\alpha^j} - B_{ij}(w+w^*)\Big]\nabla_{kl}\Psi d\Omega$$
$$- \int_\Omega C^{\lambda\mu}_{*ij}\widetilde{T}^{ij}_{\mathrm{p}}\nabla_{\lambda\mu}\Psi d\Omega. \tag{19.4}$$

From (19.4) it follows that

$$\int_\Omega C^{ik}C^{jl}w_{\alpha^i}w_{\alpha^j}\nabla_{kl}\Psi d\Omega$$
$$= 2\Big\{\|\Psi\|^2_{H_9} - \frac{1}{2}\int_\Omega C^{ik}C^{jl}[w^*_{\alpha^i}(w+w^*)_{\alpha^j} + w^*_{\alpha^j}(w+w^*)_{\alpha^i} \tag{19.5}$$
$$- 2B_{ij}(w+w^*)]\nabla_{kl}\Psi d\Omega + \int_\Omega C^{\lambda\mu}_{*ij}\widetilde{T}^{ij}_{\mathrm{p}}\nabla_{\lambda\mu}\Psi d\Omega\Big\}.$$

Substituting (19.5) into (19.3), we obtain (19.2). Lemma 19.1 is proved.

19.2. Let us consider the functional $\Pi(w) = \|\Psi\|^2_{H_9}$. It is easily seen that by (18.1) we have the representation

$$\Pi(w) = \|\Psi\|^2_{H_9} = \|\mathbf{K}_{9\kappa0} + \mathbf{K}_{9\kappa1} + \mathbf{K}_{9\kappa2}\|^2_{H_9} = \sum_{\mu=0}^{4}\Pi_\mu(w), \tag{19.6}$$
$$\Pi_0(w) = \|\mathbf{K}_{9\kappa0}\|^2_{H_9}, \quad \Pi_1(w) = 2(\mathbf{K}_{9\kappa0}\cdot\mathbf{K}_{9\kappa1})_{H_9},$$
$$\Pi_2(w) = \|\mathbf{K}_{9\kappa2}\|^2_{H_9} + 2(\mathbf{K}_{9\kappa0}\cdot\mathbf{K}_{9\kappa2})_{H_9},$$

$$\Pi_3(w) = 2\big(\mathbf{K}_{9\kappa 1} \cdot \mathbf{K}_{9\kappa 2}\big)_{H_9}, \quad \Pi_4(w) = \|\mathbf{K}_{9\kappa 2}\|^2_{H_9}, \tag{19.7}$$

where $\Pi_\mu(w)$ are homogeneous functionals of order μ in w.

Lemma 19.2. *All the functionals $\Pi_\mu(w)$ are weakly continuous in H_κ.*

To prove the claim, let us consider, for example, the functional $\Pi_2(w)$, and let $w_n \rightharpoonup w$. By Theorem 18.1, concerning complete continuity of $\mathbf{K}_{9\kappa\mu}$, we have

$$\big\|\mathbf{K}_{9\kappa\mu}(w) - \mathbf{K}_{9\kappa\mu}(w_n)\big\|_{H_9} \to 0, \qquad \mu = 1, 2,$$

while from (19.7) we have

$$\Pi_2(w_n) \to \Pi_2(w),$$

which indeed proves Lemma 19.2 for $\Pi_2(w)$. The proofs for all the other $\Pi_\mu(w)$ are completely analogous.

Corollary. For all $\Pi_\mu(w)$ we have the inequality

$$\big|\Pi_\mu(w)\big| \le m\, \|w\|^\mu_{H_\kappa}, \qquad \mu = 0,\ 1,\ 2,\ 3,\ 4. \tag{19.8}$$

Lemma 19.3. *The relation $\Pi_4(w) = 0$ under Conditions 1, 5, and 8 of Section 7 is possible if and only if*

$$w \equiv 0. \tag{19.9}$$

Let us note that if $\Pi_4(w) = 0$, then from (19.7) it follows that

$$\mathbf{K}_{9\kappa 2}(w) \equiv 0,$$

while from (18.7), which defines $\mathbf{K}_{9\kappa 2}$, we find that

$$\int_\Omega C^{ik} C^{jl} w_{\alpha^i} w_{\alpha^j} \nabla_{kl}\theta d\Omega = 0, \tag{19.10}$$

for any $\theta \in H_9$. From (17.35) and (19.10) we obtain

$$\int_\Omega C^{ik} C^{jl} \nabla_{kl} w \nabla_{ij} w \theta d\Omega = 0,$$

so that

$$\nabla_{11} w \nabla_{22} w - \nabla^2_{12} w = 0. \tag{19.11}$$

Since we are considering developable shallow shells, there are generalized coordinates in which (19.11) can be written in the form

$$w_{\alpha^1\alpha^1} w_{\alpha^2\alpha^2} - w^2_{\alpha^1\alpha^2} = 0. \tag{19.12}$$

Furthermore, after the change of variables (17.13), $w \in H_\kappa$, but under the condition (7.8) we have that on all of Γ,

$$w|_\Gamma = 0. \tag{19.13}$$

By (19.12), (19.13) we are under the conditions of Lemma 15.4, from which (19.9) follows. Lemma 19.3 has been proved.

19.3. As in Section 15, on the sphere $\Sigma(1, 0)$ in the space H_κ we introduce the set $\Sigma'(1, 0)$ of functions that satisfy the inequality

$$\|v\|^2_{H_\kappa} - \sum_{k,l=1}^{2} \frac{3}{2} m \left\| C^{ik} C^{jl} \Big(- B_{ij} v + v_{\alpha^i} w^*_{\alpha^j} + v_{\alpha^j} w^*_{\alpha^i} \Big) \right\|^2_{L_2\Omega}$$
$$- \int_\Omega \widetilde{T}^{ij}_{\mathrm{p}} v_{\alpha^i} v_{\alpha^j} d\Omega \leq \frac{1}{2}.$$

Furthermore, let $\overline{\Sigma}'(1, 0)$ be the weak closure of $\Sigma'(1, 0)$ in H_κ. We denote by $\Sigma''(1, 0)$ the complement of $\Sigma'(1, 0)$ in the sphere $\Sigma(1, 0)$. Thus

$$\Sigma(1, 0) = \Sigma'(1, 0) \bigcup \Sigma''(1, 0).$$

Clearly, on $\Sigma''(1, 0)$ we have

$$\|v\|^2_{H_\kappa} - \frac{3}{2} m \sum_{k,l=1}^{2} \left\| C^{ik} C^{jl} \Big(- B_{ij} v + v_{\alpha^i} w^*_{\alpha^j} + v_{\alpha^j} w^*_{\alpha^i} \Big) \right\|^2_{L_2\Omega}$$
$$- \int_\Omega \widetilde{T}^{ij}_{\mathrm{p}} v_{\alpha^i} v_{\alpha^j} d\Omega > \frac{1}{2}. \tag{19.14}$$

In accordance with this decomposition of $\Sigma(1, 0)$, every sphere $\Sigma(R, 0)$ in H_κ is decomposed into two parts, $\Sigma'(R, 0)$ and $\Sigma''(R, 0)$, central projections $w \mapsto Rw$ of $\Sigma'(1, 0)$ and $\Sigma''(1, 0)$ from $\Sigma(1, 0)$ onto $\Sigma(R, 0)$, respectively.

Lemma 19.4. *The set $\overline{\Sigma}'(1, 0)$ does not contain zero.*

Lemma 19.4 is proved in the same way as Lemma 15.1.

Lemma 19.5. *On $\Sigma'(R, 0)$ we have the inequality*

$$\Pi_4(w) \geq m R^4. \tag{19.15}$$

To prove the claim, we first establish the following inequality on $\Sigma'(1, 0)$:

$$\Pi_4(w) \geq m > 0. \tag{19.16}$$

Indeed, if (19.16) did not hold, then there would exist a sequence $w_n \in \Sigma'(1, 0)$ such that

$$\Pi_4(w_n) \to 0.$$

Here we can take w_n to converge weakly to w_0 in H_κ:

$$w_n \rightharpoonup w_0,$$

and by weak continuity of $\Pi_4(w)$ we have that

$$\Pi_4(w_0) = 0.$$

From Lemma 19.3 we have that

$$w_0 \equiv 0.$$

However, this is impossible, since $w_0 \in \overline{\Sigma}'(1, 0)$, and by Lemma 19.4, $\overline{\Sigma}'(1, 0)$ does not contain zero. Therefore, (19.16) has been established. But then from the homogeneity of $\Pi_4(w)$, (19.15) follows, and Lemma 19.5 has been proved.

19.4.

Theorem 19.1. *Let all the Conditions 1–8 of Section 17 be satisfied. Then on spheres of sufficiently large radius R in H_κ we have the inequality*

$$\Phi(w, t) \geq \frac{1}{3}R^2, \qquad 0 \leq t \leq 1. \tag{19.17}$$

To prove (19.17), we shall use inequalities that follow from the elementary inequality

$$|ab| \leq \frac{\epsilon}{2}a^2 + \frac{1}{2\epsilon}b^2,$$

which holds for all a, b, and $\epsilon > 0$ as well as by the Cauchy–Bunyakovskii–Schwarz inequality and the inequality (12.34):

$$\begin{aligned}
&\left|\int_\Omega C^{ik}C^{jl}\Big(- B_{ij}w + w^*_{\alpha^i} w_{\alpha^j} + w^*_{\alpha^j} w_{\alpha^i}\Big)\nabla_{kl}\Psi d\Omega\right| \\
&\quad \leq \epsilon m \,\|\Psi\|^2_{H_9} + \frac{1}{2\epsilon}\sum_{k,l=1}^{2} \left\|C^{ik}C^{jl}\Big(- B_{ij}w + w_{\alpha^i} w^*_{\alpha^j} + w_{\alpha^j} w^*_{\alpha^i}\Big)\right\|^2_{L_{2\Omega}} \\
&\quad \leq \epsilon m \,\|\Psi\|^2_{H_9} + \frac{m}{\epsilon}\,\|w\|^2_{H_\kappa},
\end{aligned} \tag{19.18}$$

$$\begin{aligned}
&\left|\int_\Omega C^{ik}C^{jl}\Big(- 2B_{ij}w^* + w^*_{\alpha^i} w^*_{\alpha^j}\Big)\nabla_{kl}\Psi d\Omega\right| \\
&\quad \leq \frac{\epsilon}{2}\sum_{k,l=1}^{2} \|\nabla_{kl}\Psi\|^2_{L_{2\Omega}} + \frac{1}{2\epsilon}\sum_{k,l=1}^{2} \left\|C^{ik}C^{jl}\Big(- 2B_{ij}w^* + w^*_{\alpha^i} w^*_{\alpha^j})\right\|^2_{L_{2\Omega}} \\
&\quad \leq \epsilon m \,\|\Psi\|^2_{H_9} + \frac{m}{\epsilon},
\end{aligned} \tag{19.19}$$

$$\big|(\overset{0}{w} \cdot w)\big|_{H_\kappa} \leq \left\|\overset{0}{w}\right\|_{H_\kappa} \|w\|_{H_\kappa}, \tag{19.20}$$

$$\begin{aligned}
&2\left|\int_\Omega C^{kl}_{*ij}\widetilde{T}^{ij}_{\text{p}}\nabla_{kl}\Psi d\Omega\right| \leq \|C^{kl}_{*ij}\|_{C_\Omega} \sum_{k,l=1}^{2}\Big(\frac{1}{\epsilon}\left\|\widetilde{T}^{ij}_{\text{p}}\right\|^2_{L_{2\Omega}} + \epsilon\,\|\nabla_{kl}\Psi\|^2_{L_{2\Omega}}\Big) \\
&\quad \leq m\Big(\frac{1}{\epsilon}\sum_{i,j=1}^{2}\left\|\widetilde{T}^{ij}_{\text{p}}\right\|^2_{L_{2\Omega}} + \epsilon\,\|\Psi\|^2_{H_9}\Big),
\end{aligned} \tag{19.21}$$

$$\begin{aligned}
&\left|\int_\Omega \widetilde{T}^{ij}_{\text{p}} w^*_{\alpha^i} w_{\alpha^j} d\Omega\right| \leq \left\|\widetilde{T}^{ij}_{\text{p}}\right\|_{L_{2\Omega}} \left\|w^*_{\alpha^i}\right\|_{L_{4\Omega}} \|w_{\alpha^j}\|_{L_{4\Omega}} \\
&\quad \leq m\sum_{i,j=1}^{2}\left\|\widetilde{T}^{ij}_{\text{p}}\right\|_{L_{2\Omega}} \|w\|_{H_\kappa} \left\|w^*\right\|_{H_\kappa},
\end{aligned} \tag{19.22}$$

$$\left|\int_\Omega \widetilde{T}^{ij}_{\mathrm{p}} w_{\alpha^i} w_{\alpha^j} d\Omega\right| \le \left\|\widetilde{T}^{ij}_{\mathrm{p}}\right\|_{L_{2\Omega}} \|w_{\alpha^i}\|_{L_{4\Omega}} \|w_{\alpha^j}\|_{L_{4\Omega}}$$

$$\le m \sum_{i,j=1}^{2} \left\|\widetilde{T}^{ij}_{\mathrm{p}}\right\|_{L_{2\Omega}} \|w\|^2_{H_\kappa}. \tag{19.23}$$

Finally, from (19.2), (19.18)–(19.23) it follows that

$$\begin{aligned}
\Phi(w,t) \ge{}& \|w\|^2_{H_\kappa} \\
&+ t\Bigg\{2\|\Psi\|^2_{H_\kappa} - \left\|\overset{0}{w}\right\|_{H_\kappa} \|w\|_{H_\kappa} \\
&\quad - \left|\int_\Omega C^{ik}C^{jl}(-B_{ij}w + w^*_{\alpha^i} w_{\alpha^j} + w^*_{\alpha^j} w_{\alpha^i})\nabla\Psi d\Omega\right| \\
&\quad - 2\left|\int_\Omega C^{kl}_{*ij} \widetilde{T}^{ij}_{\mathrm{p}} \nabla_{kl}\Psi d\Omega\right| \\
&\quad - \left|\int_\Omega C^{ik}C^{jl}(-2B_{ij}w^* + w^*_{\alpha^i} w^*_{\alpha^j})\nabla_{kl}\Psi d\Omega\right| \\
&\quad - \left|\int_\Omega \widetilde{T}^{ij}_{\mathrm{p}} w_{\alpha^i} w_{\alpha^j} d\Omega\right| - \left|\int_\Omega \widetilde{T}^{ij}_{\mathrm{p}} w^*_{\alpha^i} w_{\alpha^j} d\Omega\right|\Bigg\} \\
\ge{}& \|w\|^2_{H_\kappa} + t\Bigg\{2\|\Psi\|^2_{H_9} - \left\|\overset{0}{w}\right\|_{H_\kappa} \|w\|_{H_\kappa} - \epsilon m\|\Psi\|^2_{H_9} \\
&\quad - \frac{1}{2\epsilon}\sum_{k,l=1}^{2} \|C^{ik}C^{jl}(-B_{ij}w + w_{\alpha^i} w^*_{\alpha^j} + w_{\alpha^j} w^*_{\alpha^i})\|^2_{L_{2\Omega}} \\
&\quad - m\Big(\frac{1}{\epsilon}\sum_{i,j=1}^{2} \left\|\widetilde{T}^{ij}_{\mathrm{p}}\right\|^2_{L_{2\Omega}} + \epsilon\|\Psi\|^2_{H_9}\Big) - \epsilon m\|\Psi\|^2_{H_9} - \frac{m}{\epsilon} \\
&\quad - \left|\int_\Omega \widetilde{T}^{ij}_{\mathrm{p}} w_{\alpha^i} w_{\alpha^j} d\Omega\right| - \left|\int_\Omega \widetilde{T}^{ij}_{\mathrm{p}} w^*_{\alpha^i} w_{\alpha^j} d\Omega\right|\Bigg\} \\
={}& \|w\|^2_{H_\kappa} + t\Bigg\{(2 - 3\epsilon m)\|\Psi\|^2_{H_9} - \left\|\overset{0}{w}\right\|_{H_\kappa} \|w\|_{H_\kappa} \\
&\quad - \frac{1}{2\epsilon}\sum_{k,l=1}^{2} \left\|C^{ik}C^{jl}(-B_{ij}w + w_{\alpha^i} w^*_{\alpha^j} + w_{\alpha^j} w^*_{\alpha^i})\right\|^2_{L_{2\Omega}} \\
&\quad - \frac{m}{\epsilon}\sum_{i,j=1}^{2} \left\|\widetilde{T}^{ij}_{\mathrm{p}}\right\|^2_{L_{2\Omega}} \\
&\quad - \frac{m}{\epsilon} - \left|\int_\Omega \widetilde{T}^{ij}_{\mathrm{p}} w_{\alpha^i} w_{\alpha^j} d\Omega\right| - \left|\int_\Omega \widetilde{T}^{ij}_{\mathrm{p}} w^*_{\alpha^i} w_{\alpha^j} d\Omega\right|\Bigg\}.
\end{aligned} \tag{19.24}$$

For ϵ we take the relation

$$2 - 3\epsilon m = 1, \text{ so } \epsilon = \frac{1}{3m}. \tag{19.25}$$

Substituting (19.25) into (19.24), we obtain

$$\begin{aligned}\Phi(w,t) \geq \|w\|_{H_\kappa}^2 + t\Big\{ &\|\Psi\|_{H_9}^2 - \left\|\overset{0}{w}\right\|_{H_\kappa} \|w\|_{H_\kappa} \\ &- \frac{3m}{2}\sum_{k,l=1}^{2} \|C^{ik}C^{jl}(-B_{ij}w + w_{\alpha^i}w_{\alpha^j}^* + w_{\alpha^j}w_{\alpha^i}^*)\|_{L_{2\Omega}}^2 \\ &- 3m^2 \sum_{i,j=1}^{2} \left\|\widetilde{T}_{\mathrm{p}}^{ij}\right\|_{L_{2\Omega}}^2 - 3m^2 - \left|\int_\Omega \widetilde{T}_{\mathrm{p}}^{ij} w_{\alpha^i} w_{\alpha^j}\, d\Omega\right| \\ &- \left|\int_\Omega \widetilde{T}_{\mathrm{p}}^{ij} w_{\alpha^i}^* w_{\alpha^j}\, d\Omega\right|\Big\}.\end{aligned} \tag{19.26}$$

Finally, from (19.18), (19.22), (19.23), and (19.26) we have

$$\begin{aligned}\Phi(w,t) \geq \|w\|_{H_\kappa}^2 + t\Big\{ &\|\Psi\|_{H_9}^2 - \left\|\overset{0}{w}\right\|_{H_\kappa} \|w\|_{H_\kappa} - 3m^2 \|w\|_{H_\kappa}^2 \\ &- 3m^2 \sum_{i,j=1}^{2} \left\|\widetilde{T}_{\mathrm{p}}^{ij}\right\|_{L_{2\Omega}}^2 - 3m^2 - m \sum_{i,j=1}^{2} \left\|\widetilde{T}_{\mathrm{p}}^{ij}\right\| \|w\|_{H_\kappa} \left(\|w\|_{H_\kappa} + \|w^*\|_{H_\kappa}\right)\Big\}.\end{aligned} \tag{19.27}$$

Let us consider $\Phi(w,t)$ on $\Sigma'(R, 0)$. From (19.27) we have

$$\begin{aligned}\Phi(w,t) \geq \|w\|_{H_\kappa}^2 + t\Big\{ &\Pi_4(w) - \sum_{\mu=0}^{3} \Pi_\mu(w) - \left\|\overset{0}{w}\right\|_{H_\kappa} \|w\|_{H_\kappa} \\ &- 3m\|w\|_{H_\kappa}^2 - 3m^2 \sum_{i,j=1}^{2} \left\|\widetilde{T}_{\mathrm{p}}^{ij}\right\|_{L_{2\Omega}}^2 - 3m^2 \\ &- m \sum_{i,j=1}^{2} \left\|\widetilde{T}_{\mathrm{p}}^{ij}\right\|_{L_{2\Omega}} \|w\|_{H_\kappa} \left(\|w\|_{H_\kappa} + \|w^*\|_{H_\kappa}\right)\Big\},\end{aligned}$$

and taking into account (19.8), (19.15), we obtain from (19.27),

$$\begin{aligned}\Phi(w,t) \geq R^2 + t\Big\{ &mR^4 - m(1 + R + R^2 + R^3) - \left\|\overset{0}{w}\right\|_{H_\kappa} R - 3m^2R^2 \\ &- 3m^2 \sum_{i,j=1}^{2} \left\|\widetilde{T}_{\mathrm{p}}^{ij}\right\|_{L_{2\Omega}}^2 - 3m^2 - m \sum_{i,j=1}^{2} \left\|\widetilde{T}_{\mathrm{p}}^{ij}\right\|_{L_{2\Omega}} R\left(\|w^*\|_{H_\kappa} + R\right)\Big\}.\end{aligned} \tag{19.28}$$

For sufficiently large R the expression in the curly braces on the right-hand side of (19.28) will be positive, and then

$$\Phi(w,t) \geq R^2.$$

Thus, (19.17) has been proved for $w \in \Sigma'(R, 0)$.

From (19.26), we deduce for points $w \in \Sigma''(R, 0)$ that

$$\begin{aligned}\Phi(w,t) \geq \|w\|_{H_\kappa}^2 &- \frac{3m}{2}\sum_{k,l=1}^{2}\left\|C^{ik}C^{jl}\left(-B_{ij}w + w_{\alpha^i}w_{\alpha^j}^* + w_{\alpha^j}w_{\alpha^i}^*\right)\right\|_{L_{2\Omega}}^2 \\ &- \left|\int_\Omega \widetilde{T}_{\mathrm{p}}^{ij} w_{\alpha^i} w_{\alpha^j}\, d\Omega\right| - \left\|\overset{0}{w}\right\|_{H_\kappa} \|w\|_{H_\kappa} - 3m^2 \sum_{i,j=1}^{2}\left\|\widetilde{T}_{\mathrm{p}}^{ij}\right\|_{L_{2\Omega}} \\ &- 3m^2 - \left|\int_\Omega \widetilde{T}_{\mathrm{p}}^{ij} w_{\alpha^i}^* w_{\alpha^j}\, d\Omega\right|.\end{aligned} \tag{19.29}$$

Furthermore, taking into account the fact that $\Sigma''(R, 0)$ is the central projection of $\Sigma''(1, 0)$ onto $\Sigma(R, 0)$ and the relation (19.14), which defines $\Sigma''(1, 0)$, we have

$$\begin{aligned}\|w\|_{H_\kappa}^2 &- \frac{3m}{2}\sum_{k,l=1}^{2}\left\|C^{ik}C^{jl}\left(-B_{ij}w + w_{\alpha^i}w_{\alpha^j}^* + w_{\alpha^j}w_{\alpha^i}^*\right)\right\|_{L_{2\Omega}}^2 \\ &- \left|\int_\Omega \widetilde{T}_{\mathrm{p}}^{ij} w_{\alpha^i} w_{\alpha^j}\, d\Omega\right| > \frac{R^2}{2},\end{aligned} \tag{19.30}$$

and then from (19.29), (19.30), (19.22) it follows that

$$\begin{aligned}\Phi(w,t) \geq \frac{R^2}{2} &- \left\|\overset{0}{w}\right\|_{H_\kappa} R - 3m^2 \sum_{i,j=1}^{2}\left\|\widetilde{T}_{\mathrm{p}}^{ij}\right\|_{L_{2\Omega}} \\ &- 3m^2 - m\left\|w^*\right\|_{H_\kappa} \sum_{i,j=1}^{2}\left\|\widetilde{T}_{\mathrm{p}}^{ij}\right\|_{L_{2\Omega}} R,\end{aligned} \tag{19.31}$$

whence we have (19.17) for large R. Theorem 19.1 has been completely proved.

Theorem 19.2. *Let all the Conditions 1–8 of Section 17 hold. In this case the vector field $w - \mathbf{G}_\kappa(w)$ is homotopic to the field w on spheres of sufficiently large radius R, so that its winding number on these spheres is $+1$.*

To prove this theorem, let us construct the vector field $\mathbf{\Pi}(w, t) = w - t\mathbf{G}_\kappa(w)$. Clearly, $\mathbf{\Pi}(w, 0) = w$, and $\mathbf{\Pi}(w, 1) = w - \mathbf{G}_\kappa(w)$. Let us prove now that

$$\mathbf{\Pi}(w,t) = w - t\mathbf{G}_\kappa(w) \neq 0 \text{ for } \|w\|_{H_\kappa} = R,\ 0 \leq t \leq 1, \tag{19.32}$$

if R is sufficiently large. Indeed, if for some w_0 and $0 \leq t_0 \leq 1$ we have that

$$\mathbf{\Pi}(w_0, t_0) = w_0 - t_0\mathbf{G}_\kappa(w_0) = 0,$$

then

$$\left(\mathbf{\Pi}(w_0, t_0)\cdot w_0\right)_{H_\kappa} = \|w_0\|_{H_\kappa}^2 - t_0\left(\mathbf{G}_\kappa(w_0)\cdot w_0\right)_{H_\kappa} = \Phi(w_0, t_0) = 0,$$

which contradicts Theorem 19.1. Theorem 19.2 has been proved.

Theorem 19.3. *Assume that Conditions 1–5, 8 of Section 17 are satisfied and that Furthermore,* $\widetilde{T}_{\mathrm{p}}^{ij} \in L_{2\Omega}$. *In this case the boundary value problem 9κ is solvable if and only if Conditions 6–7 of Section 17 hold, that is, for solvability it is both necessary and sufficient that* $[\widetilde{T}_{\mathrm{p}}^{ij},\ R^3,\ \widetilde{M}^m] \in \overline{H}_\kappa$ *and* $\widetilde{w},\ \widetilde{w}_4$ *satisfy Condition 6 of 17.1, that is, they are admissible. Then a generalized solution* $w,\ \Psi \in W_{2\Omega}^{(2)} \times H_9$ *uniquely determines* w^*, *that is, the complex* $[\widetilde{T}_{\mathrm{p}}^{ij},\ R^3,\ \widetilde{M}^m]$.

Sufficiency follows from Theorem 19.2. To prove necessity let us consider the relation (17.2), whence

$$(w_{\mathrm{p}} \cdot \varphi)_{H_\kappa} = \int_\Omega (\widetilde{T}_{\mathrm{p}}^{ij} B_{ij} + R^3)\varphi d\Omega + \int_{\Gamma_2} \widetilde{M}^m \varphi_4 ds$$
$$= (w \cdot \varphi)_{H_\kappa} - \int_\Omega C^{ik} C^{jl} \big(B_{ij}\varphi - \varphi_{\alpha^i} w_{\alpha^j}\big) \nabla_{kl} \Psi d\Omega$$
$$+ \int_\Omega \widetilde{T}_{\mathrm{p}}^{ij} w_{\alpha^i} \varphi_{\alpha^j} d\Omega,$$

where $\varphi_4 = \partial\varphi/\partial m$. It is easily seen that if $w \in W_{2\Omega}^{(2)}, \varphi \in H_\kappa, \Psi \in H_9, \widetilde{T}_{\mathrm{p}}^{ij} \in L_{2\Omega}$, then the right-hand side of the above equation is a linear functional with respect to φ in H_κ. Its linearity is obvious. Let us establish its boundedness:

$$\left|(w \cdot \varphi)_{H_\kappa} - \int_\Omega C^{ik} C^{jl} \big(B_{ij}\varphi - \varphi_{\alpha^i} w_{\alpha^j}\big) \nabla_{kl} \Psi d\Omega + \int_\Omega \widetilde{T}_{\mathrm{p}}^{ij} w_{\alpha^i} \varphi_{\alpha^j} d\Omega\right|$$
$$\leq m\Big(\|w\|_{H_\kappa} + \left\|C^{ik} C^{jl} B_{ij}\right\|_{C_\Omega} \|\nabla_{kl}\Psi\|_{L_{2\Omega}}$$
$$+ \left\|C^{ik} C^{jl}\right\|_{C_\Omega} \|w_{\alpha^j}\|_{L_{4\Omega}} \|\nabla_{kl}\Psi\|_{L_{2\Omega}}$$
$$+ \sum_{i,j=1}^{2} \left\|\widetilde{T}_{\mathrm{p}}^{ij}\right\|_{L_{2\Omega}} \|w_{\alpha^i}\|_{L_{4\Omega}} \Big) \|\varphi\|_{H_\kappa}.$$

Finally, if $w \in H_\kappa$, then the corresponding complete displacement w, equal to $\overset{0}{w} + w_{\mathrm{p}} + w$, is in $W_{2\Omega}^{(2)}$, whence $\overset{0}{w} \in W_{2\Omega}^{(2)}$, and thus $\widetilde{w}, \widetilde{w}_4$ must necessarily satisfy condition 6 of 17.1. It is obvious that w_{p}^*, and therefore the complex $[\widetilde{T}_{\mathrm{p}}^{ij} B_{ij} + R^3,\ \widetilde{M}^m]$, are well-defined.

Theorem 19.4. *Let Conditions 1–8 of Section 17 be satisfied. In this case all the solutions of the boundary value problem 9κ lie in a sphere* $\Sigma_{H_\kappa}(R,\ 0)$ *of radius R in the space* H_κ.

Remark 19.1. We remind the reader that in our understanding a developable shell has the metric of the middle surface that is close to the metric of some developable surface in the sense of (8.1).

Finally, let us note that all the clarifications of Section 16 contained in remarks 16.2–16.5 also pertain to the facts established in Theorems 19.1–19.4.

19.5.

Theorem 19.5. *Assume that a shallow shell satisfies Conditions 1–5 of Section 17 and* $\widetilde{T}_{\mathrm{p}}^{ij} \in L_{2\Omega}$. *Then the boundary value Problem* 9κ (17.40), (17.41), *is solvable if and only if* $[\widetilde{T}_{\mathrm{p}}^{ij}, R^3, \widetilde{M}^m] \in \widetilde{H}_\kappa$ *and* $\widetilde{w}$, $\partial\widetilde{w}/\partial\nu$ *satisfy Condition 6 of 17.1, that is, if they are admissible. Then the generalized solution* $w \in W_{2\Omega}^{(2)}$, $\Psi \in H_9$ *is uniquely defined by the complex of loads* $[\widetilde{T}_{\mathrm{p}}^{ij}, R^3, \widetilde{M}^m]$.

19.6. Let us assume that the boundary value problem of nonlinear shell theory has a group of automorphisms $\mathcal{X}$, or, that amounts to the same, that the nonlinear operator equation (NOE) (17.4), (17.5) has that group of automorphisms. Let the subspaces $H_{\kappa s}$, H_{9s} be induced by that group.

Theorem 19.6. *Assume that Conditions 1–8 of Section 17 hold. Then the winding number of the vector field* (19.32) *for* $t = 1$ *on spheres* $\Sigma_{H_{\kappa s}}(R, 0)$ *is* $+1$ *for sufficiently large R. Furthermore, the NOE* (17.4), (17.5) *have at least one solution. All the solutions of the NOE* (17.4), (17.5) *are contained in a sphere in the subspace* $H_{\kappa s}$.

19.7. The first studies of mathematical problems in the nonlinear theory of plates and shells date back to 1955. This is the year of publication of the paper of the author [339], which used the variational approach. For certain clamping conditions for the Vlasov system of equations in displacements, solvability of Bubnov–Galerkin equations was established for every n, and the passage to the limit $n \to \infty$ was justified. Thus a solvability theorem of the boundary value problem was proved as well. The year 1957 saw the appearance of the work of Morozov [200], in which the Hildebrandt–Graves theorem [118] was applied to solvability of the equations. In 1965 appeared the paper of Berger [25], in which the first boundary problem for the von Kármán equations was considered. Subsequently, the problem of solvability of boundary value problems for the von Kármán equations for plates was treated by many authors: Dubinskii [69], Hlavaček and Naumann [119, 120], John and Nečas [126], Knightly [143], Knightly and Sather [144], Nečas and Naumann [216], Rabier [251]. Let us also note the long series of works of Morozov [201, 203, 204, 205] and the monograph of Ciarlet and Rabier [58]. The results of the above authors are based on an a priori estimate of the solution in an energy norm, which is obtained here immediately. In the case of shells, obtaining a priori estimates presents grave difficulties. Let us note here the work of Bernadou and Oden [30], where under somewhat restrictive assumptions on the loads, a solvability theorem was proved for the problem in displacements in general nonlinear coordinates.

The approach of Sections 16, 19 is due to the author and was first published in 1957 [342]. Later, these techniques were applied to a wide range of problems [342, 344, 346, 349, 354, 356, 377, 378, 359, 365, 366, 362, 180, 216]. The method is based, as can be seen above, on a priori estimates of the type of (16.25). The main difficulty is connected with obtaining these estimates. The second part of the proof is based on Leray–Schauder infinite-dimensional degree theory [162]. Subsequently, we have the pioneering works of Vishik [328, 329, 330], Browder

[41], Lions [180] (see also [84]), and other authors, where different versions of the "acute angle" lemma are presented. See also the fundamental survey of Dubinskii [70]. The "acute angle" lemma can also be used in the type of questions we are considering here. Naturally, here as well the arguments are based on a priori estimates (16.25), (19.17). However, using the degree of the mapping (where at all possible) has advantages. The point is that we obtain not only a solvability theorem but also an important geometric characteristic: the total index of all solutions of the boundary value problem. Knowing this characteristic in many cases allows us to get an indication of the number of solutions. At the same time we note that the theory of monotone operators that appeared later in the work of Vishik [329, 330], Kachurovskii [127], Vainberg [319], Minty [199], and Browder [42, 43] (for more details see the above-mentioned survey of Dubinskii, a detailed survey of Skrypnik [281], as well as [280]) cannot provide exhaustive answers to the range of questions we are dealing with, since in the nonlinear theory of shallow shells the operators $\mathbf{G}_{t\kappa}$, $\mathbf{G}_{\kappa\kappa}$ in the most interesting cases are not monotone. From among later studies of the question of solvability of the main boundary value problems of nonlinear shell theory let us note [33, 66]. Let us also note the books of Ciarlet [400, 402] and, finally, the extremely interesting monograph of S.S. Antman [396], where in addition to general problems of nonlinear theory of elasticity, the author considers many problems of nonlinear shell theory from a different point of view, which also uses variational considerations. We should also note all the papers of Antman dealing with different aspects of the so-called exact one-dimensional shell theory. He obtains a number of interesting results concerning qualitative behavior of solutions depending on the loads and their stability.

20. Differentiability Properties of Generalized Solutions of the Problems $t\kappa$ and 9κ. Conditions for the Existence of Classical Solutions

20.1. In this section we shall consider the question of improvement of differentiability properties of generalized solutions as we increase the demands on the initial data of the problem. We restrict ourselves to the case when on the entire boundary curve Γ the following conditions apply

$$w_i|_\Gamma = \widetilde{w}_i, \tag{20.1}$$

$$w|_\Gamma = \widetilde{w}, \tag{20.2}$$

$$\left.\frac{\partial w}{\partial m}\right|_\Gamma = \widetilde{w}_4. \tag{20.3}$$

Thus, as initial data of the problem we consider here the middle surface S, the boundary curve Γ, the elastic constants D_{s}^{ijkl}, D_{f}^{ijkl}, the loads R^3, R^s, and the boundary displacements $\widetilde{w}_i$, $\widetilde{w}$, and $\widetilde{w}_{\mathrm{p}}$. Consequently, we are dealing here with Problem 5.1.

The study of differentiability properties of solutions is based on the following idea. The principal linear part of the differential equations we are considering is elliptic; the dependence of differentiability properties of solutions of elliptic boundary value problems on the properties of right-hand sides is well understood. We have already proved solvability of the corresponding boundary value problems in energy spaces, and thus in Sobolev spaces. Considering now the nonlinear terms in the equation as being known, we obtain elliptic linear boundary value problems with right-hand sides in some function space. It turns out that then solutions lie in a Sobolev space with better differentiability properties. Iterating this procedure, we obtain a dependence between differentiability of solutions and differentiability properties of the load complex. Under certain conditions the generalized solution has all the necessary usual derivatives and becomes a classical one.

The study of differentiability properties of generalized solutions is of importance not only in itself but also since the better they are, the better they would be approximated, and thus the more efficient would be the majority of methods of approximate solution of boundary value problems.

A general analysis of differentiability properties of solutions of elliptic systems is dealt with in a large number of works. We mention here the studies [1, 23, 264, 289, 290] as well as [334, 116, 160]. These studies are based on the concepts of systems that are elliptic in the sense of Petrovskii [233] and of boundary value problems that satisfy complementarity and covering conditions.

Definition 20.1. A system of differential expressions

$$\sum_{j=1}^{n} \mathcal{E}_{tj}\left(\frac{\partial}{\partial x_p}, x_s\right) u_j, \qquad t = 1, \ldots, n; \quad p, s = 1, \ldots, m, \tag{20.4}$$

is called elliptic in the sense of Petrovskii [233] in a domain Ω of variables x_p, x_s if the algebraic system in ξ_p,

$$\sum_{j=1}^{n} \mathcal{E}_{tj}\left(\xi_p, x_s\right) u_j = 0, \tag{20.5}$$

has for any $x_s \in \overline{\Omega}$ only the zero real solution, $u_j = 0$.

The concept of an elliptic system has been substantially generalized in [1, 289, 290]. In the case of homogeneous differential systems (20.4), which we have to deal with, both definitions single out the same class of systems.

20.2.

Lemma 20.1. *If the material of the shell is regular, that is, it is elliptic and satisfies relation* (4.18), *then the differential system* $D_{\mathrm{s}}^{ijkl} w_{l\alpha^k\alpha^j} = f^i$ *is elliptic in the sense of Petrovskii.*

To prove the claim, let us consider an elastic plate with constants D_{s}^{ijkl} and let us make the change of variables $w_{l\alpha^k\alpha^j} \mapsto \xi_k \xi_j w_l$. As a result we obtain the polynomials $D_{\mathrm{s}}^{ijkl} \xi_k \xi_j w_l$; violation of ellipticity would mean that the system

$$D_{\mathrm{s}}^{ijkl} \xi_k \xi_j w_l = 0, \qquad i = 1, 2, \tag{20.6}$$

has a nontrivial solution w_l for some ξ_k, ξ_j. Then from (20.6) we have

$$D_{\mathrm{s}}^{ijkl}\xi_k\xi_j w_l w_i = 0. \tag{20.7}$$

Let us assume now that the plate undergoes a displacement of the form

$$w_l(\alpha^1, \alpha^2) = w_l \exp(\xi_1\alpha^1 + \xi_2\alpha^2), \qquad w \equiv 0,\ l = 1, 2, \tag{20.8}$$

where ξ_1, ξ_2 guarantee the existence of nontrivial solutions w_l in (20.6), and let us compute for (20.8) the density of energy Q_{s} using (4.12).

As a result we easily obtain

$$Q_{\mathrm{s}} = D_{\mathrm{s}}^{ijkl}\xi_k\xi_j w_l w_i = 0,$$

which contradicts (4.18). Lemma 20.1 has been proved.

Next, let us consider the homogeneous differential expression

$$\begin{aligned} D_{\mathrm{f}}^{1111}\frac{\partial^4 w}{\partial\alpha_1^4} + 4D_{\mathrm{f}}^{1112}\frac{\partial^4 w}{\partial\alpha_1^3\partial\alpha_2} + 6D_{\mathrm{f}}^{1122}\frac{\partial^4 w}{\partial\alpha_1^2\partial\alpha_2^2} \\ + 4D_{\mathrm{f}}^{1222}\frac{\partial^4 w}{\partial\alpha_1\partial_2^3} + D_{\mathrm{f}}^{2222}\frac{\partial^4 w}{\partial\alpha_2^4}, \end{aligned} \tag{20.9}$$

which defines the principal terms of the expression

$$\widetilde{\nabla}_{ij}\left(D_{\mathrm{f}}^{ijkl}\nabla_{kl}w\right).$$

Lemma 20.2. *If the material of the shell satisfies the regularity condition* (4.17), *then the differential expression* (20.9) *is elliptic in the sense of Petrovskii.*

To prove the lemma, we have to show that the relation

$$D^*(\xi_1, \xi_2) = D_{\mathrm{f}}^{1111}\xi_1^4 + 4D_{\mathrm{f}}^{1112}\xi_1^3\xi_2 + 6D_{\mathrm{f}}^{1122}\xi_1^2\xi_2^2 + 4D_{\mathrm{f}}^{1222}\xi_1\xi_2^3 + D_{\mathrm{f}}^{2222}\xi_2^4 = 0$$

is possible only for $\xi_i = 0$,

Let the middle surface S be planar, so that we are considering a plate; furthermore, let α^1, α^2 be the Cartesian coordinates on it. Let us consider a transverse displacement

$$w = \exp(\xi_1\alpha^1 + \xi_2\alpha^2). \tag{20.10}$$

It is easily seen that in this case

$$Q_{\mathrm{f}} = D_{\mathrm{f}}^{ijkl}\nabla_{ij}w\nabla_{kl}w = D_{\mathrm{f}}^{ijkl}w_{\alpha^i\alpha^j}w_{\alpha^k\alpha^l} = D^*(\xi_1, \xi_2)\exp 2(\xi_1\alpha^1 + \xi_2\alpha^2).$$

On the other hand,

$$\overset{1}{\gamma}{}_{11}^2 + \overset{1}{\gamma}{}_{22}^2 + \overset{1}{\gamma}{}_{12}^2 = (\xi_1^4 + \xi_2^4 + \xi_1^2\xi_2^2)\exp 2(\xi_1\alpha^1 + \xi_2\alpha^2). \tag{20.11}$$

If $D^*(\xi_1, \xi_2) = 0$, then by (4.18) the right-hand side of (20.11) is zero, and Lemma 20.2 is proved.

Conditions for ellipticity of the linear part of the boundary value problems under consideration guarantee the improvement of differentiability properties of solutions as we increase the smoothness of initial data.

20.3. Let us first consider the differential expression $\nabla_j\left(D_{\mathrm{s}}^{ijkl} D\nabla_k w_l\right)$, the principal part of which is $D_{\mathrm{s}}^{ijkl} w_{l\alpha^i\alpha^k}$, and the corresponding operator equation of the form

$$(\omega\cdot\chi)_{H_5}=\int_\Omega \overset{*}{R}{}^s\varphi_s d\Omega-(\overset{0}{\omega}\cdot\chi)_{H_5}, \tag{20.12}$$

in which $\omega\in H_5$, $\chi\in H_5$ are arbitrary, $\overset{*}{R}{}^s\in\overline{H}_5$.

Lemma 20.3. *Let $\tilde{\Omega}$ be a subdomain of Ω lying entirely in its interior (see Figure 20.1), and assume that the following conditions are satisfied:*

$$D_{\mathrm{s}}^{ijkl}\in C_\Omega^{1+\rho}\left(H_{\tilde{\Omega}}^{1+\rho,\lambda}\right),\qquad S\in C_{\tilde{\Omega}}^{3+\rho}\left(H_{\tilde{\Omega}}^{3+\rho,\lambda}\right), \tag{20.13}$$

$$\overset{*}{R}{}^s\in W_{p\tilde{\Omega}}^{(\rho)}\left(H_{\tilde{\Omega}}^{\rho,\lambda}\right), \tag{20.14}$$

$$\overset{0}{\omega}\in W_{p\tilde{\Omega}}^{(\rho+2)}\left(H_{\tilde{\Omega}}^{\rho+2,\lambda}\right). \tag{20.15}$$

Then in any subdomain Ω' contained in the interior of $\tilde{\Omega}$ for any solution ω of the operator equation (20.12) *we have the relations*

$$\omega\in W_{p\Omega'}^{(\rho+2)}\left(H_{\Omega'}^{\rho+2,\lambda^1}\right). \tag{20.16}$$

As was already mentioned above, in (20.16), *$\lambda^1=\lambda$ if $\lambda<1$, and λ^1 is arbitrarily close to* 1 *if $\lambda=1$.*

We note that the result of Lemma 20.3 is in no way connected with boundary conditions of the boundary value problems and therefore in fact applies to any boundary value problem $t\kappa$. For a proof of the lemma see [1, 22, 289, 290].

The study of differentiability properties of ω in a neighborhood of the boundary Γ will be based on complementarity and covering conditions, which the reader can find in [1, 289, 290]. In these papers complementarity is expressed algebraically; however, applying the algebraic criteria in concrete cases presents difficulties. We can use a different approach [32], based on the simple idea that covering and complementarity conditions are necessary to ensure unique solvability of the corresponding problem in a half-space (a half-plane in our case) in an energy space. However, we have already proved the required statement and obtained all the necessary estimates for solutions of linear problems. In particular, we have the

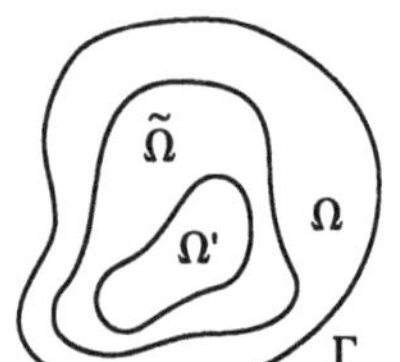

FIGURE 20.1.

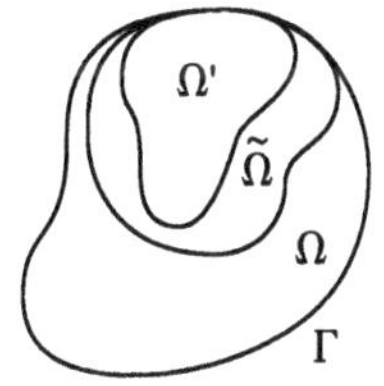

FIGURE 20.2.

inequality

$$\|\omega\|_{H_5} \leq m\Big(\sum_{s=1}^{2}\Big\|\overset{*}{R}{}^s\Big\|_{\overline{H}_5} + \Big\|\overset{0}{\omega}\Big\|_{H_5}\Big), \tag{20.17}$$

from which we immediately obtain the following:

Lemma 20.4. *The differential system* $\widetilde{\nabla}_j D_{\text{s}}^{ijkl}\nabla_k w_l$ *with the boundary conditions* (20.1) *satisfies the complementarity and covering conditions.*

Let us assume now that the subdomain $\widetilde{\Omega}$ in which conditions (20.13)–(20.15) hold can include a segment $\widetilde{\Gamma}$ of the boundary Γ (see Figure 20.2)

Lemma 20.5. *Let* $\omega \in H_5$ *and let it satisfy the operator equation* (20.12). *Assume that conditions* (20.13)–(20.15) *hold and that furthermore,*

$$\widetilde{\Gamma} \in C_{\widetilde{\Gamma}}^{(\rho+2)}\big(H_{\widetilde{\Gamma}}^{\rho+2,\lambda}\big).$$

Then (20.16) *holds in any subdomain* Ω' *that is contained in* $\widetilde{\Omega}$, *the boundary of which may intersect* $\widetilde{\Gamma}$ *(see Figure* (20.2)*).*

This lemma follows immediately from Lemmas 20.1 and 20.4.

20.4. Let us move on now to a study of boundary value problems for w. Let us consider the differential expression $\widetilde{\nabla}_{ij} D_{\text{f}}^{ijkl} D\nabla_{kl} w$, the principal part of which is $DD_{\text{f}}^{ijkl} w_{\alpha^i\alpha^j\alpha^k\alpha^l}$, and the corresponding operator equation

$$(w \cdot \varphi)_{H_1} = \int_\Omega \overset{*}{R}{}^3 \varphi \, d\Omega - (\overset{0}{w} \cdot \varphi)_{H_1}. \tag{20.18}$$

Here w, $\varphi \in H_1$, $\overset{*}{R}{}^3 \in \overline{H}_1$, and $\overset{0}{w}$ is defined by (13.1), (13.2), which in our case take the form

$$\overset{0}{w}\Big|_{\Gamma} = \widetilde{w}, \tag{20.19}$$

$$\frac{\partial \overset{0}{w}}{\partial m}\Bigg|_{\Gamma_1} = \widetilde{w}_4. \tag{20.20}$$

Lemma 20.6. *Let* $\widetilde{\Omega}$ *be a subdomain of* Ω *contained in its interior (see Figure* 20.3*) and let the following conditions be satisfied:*

$$D_{\text{f}}^{ijkl} \in C_{\widetilde{\Omega}}^{2+\rho}\big(H_{\widetilde{\Omega}}^{2+\rho,\lambda}\big), \qquad S \in C_{\widetilde{\Omega}}^{4+\rho}\big(H_{\widetilde{\Omega}}^{4+\rho,\lambda}\big), \tag{20.21}$$

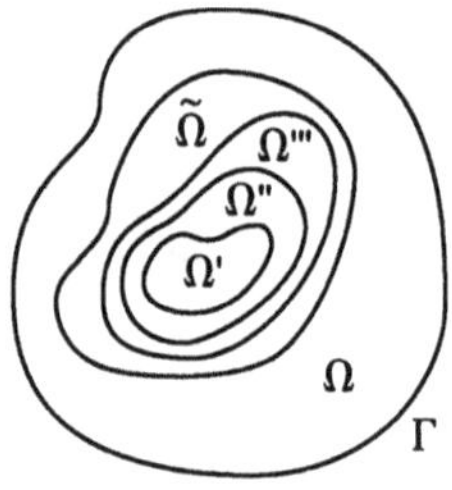

FIGURE 20.3.

$$\overset{*}{R}{}^3 \in W^{(\rho)}_{p\tilde{\Omega}}\left(H^{\rho,\lambda}_{\tilde{\Omega}}\right), \tag{20.22}$$

$$\overset{0}{w} \in W^{(4+\rho)}_{\tilde{\Omega}}\left(H^{4+\rho,\lambda}_{\tilde{\Omega}}\right). \tag{20.23}$$

Then any function $w \in H_1$ *satisfying* (20.18) *also satisfies the relation*

$$w \in W^{(4+\rho)}_{p\Omega'}\left(H^{4+\rho,\lambda^1}_{\Omega'}\right), \tag{20.24}$$

where Ω' *is any subdomain of* Ω *contained entirely in the interior of* $\tilde{\Omega}$ *(see Figure* 20.1*).*

Lemma 20.6 follows from Lemma 20.2.

Lemma 20.7. *The boundary conditions* (20.19), (20.20) *cover* [32] *the differential expression* $\tilde{\nabla}_{ij}\left(D^{ijkl}_{\mathrm{f}} D\nabla_{kl} w\right)$.

Lemma 20.7 follows from the inequality

$$\|w\|_{H_1} \le m\left(\left\|\overset{*}{R}{}^3\right\|_{\overline{H}_1} + \left\|\overset{0}{w}\right\|_{W^{(2)}_{2\Omega}}\right),$$

which in turn follows from (12.35) and (20.18).

Lemma 20.8. *Assume that conditions* (20.21)–(20.20) *hold in a subdomain* $\tilde{\Omega}$, *the boundary of which can intersect* Γ *in a segment* $\tilde{\Gamma}$, *and that furthermore,*

$$\tilde{\Gamma} \in C^{(4+\rho)}_{\tilde{\Gamma}}\left(H^{4+\rho,\lambda}_{\tilde{\Gamma}}\right).$$

Then (20.24) *holds in any subdomain* Ω' *contained in the interior of* $\tilde{\Omega}$, *the boundary of which can intersect* $\tilde{\Gamma}$ *(see Figure* 20.3*).*

20.5. Let us now analyze the differentiability properties of generalized solutions of nonlinear boundary value problems under conditions (20.1)–(20.3).

Theorem 20.1. *Assume that the conditions of one of Theorems 16.5, 16.8, 16.11 are satisfied and that moreover, relations* (20.13), (20.15), (20.21), (20.23) *hold. Furthermore, let*

$$R^3, R^s \in W^{(\rho)}_{p\tilde{\Omega}}\left(H^{\rho,\lambda}_{\tilde{\Omega}}\right).$$

Then in any subdomain Ω' *contained in the interior of* $\tilde{\Omega}$ *the corresponding generalized solution satisfies the relations* (20.16), (20.24).

To prove Theorem 20.1, we note that in the conditions of this theorem, (13.6), (13.7), which define a generalized solution, can be written in the form

$$[\omega \cdot \chi]_{H_l} = \int_\Omega \left\{ \left[\left(B_{kl} w - \frac{1}{2} w_{\alpha^k} w_{\alpha^l} \right) D_{\mathrm{s}}^{\lambda jkl} D \right]_{\alpha^i} D^{-1} \right.$$
$$\left. + \left(B_{kl} w - \frac{1}{2} w_{\alpha^k} w_{\alpha^l} \right) D_{\mathrm{s}}^{ijkl} \mathbf{G}_{ij}^{\lambda} - R^{\lambda} \right\} \varphi_\lambda d\Omega, \tag{20.25}$$

$$(w \cdot \varphi)_{H_\kappa} = \int_\Omega \left[T^{ij} (B_{ij} + \nabla_{ij} w) + R^3 - R^s w_{\alpha^s} \right] \varphi d\Omega, \tag{20.26}$$

where $\mathbf{b} = (\chi, \varphi) = (\varphi_1, \varphi_2, \varphi) \in H_{15}$ have compact support in Ω.

Let us consider first the case $\rho = 0$ and $1 < p < 2$. Comparing (20.12) (and taking into account (20.10)) and (20.25), we see that they are identical if we take

$$\begin{aligned} \overset{*}{R}{}^s = &- \left[\left(B_{kl} w - \frac{1}{2} w_{\alpha^k} w_{\alpha^l} \right) D_{\mathrm{s}}^{sjkl} D \right]_{\alpha^j} D^{-1} \\ &- \left(B_{kl} w - \frac{1}{2} w_{\alpha^k} w_{\alpha^l} \right) D_{\mathrm{s}}^{ijkl} \mathbf{G}_{ij}^{s} - R^s \end{aligned} \tag{20.27}$$

Since $w \in H_1$ and $p > 2$, we easily conclude that $\overset{*}{R}{}^s \in L_{p\overline{\Omega}}$. Then from Lemma 20.3 for $\rho = 0$ we have

$$\omega \in W_{p\Omega'''}^{(2)}, \qquad \Omega' \subset \Omega''' \subset \Omega, \tag{20.28}$$

and by the embedding Theorem 10.3,

$$\omega \in H_{\Omega'''}^{0,2-\frac{2}{p}}, W_{\frac{2p}{2-p}\Omega'''}^{(1)}.$$

But then, by (4.14), (4.15), $T^{ij} \in L_{\frac{2p}{2-p}\Omega'''}$. Next, comparing (20.25) and (20.26), we see that they are identical if we take

$$\overset{*}{R}{}^3 = T^{ij}(B_{ij} + \nabla_{ij} w) + R^3 - R^s w_{\alpha^s}. \tag{20.29}$$

From (20.29) we easily obtain

$$\overset{*}{R}{}^3 \in L_{p_1, \Omega'''},$$

where p_1 is a number smaller than p but arbitrarily close to it. From (20.26) and Lemma 20.6 we have

$$w \in W_{p_1, \Omega''}^{(4)}, \qquad \Omega' \subset \Omega'' \subset \Omega'''.$$

But in that case from (20.29) we conclude that

$$\overset{*}{R}{}^3 \in L_{p\Omega''}. \tag{20.30}$$

From (20.30) and Lemma 20.6 we have

$$w \in W_{p\Omega'}^{(4)}, \qquad \Omega' \subset \Omega'' \subset \Omega'''. \tag{20.31}$$

Relations (20.28), (20.31) complete the proof of Theorem 20.1 in the case $\rho = 0$, $1 < p < 2$, since Ω''', Ω'', Ω' are arbitrary.

Let us consider the case $\rho = 0$, $p = 2$. Here we immediately obtain

$$\omega \in W^{(2)}_{2\Omega'}, \quad H^{0,\lambda^1}_{\Omega'}; \quad W^{(1)}_{q_1\Omega'}; \quad T^{ij} \in L_{q_2\Omega'} \text{ for any } q_1, q_2 \geq 1, \tag{20.32}$$

and then (20.28), (20.31) are preserved.

If $p > 2$, by Lemma 20.3 we have from (20.27) that

$$\omega \in W^{(2)}_{p_2,\Omega'''}, \qquad \Omega' \subset \Omega''', \; p_2 < 2,$$

and by the embedding Theorem 10.3,

$$\omega \in H^{1,1-\frac{2}{p}}_{\Omega'''}.$$

But then from (20.27), (20.30), and respectively (20.31), by the embedding Theorem 10.3,

$$w \in H^{3,1-\frac{2}{p}}_{\Omega''},$$

and we obtain (20.28), (20.31). Thus Theorem 20.1 is completely proved for $\rho = 0$.

The case $\rho > 0$ is analyzed using induction. Assume that Theorem 20.1 is true for some $\rho > 0$ and let us consider the case $\rho + 1$. Then from (20.27) we conclude that $\overset{*}{R}{}^3 \in W^{(\rho+1)}_{p\Omega'''}$, and from Lemma 20.3 we have

$$\omega \in W^{(\rho+3)}_{p\Omega''}, \qquad \Omega' \subset \Omega'' \subset \Omega'''.$$

But then from (4.14), (4.15) we obtain

$$T^{ij} \in W^{(\rho+3)}_{p\Omega''},$$

and from (20.29) it follows that $\overset{*}{R}{}^3 \in W^{(\rho+1)}_{p\Omega''}$; next, from Lemma 20.5

$$w \in W^{(\rho+5)}_{p\Omega'}, \qquad \Omega' \subset \Omega''.$$

If we recall that the domains Ω', Ω'', Ω''' are arbitrary (see Figure (20.4)), we can conclude that Theorem 20.1 has been completely proved for the spaces $W^{(\rho)}_{p\Omega}$.

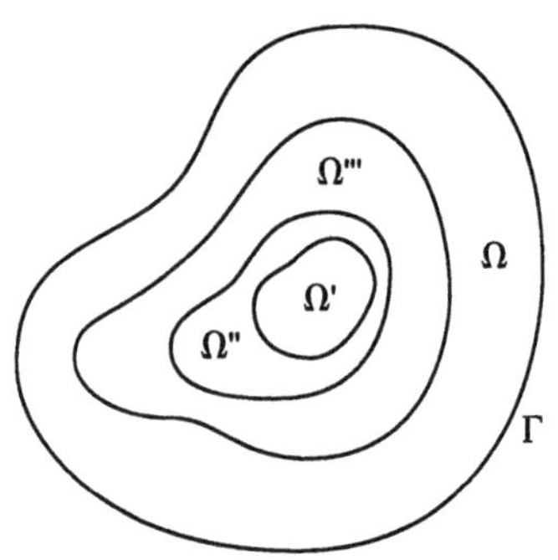

FIGURE 20.4.

In the case of spaces $H_{\Omega}^{k,\lambda}$, Theorem 20.1 is established by using the above results. Indeed, if R^s, $R^3 \in H_{\Omega'''}^{\rho,\lambda}$, then clearly R^3, $R^s \in W_{p\Omega'''}^{(\rho)}$ for any $p > 1$. Then from Theorem 20.1 for $W_{p\Omega}^{(\rho)}$ it follows that in some subdomain $\Omega'' \subset \Omega'''$ we have the relation

$$w \in W_{p\Omega''}^{(\rho+4)}, \qquad \omega \in W_{p\Omega''}^{(\rho+2)}, \quad \forall p > 1,$$

and from the embedding Theorem 10.3 we have

$$w \in H_{\Omega''}^{\rho+3,\lambda^1}, \qquad \omega \in W_{\Omega''}^{\rho+1,\lambda^1}.$$

But in that case from (20.27), (20.29), respectively, we obtain $\widetilde{R}^\lambda \in H_{\Omega''}^{\rho,\lambda^1}$, $\widetilde{R}^3 \in H_{\Omega''}^{\rho,\lambda^1}$, and by Lemmas 20.3, 20.6 we have

$$w \in H_{\Omega'}^{\rho+4,\lambda^1}, \qquad \omega \in W_{\Omega'}^{\rho+2,\lambda^1}, \Omega' \subset \Omega''.$$

Theorem 20.1 has been completely proved.

20.6. We have the following theorems.

Theorem 20.2. *Let the conditions of one of Theorems 16.5, 16.8, 16.11 be satisfied. Moreover, let conditions* (20.13), (20.15), (20.21), (20.23) *hold in a subdomain* $\widetilde{\Omega}$ *that is entirely contained in* Ω, *the boundary of which may include* $\widetilde{\Gamma}$, *a piece of the boundary curve* Γ;

$$\widetilde{\Gamma} \in C_{\widetilde{\Gamma}}^{\rho+4}(H_{\widetilde{\Gamma}}^{\rho+4,\lambda}).$$

Then in any subdomain $\Omega' \subset \widetilde{\Omega}$ *(Figure 20.2), the boundary of which may intersect* $\widetilde{\Gamma}$, *any generalized solution given by Theorems 16.5, 16.8, 16.11, will satisfy the relations* (20.16), (20.24).

The proof of Theorem 20.2 is similar to that of Theorem 20.1, with the difference that in addition to Lemmas 20.3, 20.6, we also use Lemmas 20.4, 20.7. We do not give the details.

Theorem 20.3. *Let the conditions of one of Theorems 16.5, 16.8, 16.11 be satisfied. Moreover, let conditions* (20.13), (20.15), (20.21), (20.23) *hold in the entire domain* $\overline{\Omega}$; *furthermore, let*

$$\Gamma \in C_{\Gamma}^{\rho+4}(H_{\Gamma}^{\rho+4,\lambda}).$$

Then relations (20.19), (20.24) *hold throughout* $\overline{\Omega}$.

Theorem 20.3 follows directly from Theorems 20.1, 20.2.

20.7. In this section we shall consider the differentiability properties of solutions of nonlinear boundary value problems with the Airy stress function (7.51), (7.60). Boundary conditions are taken in the form (7.24), and furthermore,

$$w|_{\Gamma} = \widetilde{w}, \qquad \left.\frac{\partial w}{\partial m}\right|_{\Gamma} = \widetilde{w}_4.$$

Lemma 20.9. *If the material of the shell is regular, so that* (4.18) *holds, then the differential form* $C_*^{\lambda\mu st}\Psi_{\alpha^\lambda\alpha^\mu\alpha^s\alpha^t}$ *is elliptic in the sense of Petrovskii.*

First of all, we note that by (7.56),

$$C_*^{\lambda\mu st} = C_{ijkl,\mathbf{s}}, \quad \lambda + i = \mu + j = s + k = t + l = 3,$$

since the shell is developable and $C^{ik} = \pm 1$.

Thus it has to be shown that the differential form $C_{ijkl,\mathbf{s}}\Psi_{\alpha^i\alpha^j\alpha^k\alpha^l}$ is elliptic in the sense of Petrovskii. In other words, we have to demonstrate that if

$$\begin{aligned} C^*(\xi_1, \xi_2) = C_{1111,\mathbf{s}}\xi_1^4 + 4C_{1112,\mathbf{s}}\xi_1^3\xi_2 + 6C_{1122,\mathbf{s}}\xi_1^2\xi_2^2 \\ + 4C_{1222,\mathbf{s}}\xi_1\xi_2^3 + C_{2222,\mathbf{s}}\xi_2^4 = 0, \end{aligned} \tag{20.33}$$

then

$$\xi_1 = \xi_2 = 0. \tag{20.34}$$

To prove (20.34) we consider an anisotropic plate, a stressed state of which is described by the Airy stress function of the form

$$\widetilde{\Psi} = \exp(\xi_1\alpha^1 + \xi_2\alpha^2),$$

and then the stresses T^{ij} according to (7.2) will have the form

$$T^{11} = \xi_2^2 \exp(\xi_1\alpha^1 + \xi_2\alpha^2), \quad T^{12} = \xi_1\xi_2 \exp(\xi_1\alpha^1 + \xi_2\alpha^2) \quad (1 \rightleftharpoons 2).$$

Furthermore, from (4.16) it follows that

$$Q_\mathbf{s} = C_\mathbf{s}^{\lambda\mu st}T_{\lambda\mu}T_{st} = C_{\lambda\mu st}T^{\lambda\mu}T^{st} = C^*(\xi_1, \xi_2)\exp(\xi_1\alpha^1 + \xi_2\alpha^2).$$

By (4.17) we have

$$C^*(\xi_1, \xi_2)\exp(\xi_1\alpha^1 + \xi_2\alpha^2) > m\left(\xi_1^4 + \xi_2^4 + \xi_1^2\xi_2^2\right)\exp(\xi_1\alpha^1 + \xi_2\alpha^2). \tag{20.35}$$

If (20.33) holds, then from (20.35) follows (20.34), and Lemma 20.9 has been proved.

Let us consider the differential expression $\widetilde{\nabla}_{ij}\left(C_*^{ijkls}\nabla_{kl}\Psi\right)$, the principal part of which is $C_*^{ijkls}\Psi_{\alpha^i\alpha^j\alpha^k\alpha^l}$, and the corresponding operator equation

$$(\Psi \cdot \theta)_{H_9} = \int_\Omega \widetilde{R} \cdot \theta d\Omega. \tag{20.36}$$

Lemma 20.10. *Let* $\widetilde{\Omega}$ *be a subdomain of* Ω *entirely contained in its interior, and assume that the following conditions are satisfied:*

$$C_*^{ijkl} \in C_\Omega^{2+\rho}\left(H_\Omega^{2+\rho,\lambda}\right), \quad S \in C_\Omega^{4+\rho}\left(H_\Omega^{4+\rho,\lambda}\right), \quad \widetilde{R} \in W_{p\widetilde{\Omega}}^{(\rho)}\left(H_\Omega^{\rho,\lambda}\right). \tag{20.37}$$

Then in any subdomain Ω' *contained in the interior of* Ω *(see Figure* 20.1*) and for any solution of* (20.36) *we have the relations*

$$\Psi \in W_{p\Omega'}^{(\rho+4)}\left(H_{\Omega'}^{\rho+4,\lambda^1}\right). \tag{20.38}$$

Lemma 20.11 follows immediately from ellipticity of $C_*^{ijkl} w_{\alpha^i\alpha^j\alpha^k\alpha^l}$ in the sense of Petrovskii (Lemma 20.9).

Lemma 20.11. *The boundary conditions of the first boundary value problem (prescription of Ψ and $\partial\Psi/\partial m$ on Γ) cover the differential expression $\widetilde{\nabla}_{ij} C_*^{ijkl} \nabla_{kl} \Psi$.*

This lemma is proved in the same way as Lemma 20.7.

Lemma 20.12. *Let $\widetilde{\Omega}$ be a subdomain of Ω, the boundary of which can include $\widetilde{\Gamma}$, a segment of the boundary curve Γ. Assume that conditions* (20.37) *are satisfied in $\widetilde{\Omega}$ and that moreover,*

$$\widetilde{\Gamma} \in C_{\widetilde{\Gamma}}^{\rho+4}\left(H_{\widetilde{\Gamma}}^{\rho+4,\lambda}\right).$$

Then any solution of (20.36) *satisfies* (20.38) *in any subdomain $\Omega' \subset \widetilde{\Omega}$.*

Lemmas 20.9–20.12 allow us to state the main results concerning the differentiability properties of generalized solutions of problem 91.

Theorem 20.4. *Let Conditions 1–8 of Section 17 be satisfied and assume, furthermore, that in a subdomain $\widetilde{\Omega}$ contained in the interior of Ω we have the relations*

$$D_{\mathrm{f}}^{ijkl} \in C_{\widetilde{\Omega}}^{2+\rho}\left(H_{\widetilde{\Omega}}^{2+\rho,\lambda}\right), \quad S \in C_{\widetilde{\Omega}}^{4+\rho}\left(H_{\widetilde{\Omega}}^{4+\rho,\lambda}\right), \quad R^3 \in W_{p\widetilde{\Omega}}^{(\rho)}\left(H_{\Omega}^{\rho,\lambda}\right),$$
$$\overset{0}{w} \in W_{p\widetilde{\Omega}}^{(4+\rho)}\left(H_{\Omega}^{4+\rho,\lambda}\right), \quad C_*^{ijkl} \in C_{\widetilde{\Omega}}^{2+\rho}\left(H_{\widetilde{\Omega}}^{2+\rho,\lambda}\right),$$
$$C_{*ij}^{\lambda\mu} \widetilde{T}_{\mathrm{p}}^{ij} \in W_{p\widetilde{\Omega}}^{(\rho)}\left(H_{\widetilde{\Omega}}^{\rho,\lambda}\right); \quad p > 1. \tag{20.39}$$

Then any generalized solution of Problem 91 satisfies

$$w,\ \Psi \in W_{p\Omega'}^{(\rho+4)}\left(H_{\Omega'}^{\rho,\lambda}\right) \tag{20.40}$$

in any domain Ω' contained in the interior of $\widetilde{\Omega}$ (see Figure 20.1*).*

Theorem 20.4 follows from Lemmas 20.6, 20.10.

Theorem 20.5. *Let Conditions 1–8 of Section 17 be satisfied as well as condition* (20.39) *and let the boundary of a subdomain Ω' include $\widetilde{\Gamma}$, a segment of the boundary curve Γ. Moreover, let*

$$\widetilde{\Gamma} \in C_{\widetilde{\Gamma}}^{\rho+4}\left(H_{\widetilde{\Gamma}}^{\rho+4,\lambda}\right). \tag{20.41}$$

Then (20.40) *will hold for any generalized solution of Problem 91 in any subdomain Ω' contained in $\widetilde{\Omega}$, the boundary of which may intersect $\widetilde{\Gamma}$ (see Figure* 20.2*).*

Theorem 20.5 follows immediately from Lemmas 20.8, 20.12.

Theorem 20.6. *Let Conditions 1–8 of Section 17 and* (20.39) *be satisfied in the entire domain $\overline{\Omega}$, while* (20.41) *holds for the entire boundary curve Γ. Then*

$$w,\ \Psi \in W_{p\Omega}^{(\rho+4)}\left(H_{\overline{\Omega}}^{\rho+4,\lambda}\right). \tag{20.42}$$

Theorem 20.6 follows from Theorem 20.5.

Let us clarify the meaning of the results dealing with differentiability properties of solutions. Roughly speaking, the dependence of differentiability properties of solutions of all the nonlinear problems on the smoothness of exterior loads (for a sufficiently smooth shell) is that same as for the linear problems obtained by neglecting the nonlinear terms. Smoothness conditions may be violated in a neighbourhood of points at which boundary conditions change, which necessitated the somewhat cumbersome way of stating results.

Qualitatively, the differentiability properties of solutions for all types of problems are the same as those of solutions of Laplace's equation for w_1, w_2 and as those of the corresponding solution of the biharmonic equation for w, Ψ.

CHAPTER V

The Variational Approach to the Problem of Solvability of Boundary Value Problems of Nonlinear Shallow Shell Theory

21. The Variational Approach to the Problem of Solvability of Boundary Value Problems of Nonlinear Shallow Shell Theory in Displacements

21.1. Up till now we have been considering applications of topological methods to the problems of nonlinear shallow shell theory. In this chapter we will analyze the same problems from a different point of view, a variational one. The derivation of the governing equations of the two versions of the theory we are considering is based on variational arguments, which makes a variational analysis of these problems very natural. In this section we shall use this approach to study the problems of shallow shell theory in displacements, since in this way we obtain results that are independent and complementary to the ones obtained by topological methods.

We start by studying the structure and properties of the total energy functional. We shall show how to represent it in a useful form. Then we shall show that the vanishing of the first variation of the total energy functional (equation (21.20)) coincides with the equations of generalized solutions of the problem introduced above. After that, we shall consider the problem of minimizing the total energy functional and properties of minimizing sequences.

A similar approach will be applied in Section 22 to problems with the Airy stress function, with the obvious replacement of the total energy functional by the Alumyae functional.

We shall assume that Conditions 1–8 of Section 13 hold. Let us introduce a new vector of displacements, $\mathbf{a} = (w_1, w_2, w)$, using (13.10) and let us consider the total energy functional defined by (6.38),

$$\mathcal{I}_{t\kappa}(\mathbf{a}+\mathbf{a}^*) = \frac{1}{2}\left\|w+w^*\right\|_{H_\kappa}^2 + \frac{1}{2}\left\{\int_\Omega D_{\mathrm{s}}^{ijkl}\overset{0}{\epsilon}_{ij}(\mathbf{a}+\mathbf{a}^*)\overset{0}{\epsilon}_{kl}(\mathbf{a}+\mathbf{a}^*)d\Omega\right.$$
$$+\int_{\Gamma_6} k_{\mathrm{s}}^{\tau\tau}(w_\tau+w_\tau^*)^2\,ds + \int_{\Gamma_7} k_{\mathrm{s}}^{mm}(w_m+w_m^*)^2\,ds$$
$$\left.+\int_{\Gamma_8} k_{\mathrm{s}}^{ij}(w_i+w_i^*)(w_j+w_j^*)\,ds\right\}$$
$$-(\boldsymbol{\omega}_{\mathrm{p}}\cdot\boldsymbol{\omega})_{H_t} - (w_{\mathrm{p}}\cdot w)_{H_\kappa}. \tag{21.1}$$

(There is no summation on m and τ.) In (21.1) we took into account the definition (12.5) of the space H_κ. For the considerations to follow, it is useful to transform (21.1) using (11.5). Then we obtain

$$\mathcal{I}_{t\kappa}(\mathbf{a}+\mathbf{a}^*) = \frac{1}{2}\left\|w+w^*\right\|_{H_\kappa}^2 + \frac{1}{2}\left\|\boldsymbol{\omega}+\boldsymbol{\omega}^*\right\|_{H_t}^2$$
$$+\int_\Omega D_{\mathrm{s}}^{ijkl}\overset{0}{e}_{ij}(\mathbf{a}+\mathbf{a}^*)\Big[-B_{kl}(w+w^*)$$
$$+\frac{1}{2}(w+w^*)_{\alpha^k}(w+w^*)_{\alpha^l}\Big]d\Omega$$
$$+\frac{1}{2}\int_\Omega D_{\mathrm{s}}^{ijkl}\Big[-B_{ij}(w+w^*)+\frac{1}{2}(w+w^*)_{\alpha^i}(w+w^*)_{\alpha^j}\Big]$$
$$\times\Big[-B_{kl}(w+w^*)+\frac{1}{2}(w+w^*)_{\alpha^k}(w+w^*)_{\alpha^l}\Big]d\Omega$$
$$-(\boldsymbol{\omega}_{\mathrm{p}}\cdot\boldsymbol{\omega})_{H_t} - (w_{\mathrm{p}}\cdot w)_{H_\kappa}. \tag{21.2}$$

Lemma 21.1. *The functional $\mathcal{I}_{t\kappa}(\mathbf{a}+\mathbf{a}^*)$ is well-defined everywhere in $H_{t\kappa}$.*

To prove the lemma, we note that by Conditions 5, 6 of Section 13, $\boldsymbol{\omega}^* \in W_{2\Omega}^{(1)}$, $w^* \in W_{2\Omega}^{(2)}$. If furthermore $w \in H_\kappa$, then it can be seen from (3.16) that $\overset{0}{\epsilon}_{ij}(\mathbf{a}+\mathbf{a}^*) \in L_{2\Omega}$. In view of these facts, the second surface integral on the right-hand side of (21.1) is defined. Furthermore, since $w^* \in W_{2\Omega}^{(2)}$ and $w \in H_\kappa$, $\overset{1}{e}_{ij}(w+w^*) \in L_{2\Omega}$ by (3.17), so that the first surface integral on the right-hand side of (21.1) is defined. The line integrals on the right-hand side of (21.1) are defined by the embedding Theorems 11.3, (12.3), and condition 8 of Section 13. Lemma 21.1 is proved.

In the sequel we shall need a representation of the functional $\mathcal{I}_{t\kappa}$ in terms of the argument $\mathbf{a}+\mathbf{a}^* +\gamma\mathbf{b}$, where $\mathbf{b} = (\varphi_1, \varphi_2, \varphi) \in H_t$. It is easily seen that we have the representation

$$\mathcal{I}_{t\kappa}(\mathbf{a}+\mathbf{a}^*+\gamma\mathbf{b}) = \mathcal{I}_{t\kappa}(\mathbf{a}+\mathbf{a}^*) + \sum_{\mu=1}^{4} \gamma^{\mu}\mathcal{I}_{t\kappa\mu}(\mathbf{a}+\mathbf{a}^*,\ \mathbf{b}), \tag{21.3}$$

where $\mathcal{I}_{t\kappa\mu}(\mathbf{a}+\mathbf{a}^*,\ \mathbf{b})$ are some functionals in $H_{t\kappa}$ in $\mathbf{a}+\mathbf{a}^*$ and $\mathbf{b}$. $\mathcal{I}_{t\kappa\mu}(\mathbf{a}+\mathbf{a}^*,\ \mathbf{b})$ is a homogeneous functional of order μ in $\mathbf{b}$ for a fixed $\mathbf{a}+\mathbf{a}^*$. Thus $\mathcal{I}_{t\kappa}(\mathbf{a}+\mathbf{a}^*+\gamma\mathbf{b})$ is a fourth degree polynomial in γ. The relation (21.3) follows immediately from (21.1) if we take into account the structure of representation of $\overset{0}{\epsilon}_{ij}$, $\overset{1}{e}_{ij}$ in terms of displacements, as given by (3.16), (3.17). We will obtain now explicit expressions for $\mathcal{I}_{t\kappa\mu}(\mathbf{a}+\mathbf{a}^*,\ \mathbf{b})$. We have

$$\mathcal{I}_{t\kappa 1}(\mathbf{a}+\mathbf{a}^*,\ \mathbf{b}) = \frac{d}{d\gamma}\mathcal{I}_{t\kappa}(\mathbf{a}+\mathbf{a}^*+\gamma\mathbf{b})\ |_{\gamma=0},$$

and furthermore,

$$\frac{d}{d\gamma}\frac{1}{2}\ \|w+w^*+\gamma\varphi\|^2_{H_\kappa}\Big|_{\gamma=0} = \big((w+w^*)\cdot\varphi\big)_{H_\kappa}, \tag{21.4}$$

$$\frac{d}{d\gamma}\frac{1}{2}\ \|\omega+\omega^*+\gamma\chi\|^2_{H_t}\Big|_{\gamma=0} = \big((\omega+\omega^*)\cdot\chi\big)_{H_t}. \tag{21.5}$$

For the derivative of the third term on the right-hand side of (21.2) we obtain

$$\begin{aligned}
&\frac{d}{d\gamma}\int_\Omega D_{\mathrm{s}}^{ijkl}\overset{\widetilde{0}}{e}_{ij}(\mathbf{a}+\mathbf{a}^*+\gamma\mathbf{b})\Big[-B_{kl}(w+w^*+\gamma\varphi)\\
&\qquad+\frac{1}{2}(w+w^*+\gamma\varphi)_{\alpha^k}(w+w^*+\gamma\varphi)_{\alpha^l}\Big]d\Omega\ |_{\gamma=0}\\
&=\int_\Omega D_{\mathrm{s}}^{ijkl}\overset{\widetilde{0}}{e}_{ij}(\mathbf{b})\Big[-B_{kl}(w+w^*)+\frac{1}{2}(w+w^*)_{\alpha^k}(w+w^*)_{\alpha^l}\Big]d\Omega\\
&\quad+\int_\Omega D_{\mathrm{s}}^{ijkl}\overset{\widetilde{0}}{e}_{ij}(\mathbf{a}+\mathbf{a}^*)\Big[-B_{kl}\varphi+(w+w^*)_{\alpha^k}\varphi_{\alpha^l}\Big]d\Omega.
\end{aligned} \tag{21.6}$$

Finally, for the derivative of the fourth term of the right-hand side of (21.2) we derive

$$\begin{aligned}
&\frac{d}{d\gamma}\frac{1}{2}\int_\Omega D_{\mathrm{s}}^{ijkl}\Big[-B_{ij}(w+w^*+\gamma\varphi)+\frac{1}{2}(w+w^*+\gamma\varphi)_{\alpha^k}(w+w^*+\gamma\varphi)_{\alpha^l}\Big]\\
&\quad\times\Big[-B_{kl}(w+w^*+\gamma\varphi)+\frac{1}{2}(w+w^*+\gamma\varphi)_{\alpha^k}(w+w^*+\gamma\varphi)_{\alpha^l}\Big]d\Omega\ |_{\gamma=0}\\
&=\int_\Omega D_{\mathrm{s}}^{ijkl}\Big[-B_{ij}\varphi+\varphi_{\alpha^i}(w+w^*)_{\alpha^j}\Big]\\
&\quad\times\Big[-B_{kl}(w+w^*)+\frac{1}{2}(w+w^*)_{\alpha^k}(w+w^*)_{\alpha^l}\Big]d\Omega,
\end{aligned} \tag{21.7}$$

and furthermore

$$\frac{d}{d\gamma}(\omega_p \cdot (\omega + \gamma\chi))_{H_t}\Big|_{\gamma=0} = (\omega_p \cdot \chi)_{H_t}, \tag{21.8}$$

$$\frac{d}{d\gamma}(w_p \cdot (w + \gamma\varphi))_{H_\kappa}\Big|_{\gamma=0} = (w_p \cdot \varphi)_{H_\kappa}. \tag{21.9}$$

From (21.3)–(21.9) we obtain

$$\begin{aligned}
\mathcal{I}_{t\kappa 1}(\mathbf{a} + \mathbf{a}^*, \mathbf{b}) =& ((w + w^*) \cdot \varphi)_{H_\kappa} + ((\omega + \omega^*) \cdot \chi)_{H_t} \\
&+ \int_\Omega D_s^{ijkl} \Big\{ \tilde{\overset{0}{e}}_{ij}(\mathbf{b}) \Big[- B_{kl}(w + w^*) \\
&\qquad + \frac{1}{2}(w + w^*)_{\alpha^k}(w + w^*)_{\alpha^l} \Big] \\
&+ \tilde{\overset{0}{e}}_{ij}(\mathbf{a} + \mathbf{a}^*) \Big[- B_{kl}\varphi + (w + w^*)_{\alpha^i}\varphi_{\alpha^j} \Big] - \Big[B_{ij}(w + w^*) \\
&+ \frac{1}{2}(w + w^*)_{\alpha^i}(w + w^*)_{\alpha^j} \Big] \Big[- B_{kl}\varphi + (w + w^*)_{\alpha^k}\varphi_{\alpha^l} \Big] \Big\} d\Omega \\
&- \omega_p \cdot \omega)_{H_t} - (w_p \cdot \varphi)_{H_\kappa}.
\end{aligned} \tag{21.10}$$

Using (3.16) and (13.8), (13.12), (21.10) can be written in the form

$$\begin{aligned}
\mathcal{I}_{t\kappa 1} =& (w \cdot \varphi)_{H_\kappa} + (\overset{0}{w} \cdot \varphi)_{H_\kappa} + \int_\Omega T^{kl}(\mathbf{a} + \mathbf{a}^*)[- B_{kl}\varphi + \varphi_{\alpha^k}(w + w^*)_{\alpha^l}] d\Omega \\
&+ (\omega \cdot \chi)_{H_t} + (\overset{0}{\omega} \cdot \chi)_{H_t} \\
&+ \int_\Omega D_s^{ijkl} \nabla_i \varphi_j \Big[- B_{kl}(w + w^*) + \frac{1}{2}(w + w^*)_{\alpha^k}(w + w^*)_{\alpha^l} \Big] d\Omega.
\end{aligned} \tag{21.11}$$

Finally, if we take into account the definition of the operator $\mathbf{G}_{t\kappa j}$ by (13.30), (21.11) can be written as

$$\mathcal{I}_{t\kappa 1}(\mathbf{a} + \mathbf{a}^*, \mathbf{b}) = ((\mathbf{a} - \mathbf{G}_{t\kappa}(\mathbf{a} + \mathbf{a}^*)) \cdot \mathbf{b})_{H_t \kappa}. \tag{21.12}$$

Clearly, $\mathcal{I}_{t\kappa 1}(\mathbf{a} + \mathbf{a}^*, \mathbf{b})$ defines the first variation of the total energy functional (where **b** is the variation of the vector of displacements). To define $\mathcal{I}_{t\kappa 2}$, the second variation of the functional, we have

$$\mathcal{I}_{t\kappa 2} = \frac{1}{2}\frac{d^2}{d\gamma^2}\mathcal{I}_{t\kappa}(\mathbf{a} + \mathbf{a}^* + \gamma\mathbf{b})\mid_{\gamma=0};$$

next,

$$\frac{1}{2}\frac{d^2}{d\gamma^2}\left\| w + w^* + \gamma\varphi \right\|_{H_\kappa}^2 = \|\varphi\|_{H_\kappa}^2, \tag{21.13}$$

and furthermore,

$$\frac{d^2}{d\gamma^2}\frac{1}{2}\int_\Omega D_{\mathrm{s}}^{ijkl}\overset{0}{\epsilon}_{ij}(\mathbf{a}+\mathbf{a}^*+\gamma\mathbf{b})\overset{0}{\epsilon}_{kl}(\mathbf{a}+\mathbf{a}^*+\gamma\mathbf{b})d\Omega\mid_{\gamma=0}$$
$$=\frac{d}{d\gamma}\int_\Omega T^{ij}(\mathbf{a}+\mathbf{a}^*+\gamma\mathbf{b})\frac{d}{d\gamma}\overset{0}{\epsilon}_{ij}(\mathbf{a}+\mathbf{a}^*+\gamma\mathbf{b})d\Omega\mid_{\gamma=0} \tag{21.14}$$
$$=\int_\Omega\big\{D_{\mathrm{s}}^{ijkl}[\overset{0}{\epsilon}_{ij}(\mathbf{b})+(w+w^*)_{\alpha^i}\varphi_{\alpha^j}][\overset{0}{\epsilon}_{kl}(\mathbf{b})+(w+w^*)_{\alpha^k}\varphi_{\alpha^l}]$$
$$+\big[T^{ij}(\mathbf{a}+\mathbf{a}^*)\varphi_{\alpha^i}\varphi_{\alpha^j}\big]\big\}d\Omega.$$

Moreover,

$$\frac{d^2}{d\gamma^2}\frac{1}{2}\Big[\int_{\Gamma_6}k_{\mathrm{s}}^{\tau\tau}(w_\tau+w_\tau^*+\gamma\varphi_\tau)^2ds+\int_{\Gamma_7}k_{\mathrm{s}}^{mm}(w_m+w_m^*+\gamma\varphi_m)^2ds$$
$$+\int_{\Gamma_8}k_{\mathrm{s}}^{ij}(w_i+w_i^*+\gamma\varphi_i)(w_j+w_j^*+\gamma\varphi_j)ds\Big] \tag{21.15}$$
$$=\int_{\Gamma_6}k_{\mathrm{s}}^{\tau\tau}\varphi_\tau^2ds+\int_{\Gamma_7}k_{\mathrm{s}}^{mm}\varphi_m^2ds+\int_{\Gamma_8}k_{\mathrm{s}}^{ij}\varphi_i\varphi_jds.$$

From (21.13)–(21.15) we obtain

$$\mathcal{I}_{\iota\kappa 2}(\mathbf{a}+\mathbf{a}^*,\,\mathbf{b})=\|\chi\|_{H_\kappa}^2+2\int_\Omega\big\{D_{\mathrm{s}}^{ijkl}[\overset{0}{e}_{ij}(\mathbf{b})(w+w^*)_{\alpha^k}\varphi_{\alpha^l}$$
$$+(w+w^*)_{\alpha^i}(w+w^*)_{\alpha^j}\varphi_{\alpha^k}\varphi_{\alpha^l}]$$
$$+\overset{0}{\epsilon}_{kl}(\mathbf{a}+\mathbf{a}^*)\varphi_{\alpha^i}\varphi_{\alpha^j}\big\}d\Omega.$$

For $\mathcal{I}_{\iota\kappa 3}(\mathbf{a}+\mathbf{a}^*,\,\mathbf{b})$ we have

$$\mathcal{I}_{\iota\kappa 3}(\mathbf{a}+\mathbf{a}^*,\,\mathbf{b})=\frac{1}{6}\frac{d^3}{d\gamma^3}\mathcal{I}_{\iota\kappa}(\mathbf{a}+\gamma\mathbf{b})\mid_{\gamma=0}$$
$$=\frac{1}{6}\frac{d^3}{d\gamma^3}\int_\Omega D_{\mathrm{s}}^{ijkl}\overset{0}{\epsilon}_{ij}(\mathbf{a}+\mathbf{a}^*+\gamma\mathbf{b})\overset{0}{\epsilon}_{kl}(\mathbf{a}+\mathbf{a}^*+\gamma\mathbf{b})d\Omega$$
$$=2\int_\Omega D_{\mathrm{s}}^{ijkl}\varphi_{\alpha^i}\varphi_{\alpha^j}\varphi_{\alpha^k}(w+w^*)_{\alpha^l}d\Omega,$$
$$\mathcal{I}_{\iota\kappa 4}=\frac{1}{8}\int_\Omega D_{\mathrm{s}}^{ijkl}\varphi_{\alpha^i}\varphi_{\alpha^j}\varphi_{\alpha^k}\varphi_{\alpha^l}d\Omega.$$

Lemma 21.2. *We have the inequalities*

$$|\mathcal{I}_{\iota\kappa 1}(\mathbf{a}+\mathbf{a}^*,\,\mathbf{b})|\le m\big(\,\|\mathbf{a}+\mathbf{a}^*\|_{H_{\iota\kappa}}^2+\|\mathbf{a}+\mathbf{a}^*\|_{H_{\iota\kappa}}$$
$$+\|\omega_{\mathrm{f}}^*\|_{H_\iota}+\|w^*\|_{H_\kappa}\big)\,\|\mathbf{b}\|_{H_{\iota\kappa}}, \tag{21.16}$$
$$|\mathcal{I}_{\iota\kappa 2}(\mathbf{a}+\mathbf{a}^*,\,\mathbf{b})|\le m\big(\,\|w+w^*\|_{L_{4\Omega}}\,\|w+w^*\|_{L_{4\Omega}}^2$$

$$+\left\|\overset{0}{\epsilon}_{kl}(\mathbf{a}+\mathbf{a}^*)\right\|_{L_{2\Omega}}\Big)\,\|\mathbf{b}\|^2_{H_{t\kappa}}, \tag{21.17}$$

$$|\mathcal{I}_{t\kappa 3}(\mathbf{a}+\mathbf{a}^*,\ \mathbf{b})| \le \left\|w+w^*\right\|_{H_\kappa}\|\mathbf{b}\|^3_{H_{t\kappa}}, \tag{21.18}$$

$$|\mathcal{I}_{t\kappa 4}(\mathbf{b})| \le m\,\|\mathbf{b}\|^4_{H_{t\kappa}}. \tag{21.19}$$

Relations (21.16)–(21.19) follow directly from the embedding Theorems 11.3, 12.3 and the structure of the relations (4.14) for T^{kl}, (3.16) for $\overset{0}{\epsilon}_{kl}$, and (3.11) for $\overset{0}{e}_{ij}$.

Let us recall the concept of a gradient. Let $J(x)$ be a smooth (in the sense that all the differentiation operations below are justified in the classical sense) functional defined on a Hilbert space H. For an element $x \in X$, let us consider $J(x+\alpha y)$, where $\alpha \in \mathbf{R}$, $y \in X$. If for all $y \in X$ there is an element z such that

$$\left.\frac{dJ(x+\alpha y)}{d\alpha}\right|_{\alpha=0} = (z,\ y)_X,$$

then z is called the *gradient of the functional* $J(x)$ at the point x. Obviously, z depends on x and therefore, as x is varied, defines an operator in x. This operator is denoted by $\operatorname{grad}_X J(x)$. A point x where $\operatorname{grad}_X J(x) = 0$ is called a *critical point* of the functional.

Lemma 21.3. *We have*

$$\operatorname{grad}_{H_{t\kappa}} \mathcal{I}_{t\kappa}(\mathbf{a}+\mathbf{a}^*) = \mathbf{a} - \mathbf{G}_{t\kappa}(\mathbf{a}+\mathbf{a}^*), \tag{21.20}$$

where the operator $\mathbf{G}_{t\kappa}(\mathbf{a}+\mathbf{a}^*)$ *is given by* (13.31).

In (21.20) "grad" stands for the gradient in $H_{t\kappa}$.

To prove (21.20) we note that for $\gamma = 1$ we have from (21.4),

$$\mathcal{I}_{t\kappa}(\mathbf{a}+\mathbf{a}^*+\mathbf{b}) - \mathcal{I}_{t\kappa}(\mathbf{a}+\mathbf{a}^*) = \mathcal{I}_{t\kappa 1}(\mathbf{a}+\mathbf{a}^*,\ \mathbf{b})\sum_{\mu=2}^{4}\mathcal{I}_{t\kappa\mu}(\mathbf{a}+\mathbf{a}^*,\ \mathbf{b}).$$

If we now take into account (21.12) and the estimates of Lemma 21.2, we deduce the claim of Lemma 21.3.

From (21.20) follows the following important theorem.

Theorem 21.1. *The set of critical points of the functional* $\mathcal{I}_{t\kappa}(\mathbf{a}+\mathbf{a}^*)$ *in* $H_{t\kappa}$ *coincides with the set of generalized solutions of problem* $t\kappa$.

21.2. Since all the critical points of the functional $\mathcal{I}_{t\kappa}(\mathbf{a}+\mathbf{a}^*)$ are generalized solutions of problem $t\kappa$, they satisfy (13.31) and (13.32), from which it follows that these critical points lie on a hypersurface HS1 in $H_{t\kappa}$ given by (13.33). Therefore, in the following we shall consider $\mathcal{I}_{t\kappa}(\mathbf{a}+\mathbf{a}^*)$ on HS1, that is, taking (13.33) into account. Thus we are considering $\mathcal{I}_{t\kappa}(\mathbf{a}+\mathbf{a}^*)$ already in H_κ, since by (13.33), ω is expressed in terms of w. Below, if (13.33) has been taken ito account, we shall denote $\mathcal{I}_{t\kappa}(\mathbf{a}+\mathbf{a}^*)$ by $\mathcal{I}_{\kappa\kappa}(w)$.

Lemma 21.4. *We have the representation*

$$\mathcal{I}_{\kappa\kappa}(w) = \frac{1}{2}\,\|w\|^2_{H_\kappa} + \widetilde{\mathcal{I}}_{\kappa\kappa}(w), \tag{21.21}$$

where $\widetilde{\mathcal{I}}_{\kappa\kappa}(w)$ is a weakly continuous functional in H_κ.

The representation (21.21) follows directly from (21.2), and furthermore,

$$\begin{aligned}
\widetilde{\mathcal{I}}_{\kappa\kappa}(w) &= \frac{1}{2}\int_\Omega D_{\rm f}^{ijst}[\overset{1}{e}_{ij}(w)\overset{1}{e}_{st}(w^*) + \overset{1}{e}_{ij}(w^*)\overset{1}{e}_{st}(w)]d\Omega \\
&+ \frac{1}{2}\int_\Omega D_{\rm f}^{ijst}\overset{1}{e}_{ij}(w^*)\overset{1}{e}_{st}(w^*)d\Omega \\
&+ \frac{1}{2}\int_\Omega D_{\rm s}^{ijst}\overset{0}{\epsilon}_{ij}(\mathbf{a}+\mathbf{a}^*)\overset{0}{\epsilon}_{st}(\mathbf{a}+\mathbf{a}^*)d\Omega \\
&+ \int_{\Gamma_6} k_{\rm s}^{\tau\tau} w_\tau^2 ds + \int_{\Gamma_7} k_{\rm s}^{mm} w_m^2 ds + \int_{\Gamma_8} k_{\rm s}^{ij} w_i w_j \,|_{i,j=1,2}\, ds \\
&- (\omega_{\rm p}^* \cdot \omega)_{H_t} - (w_{\rm p}^* \cdot w)_{H_\kappa}.
\end{aligned} \tag{21.22}$$

Let us comment only on weak continuity of $\widetilde{\mathcal{I}}_{\kappa\kappa}(w)$ in H_κ. Let

$$w_n \rightharpoonup w \text{ in } H_\kappa.$$

It is easily seen that the first area integral on the right-hand side of (21.6) is a linear functional in H_κ, as is the last term on the right-hand side of (21.22). Furthermore, by complete continuity of $\mathbf{K}_{t\kappa}(w)$ (Theorem 14.1), we have

$$\mathbf{K}_{t\kappa}(w_n) \to \mathbf{K}_{t\kappa}(w) \text{ in } H_t,$$

which implies

$$\overset{0}{\epsilon}_{ij}(w_n) \to \overset{0}{\epsilon}_{ij}(w) \text{ in } L_{2\Omega} \tag{21.23}$$

by (14.32). From (21.16) we have

$$\begin{aligned}
&\int_\Omega D_{\rm s}^{ijkl}\overset{0}{\epsilon}_{ij}(\mathbf{a}(w_n)+\mathbf{a}^*)\overset{0}{\epsilon}_{kl}(\mathbf{a}(w_n)+\mathbf{a}^*)d\Omega \\
&\qquad \to \int_\Omega D_{\rm s}^{ijkl}\overset{0}{\epsilon}_{ij}(\mathbf{a}(w)+\mathbf{a}^*)\overset{0}{\epsilon}_{kl}(\mathbf{a}(w)+\mathbf{a}^*)d\Omega.
\end{aligned} \tag{21.24}$$

Finally, the line integrals on the right-hand side of (21.22) are weakly continuous functionals, since by complete continuity of $\mathbf{K}_{t\kappa}(w)$ and the embedding Theorem 11.3 we have

$$\begin{aligned}
&\int_{\Gamma_6} k_{\rm s}^{\tau\tau} w_\tau^2(w_n)ds + \int_{\Gamma_7} k_{\rm s}^{mm} w_m^2(w_n)ds + \int_{\Gamma_8} k_{\rm s}^{ij} w_i(w_n)w_j(w_n)ds \\
&\quad \to \int_{\Gamma_6} k_{\rm s}^{\tau\tau} w_\tau^2(w)ds + \int_{\Gamma_7} k_{\rm s}^{mm} w_m^2(w)ds + \int_{\Gamma_8} k_{\rm s}^{ij} w_i(w)w_j(w)ds.
\end{aligned} \tag{21.25}$$

Relations (21.24), (21.25) complete the proof of weak continuity of $\widetilde{\mathcal{I}}_{t\kappa}(w)$, and thus of Lemma 21.4.

Lemma 21.5. *We have*

$$\operatorname{grad}_{H_\kappa} \mathcal{I}_{\kappa\kappa}(w) = w - \mathbf{G}_{\kappa\kappa}(w),$$

where the operator $\mathbf{G}_{\kappa\kappa}(w)$ *is given by* (13.36).

We do not present a proof of Lemma 21.5, as it is analogous to the proof of Lemma 21.3.

From Lemma 21.5 follows the following important theorem.

Theorem 21.2. *A vector function* $\mathbf{a} = (w_1, w_2, w)$ *is a generalized solution of problem* $t\kappa$ *if and only if* w *is a critical point of the functional* $\mathcal{I}_{\kappa\kappa}$ *in* H_κ, *while* $\omega = (w_1, w_2)$ *is determined from* (13.33).

Theorem 21.3. *Assume that all conditions of Theorem 16.1 hold. Then the functional* $\mathcal{I}_{\kappa\kappa}(w) = \mathcal{I}_{t\kappa}(\mathbf{a}(w) + \mathbf{a}^*)$ *defined by* (21.21) *is an increasing functional in* H_κ, *that is,* $\mathcal{I}_{\kappa\kappa}(w) \to \infty$ *as* $\|w\|_{H_\kappa} \to \infty$.

In fact, we have a stronger claim: *On spheres of sufficiently large radius in* H_κ *we have the inequality*

$$\mathcal{I}_{\kappa\kappa}(w) \ge mR^2, \quad m > 0. \tag{21.26}$$

We shall prove (21.26) using the scheme of the proof of Theorems 16.1 and 19.1. For this let us introduce a subdivision of the unit sphere $\Sigma_{H_\kappa}(1, 0)$ into two parts, $\Sigma'_{H_\kappa}(1, 0)$ and $\Sigma''_{H_\kappa}(1, 0)$. Let $\Sigma'_{H_\kappa}(1, 0)$ be the set of points v that satisfy the inequality

$$\|v\|^2_{H_\kappa} - \left(\omega^*_{\mathrm{p}} \cdot \omega_2(v)\right)_{H_t} < \frac{1}{2},$$

while $\Sigma''_{H_\kappa}(1, 0)$ is the set of points v for which

$$\|v\|^2_{H_\kappa} - \left(\omega^*_{\mathrm{p}} \cdot \omega_2(v)\right)_{H_t} \ge \frac{1}{2}. \tag{21.27}$$

By the radial projection $v \mapsto Rv$, this subdivision of $\Sigma_{H_\kappa}(1, 0)$ generates a subdivision of $\Sigma_{H_\kappa}(R, 0) = \Sigma'_{H_\kappa}(R, 0) \cup \Sigma''_{H_\kappa}(R, 0)$. Furthermore, on $\Sigma''_{H_\kappa}(R, 0)$ we have the inequality

$$\|w\|^2_{H_\kappa} - \left(\omega^*_{\mathrm{p}} \cdot \omega_2(w)\right)_{H_t} \ge \frac{1}{2} \|w\|^2_{H_\kappa}. \tag{21.28}$$

Lemma 21.6. *The weak closure of* $\Sigma'_{H_\kappa}(1, 0)$ *in* H_κ *does not contain zero.*

The lemma follows from the fact that were the weak closure of $\Sigma'_{H_\kappa}(1, 0)$ in H_κ to contain zero, then there would exist a sequence in $\Sigma'_{H_\kappa}(1, 0)$ such that

$$v_n \rightharpoonup 0 \text{ in } H_\kappa.$$

By Theorem 14.1 on complete continuity of $\mathbf{K}_{t\kappa\mu}(w)$ we would have then

$$\|\omega_2(v_n)\|_{H_t} \to 0,$$

which is impossible because $\|v_n\|_{H_\kappa} = 1$.

To estimate $\mathcal{I}_{\kappa\kappa}(w)$ on $\Sigma'_{H_\kappa}(R,\ 0)$ we note that by Theorem 16.1 inequalities (16.12)–(16.15) hold, from which we obtain

$$\begin{aligned}\mathcal{I}_{\kappa\kappa}(w) &\geq \frac{1}{2}\Pi(w) - \left|(\omega_{\mathrm{p}}^* \cdot \omega)_{H_\iota}\right| - \left|(w_{\mathrm{p}}^* \cdot w)_{H_\kappa}\right| \\ &\geq \frac{1}{2}\left[\Pi_4(w) - \sum_{\mu=0}^{3}\left|\Pi_\mu(w)\right|\right] - \left\|\omega_{\mathrm{p}}^*\right\|_{H_\iota}\|\omega\|_{H_\iota} - \left\|w_{\mathrm{p}}^*\right\|_{H_\kappa}\|w\|_{H_\kappa} \\ &\geq \frac{1}{2}m\left(\|w\|_{H_\kappa}^4 - \sum_{\mu=0}^{3}\|w\|_{H_\kappa}^\mu\right) - \left\|\omega_{\mathrm{p}}^*\right\|_{H_\iota}\|\omega\|_{H_\iota} - \left\|w_{\mathrm{p}}^*\right\|_{H_\kappa}\|w\|_{H_\kappa}.\end{aligned} \tag{21.29}$$

If we take (14.3) into account, we have

$$\|\omega\|_{H_\iota} \leq m\left(1 + \|w\|_{H_\kappa} + \|w\|_{H_\kappa}^2\right), \tag{21.30}$$

while from (21.29), (21.30) we have for elements of $\Sigma'_{H_\kappa}(R,\ 0)$ the estimate

$$\mathcal{I}_{\kappa\kappa}(w) \geq m\,\|w\|_{H_\kappa}^4. \tag{21.31}$$

For points in $\Sigma''_{H_\kappa}(R,\ 0)$, by neglecting all the nonnegative terms on the right-hand side of (21.22) we have

$$\begin{aligned}\mathcal{I}_{\kappa\kappa}(w) \geq \|w\|_{H_\kappa}^2 &- \left|\frac{1}{2}\int_\Omega D_{\mathrm{f}}^{ijst}\left[\overset{1}{e}_{ij}(w)\overset{1}{e}_{st}(w^*) + \overset{1}{e}_{ij}(w^*)\overset{1}{e}_{st}(w)\right]d\Omega\right| \\ &- \left|(\omega_{\mathrm{p}}^* \cdot \omega_2(w))_{H_\iota}\right| - \left|(\omega_{\mathrm{p}}^* \cdot (\omega_1(w) + \omega_0))_{H_\iota}\right| - \left\|w_{\mathrm{p}}^*\right\|_{H_\kappa}\|w\|_{H_\kappa}.\end{aligned} \tag{21.32}$$

Now we take into account the inequality

$$\left|\frac{1}{2}\int_\Omega D_{\mathrm{f}}^{ijst}\left[\overset{1}{e}_{ij}(w)\overset{1}{e}_{st}(w^*) + \overset{1}{e}_{ij}(w^*)\overset{1}{e}_{st}(w)\right]d\Omega\right| \leq m\,\|w\|_{H_\kappa}, \tag{21.33}$$

which follows from (12.25), (3.17), and the obvious inequality that follows from (14.2),

$$\left|(\omega_{\mathrm{p}}^* \cdot (\omega_1(w) + \omega_0))_{H_\iota}\right| \leq m(1 + \|w\|_{H_\kappa}). \tag{21.34}$$

If, moreover, we take into consideration (21.28), then from (21.32)–(21.34) on $\Sigma''_{H_\kappa}(R,\ 0)$ we obtain

$$\mathcal{I}_{\kappa\kappa}(w) \geq m\,\|w\|_{H_\kappa}^2 \tag{21.35}$$

for sufficiently large $\|w\|_{H_\kappa}$. Then (21.26) follows from (21.31), (21.35). Theorem 21.3 is proved.

Theorem 21.4. *Let all the conditions of Theorem 16.1 be satisfied. Then the boundary value problem $t\kappa$ has at least one generalized solution w that furnishes the absolute minimum of the functional $\mathcal{I}_{\kappa\kappa}(w)$; $\omega = (w_1,\ w_2)$ is determined from* (13.33).

We remind the reader that in our terminology the fact that w_0 furnishes the absolute minimum means that in all of H_κ,

$$\mathcal{I}_{\kappa\kappa}(w) \geq \mathcal{I}_{\kappa\kappa}(w_0).$$

Having established (21.26), we can now prove Theorem 21.4.

Indeed, by (21.22), Lemma 21.4, and (21.26) we conclude that the functional $\mathcal{I}_{\kappa\kappa}(w)$ is bounded from below in H_κ. Let d be the greatest lower bound of $\mathcal{I}_{\kappa\kappa}(w)$ and let w_n be a sequence such that

$$\mathcal{I}_{\kappa\kappa}(w_n) \to d. \tag{21.36}$$

Below we shall call such a sequence w_n an *absolutely minimizing sequence*.

It is easily seen that the entire sequence w_n is bounded. Indeed, if it is not bounded, there would exist a subsequence (which we again denote by w_n) such that

$$\|w_n\|_{H_\kappa} \to \infty \text{ as } n \to \infty,$$

but then (21.36) would contradict Theorem 21.3 (relation (21.26)). Since w_n is bounded, it can be taken to converge weakly to an element w_0. Furthermore, we have

$$\mathcal{I}_{\kappa\kappa}(w_n) = \frac{1}{2}\|w_n\|^2_{H_\kappa} + \widetilde{\mathcal{I}}_{\kappa\kappa}(w_n) \to d \text{ as } w_n \to w_0,$$

and moreover,

$$\left\|\frac{w_m - w_n}{2}\right\|^2_{H_\kappa} + \left\|\frac{w_m + w_n}{2}\right\|^2_{H_\kappa} = \frac{1}{2}\left(\|w_m\|^2_{H_\kappa} + \|w_n\|^2_{H_\kappa}\right) \tag{21.37}$$

for any two terms w_n, w_m of the weakly convergent subsequence that we chose from the absolutely minimizing sequence.

From (21.37) it follows that

$$\begin{aligned}\left\|\frac{w_m - w_n}{2}\right\|^2_{H_\kappa} &= \frac{1}{2}\left(\|w_m\|^2_{H_\kappa} + \|w_n\|^2_{H_\kappa}\right) - \left\|\frac{w_m + w_n}{2}\right\|^2_{H_\kappa} + \frac{1}{2}\widetilde{\mathcal{I}}_{\kappa\kappa}(w_m)\\ &\quad + \frac{1}{2}\widetilde{\mathcal{I}}_{\kappa\kappa}(w_n) - \widetilde{\mathcal{I}}_{\kappa\kappa}\left(\frac{w_m + w_n}{2}\right) - \frac{1}{2}\widetilde{\mathcal{I}}_{\kappa\kappa}(w_m)\\ &\quad - \frac{1}{2}\widetilde{\mathcal{I}}_{\kappa\kappa}(w_n) + \widetilde{\mathcal{I}}_{\kappa\kappa}\left(\frac{w_m + w_n}{2}\right).\end{aligned} \tag{21.38}$$

Furthermore, since w_m, $w_n \rightharpoonup w_0$,

$$\frac{1}{2}\widetilde{\mathcal{I}}_{\kappa\kappa}(w_m) - \frac{1}{2}\widetilde{\mathcal{I}}_{\kappa\kappa}(w_n) + \widetilde{\mathcal{I}}_{\kappa\kappa}\left(\frac{w_m + w_n}{2}\right) \to 0, \tag{21.39}$$

and

$$\frac{d}{2} \leq \frac{1}{2}\left(\|w_m\|^2_{H_\kappa} + \widetilde{\mathcal{I}}_{\kappa\kappa}(w_m)\right) \leq \frac{d}{2} + \epsilon_m,$$
$$\frac{d}{2} \leq \frac{1}{2}\left(\|w_n\|^2_{H_\kappa} + \widetilde{\mathcal{I}}_{\kappa\kappa}(w_n)\right) \leq \frac{d}{2} + \epsilon_n, \tag{21.40}$$

Besides, we have

$$d \leq \left\|\frac{w_m + w_n}{2}\right\|^2_{H_\kappa} + \widetilde{\mathcal{I}}_{\kappa\kappa}\left(\frac{w_m + w_n}{2}\right). \tag{21.41}$$

In (21.40), (21.41), ϵ_m, $\epsilon_n \to 0$ as m, $n \to \infty$. From (21.38)–(21.41) it follows that

$$\left\|\frac{w_m - w_n}{2}\right\|^2_{H_\kappa} \leq \frac{d}{2} + \epsilon_m + \frac{d}{2} + \epsilon_n - d \leq \epsilon_m + \epsilon_n \to 0,\ m,\ n \to \infty.$$

Thus w_m is strongly convergent, and

$$w_m \to w_0,\ \mathcal{I}_{\kappa\kappa}(w_0) = d.$$

Therefore, w_0 is an element that furnishes the absolute minimum of $\mathcal{I}_{\kappa\kappa}(w)$, so that, as follows from Lemma 21.5, w_0 generates a generalized solution of problem $t\kappa$. Theorem 21.4 is proved.

Theorem 21.5. *Let all the conditions of Theorem 16.2 be satisfied. Then the boundary value Problem $t\kappa$ has at least one generalized solution w that furnishes the absolute minimum of the functional $\mathcal{I}_{\kappa\kappa}(w)$; here ω is given by* (13.33).

Theorem 21.6. *Let all the conditions of Theorem 16.3 be satisfied. Then the boundary value Problem $t\kappa$ has at least one generalized solution w that furnishes the absolute minimum of the functional $\mathcal{I}_{\kappa\kappa}(w)$; here ω is given by* (13.33).

21.3. In Section 16 we also established solvability of boundary value problems of nonlinear shallow shell theory in displacements. It is of interest to compare the two methods of proving solvability that we have presented. In this respect, let us indicate some of their common origins. The reader has already probably understood that both these methods of proof are based on the inequality (16.18) in Lemma 16.3. At the same time, the two methods are essentially different. More than that, similarities in the general formulation of the results notwithstanding, there is a significant difference in the information on the solution that is being provided by each method of proof. The proofs of Section 16 are based on the computation of the winding number of the vector field $w - \mathbf{G}_{\kappa\kappa}(w)$. As a result, we obtain the existence of a solution with a nonzero index. At the same time, in Section 21, using variational principles, we obtain a solution that furnishes the absolute minimum of $\mathcal{I}_{\kappa\kappa}(w)$. Very frequently, these will actually be qualitatively different solutions, and such cases will occur particularly often in the case of nonunique solutions of the problem. Therefore, it makes no sense to counterpose the two methods. Furthermore, their use in tandem will frequently help us to estimate the number of solutions of the boundary value problem under consideration.

21.4. Let us study in more detail the structure of absolutely minimizing sequences. First of all, let us note that in theorems 21.4–21.6 we have actually established the following important fact.

Theorem 21.7. *Under the conditions of Theorems 16.1–16.3 any absolutely minimizing sequence w_n for the corresponding functional $\mathcal{I}_{\kappa\kappa}(w)$ is strongly compact, and every strong limit of that sequence w_0 furnishes an absolute minimum of $\mathcal{I}_{\kappa\kappa}(w)$.*

Let us introduce some definitions.

Definition 21.1. Assume that everywhere in H_κ we have the relation

$$\mathcal{I}_{\kappa\kappa}(w) > \mathcal{I}_{\kappa\kappa}(w_0), \quad w \neq w_0. \tag{21.42}$$

Then we call the point w_0 a point of *strict absolute minimum.*

Theorem 21.8. *Let w_0 be a strict absolute minimum of $\mathcal{I}_{\kappa\kappa}(w)$. Then every absolutely minimizing sequence w_n converges strongly to w_0.*

Theorem 21.8 follows directly from Theorem 21.6.

Theorem 21.9. *Let $w_{01}, \ldots, w_{0N}$ be absolute minima of the functional $\mathcal{I}_{\kappa\kappa}(w)$, the number of which we assume to be finite, so that*

$$\mathcal{I}_{\kappa\kappa}(w_{01}) = \mathcal{I}_{\kappa\kappa}(w_{02}) = \cdots = \mathcal{I}_{\kappa\kappa}(w_{0N}) = d.$$

Then any absolutely minimizing sequence w_n can be decomposed into a finite number (not exceeding N) of strongly convergent sequences each of which converges to a different element w_{0k}, $k = 1, \ldots, N$.

To prove this theorem, we note that since the number of the elements w_{0k} is finite, each of them can be taken to lie in a ball $B_k(\rho, w_{0k})$ of radius ρ such that the $\overline{B}_k$ have no common points. Next, it is easily seen that there is only a finite number of elements of the sequence w_n not contained in the balls $\overline{B}_k$. Indeed, were there to be an infinite number of such elements, that would mean that there is an absolutely minimizing sequence not contained in $\overline{B}_k$. It would obviously be strongly compact in H_κ and then outside of $\cup B_k$ there would be another element that furnishes an absolute minimum for the functional $\mathcal{I}_{\kappa\kappa}(w)$, which is impossible. Let us now consider the balls B_k that contain an infinite number of elements of w_n. Clearly, these elements again form an absolutely minimizing sequence. By the arguments above, it will be strongly compact, and thus any of its strong limit points is a point of absolute minimum of $\mathcal{I}_{\kappa\kappa}$. But only w_{0k} can be such a point in B_k, so that the infinite sequence of elements in B_k is a sequence that strongly converges to w_{0k}. Thus the entire absolutely minimizing sequence is composed of a finite number of convergent sequences and a finite number of elements lying outside of $\cup B_k$ and in some of these balls. These can be said to belong to any convergent sequence. Theorem 21.9 is proved.

21.5. We introduce the following definitions.

Definition 21.2. Assume that on a sphere $\Sigma_{H_\kappa}(r,\ w^{**})$ the following inequality holds:

$$\mathcal{I}_{\kappa\kappa}(w) > \mathcal{I}_{\kappa\kappa}(w^*) \text{ (respectively } \mathcal{I}_{\kappa\kappa}(w) < \mathcal{I}_{\kappa\kappa}(w^*)),$$

where w^* is inside $\Sigma_{H_\kappa}(r,\ w^{**})$. Then we shall call $\Sigma_{H_\kappa}(r,\ w^{**})$ a *sphere of relative minimum (maximum)*.

Definition 21.3. Suppose for a point w_0 there is a ball $B(r,\ w_0)$ where the following inequality holds:

$$\mathcal{I}_{\kappa\kappa}(w) \geq \mathcal{I}_{\kappa\kappa}(w_0)\ (\mathcal{I}_{\kappa\kappa}(w) \leq \mathcal{I}_{\kappa\kappa}(w_0)),\quad w \in B(r,\ w_0). \tag{21.43}$$

Then we shall call w_0 a *relative minimum (maximum)* of the functional $\mathcal{I}_{\kappa\kappa}(w)$, while $B(r,\ w_0)$ will be called a *ball of relative minimum (maximum)*. If in (21.43) we have a strict inequality, then w_0 will be a *strict relative minimum (maximum)* of the functional $\mathcal{I}_{\kappa\kappa}(w)$. Similarly, $B(r,\ w_0)$ will then be called a *ball of strict relative minimum (maximum)*.

Theorem 21.10. *Inside each sphere $\Sigma_{H_\kappa}(r,\ w^{**})$ of relative minimum (maximum) there is at least one generalized solution w_0 of the boundary value problem $t\kappa$ that is a relative minimum (maximum) of $\mathcal{I}_{\kappa\kappa}$ for a ball $B(r_1,\ w_0)$, $r_1 > r$.*

To prove the theorem, we note that by Lemma 21.4, in $\overline{B}(r,\ w^{**})$ the functional is bounded from below (above). Therefore, its sharp lower bound d_{rel} (sharp upper bound $\mathcal{D}_{\text{rel}}$) is defined in $\overline{B}(r,\ w^{**})$, and let $w_n \in \overline{B}(r,\ w^{**})$ be a sequence such that

$$\mathcal{I}_{\kappa\kappa}(w_n) \to d_{\text{rel}}\ (\mathcal{D}_{\text{rel}}).$$

We shall call such a sequence a relatively minimizing (maximizing) sequence. Clearly, w_n is bounded and weakly compact; let w_0 be one of its weak limits.

Repeating verbatim the arguments of Theorem 21.3 for the case of a relatively minimizing sequence w_n, we obtain the existence of an element w_0, for which

$$\mathcal{I}_{\kappa\kappa}(w_0) = d_{\text{rel}}.$$

Next, it is obvious that by (21.43), w_0 is necessarily contained inside $\Sigma_{H_\kappa}(r,\ w^{**})$ and therefore is a critical point of $\mathcal{I}_{\kappa\kappa}(w)$ and thus is a generalized solution of the problem $t\kappa$. If w_n is a relatively maximizing sequence, we use the functional $-\mathcal{I}_{\kappa\kappa}(w)$, for which this sequence will already be relatively minimizing, and we also deduce that w_0 is a critical point of $\mathcal{I}_{\kappa\kappa}(w)$. Theorem 21.10 is proved.

In the process of proof we have established a number of facts that allow us to state the following result.

Theorem 21.11. *Under the conditions of Theorems 16.1–16.3 every relatively minimizing (maximizing) sequence w_n is strongly compact in H_κ, and every strong limit w_0 of that sequence is a point of relative minimum (maximum) of $\mathcal{I}_{\kappa\kappa}(w)$.*

21.6. In this section we shall study the functional $\mathcal{I}_{t\kappa}(\mathbf{a} + \mathbf{a}^*)$ in the space $H_{t\kappa}$. Recall that earlier, the functional $\mathcal{I}_{t\kappa}(\mathbf{a} + \mathbf{a}^*)$ was considered on HS1 defined by (13.33).

Lemma 21.7. *The functional $\mathcal{I}_{t\kappa}(\mathbf{a}+\mathbf{a}^*)$ is bounded from below for arbitrary* $\mathbf{a}$ *in $H_{t\kappa}$.*

To prove Lemma 21.7 we write $\mathcal{I}_{t\kappa}(\mathbf{a}+\mathbf{a}^*)$ in the form

$$\mathcal{I}_{t\kappa}(\mathbf{a}+\mathbf{a}^*) = \frac{1}{2}\,\|\omega\|^2_{H_\kappa} + \mathcal{I}_{1t\kappa} + \mathcal{I}_{0t\kappa}, \tag{21.44}$$

where

$$\mathcal{I}_{1t\kappa} = \int_\Omega D_{\mathrm{s}}^{ijkl}\overset{\widetilde{0}}{\epsilon}_{kl}(\omega)\overset{0}{\epsilon}_{ij}(\mathbf{a}^*)\,d\Omega + (\omega^*_{\mathrm{p}}\cdot\omega)_{H_t}.$$

The functional $\mathcal{I}_{0t\kappa}$ depends only on w. The structure of (21.44) follows directly from (21.2). Next, it is easily seen that $\mathcal{I}_{1t\kappa}$ is a functional that is linear with respect to ω in H_t if w is held fixed, while the functional $\mathcal{I}_{t\kappa}$ is quadratic in ω. By well-known results [196], for fixed w, $\mathcal{I}_{t\kappa}(\mathbf{a}+\mathbf{a}^*)$ has a unique minimum in H_t. At the pont of minimum, as is easily seen, ω and w are related by (13.33). In other words, the minimum of $\mathcal{I}_{t\kappa}$ for a fixed w is attained on HS1 (13.33). However, on HS1 (13.33) $\mathcal{I}_{t\kappa}(\mathbf{a}+\mathbf{a}^*)$ becomes $\mathcal{I}_{\kappa\kappa}(w)$, and by Theorems 21.4–21.6, $\mathcal{I}_{\kappa\kappa}(w)$ has an absolute minimum in all of H_κ. Therefore, we have the inequality

$$\mathcal{I}_{t\kappa}(\mathbf{a}+\mathbf{a}^{**}) \geq \mathcal{I}_{\kappa\kappa}(w) = d,$$

which concludes the proof of Lemma 21.7.

Lemma 21.8. *Every absolutely minimizing sequence $\mathbf{a}_n$ is bounded in $H_{t\kappa}$.*

To prove the lemma, we introduce a sequence $\widetilde{\mathbf{a}}_n$ with components $\omega_n = \mathbf{K}_{t\kappa}(w_n)$, w_n. We clearly have

$$d \leq \mathcal{I}_{t\kappa}(\widetilde{\mathbf{a}}_n) = \mathcal{I}_{\kappa\kappa}(w_n) \leq \mathcal{I}_{t\kappa}(\mathbf{a}_n) \leq \mathcal{D}, \tag{21.45}$$

where $\mathcal{D}$ is some constant. Thus, $\widetilde{\mathbf{a}}_n$ is a minimizing sequence on HS1 (13.33). But then in the proof of Theorem 21.4 we obtained

$$\|w_n\|_{H_\kappa} \leq R, \tag{21.46}$$

where R does not depend on n. If (21.45) is satisfied, it follows from (21.44) that

$$\mathcal{D} \geq \mathcal{I}_{t\kappa}(\mathbf{a}_n) \geq \frac{1}{2}\,\|\omega_n\|^2_{H_t} - m(\|\omega_n\|_{H_t} + 1), \tag{21.47}$$

whence

$$\|\omega_n\|_{H_t} \leq R.$$

Lemma 21.8 follows from (21.46), (21.47).

Theorem 21.12. *Assume that the conditions of Theorems 16.1–16.3 are satisfied. Then every absolutely minimizing sequence $\mathbf{a}_n$ is strongly compact in $H_{t\kappa}$, and every limit of sequences in $\mathbf{a}_n$ furnishes an absolute minimum of $\mathcal{I}_{t\kappa}$ in $H_{t\kappa}$.*

To prove this theorem, we note also that the functional $\mathcal{I}_{t\kappa}(\mathbf{a}+\mathbf{a}^*)$ can be written in the form

$$\mathcal{I}_{t\kappa}(\mathbf{a}+\mathbf{a}^*) = \frac{1}{2}\,\|\mathbf{a}\|^2_{H_{t\kappa}} + \widetilde{\mathcal{I}}_{t\kappa}(\mathbf{a}), \tag{21.48}$$

where the functional $\widetilde{\mathcal{I}}_{t\kappa}(\mathbf{a})$ is weakly continuous in $H_{t\kappa}$.

The structure of (21.48) also directly follows from (21.1). Weak continuity of $\widetilde{\mathcal{I}}_{t\kappa}(\mathbf{a})$ follows immediately from complete continuity of the operators of embedding H_t into $L_{q\Omega}$, and of H_κ into $W^{(1)}_{q\Omega}$, C_Ω for any $q \geq 1$. Furthermore, by Lemma 21.8 the whole set $\mathbf{a}_n$ is bounded and therefore weakly compact. Thus from any absolutely minimizing sequence we can choose a weakly convergent sequence $\mathbf{a}_n$. However, the structure of (21.48) allows us to repeat the arguments of Theorem 21.4 and to establish strong convergence of $\mathbf{a}_n$ and thus strong compactness of any absolutely minimizing sequence. Theorem 21.12 is proved.

21.7. We state the following results without proof.

Theorem 21.13. *Let $\mathbf{a}_0$ be a strict absolute minimum of $\mathcal{I}_{t\kappa}(\mathbf{a})$ in $H_{t\kappa}$. Then any absolutely minimizing sequence converges strongly to $\mathbf{a}_0$.*

Theorem 21.14. *Let $\mathbf{a}_{01}, \ldots, \mathbf{a}_{0N}$ be absolute minima of the functional $\mathcal{I}_{t\kappa}$ in $H_{t\kappa}$, the number of which we assume to be finite. Then any absolutely minimizing sequence w_n can be decomposed into a finite number (not exceeding N) of convergent sequences each of which converges to its absolute minimum $\mathbf{a}_{0k}$.*

Theorem 21.15. *Inside each sphere $\Sigma_{H_{t\kappa}}(r, \mathbf{a}^{**})$ of relative minimum (maximum) there is at least one generalized solution $\mathbf{a}_0$ of the boundary value problem $t\kappa$ that is a relative minimum (maximum) of $\mathcal{I}_{t\kappa}$ for a ball $B(r_1, \mathbf{a}_0)$, $r_1 < r$.*

Theorem 21.16. *Under the conditions of Theorems 16.1–16.3 every relatively minimizing (maximizing) sequence $\mathbf{a}_n$ is strongly compact in $H_{t\kappa}$, and every strong limit $\mathbf{a}_0$ of that sequence is a generalized solution of the problems $t\kappa$ that is a point of relative minimum (maximum) of $\mathcal{I}_{t\kappa}$.*

We do not present proofs of Theorems 21.13–21.16, as they are identical to those of Theorems 21.8–21.11.

22. The Variational Approach to the Problem of Solvability of Boundary Value Problems of Nonlinear Shallow Shell Theory with an Airy Stress Function

22.1. In this section we shall study the problem of solvability of boundary value problems of shallow shell theory using a variational approach. As has been mentioned, the results of this approach complement the ones obtained by topological methods. The variational approach gives us important information on the convergence of (absolutely and relatively) minimizing sequences that are used in the numerical solution of corresponding problems.

We shall assume that Conditions 1–8 of Section 17 are satisfied; we start with the functional (7.25), by variation of which we obtained equations (7.51), (7.60), the defining equations of nonlinear shallow shell theory with an Airy stress function. From (7.25), taking into account (17.13), we have ($w_4 = \partial w/\partial m$)

$$
\begin{aligned}
&\mathcal{I}_{9\kappa}(\Psi, w+w^*)\\
&\quad= \frac{1}{2}\Big\{ \int_\Omega [D_{\mathrm{f}}^{ijkl}\nabla_{ij}(w+w^*)\nabla_{kl}(w+w^*)\\
&\qquad\qquad - 2(w+w^*)(C^{ik}C^{jl}\nabla_{kl}\Psi + \widetilde{T}_{\mathrm{p}}^{ij})B_{ij}\\
&\qquad\qquad + (C^{ik}C^{jl}\nabla_{kl}\Psi + \widetilde{T}_{\mathrm{p}}^{ij})(w+w^*)_{\alpha^i} \times (w+w^*)_{\alpha^j}\\
&\qquad\qquad - C_{ijkl,\mathrm{s}}(C^{k\lambda}C^{l\mu}\nabla_{\lambda\mu}\Psi + \widetilde{T}_{\mathrm{p}}^{kl})(C^{is}C^{jt}\nabla_{st}\Psi + \widetilde{T}_{\mathrm{p}}^{ij})]\,d\Omega\\
&\qquad\qquad + \int_{\Gamma_2} k_{\mathrm{f}}^{44}(w_4+w_4^*)^2\,ds - \int_{\Gamma_2}(w_4+w_4^*)\widetilde{M}^m\,ds\\
&\qquad\qquad - \int_\Omega (w+w^*)R^3\,d\Omega\Big\}.
\end{aligned}
\tag{22.1}
$$

Lemma 22.1. *The functional $\mathcal{I}_{9\kappa}$ (22.1) is defined everywhere in $H_{9\kappa}$.*

Indeed, if $w,\ w^* \in H_\kappa$, then as we have already remarked more than once, $\nabla_{ij}w,\ \nabla_{ij}w^* \in L_{2\Omega}$, and the first integral on the right-hand side of (7.25) is defined. Next, since $\nabla_{ij}\Psi \in L_{2\Omega}$ and by the embedding Theorem 12.3 (see (12.28)), $w,\ w^* \in L_{2\Omega}$, by condition 1 of Section 17, $B_{ij} \in C_\Omega$, and by condition 7, $\widetilde{T}_{\mathrm{p}}^{ij} \in L_{2\Omega}$. From all this it follows that the second and the fourth terms on the right-hand side of (7.25) are defined. If we note that by the embedding Theorem 12.3 (relation (12.26), $w_{\alpha^i},\ w^*_{\alpha^i} \in L_{q\Omega}$ for any $q \geq 1$, we can claim that the third term in (7.25) is also defined. Finally, $\int_{\Gamma_2} k_{\mathrm{f}}^{44}(w_4 + w_4^*)^2\,ds$ is computable, as (Theorem 12.3, (12.26)) $w_4,\ w_4^* \in L_{q\Omega}$ for any $q \geq 1$. The last two integrals in (7.25) are defined by the condition (17.7) on loading terms. Lemma 22.1 is proved.

Below we shall need a representation of the functional $\mathcal{I}_{9\kappa}$ in terms of arguments $\Psi + \gamma\Theta$, and $w + w^* + \gamma\varphi$. Directly from (7.25) we obtain

$$
\mathcal{I}_{9\kappa}(\Psi+\gamma\Theta,\ w+w^*+\gamma\varphi) = \mathcal{I}_{9\kappa}(\Psi,\ w+w^*) + \sum_{\mu=1}^{4}\gamma^\mu \mathcal{I}_{9\kappa\mu}(\psi,\ w+w^*,\ \mathbf{d}),
$$

where the vector **d** has components $(\Theta,\ \varphi)$, and furthermore,

$$\begin{aligned}
&\mathcal{I}_{9\kappa 1}(\Psi,\ w+w^*,\ \mathbf{d}) \\
&= (w\cdot\varphi)_{H_\kappa} - (\Psi\cdot\Theta)_{H_9} + (w^*\cdot\varphi)_{H_\kappa} \\
&\quad - \int_\Omega [(C^{ik}C^{jl}\nabla_{kl}\Psi + \widetilde{T}^{ij}_{\text{p}})(B_{ij}\varphi - \varphi_{\alpha^i}(w+w^*)_{\alpha^j} + R^3\varphi)]\,d\Omega \\
&\quad - \int_{\Gamma_2} \widetilde{M}^m \frac{\partial\varphi}{\partial m}\,ds - \int_\Omega C^{ik}C^{jl} \\
&\quad \times \Big[\frac{1}{2}(w+w^*)_{\alpha^i}(w+w^*)_{\alpha^j} - (w+w^*)B_{ij}\Big]\nabla_{kl}\Theta\,d\Omega \\
&\quad + \int_\Omega C^{\lambda\mu}_{*ij}\widetilde{T}^{ij}_{\text{p}}\nabla_{\lambda\mu}\Theta\,d\Omega \\
&= ((w - \mathbf{G}_{\kappa\kappa}(w))\cdot\varphi)_{H_\kappa} + ((\mathbf{K}_{9\kappa}(w) - \Psi)\cdot\Theta)_{H_9} \\
&= ((\mathbf{c} - \mathbf{G}_{9\kappa}(\mathbf{c}))\cdot\mathbf{d})_{H_{9\kappa}}.
\end{aligned} \tag{22.2}$$

In (22.2) the vector **c** has components $(\Psi,\ w)$; the operator $\mathbf{G}_{9\kappa}$ is defined by relation (17.20). The vector **d** has components $(\Theta,\ \varphi \in H_{9\kappa})$. Equation (22.2) itself has been obtained using (17.16), (17.18). Next,

$$\begin{aligned}
\mathcal{I}_{9\kappa 2}(\Psi,\ w+w^*,\ \mathbf{d}) &= \frac{1}{2}\,\|\varphi\|^2_{H_\kappa} - \int_\Omega \varphi B_{ij}C^{ik}C^{jl}\nabla_{kl}\Theta\,d\Omega \\
&\quad + \frac{1}{2}\int_\Omega \Big\{C^{ik}C^{jl}\nabla_{kl}\Theta[\varphi_{\alpha^i}(w+\overset{0}{w})_{\alpha^j} + \varphi_{\alpha^j}(w+\overset{0}{w})_{\alpha^i}] \\
&\quad + (C^{ik}C^{jl}\nabla_{kl}\Psi + \widetilde{T}^{ij}_{\text{p}})\varphi_{\alpha^i}\varphi_{\alpha^j}\Big\}\,d\Omega - \frac{1}{2}\,\|\Psi\|^2_{H_9}\,,\ \mathcal{I}_{9\kappa 3} \\
&= \frac{1}{2}\int_\Omega C^{ik}C^{jl}\nabla_{kl}\Theta\varphi_{\alpha^i}\varphi_{\alpha^j}\,d\Omega.
\end{aligned}$$

Theorem 22.1. *Assume that all the Conditions 1–8 of Section 17 are satisfied. Then we have that*

$$\operatorname{grad}_{H_{9\kappa}}\mathcal{I}_{9\kappa}(\Psi,\ w+w^*) = \mathbf{c} - \mathbf{G}_{9\kappa}(\mathbf{c}),$$

and therefore the set of critical points of the functional $\mathcal{I}_{9\kappa}$ in $H_{9\kappa}$ coincides with the set of generalized solutions of Problem 9κ (see (17.20)*).*

Theorem 22.1 provides a rigorous justification for the mixed variational principle of Alumyae, which we used in the derivation of the main equations (7.51), (7.60).

Once we have established (22.2), which follows by direct comparison of its left-hand side with (17.16), (17.18), the proof of Theorem 22.1 is based on the estimates

$$|\mathcal{I}_{9\kappa 2}| \le m\,\|\mathbf{d}\|^2_{H_{9\kappa}}\,,\quad |\mathcal{I}_{9\kappa 3}| \le m\,\|\mathbf{d}\|^3_{H_{9\kappa}}\,,$$

which follow immediately from (12.34) and the embedding Theorem 12.3 (see (12.25), (12.28)).

From Theorem 22.1 we immediately obtain the following result.

Theorem 22.2. *A vector* $\mathbf{c} = (\Psi,\ w)$ *is a generalized solution of Problem* 9κ *if and only if* $\mathbf{c}$ *is a critical point of the functional* $\mathcal{I}_{9\kappa}$ *in* $H_{9\kappa}$.

22.2. An important part in the variational approach to boundary value problems of nonlinear shell theory is played by a different functional,

$$\begin{aligned}\mathcal{I}_\kappa(w+w^*) &= \frac{1}{2}\left\|w+w^*\right\|^2_{H_\kappa} + \frac{1}{2}\left\|\Psi\right\|^2_{H_9} \\ &\quad + \int_\Omega \widetilde{T}^{ij}_{\text{p}}\Big[\frac{1}{2}(w+w^*)_{\alpha^i}(w+w^*)_{\alpha^j} - B_{ij}(w+w^*)\Big]\,d\Omega \\ &\quad - \int_{\Gamma_2} \widetilde{M}^m w_4\,ds - \int_\Omega R^3 w\,d\Omega.\end{aligned} \tag{22.3}$$

In (22.3) Ψ is taken to be expressed in terms of w by (17.17), and the functional is considered on the hypersurface HS2; w ranges through the entire space H_κ.

From Theorem 22.2 we obtain the following.

Theorem 22.3. *We have the relation*

$$\operatorname{grad}_{H_\kappa} \mathcal{I}_\kappa(w) = w - \mathbf{G}_\kappa(w). \tag{22.4}$$

To prove this, let us consider $\mathcal{I}_\kappa(w+\gamma\varphi)$. Since, as is easily seen, $\mathcal{I}_\kappa(w+\gamma\varphi)$ is a fourth-degree polynomial in γ, we have the representation

$$\mathcal{I}_\kappa(w+\gamma\varphi) = \mathcal{I}_\kappa(w) + \sum_{\mu=1}^{4} \gamma^\mu \mathcal{I}_{\kappa\mu}(w,\ \varphi),$$

where

$$\mathcal{I}_{\kappa\mu}(w,\ \varphi) = \frac{1}{\mu!}\frac{d^\mu \mathcal{I}_\kappa}{d\gamma^\mu}\bigg|_{\gamma=0}.$$

From (22.3) we obtain

$$\begin{aligned}\mathcal{I}_{\kappa 1} = \frac{d\mathcal{I}_\kappa}{d\gamma}\bigg|_{\gamma=0} &= ((w+w^*)\cdot\varphi)_{H_\kappa} + \left(\Psi\cdot\frac{d\Psi}{d\gamma}\right)_{H_9}\bigg|_{\gamma=0} \\ &\quad + \int_\Omega \big[\widetilde{T}^{ij}_{\text{p}}(w+w^*)_{\alpha^i}\varphi_{\alpha^j} - B_{ij}\varphi\big]\,d\Omega \\ &\quad - \int_{\Gamma_2} \widetilde{M}^m \varphi_4\,ds - \int_\Omega R^3\varphi\,d\Omega,\quad \varphi_4 = \frac{\partial\varphi}{\partial m}.\end{aligned} \tag{22.5}$$

From (17.15), (17.34) we have

$$\left(\frac{d\Psi}{d\gamma}\bigg|_{\gamma=0}\cdot\Theta\right)_{H_9} = \int_\Omega C^{ik}C^{jl}[(w+w^*)_{\alpha^i}\varphi_{\alpha^j} - B_{ij}\varphi]\nabla_{kl}\Theta\,d\Omega. \tag{22.6}$$

Relation (22.6) is obtained if in (17.15) we change variables $w \mapsto w + w^* + \gamma\varphi$ and differentiate with respect to γ. Setting $\Theta = \Psi$ in (22.6), we obtain

$$\left(\left. \frac{d\Psi}{d\gamma} \right|_{\gamma=0} \cdot \Theta \right)_{H_9} = \int_\Omega C^{ik} C^{jl} [(w + \overset{0}{w})_{\alpha^i} \varphi_{\alpha^j} - B_{ij}\varphi] \nabla_{kl} \Psi \, d\Omega. \quad (22.7)$$

From (22.5), (22.7) it follows that

$$\mathcal{I}_{9\kappa 1} = \left. \frac{d\mathcal{I}_\kappa}{d\gamma} \right|_{\gamma=0}$$

$$= ((w + w^*) \cdot \varphi)_{H_\kappa} + \int_\Omega C^{ik} C^{jl} [(w + w^*)_{\alpha^i} \varphi_{\alpha^j} - B_{ij}\varphi] \nabla_{kl} \Psi \, d\Omega$$

$$+ \int_\Omega \widetilde{T}^{ij}_{\text{p}} [(w + w^*)_{\alpha^i} \varphi_{\alpha^j} - B_{ij}\varphi] \, d\Omega - \int_{\Gamma_2} \widetilde{M}^m \varphi_4 \, ds - \int_\Omega R^3 \varphi \, d\Omega. \quad (22.8)$$

On the other hand, by (17.18),

$$((w - \mathbf{G}_\kappa(w)) \cdot \varphi)_{H_\kappa} = ((w + w^*) \cdot \varphi)_{H_\kappa}$$

$$- \int_\Omega \big[(C^{ik} C^{jl} \nabla_{kl} \Psi + \widetilde{T}^{ij}_{\text{p}})$$

$$\times (B_{ij}\varphi - \varphi_{\alpha^i} (w + w^*)_{\alpha^j}) + R^3 \varphi \big] \, d\Omega$$

$$- \int_{\Gamma_2} \widetilde{M}^m \varphi_4 \, ds.$$

To finally establish (22.4), we need to prove for small $\|\varphi\|_{H_\kappa}$ the estimate

$$\left| \sum_{\mu=2}^{4} \mathcal{I}_{9\kappa\mu}(w, \varphi) \right| \le m \, \|\varphi\|^2_{H_\kappa}. \quad (22.9)$$

From (22.3) we have

$$\frac{d^2}{d\gamma^2} \mathcal{I}_\kappa \, (w + \gamma\varphi)|_{\gamma=0} = \|\varphi\|^2_{H_\kappa} + \left. \frac{d}{d\gamma} \left(\Psi \cdot \frac{d\Psi}{d\gamma} \right)_{H_9} \right|_{\gamma=0} + \int_\Omega \widetilde{T}^{ij}_{\text{p}} \varphi_{\alpha^i} \varphi_{\alpha^j} \, d\Omega. \quad (22.10)$$

Let us compute the middle term in (22.10):

$$\frac{d}{d\gamma} \left(\Psi \cdot \frac{d\Psi}{d\gamma} \right)_{H_9} = \left\| \frac{d\Psi}{d\gamma} \right\|^2_{H_9} + \left(\Psi \cdot \frac{d^2\Psi}{d\gamma^2} \right)_{H_9}. \quad (22.11)$$

Next, we have from (22.6), since $\Theta \in H_9$ is arbitrary and

$$\|\nabla_{kl}\Theta\|_{L_{2\Omega}} \le m \, \|\Theta\|_{H_9},$$

the inequality

$$\left. \left\| \frac{d\Psi}{d\gamma} \right\|_{H_9} \right|_{\gamma=0} \le \sum_{k,l=1}^{2} \left\| C^{ik} C^{jl} (w + w^*)_{\alpha^i} \varphi_{\alpha^j} - B_{ij}\varphi \right\|_{L_{2\Omega}} \le m \, \|\varphi\|_{H_\kappa}. \quad (22.12)$$

If we set $w \mapsto w + \gamma\varphi$ in (17.15), we obtain immediately

$$\left(\Theta \cdot \frac{d^2\Psi}{d\gamma^2}\right)_{H_9}\bigg|_{\gamma=0} = \frac{1}{2}\int_\Omega C^{ik}C^{jl}\varphi_{\alpha^i}\varphi_{\alpha^j}\nabla_{kl}\Theta\, d\Omega,$$

whence

$$\left\|\frac{d^2\Psi}{d\gamma^2}\right\|_{H_9}\bigg|_{\gamma=0} \le m\sum_{i,j=1}^{2}\|\varphi_{\alpha^i}\varphi_{\alpha^j}\|_{L_{2\Omega}} \le m_1\|\varphi\|^2_{H_\kappa}. \tag{22.13}$$

The final calculations in (22.12), (22.13) use (12.26), (12.28). From (22.10), (22.11) we have

$$\begin{aligned}|\mathcal{I}_{\kappa 2}(w,\varphi)| &\le \frac{1}{2}\left|\frac{d^2}{d\gamma^2}\mathcal{I}_\kappa(w+\gamma\varphi)\bigg|_{\gamma=0}\right| \\ &\le \frac{1}{2}\|\varphi\|^2_{H_\kappa} + \frac{1}{2}\left\|\left(\frac{d\Psi}{d\gamma}\right)^2\right\|_{H_9}\bigg|_{\gamma=0} \\ &\quad + \frac{1}{2}\left|\left(\Psi\cdot\frac{d^2\Psi}{d\gamma^2}\right)_{H_9}\bigg|_{\gamma=0}\right| + \left|\int_\Omega \widetilde{T}^{ij}_{\mathrm{p}}\varphi_{\alpha^i}\varphi_{\alpha^j}\, d\Omega\right|.\end{aligned} \tag{22.14}$$

Finally, from (22.12), (22.13), (22.14) and since $\widetilde{T}^{ij}_{\mathrm{p}} \in L_{2\Omega}$, we have

$$|\mathcal{I}_{\kappa 2}(w,\varphi)| \le m\|\varphi\|^2_{H_\kappa}. \tag{22.15}$$

Next, for $\mathcal{I}_{\kappa 3}(w,\varphi)$ we have

$$\begin{aligned}\mathcal{I}_{\kappa 3}(w,\varphi) &= \frac{1}{3!}\frac{d^3}{d\gamma^3}\mathcal{I}_\kappa(w+\gamma\varphi)\bigg|_{\gamma=0} = \frac{1}{6}\frac{d^3}{d\gamma^3}\|\Psi\|^2_{H_9}\bigg|_{\gamma=0} \\ &= \left(\frac{d\Psi}{d\gamma}\cdot\frac{d^2\Psi}{d\gamma^2}\right)_{H_9}\bigg|_{\gamma=0} + \frac{1}{3}\left(\Psi\cdot\frac{d^3\Psi}{d\gamma^3}\right)_{H_9}\bigg|_{\gamma=0}\end{aligned}$$

From (17.16), (17.34) we have

$$\frac{d^3\Psi}{d\gamma^3}\bigg|_{\gamma=0} = 0, \tag{22.16}$$

and then from (22.12), (22.13), (22.16) it follows that

$$|\mathcal{I}_{\kappa 3}(w,\varphi)| \le m\|\varphi\|^3_{H_\kappa}. \tag{22.17}$$

Finally,

$$|\mathcal{I}_{\kappa 4}(w,\,\varphi)||_{\gamma=0} = \frac{1}{4!}\left|\frac{d^4}{d\gamma^4}\mathcal{I}_\kappa(w+\gamma\varphi)\right|\Bigg|_{\gamma=0}$$

$$= \frac{1}{48}\left|\frac{d^4}{d\gamma^4}\|\Psi\|^2_{H_9}\right|\Bigg|_{\gamma=0} = \frac{1}{4}\left\|\frac{d^2\Psi}{d\gamma^2}\right\|^2_{H_9}\Bigg|_{\gamma=0} \le m\,\|\varphi\|^4_{H_\kappa}\,. \tag{22.18}$$

Inequality (22.9) follows from (22.15), (22.17), (22.18). Theorem 22.3 is proved.

Theorem 22.4. *A vector function* $\mathbf{c} = (\Psi,\, w)$ *will be a generalized solution of Problem 9κ if and only if the point* $w \in H_\kappa$ *is a critical point of the functional* $\mathcal{I}_\kappa$ *and* Ψ *is determined by* (17.17).

22.2. We have the following theorem.

Theorem 22.5. *The functional* $\mathcal{I}_\kappa(w)$ *can be represented in the form*

$$\mathcal{I}_\kappa(w) = \frac{1}{2}\,\|w\|^2_{H_\kappa} + \widetilde{\mathcal{I}}_\kappa(w), \tag{22.19}$$

where $\widetilde{\mathcal{I}}_\kappa(w)$ *is a weakly continuous functional in* H_κ. *Furthermore, we have*

$$\mathcal{I}_\kappa(w) \to \infty \text{ as } \|w\|_{H_\kappa} \to \infty, \tag{22.20}$$

that is, $\mathcal{I}_\kappa$ *is increasing.*

To prove the theorem, we note that (22.19) follows immediately from (22.3) and that in addition

$$\widetilde{\mathcal{I}}_\kappa(w) = (w\cdot w^*)_{H_\kappa} + \frac{1}{2}\,\|w^*\|^2_{H_\kappa} + \frac{1}{2}\,\|\Psi\|^2_{H_9}$$

$$+ \int_\Omega \widetilde{T}^{ij}_{\text{p}}\Big[\frac{1}{2}(w+w^*)_{\alpha^i}(w+w^*)_{\alpha^j} - B_{ij}(w+w^*)\Big]\,d\Omega \tag{22.21}$$

$$- \int_\Omega R^3 w\,d\Omega - \int_{\Gamma_2} \widetilde{M}^m w_4\,ds.$$

It is easily seen that $\widetilde{\mathcal{I}}_\kappa(w)$ is a weakly continuous functional in H_κ. Indeed, the fifth and the sixth terms on the right-hand side of $\to \mathcal{I}_\kappa(w)$ (22.21) are linear and bounded in H_κ. The functional $\frac{1}{2}\,\|\Psi\|^2_{H_9}$ is weakly continuous by Theorem 18.1. Finally, weak continuity of the fourth term follows from complete continuity of the operator of embedding H_κ into $W^{(1)}_{4\Omega}$, Theorem 12.3. Thus, weak continuity of $\widetilde{\mathcal{I}}_\kappa(w)$ has been established.

Let us now prove (22.20). For this, we use the scheme of the arguments of Lemma 16.5 and Theorem 21.2. On $\Sigma_{H_\kappa}(1,\,0)$ let us introduce the set $\Sigma'_{H_\kappa}(1,\,0)$ of elements v for which

$$\|v\|^2_{H_\kappa} - 2\left|\int_\Omega \widetilde{T}^{ij}_{\text{p}}\,v_{\alpha^i}v_{\alpha^j}\,d\Omega\right| < \frac{1}{2}.$$

The complement of $\Sigma'_{H_\kappa}(1, 0)$ in $\Sigma_{H_\kappa}(1, 0)$ will be called $\Sigma''_{H_\kappa}(1, 0)$. Clearly, on $\Sigma''_{H_\kappa}(1, 0)$ we have

$$\|v\|^2_{H_\kappa} - 2\left|\int_\Omega \widetilde{T}^{ij}_{\mathrm{p}} v_{\alpha^i} v_{\alpha^j}\, d\Omega\right| \geq \frac{1}{2}. \tag{22.22}$$

Next, such a subdivision of the sphere $\Sigma_{H_\kappa}(1, 0)$ generates a subdivision of the sphere $\Sigma_{H_\kappa}(R, 0)$ by the radial projection onto $\Sigma'_{H_\kappa}(R, 0)$ and $\Sigma''_{H_\kappa}(R, 0)$. By weak continuity of $\int_\Omega \widetilde{T}^{ij}_{\mathrm{p}} v_{\alpha^i} v_{\alpha^j}\, d\Omega$ in H_κ, $\overline{\Sigma}'_{H_\kappa}(1, 0)$ does not contain zero. Next, we use the representation (19.6) for $\|\Psi(w)\|^2_{H_9}$. Here clearly, inequalities (19.8), (19.15) will be satisfied. Using the scheme of proof of Theorem 21.2, on $\Sigma'_{H_\kappa}(R, 0)$ we have that

$$\begin{aligned}\mathcal{I}_\kappa(w) \geq \frac{1}{2}\|w\|^2_{H_\kappa} &+ \frac{1}{2}(\Pi_4(w) - |\Pi_3(w)| - |\Pi_2(w)| \\ &- |\Pi_1(w)| - |\Pi_0(w)|) - \left|\int_\Omega \widetilde{T}^{ij}_{\mathrm{p}}\Big[\frac{1}{2}(w + w^*)_{\alpha^i}(w + w^*)_{\alpha^j}\right. \\ &\left. - B_{ij}(w + w^*)\Big]\, d\Omega\right| - \left|\int_\Omega R^3 w\, d\Omega + \int_{\Gamma_2} \widetilde{M}^m \frac{dw}{dm}\, ds\right|.\end{aligned} \tag{22.23}$$

Since by assumption $\widetilde{T}^{ij}_{\mathrm{p}} \in L_{2\Omega}$, we have

$$\begin{aligned}&\left|\int_\Omega \widetilde{T}^{ij}_{\mathrm{p}}\Big[\frac{1}{2}(w + w^*)_{\alpha^i}(w + w^*)_{\alpha^j} - B_{ij}(w + w^*)\Big]\, d\Omega\right| \\ &\leq \left\|\widetilde{T}^{ij}_{\mathrm{p}}\right\|_{L_{2\Omega}} \left(\left\|\frac{1}{2} w_{\alpha^i} w_{\alpha^j}\right\|_{L_{2\Omega}} + \left\|\frac{1}{2}\left(w_{\alpha^i} w^*_{\alpha^j} + w_{\alpha^j} w^*_{\alpha^i}\right) - B_{ij} w\right\|_{L_{2\Omega}}\right. \\ &\qquad \left. + \left\|\frac{1}{2} w^*_{\alpha^i} w^*_{\alpha^j} - B_{ij} w^*\right\|_{L_{2\Omega}}\right).\end{aligned} \tag{22.24}$$

By inequalities (13.16), (13.19), we easily obtain

$$\begin{aligned}&\left|\int_\Omega \widetilde{T}^{ij}_{\mathrm{p}}\Big[\frac{1}{2}(w + w^*)_{\alpha^i}(w + w^*)_{\alpha^j} - B_{ij}(w + w^*)\Big]\, d\Omega\right| \\ &\leq m\left(1 + \|w\|_{H_\kappa} + \|w\|^2_{H_\kappa}\right).\end{aligned} \tag{22.25}$$

Finally, by Condition 7 of Section 17,

$$\left|\int_\Omega R^3 w\, d\Omega + \int_{\Gamma_2} \widetilde{M}^m \frac{dw}{dm}\, ds\right| \leq m\, \|w\|_{H_\kappa}. \tag{22.26}$$

Taking into account (19.8), (19.16), (22.23)–(22.26), we have for all w,

$$\mathcal{I}_\kappa(w) \geq \frac{1}{2}R^2 + m(R^4 - R^3 - R^2 - R - 1),$$

and for sufficiently large R,

$$\mathcal{I}_\kappa(w) \geq \frac{1}{2}R^2. \tag{22.27}$$

For elements $w \in \Sigma''_{H_\kappa}(R, 0)$ we have

$$\|w\|^2_{H_\kappa} - 2\left|\int_\Omega \widetilde{T}^{ij}_{\mathrm{p}} w_{\alpha^i} w_{\alpha^j}\, d\Omega\right| \geq \frac{1}{2}R^2.$$

From (22.3) we have for these elements

$$\begin{aligned}\mathcal{I}_\kappa(w) \geq \frac{1}{2}\|w\|^2_{H_\kappa} &- \frac{1}{2}\left|\int_\Omega \widetilde{T}^{ij}_{\mathrm{p}} w_{\alpha^i} w_{\alpha^j}\, d\Omega\right| \\ &- \left|\int_\Omega \widetilde{T}^{ij}_{\mathrm{p}}\left(w_{\alpha^i} w^*_{\alpha^j} - B_{ij} w\right) d\Omega\right| - \left|\int_\Omega \widetilde{T}^{ij}_{\mathrm{p}} B_{ij} w^*\, d\Omega\right| \\ &- \left|\int_\Omega R^3 w\, d\Omega + \int_{\Gamma_2} \widetilde{M}^m w_4\, ds\right| \\ \geq \frac{1}{4}R^2 - m(R+1),\end{aligned}$$

and for sufficiently large R, from (22.8),

$$\mathcal{I}_\kappa(w) \geq \frac{1}{8}R^2. \tag{22.28}$$

Finally, from (22.27), (22.28) we have that for sufficiently large R everywhere on the sphere,

$$\mathcal{I}_\kappa \geq \frac{1}{8}R^2.$$

Theorem 22.5 is completely proved.

Theorem 22.6. *Let Conditions 1–8 of Section 17 be satisfied. Then the functional $\mathcal{I}_\kappa(w)$ has at least one point of absolute minimum, which together with a function $\Psi \in H_9$ defined by* (17.7) *provides a generalized solution of the boundary value problem of nonlinear shallow shell theory with an Airy stress function* (7.51), (7.60), (6.1)–(6.3), (6.20).

To prove the claim, we note that if Ψ is defined by (17.17) then (17.15) and thus (17.39) have to hold true. Next, since $\mathcal{I}_\kappa(w)$ is an increasing functional in H_κ, existence of at least one absolute minimum is proved as in Theorems 21.4–21.6. But at that point we must have

$$\operatorname{grad}_{H_\kappa} \mathcal{I}_\kappa(w) = 0,$$

and from (22.4) we have

$$w = \mathbf{G}_\kappa(w),$$

by which (17.14) must be satisfied, so that (17.38) holds. Theorem 22.6 is proved.

22.4. Thus the existence theorem for a generalized solution of problem 9κ defined by (7.51), (7.60), (6.1)–(6.3), (6.20), $\Gamma = \Gamma_1 + \Gamma_2$, has been proved by two different methods that led to Theorems 19.3 and 22.6. However, since the theorems were proved under correspondingly identical conditions, their results cannot be taken

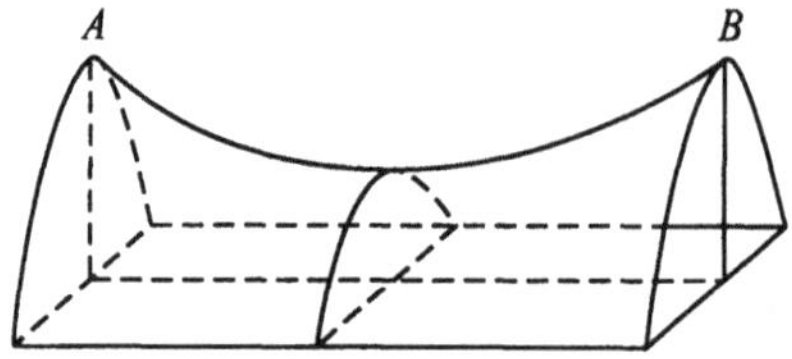

FIGURE 22.1.

to be identical. The point is that the generalized solution of Theorem 19.3 does not have to coincide with the solution provided by Theorem 22.6. These can be completely different solutions. In addition, the use of variational and topological approaches in tandem allows us in a number of cases to estimate the number of solutions and to establish their nonuniqueness. Let us also observe that Remark 16.1 applies to Theorem 22.6.

22.5. The generalized solution w, Ψ of Problem 9κ provided by Theorem 22.6, which furnishes the absolute minimum of the functional $\mathcal{I}_\kappa(w)$ in H_κ, will also be, by Theorem 22.1, a critical point of the functional $\mathcal{I}_{9\kappa}(\Psi, w)$ in $H_{9\kappa}$. However, w, Ψ will not always furnish a minimum of $\mathcal{I}_{9\kappa}(\Psi, w)$. Moreover, it can only be asserted that Ψ, w is an extremal point, which does not have to be either a maximum or a minimum. Using analogy, the situation with the functional $\mathcal{I}_{9\kappa}$ is similar to the problem of finding critical points of a function $\mathcal{I}$ on a hyperboloid in three-dimensional space (see Figure 22.1). When we restrict ourselves to HS2 by (17.17), this fixes a curve $\mathcal{AB}$ on which the function $\mathcal{I}$ already has a minimum. This corresponds to Theorem 22.6.

22.6. In the case of Problems 9κ that we are considering, we have the analogues of Theorems 21.8–21.11. Due to complete identity in statements and proofs, we do not present them here. Let us note that Theorem 22.5 essentially contains the following theorem.

Theorem 22.7. *Under the conditions of Theorems 19.1 and 22.6, every absolutely minimizing sequence w_n for the functional $\mathcal{I}_\kappa(w)$ in H_κ is strongly compact, and every strong limit w_0 of that sequence furnishes an absolute minimum of $\mathcal{I}_\kappa(w)$.*

22.7. Let us make some general comments concerning Chapter V. For the first time, variational considerations in nonlinear shell theory were used to prove solvability of boundary value problems by Vorovich [339, 342]. These were followed by [366]. Applications of the variational approach to plates can be found in [58]. Let us also note [30, 413, 414]. The scheme of arguments of Section 21–22 for functionals of nonlinear shallow shell theory is published here for the first time. The arguments are based, as the reader has probably noticed, on the inequalities (21.26) (in Theorem 21.3) and (22.28) (Theorem 22.5). Once these are established, Theorems 21.4–21.7, 22.6 on the existence of absolute minima of functionals $\mathcal{I}_{\kappa\kappa}(w)$, $\mathcal{I}_\kappa(w)$ follow immediately from the results of Krasnosel'skii [162], who introduced the concept of an increasing functional, or of Vainberg and Kachurovskii [320, 318, 319]. The concluding scheme of arguments of Theorems 21.4–21.7,

22.6 used by the present author is also not without independent interest. Let us also note that in problems of nonlinear shallow shell theory the functionals $\mathcal{I}_{t\kappa}(\mathbf{a})$, $\mathcal{I}_{9\kappa}(\mathbf{c})$, $\mathcal{I}_{\kappa\kappa}(w)$, $\mathcal{I}_{\kappa}(w)$ are not convex, which precludes us from using the theory for convex functionals, which is surveyed in [319].

CHAPTER VI

Numerical-Analytical Methods in the Nonlinear Theory of Shallow Shells

23. Expansion in Powers of a Small-parameter (Nonsingular Solutions)

23.1. Up till now we have mainly studied the problem of solvability of nonlinear boundary value problems of shallow shell theory. The variational approach to the study of these problems gives us a clue for the analysis of variational methods of approximate solution of these problems. We turn now to a detailed treatment of some methods of numerical and analytical solution of problems in shallow shell theory.

Let us recall first some general definitions. Let T be a continuous nonlinear operator acting in a Hilbert space H. Let us assume that at a point x_0 we have the representation

$$T(x_0 + y) - T(x_0) = S(x_0)y + o(\|y\|)$$

for all $y \in H$, where $S(x_0)$ is a linear (continuous) operator in y. Then we say that T is Fréchet-differentiable at x_0, and the operator $S(x_0)$ defined by that equality is the gradient of the operator T at the point x_0 and will be denoted by $S(x_0) = \mathrm{grad}_H T(x_0)$. The gradient of an operator plays an important part in the study of the properties of the operator itself. Basically, this is an extension of the concept of the gradient of a mapping on an n-dimensional Euclidean space to Hilbert spaces.

Let w_0 be the solution of the operator equation (13.36),

$$w_0 = \mathbf{G}_{\kappa\kappa}(w_0). \tag{23.1}$$

The operator $\mathbf{G}_{\kappa\kappa}$ is Fréchet-differentiable [162] at the point w_0. This fact was essentially established in Section 14 and will not be justified further. Furthermore, assume that for the equation

$$v = \sigma \operatorname{grad}_{H_\kappa} \mathbf{G}_{\kappa\kappa}(w_0)v, \tag{23.2}$$

$\sigma = 1$ is not an eigenvalue. In this case we call w_0 a nonsingular solution of equation (23.1).

Lemma 23.1. *Any nonsingular solution of equation* (23.1) *is isolated* [162].

Indeed, let w_n be a sequence of solutions of

$$w_n = \mathbf{G}_{\kappa\kappa}(w_n) \text{ such that } w_n \to w_0. \tag{23.3}$$

Then for the sequence $v_n = w_n - w_0$ we obtain

$$v_n = \mathbf{G}_{\kappa\kappa}(w_0 + v_n) - \mathbf{G}_{\kappa\kappa}(w_0) = \operatorname{grad}_{H_\kappa} \mathbf{G}_{\kappa\kappa}(v_n) + \mathbf{R}(w_0, v_n), \tag{23.4}$$

where

$$\frac{\|\mathbf{R}(w_0, v_n)\|_{H_\kappa}}{\|v_n\|^2_{H_\kappa}} \leq M. \tag{23.5}$$

Since w_0 is a nonsingular solution, from (23.4) we have

$$v_n = \mathbf{KR}(w_0, v_n), \tag{23.6}$$

where

$$\mathbf{K} = \left(\mathbf{I} - \operatorname{grad}_{H_\kappa} \mathbf{G}_{\kappa\kappa}\right)^{-1}. \tag{23.7}$$

The operator $\mathbf{K}$ exists and is bounded, since $\operatorname{grad}_{H_\kappa} \mathbf{G}_{\kappa\kappa}$ is a completely continuous operator and $\sigma = 1$ is not its eigenvalue. From (23.6) we have

$$\|v_n\|_{H_\kappa} \leq \|\mathbf{K}\| \, \|\mathbf{R}(w_0, v)\|_{H_\kappa}, \tag{23.8}$$

which is impossible due to (23.5). Lemma 23.1 is proved.

Using an example, let us explain the mechanical contents of the concept of a "nonsingular solution." Let us consider a spherical rigidly hinged shell, acted upon by a negative pressure p. This problem has been studied by many authors [306, 308, 322]. A typical characteristic of the loading of the dome, the dependence $p(f)$ of the pressure on the vertical deflection, is shown in Figure 23.1 for an intermediate value of elevation. All the solutions of this problem, apart from the points p_i, f_i, $i = 1, 2$, are nonsingular.

23.2. Let us consider the NOE (13.36), to which we reduce the boundary value problems $t\kappa$, and let us indicate the general scheme of using the method of small-parameters in such problems. This scheme covers virtually all the cases encountered in nonlinear shell theory. Let us assume that w_0 is a nonsingular solution of (13.36) and let us set

$$w \to w_0 + w. \tag{23.9}$$

Then for the new w we obtain

$$w = \mathbf{G}_{\kappa\kappa}(w_0 + w) - \mathbf{G}_{\kappa\kappa}(w_0) + \mathbf{R}(w_0 + w). \tag{23.10}$$

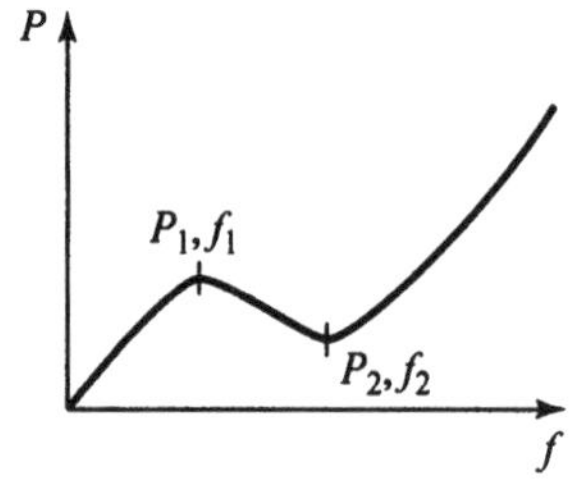

FIGURE 23.1.

On the right hand-side of (23.10) the term $\mathbf{R}(w_0 + w)$ represents a perturbation of the problem that led to the appearance of w. This term could include both a change in the external loading as well as a change in the parameters of the shell itself. Choosing it in an appropriate form, we can cover a wide class of problems. From the results of Sections 14 and 17 it is natural to put

$$\mathbf{G}_{\kappa\kappa}(w_0+w)-\mathbf{G}_{\kappa\kappa}(w_0)=\mathbf{G}_{\kappa\kappa 1}(w_0)w+\mathbf{G}_{\kappa\kappa 2}(w_0,\ w)+\mathbf{G}_{\kappa\kappa 3}(w_0,\ w), \quad (23.11)$$

where the $\mathbf{G}_{\kappa\kappa\rho}(w_0,\ w)$ are operators in H_κ that are homogeneous in w of order ρ. Finally, instead of (23.11) we shall consider a more general equation with a parameter μ:

$$w=\mathbf{G}_{\kappa\kappa 1}(w_0)w+\mathbf{G}_{\kappa\kappa 2}(w_0,\ w)+\mathbf{G}_{\kappa\kappa 3}(w_0,\ w)+\mu\mathbf{R}(w_0+\mu w). \quad (23.12)$$

Assuming that for $\mathbf{R}(w_0 + \mu w)$ we have the expansion

$$\begin{gathered}\mathbf{R}(w_0+\mu w)=\sum_{k=0}^{\infty}\mathbf{R}_k(w_0,\ w)\mu^k,\\ \|\mathbf{R}_k(w_0,\ w)\|_{H_\kappa}\le\|\mathbf{R}_k\|\ \|w\|^k_{H_\kappa},\end{gathered} \quad (23.13)$$

which converges in H_κ at $\mu = 1$, we shall seek the solution of (23.12) in the form

$$w=\sum_{k=0}^{\infty} w_{1k}\mu^k, \quad (23.14)$$

the coefficients of which are found consecutively from (23.12). This method was quite frequently used in the earlier stages of the development of the nonlinear theory of shells. Let us note in this connection the survey of Chien Wei-Zang [53] as well as his papers [55, 54], in which expansions of the form (23.13) were proposed and the parameter μ was assumed to be proportional to the maximal deflection. In a number of papers expansions in powers of the load were used. These arguments were developed in the paper of E. Kai-Yuan [129], where expansions in powers of the maximal deflection were used for various types of plate support. A characteristic feature of that paper is the computation of a large number (three or more) of expansion terms. It has to be said that due to the lack of an efficient expansion procedure, the popularity of small-parameter expansions in nonlinear shell theory waned relatively quickly. As will be explained below, this is not justified, as modifications of the method make it more efficient.

23.3. To justify the expansions (23.14) in nonlinear shell theory, we take into account the fact that

$$\mathbf{G}_{\kappa\kappa 1}(w_0, w) = \mathrm{grad}_{H_\kappa} \mathbf{G}_{\kappa\kappa}(w_0)w, \tag{23.15}$$

at the point w_0 and let us consider in more detail the structure of the operators $\mathbf{G}_{\kappa\kappa 2}(w_0, w)$, $\mathbf{G}_{\kappa\kappa 3}(w_0, w)$.

Lemma 23.2. *The homogeneous operator* $\mathbf{G}_{\kappa\kappa 2}(w_0, w)$ *admits the representation*

$$\mathbf{G}_{\kappa\kappa 2}(w_0, w) = \sum_{i=1}^{n_2} \mathcal{B}_i \left\{ \mathcal{C}_{i1}(w), \mathcal{C}_{i2}(w) \right\}, \tag{23.16}$$

where $\mathcal{C}_{i1}$, $\mathcal{C}_{i2}$ *are additive homogeneous first-order operators.* $\mathcal{C}_{i1}(w)$, $\mathcal{C}_{i2}(w)$ *act from* H_κ *into* $L_{q_1\Omega}$, *for some* $q_1 > 1$, *while* $\mathcal{B}_i$ *acts from* $L_{q_2\Omega}$ *into* H_κ; *these operators are linear in each of their variables. Furthermore, the operators* $\mathcal{C}_{i1}$, $\mathcal{C}_{i2}$, $\mathcal{B}_i$ *are bounded, so that we have the relation*

$$\left\| \mathcal{B}_i \left\{ \mathcal{C}_{i1}(\varphi), \mathcal{C}_{i2}(\psi) \right\} \right\|_{H_\kappa} \le m \left\| \varphi \right\|_{H_\kappa} \left\| \psi \right\|_{H_\kappa}, \tag{23.17}$$

for all $\varphi, \psi \in H_\kappa$.

Lemma 23.3. *The homogeneous operator* $\mathbf{G}_{\kappa\kappa 3}(w_0, w)$ *admits the representation*

$$\mathbf{G}_{\kappa\kappa 3}(w_0, w) = \sum_{i=1}^{n_3} \mathcal{E}_i \left\{ \mathcal{M}_{i1}(w), \mathcal{M}_{i2}(w) \mathcal{M}_{i3}(w) \right\}, \tag{23.18}$$

where $\mathcal{M}_{i1}$, $\mathcal{M}_{i2}$, $\mathcal{M}_{i3}$ *are additive homogeneous first-order operators. The* $\mathcal{M}_{ik}$ *act from* H_κ *into* $L_{q\Omega}$, $q > 1$, *while* $\mathcal{E}_i$ *is a homogeneous additive operator acting from* $L_{q\Omega}$ *into* H_κ. *The operators* $\mathcal{E}_i$, $\mathcal{M}_{ik}$ *are bounded, and we have the relation*

$$\left\| \mathcal{E}_i \left\{ \mathcal{M}_{i1}(\varphi), \mathcal{M}_{i2}(\psi), \mathcal{M}_{i3}(\theta) \right\} \right\|_{H_\kappa} \le m \left\| \varphi \right\|_{H_\kappa} \left\| \psi \right\|_{H_\kappa} \left\| \theta \right\|_{H_\kappa}. \tag{23.19}$$

Lemmas 23.2, 23.3 are easily proved using the structure of the operators $\mathbf{K}_{t\kappa}(w)$, $\mathbf{G}_{\kappa\kappa}(w)$, which was studied in detail in Sections 14, 17; it will not be considered further here. Passing to the justification of (23.14), we note that (23.12) can be written in the following form:

$$w = \mathbf{K}\mathbf{G}_{\kappa\kappa 2}(w_0, w) + \mathbf{K}\mathbf{G}_{\kappa\kappa 3}(w_0, w) + \mu\mathbf{K}\mathbf{R}(w_0 + \mu w), \tag{23.20}$$

where $\mathbf{K} = (\mathbf{I} - \mathbf{G}_{\kappa\kappa 1})^{-1}$ exists and is bounded, since w_0 is a nonsingular solution and $G_{\kappa\kappa 1}$ is a completely continuous operator on H_κ. The expansion (23.14) can be constructed if we substitute (23.14) into (23.20) and take into account (23.13). Furthermore, it is easy to see that if the series (23.14) converges for μ_0, then it converges for all μ such that

$$|\mu| \le |\mu_0|; \tag{23.21}$$

it satisfies (23.20) and therefore (23.12). To estimate the radius of convergence of the series (23.14), we introduce in the usual way the majorizing algebraic equation

$$x = \|\mathbf{K}\|\, m(x^2 + x^3) + \mu \|\mathbf{K}\| \sum_{k=0}^{\infty} \|\mathbf{R}_k\|\, x^k \mu^k, \tag{23.22}$$

where m is determined by the constants in the inequalities (23.17), (23.19). For the solution of (23.22) we also assume the ansatz

$$x = \sum_{k=0}^{\infty} x_k \mu^k, \tag{23.23}$$

in which the x_k are to be found from (23.22). Since (23.22) majorizes (23.20), the radius of convergence of (23.14) is no less than that of (23.23). The latter is determined by the value of μ, for which equation (23.22) has a multiple root.

Thus we have established the following result.

Theorem 23.1. *Assume that conditions of Theorem* 16.1 *(respectively,* 16.2, 16.3*) are satisfied and* w_0 *is a nonsingular solution. In this case the series* (23.14) *has a finite radius of convergence in* H_κ*, determined by* $\|\mathbf{K}\|$, $\|\mathbf{R}_k\|$, $\|w_0\|$, *and* m.

Let us remark here on an important property of the expansions (23.14): They are not that sensitive to the smoothness of initial data. Thus, they converge even in the case when R^3 contains δ-functions as components, that is, concentrated forces of sufficiently small intensivness. At the same time, the expansion (23.14) will converge at an equally fast rate for a sufficiently small uniform load p. There is no significant difference between the two cases. When other methods of approximation (Bubnov–Galerkin, Ritz, finite differences, finite element) are used, one immediately sees a large difference in the quality of approximation that depends on the smoothness of the load. Some of these methods (finite differences, finite element) cannot be used at all directly if the load contains discontinuities, as in the case of concentrated forces. Then a preliminary numerical-analytical preprocessing of the problem is required. Many examples of the use of the method of expansion in a small-parameter in mechanics can be found in [4, 72, 137, 139].

23.4. A reason for limited efficiency of the small-parameter method lies in the fact that w, considered as a function of μ, has singularities. Understanding these singularities would open up vast possibilities of extending the domain of applicability of this method. Even though other methods of solution of problems of nonlinear theory of shallow shells are developing apace, this would be very significant. Moreover, the small-parameter method itself can be easily automated. Therefore, there arises the problem of a priori determination of singularities of solutions of the NOE (23.12) considered as analytic functions of the loading parameter. The difficulties of this problem are obvious; at this stage we have to work with only partial information. There are two ways to proceed. One could go around the obstacles (that is, singularities of the solution) by reexpanding the solution in new Taylor series in the parameter μ. This approach has been effectively used by Kayuk and his coworkers [137, 134, 135, 136]. In the reexpansion it is natural to move along the real axis of μ. Using this method, Kayuk has solved a number of problems of the nonlinear theory of plates and shells; see the above references for details.

23.5. If we know the domain π of analyticity of w in μ, (23.1), then to improve convergence of the expansions we can proceed as follows. Let $\sigma = \sigma(\mu)$ be the

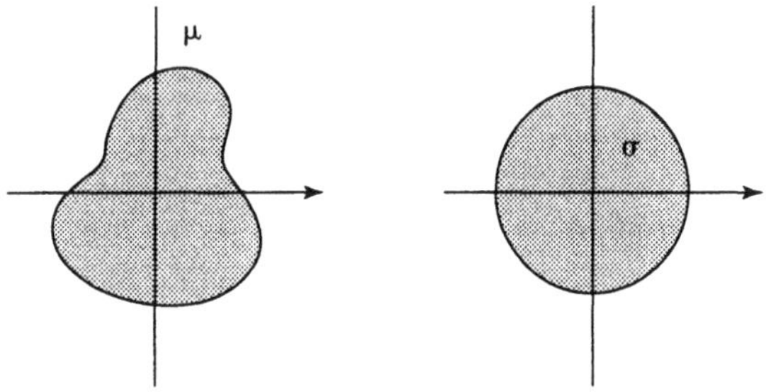

FIGURE 23.2.

mapping of π onto the unit disk and

$$\mu = \sum_{k=0}^{\infty} \mu_k \sigma^k. \tag{23.24}$$

In this case it is natural to seek w in the form

$$w = \sum_{k=0}^{\infty} \widetilde{w}_{1k} \sigma^k. \tag{23.25}$$

Clearly, the series (23.25) will converge in the unit circle and will give us a representation of the solution (13.36) in the entire domain of analyticity of $w(\mu)$. Practically, (23.25) can be obtained by an elementary recomputation of the expansion (23.14) using (23.24). However, determining the domain of analyticity π can be quite complicated. On the other hand, it is useful to have even an approximate idea of the structure of π, for example, if it is known that a domain $\widetilde{\pi}$ does not contain any singularities of $w(\mu)$, so that $\widetilde{\pi} \subset \pi$. Furthermore, let $\sigma = \sigma(\mu)$ map $\widetilde{\pi}$ into the unit disk in the σ-plane. Then, obviously, it makes sense to expand w in integer powers of σ. This will give us expansions that converge for all $\mu \in \widetilde{\pi}$. The effectiveness of this approach depends on the closeness of $\widetilde{\pi}$ and π. Construction of the expansion of w itself in σ is not hard, if we either construct an expansion of w in μ and expand σ in powers of μ or construct an inverse expansion. This approach was used already by Euler; for more details see [134, 172].

23.6. However, this approach still does not exhaust the problem. For example, in a number of cases the characteristic of the loading of a shell $f = f(p)$ (the curve "deflection at the center — load"), as is well known, is not one-to-one (see Figure 23.1). Therefore already on the real axis we have branch-point singularities. Thus, we are led to a different approach: taking into account singularities of the solution in the process of approximating it. Some experience in doing just that is available, for example in the utilization of Padé approximants [17]. The point of this approach is that if the structure of the function $f(p)$ is known *a priori*, for example, if it is known that it is a meromorphic function, then it can be approximated using the relation

$$f(p) \sim \frac{\sum_{k=0}^{N} a_k p^k}{\sum_{k=0}^{N} b_k p^k}, \tag{23.26}$$

by choosing the coefficients a_k, b_k by demanding that the maximal number of Taylor coefficients at zero for $f(p)$ and the approximating rational function (23.26) be equal. The properties of the approximant (23.26) have been studied in a number of papers [324, 98, 99, 100, 252]. It appears to be effective when the loading characteristic of the shell $f = f(p)$ is monotone. In the case of a nonmonotone characteristic of a shell, $f(p)$ is clearly not a meromorphic function. Already on the real axis it has at least two branch points: f_1, p_1, f_2, p_2 (see Figure 23.1). In the general case, in a neighborhood of each of these points we have the representation

$$f - f_i \approx \pm(p - p_i)^m.$$

Usually, one would expect that $m = \frac{1}{2}$. Because of this, in the case of a non-monotone characteristic instead of (23.26) one should use the approximation

$$f(p) \sim \frac{\sum_{k=0}^{N} a_k p^k}{\sum_{k=0}^{N} b_k p^k g(p)}, \tag{23.27}$$

where $g(p)$ is the function defined by the relation

$$p = g^3 + \mathcal{A}g^2 + \mathcal{B}g + \mathcal{C}. \tag{23.28}$$

The constants a_k, b_k, $\mathcal{A}$, $\mathcal{B}$, $\mathcal{C}$ are determined by requiring equality with a certain number of the Taylor series expansion of $f(p)$ in a neighborhood of $p = a$. These coefficients are to be found directly from (23.12), (23.20). In a number of cases the nonmonotonicity of $f(p)$ may have a more complicated structure [316, 322], and then on the right-hand side of (23.27) one has to take polynomials of higher order.

Thus, the ways to increase the effectivity of the method of small parameters are, in our opinion, far from having been exhausted. New venues are suggested if one considers the great number of deep results in the theory of analytic functions, in which their global properties are established based on the properties of their Taylor series expansions, and the possibility of automating algebraic and other computations on a computer. Approximations of the form (23.26) are also quite useful in the analysis of results of experiments with thin-walled structures. They can be used to develop high-precision methods of prediction of the upper critical pressure of a shell.

24. Expansion in Powers of a Small-parameter (Singular Solutions). The Liapunov–Schmidt Method

24.1. Let w_0 be a singular solution. In other words, we are assuming that the operator equation (23.2) has an eigenvalue $\sigma = 1$. A general analysis of such problems has been initiated in the fundamental work of Liapunov [179], Schmidt [268], and Poincaré [244]. A detailed treatment of this question can be found in [321]. Applications of the method of expansion in a small-parameter in the case of a singular solution to a number of problems in continuum mechanics can be found in [59, 171, 215]. In boundary value problems of nonlinear plate theory bifurcation theory was first used in the work of Polubarinova-Kochina [247], where the postcritical behavior of a rectangular hinged plate was considered. In 1939 Friedrichs and Stoker considered the postcritical behavior or a buckled plate [83].

A new stage in the application of methods of bifurcation theory began with the fundamental work of Koiter [146]. Postcritical behavior of a circular plate had been considered by Bodner [36] and Grigolyuk (using the Bubnov–Galerkin method) [108]. A detailed analysis of the problem under the simultaneous action of longitudinal and transversal loadings is contained in the work of Vorovich [340].

Let us first present some general considerations applied to the operator equation (13.36). First of all, let us note that the operator equation

$$w - \mathbf{G}_{\kappa\kappa 1}(w_0)w = f \tag{24.1}$$

in our case has for $f = 0$ by complete continuity of $\mathbf{G}_{\kappa\kappa 1}$ a finite number r of linearly independent solutions, eigenfunctions of the operator $\mathbf{G}_{\kappa\kappa 1}$, which we shall call the rank of w_0. They form a subspace H_κ^+ and can be taken to be orthonormalized. We shall denote the orthogonal complement of H_κ^+ in H_κ by H_κ^-, so that

$$H_\kappa = H_\kappa^- \otimes H_\kappa^+. \tag{24.2}$$

It is well known [321] that there is an operator

$$\mathbf{K} = [\mathbf{I} - \mathbf{G}_{\kappa\kappa 1}(w_0)]^{-1} \tag{24.3}$$

acting on H_κ^-.

Let us consider the operator equation (23.10). By the argument above, it is natural to look for its solution w in the set of elements of H_κ for which

$$\mathbf{G}_{\kappa\kappa 2}(w) + \mathbf{G}_{\kappa\kappa 3}(w) + \mu\mathbf{R}(w_0 + \mu w) \in H_\kappa^-, \tag{24.4}$$

and then

$$\begin{aligned} w &= w^+ + w^-, \ w^+ = l\theta, \\ w^- &= \mathbf{K}(\mathbf{G}_{\kappa\kappa 2}(w) + \mathbf{G}_{\kappa\kappa 3}(w) + \mu\mathbf{R}(w_0 + \mu w)), \end{aligned} \tag{24.5}$$

where

$$\theta \in H_\kappa^+, \ \|\theta\|_{H_\kappa} = 1 \tag{24.6}$$

and l is some constant. Thus the problem has been reduced to the operator equation (24.5) in H_κ^- and to the finite-dimensional problem of finding θ and l. It can be written down explicitly, if we take into account (24.4), so that

$$((\mathbf{G}_{\kappa\kappa 2}(w) + \mathbf{G}_{\kappa\kappa 3}(w) + \mu\mathbf{R}(w_0 + \mu w)) \cdot \theta_k) = 0, \ k = 0, 1, \dots, r, \tag{24.7}$$

where θ_k are the eigenfunctions of the operator $\mathbf{G}_{\kappa\kappa 1}$. In the theory of bifurcation of solutions of operator equations, these relations are called the bifurcation equations. Solution of the system (24.5)–(24.7) can be constructed in the following manner. We have

$$w^- = \mathbf{K}(\mathbf{G}_{\kappa\kappa 2}(w^- + l\theta) + \mathbf{G}_{\kappa\kappa 3}(w^- + l\theta) + \mu\mathbf{R}(w_0 + \mu w^- + \mu l\theta)), \tag{24.8}$$

and w^- can be determined from (24.8) for given l, θ as a series:

$$w^- = \sum_{q=0}^{\infty}\sum_{n=0}^{\infty} w_{01qn}\mu^q l^n, \quad w_{0100} = 0.$$

Its convergence for sufficiently small l and μ is easily established, since the algebraic equation

$$x = m\,\|\mathbf{K}\|\,(x^2 + |l|x + |l|^2 + x^3 + |l|^2 x + |l|x^2 + |l|^3)$$
$$+ \mu\,\|\mathbf{K}\| \sum_{k=1}^{\infty} \|\mathbf{R}_k\|\,(\mu x + |l|)^k,$$

where the constant m is taken from the inequalities (23.17), (23.19), is majorizing with respect to (24.8). Therefore, (24.8) establishes a mapping

$$\theta \mapsto w^- \tag{24.9}$$

from a neighborhood of zero in H_κ^+ into a neighborhood of zero in H_κ^-. Substituting in (24.7) the mapping (24.9), we obtain r equations, which must determine the constant l and the coefficients of the expansion of θ in some basis in H_κ^+, for example, θ_k.

The point of the above considerations is that the description of the set of solutions of the operator equation (23.10) in a neighborhood of a singular solution w_0 has been reduced to a study of a nonlinear finite-dimensional problem. However, it is this problem that proves to be the hardest. Different methods of formulating bifurcation equations in a general operator setting, and an analysis of the possibilities that arise here, can be found in [321]. There are not many cases where the analysis is carried out without a priori assumptions on the coefficients of the bifurcation equation (see, for example, [340]). As a rule, such an analysis is based on some a priori assumptions that are subsequently verified by computations.

24.2. As an example, let us consider the problem of postcritical axisymmetric behavior of a circular plate of radius a, hinged on the boundary. Let us assume that the plate is compressed along the boundary by a constant lateral load T and that a transverse load $q(r)$ is also acting on it (see Figure 24.1). Assume furthermore that the plate is supported on a nonlinear elastic foundation with reaction defined

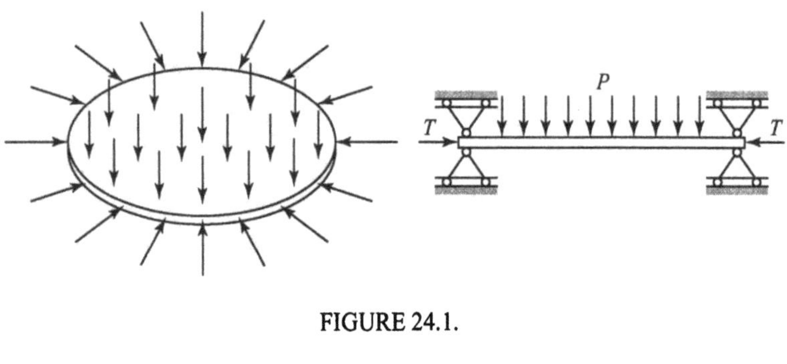

FIGURE 24.1.

by the relation

$$Q = -\lambda w^3.$$

The main equations and boundary conditions of the problem have the form

$$D_{f1} \nabla^4 w = q(r) - \lambda w^3 + \frac{2h}{r}\frac{d}{dr}\left(\frac{dw}{dr}\frac{d\Psi}{dr}\right), \tag{24.10}$$

$$\nabla^4 \Psi = -E\frac{1}{r}\frac{dw}{dr}\frac{d^2 w}{dr^2}. \tag{24.11}$$

In (24.11), $D_{f1} = \frac{2Eh^3}{3(1-\nu^2)}$ is the bending stiffness of the plate, $2h$ is its thickness, ν is the Poisson coefficient. For equations (24.10), (24.11) we have the boundary conditions

$$w(a) = \left(\frac{d^2 w}{dr^2} + \frac{\nu}{r}\frac{dw}{dr}\right)\bigg|_{r=a} = 0, \tag{24.12}$$

$$\Psi(a) = 0, \quad \left(\frac{d\Psi}{dr}\frac{1}{r}\right)\bigg|_{r=a} = -\frac{T}{2h}, \tag{24.13}$$

In this problem, we shall not exclude Ψ from consideration and pass to an operator equation in w; instead, we shall use the system (24.10)–(24.13) throughout. We shall assume that in (24.13),

$$T = T_0 + T_1,$$

where T_0 is the critical lateral compression stress of the plate, and set furthermore,

$$\Psi = \Psi^* + \frac{T_0}{4h}(a^2 - r^2). \tag{24.14}$$

Then the system (24.10)–(24.13) assumes the form

$$D_{f1} \nabla^4 w + T_0 \nabla^2 w = \mu\left[q(r) - \lambda w^3 + \frac{2h}{r}\frac{d}{dr}\left(\frac{dw}{dr}\frac{d\Psi^*}{dr}\right)\right], \tag{24.15}$$

$$\nabla^4 \Psi^* = -\frac{E}{r}\frac{dw}{dr}\frac{d^2 w}{dr^2}. \tag{24.16}$$

$$\Psi^*(a) = 0, \tag{24.17}$$

$$\left(\frac{d\Psi^*}{dr}\frac{1}{r}\right)\bigg|_{r=a} = -\frac{T_1}{2h}. \tag{24.18}$$

In (24.15), μ is the expansion parameter, which we shall later take to be $\mu = 1$. The boundary conditions (24.12) remain unchanged.

Let us next set

$$w = w_0 + \mu w_1 + \cdots, \quad \Psi^* = \Psi_0^* + \mu \Psi_1^* + \cdots. \tag{24.19}$$

Substituting (24.19) into (24.15)–(24.18), we have

$$D_{f1} \nabla^4 w_0 + T_0 \nabla^2 w_0 = 0, \tag{24.20}$$

$$w_0(a) = \left(\frac{d^2 w_0}{dr^2} + \frac{\nu}{r}\frac{dw_0}{dr}\right)\bigg|_{r=a} = 0, \tag{24.21}$$

$$\nabla^4\Psi_0^* = -\frac{E}{r}\frac{dw_0}{dr}\frac{d^2w_0}{dr^2}, \tag{24.22}$$

$$\Psi_0^*(a) = 0, \quad \left(\frac{d\Psi_0^*}{dr}\frac{1}{r}\right)\bigg|_{r=0} = -\frac{T_1}{2h}, \tag{24.23}$$

$$D_{f1}\nabla^4 w_1 + T_0\nabla^2 w_1 = q(r) - \lambda w_0^3 + \frac{2h}{r}\frac{d}{dr}\left(\frac{dw_0}{dr}\frac{d\Psi_0^*}{dr}\right), \tag{24.24}$$

and so on. From (24.20)–(24.21) it can be seen that this system corresponds to equation (24.1) for $\mu = 1$. The solution (24.14) will be singular if the boundary value problem "sits on the spectrum" and has therefore a nontrivial solution. This problem is easily analyzed, and furthermore, it turns out that $r = 1$ and we have one eigensolution,

$$w' = l\theta + \cdots, \quad \theta = \mathcal{I}_0(\kappa) - \mathcal{I}_0\left(\kappa\frac{r}{a}\right), \quad T_0 = \frac{\kappa D}{a^2}, \quad \kappa = 2.050. \tag{24.25}$$

Therefore, the system of bifurcation equations (24.7) consists of only one equation. To write it down, we need to determine Ψ_0^* from (24.22), (24.23). It is easily seen that we have

$$\Psi_0^* = Z + \frac{T_1}{4h}(a^2 - r^2) + \cdots, \tag{24.26}$$

where Z is the deflection of a fixed plate under the load

$$-\frac{El^2}{r}\frac{d\theta}{dr}\frac{d^2\theta}{dr^2}.$$

In our case the bifurcation equation assumes the form

$$\int_0^a \left\{q(r) - \lambda l^3\theta^3 + \frac{2h}{r}\frac{d}{dr}\left[l\frac{d\theta}{dr}\left(l^2\frac{dZ}{dr} - \frac{T_1}{2h}r\right)\right]\right\}\theta r\, dr = 0. \tag{24.27}$$

It is easy to see that in the approximation we are making, (24.27) means that the right-hand side of (24.24) is orthogonal to the eigenfunction $\theta(r)$. Equation (24.27) is easily written in the form

$$\sigma_1 + \sigma_2 l T_1 = \sigma_3 l^3 \ldots, \tag{24.28}$$

where

$$\sigma_1 = \int_0^a q(r)\theta(r)r\, dr,$$

$$\sigma_2 = \int_0^a r\left(\frac{d\theta}{dr}\right)^2 dr,$$

$$\sigma_3 = \int_0^a \left[\lambda\theta^4 r + 2h\frac{dZ}{dr}\left(\frac{d\theta}{dr}\right)^2\right] dr.$$

Lemma 24.1. *All* $\sigma_i > 0$, $i = 1, 2, 3$.

For σ_1, σ_2 this is obvious, as $\theta > 0$.

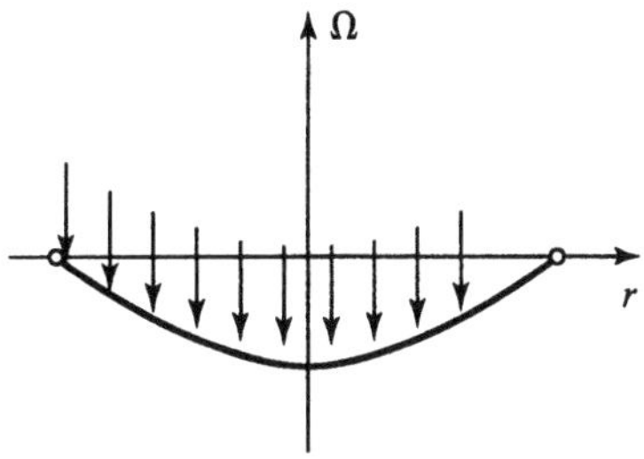

FIGURE 24.2.

Clearly, the lemma will be proved if we establish the inequality

$$\frac{dZ}{dr} \geq 0, \quad 0 \leq r \leq a. \tag{24.29}$$

As can be seen from (24.26), (24.22), Z is determined from the boundary value problem

$$\nabla^4 Z = -\frac{El^2}{r}\frac{d\theta}{dr}\frac{d^2\theta}{dr^2}, \quad Z|_{r=a} = \left.\frac{dZ}{dr}\right|_{r=a} = 0. \tag{24.30}$$

Furthermore, it follows from (24.25) that

$$\frac{d\theta}{dr} \geq 0, \quad \frac{d^2\theta}{dr^2} \geq 0, \quad \text{for } 0 \leq r \leq a. \tag{24.31}$$

Therefore, relations (24.30)–(24.31) show us that Z can be considered as the deflection of a plate of unit stiffness under a negative axisymmetric load. Then clearly, (24.29) will hold (Figure 24.2). This argument allows us to establish the number of zeros of (24.28) for different relations between T_1 and σ_i. Namely, it turns out that for

$$T_1 < 1.895\frac{\sigma_1^{2/3}\sigma_3^{1/3}}{\sigma_2},$$

the plate will have a unique equilibrium configuration. If

$$T_1 > 1.895\frac{\sigma_1^{2/3}\sigma_3^{1/3}}{\sigma_2}$$

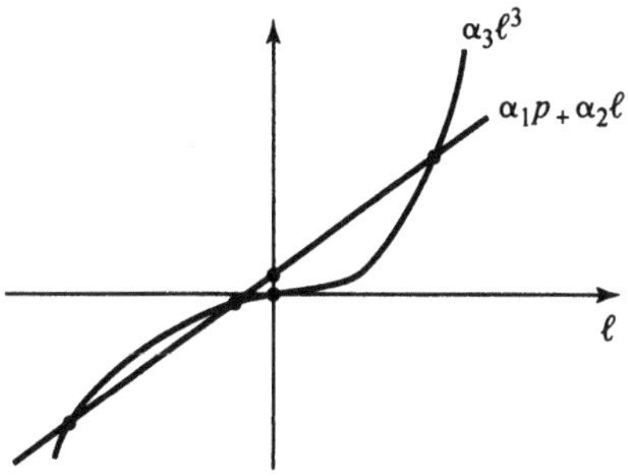

FIGURE 24.3.

then we have exhibited three equilibrium configurations: one with a positive deflection, and two others with negative ones (Figure 24.3). A more detailed study shows that to the middle (in modulus) root of (24.28) corresponds an unstable equilibrium configuration, while the two other ones are stable.

If

$$T_1 = \frac{\sigma_1^{2/3}\sigma_3^{1/3}}{\sigma_2} 1.895, \tag{24.32}$$

then the two negative roots collapse into one, and the plate will have two equilibrium configurations; the one corresponding to the double negative root of (24.28) will be unstable.

The problem of a circular plate in postcritical state has been studied by many authors. However, the analysis presented here (Theorem 24.2) allows us to state that apart from the three cases above, nothing else can happen in a neighborhood of a singular solution of (24.20), (24.21). Figure 24.3 clearly illustrates the situations possible here. The same problem has been considered in [321].

24.3. The work of Koiter [146], which deals with the influence of an initial imperfection, has led to a torrent of papers in which bifurcation theory was developed in applications to nonlinear shell theory. Here we need to note first of all the papers of Koiter himself [146, 147, 149, 150, 151, 153, 152], Budiansky [47], Budiansky and Hutchinson [48], Hutchinson and Koiter [122], and other authors [171]. The same period saw an intensive development of applications of bifurcation theory to analysis of particular problems. In detail were studied cases of bifurcation for $p = 2$, that is, in the case of a double eigenvalue. Here we should note the work of Srubshchik [293, 294, 295, 296]. The reader can find more details on this topic in the monograph [295]. The same problem is also dealt with in [117, 246].

Of course, it would be desirable to obtain rigorous results on the structure of the bifurcation equations (24.7) in a sufficiently general case, which would only use the properties of the operator equation (13.36). However, this is a general problem. Many questions in the natural science and technology reduce to this problem, united under the heading of "catastrophe theory." It has to be noted that the main concepts of this theory, its general methods and results, were created under the influence of problems of stability of elastic systems, and it could be that the first catastrophe that mankind was aware of was the loss of stability of the Euler strut. At present, general methods in catastrophe theory are being developed, and in this connection we should mention the work of Arnol'd, Varchenko, and Gusein-Zade [13, 14, 397], as well as the work of Thompson [307, 306, 308], which will be mentioned in Section 25.

24.4. Recent times have seen a fruitful synthesis of methods of bifurcation theory and of computer-based mathematics. Here one constructs numerical algorithms to find the basic solution and determines its type (singular, nonsingular). If w_0 is a singular solution, then the collection of eigenfunctions θ_k is constructed, followed by the bifurcation equation. Then one proceeds along the branch of solutions till the next singular solution is encountered. In this area, we should note the work of

Grigolyuk and Shalashilin [109, 110, 111], Shalashilin [271, 272, 273, 274, 275, 276], Gulyaev, Bazhenov, and Gotsulyak [114], Gulyaev and Mel'nichenko [115], Randhamohan [253], Kuznetsov [175], Na and Turski [214]. We should also note that the bifurcation equation simplifies if the problem has symmetries. In this case the study of bifurcation can be reduced to a system of equations of dimension less than p, the number of independent solutions of equation (24.2). For more details see [182, 183, 206, 386].

25. The Newton–Kantorovich Method

25.1. The Newton–Kantorovich method derives from the method of small-parameters considered in Sections 23, 24.

Let us recall the procedure of introducing derivatives of sufficiently smooth operator functions $P(x)$. The first (Gâteaux) derivative $P'(x)$ at a point x_0 is defined by the formula

$$P'(x_0)x = \lim_{t\to 0} \frac{P(x_0+tx)-P(x_0)}{t}.$$

$P'(x_0)x$ is a linear operator in x. The second Gâteaux derivative $P''(x)$ of the operator $P(x)$is defined by

$$P''(x_0,\, x_1,\, x_2) = \lim_{t\to 0} \frac{P'(x_0+tx_1)x_2 - P'(x_0)x_2}{t}.$$

The operator $P''(x_0,\, x_1,\, x_2)$ is linear in the variables x_1 and x_2. These variables will be dropped in the notation below.

Let us first present the general scheme of the Newton–Kantorovich method (hereafter abbreviated to NK) [130]. We consider operator equations of the form

$$\mathcal{P}(w) = 0,$$

which is solved using the following iteration scheme:

$$w_{n+1} = w_n - [\mathcal{P}'(w_n)]^{-1}\mathcal{P}(w_n). \tag{25.1}$$

Recall that in a Hilbert space H, for a Fréchet-differentiable operator we have grad $P(w_0)v = P'(w_0)v$. In practical terms the operation of differentiation of an operator is introduced using Gâteaux differentiation, which in the case of Fréchet-differentiable operators gives the same result. Namely, $P'(w_0)v = d/dt\; P(w_0 + tv)\mid_{t=0}$. The second and all subsequent derivatives are obtained in a similar way. This is the way of obtaining derivatives that we shall be using below.

The main difficulty in the applications of the NK method lies in the construction of the operator $[\mathcal{P}'(w_n)]^{-1}$. In the theory of plates and shells this operator has to be constructed numerically using some approximation scheme; both finite difference methods and finite element methods are frequently used. The iteration scheme (25.1) can be simplified:

$$w_{n+1} = w_n - [\mathcal{P}'(w_0)]^{-1}\mathcal{P}(w_n).$$

This form is called the modified NK method.

Theorem 25.1 (Kantorovich [130]). *Let the operator $\mathcal{P}$ be defined on a Banach space, and let it satisfy the following three conditions:*

1. *It has a second Gâteaux derivative in a ball $B(r, w_0)$.*
2. *The operator $\mathcal{P}'(w_0)$ is invertible.*
3. *The inequality*

$$\eta_1 \eta_2 < \frac{1}{2} \tag{25.2}$$

holds. Here

$$\eta_1 = \left\| \mathcal{P}'(w_0)^{-1} \mathcal{P}(w_0) \right\|, \quad \eta_2 = \left\| \mathcal{P}'(w_0)^{-1} \mathcal{P}''(w_0) \right\|. \tag{25.3}$$

Then there exists a solution w_* such that

$$\|w_0 - w_*\| \le r_0,$$

where

$$r_0 = \frac{1 - \sqrt{1 - 2\eta_1\eta_2}}{\eta_2}.$$

Furthermore, w_* is unique in the ball

$$\|w_* - w_0\| \le \frac{1 - \sqrt{1 - 2\eta_1\eta_2}}{\eta_2} \tag{25.4}$$

if condition (25.2) holds and in the ball (25.4) if $\eta_1\eta_2 = \frac{1}{2}$.

The rate of convergence of the NK method is characterized by

$$\|w_* - w_n\| \le 2^{2^n - n} (\eta_1\eta_2)^{2^n} \eta_2^{-1}, \tag{25.5}$$

while for the modified method we have for $\eta_1\eta_2 < 1/2$ the relation

$$\|w_* - w_n\|_{\le} \left(1 - \sqrt{1 - 2\eta_1\eta_2}\right)^{n+1} \eta_2^{-1}. \tag{25.6}$$

By construction of the NK method it is clear that it can be used if the desired solution w_* is nonsingular. The modified NK method requires w_0 to be a nonsingular solution. From (25.2), (25.3), (25.5), (25.6) it follows that the convergence rate of the NK method and of the modified NK method depends on the norms of $\mathcal{P}(w_0)$, $\mathcal{P}''(w_0)$ and $\mathcal{P}'(w_0)^{-1}$. Here it is clear that the closer to the required solution w_* we choose w_0, the smaller η_1 will be, and thus the better will be the convergence rate of the method. However, this is true only if the solution w_* is nonsingular or far from a singular one. If w_* is close to a singular solution, then we shall have a competition of two opposing effects:

$$\left\| \mathcal{P}'(w_0)^{-1} \right\| \to \infty, \quad \mathcal{P}(w_0) \to 0,$$

and the convergence rates of both the NK method and of the modified NK method can decrease drastically, which indeed is often observed in applications. Therefore, in order to avoid singular solutions (see Section 23, Figure 23.1) one needs to resort to the special techniques discussed in Sections 23, 24.

25.2. To illustrate the issues arising here, let us consider problem 51 ($t = 5$, $\kappa = 1$), to which correspond boundary conditions

$$w_1|_\Gamma = w_2|_\Gamma = w|_\Gamma = \left.\frac{\partial w}{\partial m}\right|_\Gamma = 0.$$

Furthermore, let us consider an isotropic plate, on which we act only with R^3. The NOE (13.6), (13.7) assumes the form

$$(\omega \cdot \chi)_{H_5} = -\frac{D_{s1}}{2}\int_\Omega \Big[\left(w_{\alpha^1}^2 + \nu w_{\alpha^2}^2\right)\varphi_{1\alpha^1}$$

$$+(1-\nu)w_{\alpha^1}w_{\alpha^2}\left(\varphi_{1\alpha^2} + \varphi_{2\alpha^1}\right) + \left(w_{\alpha^2}^2 + \nu w_{\alpha^1}^2\right)\varphi_{2\alpha^2}\Big]\, d\Omega \tag{25.7}$$

$$(w \cdot \varphi)_{H_1} = \int_\Omega \left(T^{ij} w_{\alpha^i\alpha^j} + R^3\right)\varphi d\Omega, \tag{25.8}$$

$$T^{11} = D_{s1}\left[w_{1\alpha^1} + \nu w_{2\alpha^2} + \frac{1}{2}\left(w_{\alpha^1}^2 + \nu w_{\alpha^2}^2\right)\right], \quad 1 \rightleftharpoons 2;$$

$$T^{12} = \frac{D_{s1}}{2}(1-\nu)\left(w_{1\alpha^2} + w_{2\alpha^1} + w_{\alpha^1}w_{\alpha^2}\right). \tag{25.9}$$

We define the operator $\mathcal{P}(w)$ by the relation

$$(\mathcal{P}(w)\cdot\varphi)_{H_1} = (w\cdot\varphi)_{H_1} - \int_\Omega \left(T^{ij} w_{\alpha^i\alpha^j} + R^3\right)\varphi d\Omega. \tag{25.10}$$

For an initial approximation w_0, let us present $\mathcal{P}(w_0)$ in a more explicit form. For this, let us substitute w_0 into the right-hand side of (25.7). Then, solving the resulting linear problem, we can determine ω_0. Next, from (25.9) we find $T^{ij}(w_0)$, and then from (25.8) we compute the corresponding load R_0^3. Then the following relations hold:

$$(\mathcal{P}(w_0)\cdot\varphi)_{H_1} = (w_0\cdot\varphi)_{H_1} - \int_\Omega \left(T^{ij}(w_0) w_{0\alpha^i\alpha^j} + R^3\right)\varphi d\Omega, \tag{25.11}$$

$$(w_0\cdot\varphi)_{H_1} - \int_\Omega \left(T^{ij}(w_0) w_{0\alpha^i\alpha^j} + R_0^3\right)\varphi d\Omega = 0; \tag{25.12}$$

from (25.11), (25.12) we have

$$(\mathcal{P}(w_0)\cdot\varphi)_{H_1} = \int_\Omega (R_0^3 - R^3)\varphi d\Omega. \tag{25.13}$$

Now let w_{**} satisfy the integral identity

$$\int_\Omega (R_0^3 - R^3)\varphi d\Omega = (w_{**}\cdot\varphi)_{H_1}. \tag{25.14}$$

Then clearly,

$$\mathcal{P}(w_0) = w_{**}.$$

It is easy to understand the physical meaning of w_{**}. From (25.14) it is seen that w_{**} is the difference in the deflection of a linear isotropic plate with the same elastic constants and the same Ω as the nonlinear plate being considered. Therefore, w_{**} describes the closeness of the initial point w_0 and the desired solution w.

To determine the convergence conditions (25.2) for our problem, we have to estimate η_1, η_2, for which first of all we need to construct $\mathcal{P}'(w_0)$. From the relation (25.11), which defines $\mathcal{P}$, we have

$$((\mathcal{P}'(w_0)\widetilde{\varphi}) \cdot \varphi)_{H_1} = (\widetilde{\varphi} \cdot \varphi)_{H_1} - \int_{\Omega} \left[T^{ij\prime}(w_0,\, \widetilde{\varphi}) w_{0\alpha^i\alpha^j} + T^{ij}(w_0) \widetilde{\varphi}_{\alpha^i\alpha^j} \right] \varphi \, d\Omega, \tag{25.15}$$

where $\widetilde{\varphi}$ determines the direction of differentiation; to compute $T^{ij\prime}(w_0,\, \widetilde{\varphi})$ we differentiate (25.7). We have

$$\begin{aligned}(\omega' \cdot \chi)_{H_5} = -D_{s1} \int_{\Omega} \Big[& (w_{0\alpha^1} \widetilde{\varphi}_{\alpha^1} + \nu w_{0\alpha^2} \widetilde{\varphi}_{\alpha^2}) \varphi_{1\alpha^1} \\ & + (1-\nu)(w_{0\alpha^1} \widetilde{\varphi}_{\alpha^2} + w_{0\alpha^2} \widetilde{\varphi}_{\alpha^1})(\varphi_{1\alpha^2} + \varphi_{2\alpha^1}) \\ & + (w_{0\alpha^2} \widetilde{\varphi}_{\alpha^2} + \nu w_{0\alpha^1} \widetilde{\varphi}_{\alpha^1}) \varphi_{2\alpha^2} \Big] d\Omega.\end{aligned} \tag{25.16}$$

Equation (25.16) defines ω' in terms of $w_0,\, \widetilde{\varphi}$. Here we take w_0 to be a fixed element of H_1. Furthermore, in (25.16), $\chi = (\varphi_1,\, \varphi_2)$ is an arbitrary element of H_5. Therefore, it follows from (25.16) that

$$\omega' = \omega'(w_0,\, \widetilde{\varphi}).$$

But then we have from (25.9) that

$$T^{11\prime} = D_{s1} \left(w'_{1\alpha^1} + \nu w'_{2\alpha^2} + w_{0\alpha^1} \widetilde{\varphi}_{\alpha^1} + \nu w_{0\alpha^2} \widetilde{\varphi}_{\alpha^2} \right), \tag{25.17}$$

and similar relations can be obtained for $T^{22\prime}$, $T^{12\prime}$. In (25.17), w'_i are the components of ω'. Therefore, we have constructed the operator $\mathcal{P}'(w_0)\widetilde{\varphi}$, since in (25.15) all the elements are expressed in terms of φ; w_0 enters as a parameter.

Let us consider the second derivative $\mathcal{P}''(w_0,\, \widetilde{\varphi},\, \widetilde{\widetilde{\varphi}})$, where $\widetilde{\widetilde{\varphi}}$ determines the direction of second differentiation. From (25.15) we have

$$\left(\mathcal{P}''(w_0,\, \widetilde{\varphi},\, \widetilde{\widetilde{\varphi}}) \cdot \varphi \right)_{H_1} = - \int_{\Omega} \Big[T^{ij\prime\prime}(w_0,\, \widetilde{\varphi},\, \widetilde{\widetilde{\varphi}}) w_{0\alpha^i\alpha^j} + T^{ij\prime}(w_0,\, \widetilde{\varphi}) \widetilde{\widetilde{\varphi}}_{\alpha^i\alpha^j} + T^{ij\prime}(w_0,\, \widetilde{\widetilde{\varphi}}) \widetilde{\varphi}_{\alpha^i\alpha^j} \Big] \varphi \, d\Omega. \tag{25.18}$$

To determine $T^{ij\prime\prime}$ we differentiate (25.16) and (25.17) again:

$$\begin{aligned}(\omega'' \cdot \chi)_{H_5} = -D_{s1} \int_{\Omega} \Big[& \left(\widetilde{\widetilde{\varphi}}_{\alpha^1} \widetilde{\varphi}_{\alpha^1} + \nu \widetilde{\widetilde{\varphi}}_{\alpha^2} \widetilde{\varphi}_{\alpha^2} \right) \varphi_{1\alpha^1} \\ & + (1-\nu) \left(\widetilde{\widetilde{\varphi}}_{\alpha^1} \widetilde{\varphi}_{\alpha^2} + \widetilde{\widetilde{\varphi}}_{\alpha^2} \widetilde{\varphi}_{\alpha^1} \right) (\varphi_{1\alpha^2} + \varphi_{2\alpha^1}) \\ & + \left(\widetilde{\widetilde{\varphi}}_{\alpha^2} \widetilde{\varphi}_{\alpha^2} + \nu \widetilde{\widetilde{\varphi}}_{\alpha^1} \widetilde{\varphi}_{\alpha^1} \right) \varphi_{2\alpha^2} \Big] d\Omega,\end{aligned} \tag{25.19}$$

$$T^{11\prime\prime} = D_{s1} \left(w''_{1\alpha^1} + \nu w''_{2\alpha^2} + \widetilde{\widetilde{\varphi}}_{\alpha^1} \widetilde{\varphi}_{\alpha^1} + \nu \widetilde{\widetilde{\varphi}}_{\alpha^2} \widetilde{\varphi}_{\alpha^2} \right). \tag{25.20}$$

In (25.20), the w_i'' are the components of ω''; similar formulae can be obtained for $T^{22''}$ and $T^{12''}$.

Let us now estimate the parameters η_1 and η_2. From (25.3) it is easy to see that to obtain an estimate for η_1, we need to estimate the solution of the LOE (linear operator equation)

$$(\widetilde{\varphi}\cdot\varphi)_{H_1} - \int_\Omega \left[T^{ij''}(w_0, \widetilde{\varphi})w_{0\alpha^i\alpha^j} + T^{ij}(w_0)\widetilde{\varphi}_{\alpha^i\alpha^j}\right]\varphi d\Omega = (w_{**}\cdot\varphi)_{H_1}. \quad (25.21)$$

From (25.21) it is easy to estimate $\widetilde{\varphi}$ in terms of w_{**}. It is easily seen that the second term in (25.12) is a completely continuous, self-adjoint operator with respect to $\widetilde{\varphi}$ in H_1. Therefore, the following functional will be weakly continuous in H_1:

$$\mathcal{E}(\widetilde{\varphi}) = \frac{1}{2}\int_\Omega \left[T^{ij'}(w_0, \widetilde{\varphi})w_{0\alpha^i\alpha^j} + T^{ij}(w_0)\widetilde{\varphi}_{\alpha^i\alpha^j}\right]\widetilde{\varphi}d\Omega. \quad (25.22)$$

Next, let $\lambda_1(w_0)$ be the maximum of $\mathcal{E}(\widetilde{\varphi})$ on the sphere $\|\widetilde{\varphi}\|_{H_1} = 1$. This maximum is attained by Theorem 9.4.

We shall distinguish the following two cases:

Case 1: $\lambda_1(w_0) < 1$,
Case 2: $\lambda_1(w_0) \geq 1$.

In the first case, we obviously have from (25.21)

$$\|\widetilde{\varphi}\|^2_{H_1}(1-\lambda_1(w_0)) \leq \|w_{**}\|_{H_1}\|\widetilde{\varphi}\|_{H_1} \text{ and } \|\widetilde{\varphi}\|_{H_1} \leq \frac{\|w_{**}\|_{H_1}}{1-\lambda_1(w_0)},$$

so that we can take

$$\eta_1 \leq \frac{\|w_{**}\|_{H_1}}{1-\lambda_1(w_0)}. \quad (25.23)$$

Let us describe (25.23) in more detail. For definiteness, let us assume that $R_0^3 - R^3 \in L_{p\Omega}$. Then from (25.19) we easily find that

$$\|w_{**}\|_{H_1} \leq m(\nu, \Gamma)\mathcal{L}^{3-2/p}D_{f1}^{-1/2}\left\|R_0^3 - R^3\right\|_{L_{p\Omega}}, \quad (25.24)$$

where $\mathcal{L}$ is a characteristic length of the shell at the middle surface. The constant m on the right hand-side of (25.24) is dimensionless; it depends only on the Poisson coefficient ν and the shape of the domain Ω. This is reflected by explicitly stating the dependence of m on the boundary curve Γ. Using (25.24), in this case we obtain for η_1

$$\eta_1 = \frac{m(\nu, \Gamma)\mathcal{L}^{3-2/p}}{1-\lambda_1(w_0)}D_{f1}^{-1/2}\left\|R_0^3 - R^3\right\|_{L_{p\Omega}}. \quad (25.25)$$

Next, let us find an estimate for η_2. This can be done using the operator equation (25.21), where instead we have on the right-hand side $(\mathcal{P}''(w_0)\cdot\varphi)_{H_1}$. Then for η_2 we have

$$\eta_2 \leq \frac{\left\|\mathcal{P}''(w_0)\right\|_{H_1}}{1-\lambda_1(w_0, \nu)}, \quad (25.26)$$

and therefore we must estimate $\|\mathcal{P}''(w_0)\|_{H_1}$, which can be done using (25.18), (25.19), (25.21). Indeed, from (25.18) we have

$$(\mathcal{P}''(w_0, \widetilde{\varphi}, \widetilde{\widetilde{\varphi}}) \cdot \varphi)_{H_1} \leq \Big(\Big\| T^{ij''}(w_0, \widetilde{\varphi}, \widetilde{\widetilde{\varphi}}) \Big\|_{L_{2\Omega}} \| w_{0\alpha^i\alpha^j} \|_{L_{2\Omega}} + \\ + \Big\| T^{ij'}(w_0, \widetilde{\widetilde{\varphi}}) \Big\|_{L_{2\Omega}} \| \widetilde{\varphi}_{\alpha^i\alpha^j} \|_{L_{2\Omega}} \\ + \Big\| T^{ij'}(w_0, \widetilde{\varphi}) \Big\|_{L_{2\Omega}} \| \widetilde{\widetilde{\varphi}}_{\alpha^i\alpha^j} \|_{L_{2\Omega}} \Big) \| \varphi \|_{C_\Omega} . \tag{25.27}$$

In order to extract the necessary estimate of $\|\mathcal{P}''(w_0, \widetilde{\varphi}, \widetilde{\widetilde{\varphi}})\|_{H_1}$ from (25.27), we first obtain some auxiliary inequalities. From (25.19) we easily deduce

$$\|\omega''\|_{H_5} \leq m(\nu, \Gamma) D_{s1}^{1/2} D_{f1}^{-1} \mathcal{L} \|\widetilde{\varphi}\|_{H_1} \|\widetilde{\widetilde{\varphi}}\|_{H_1} . \tag{25.28}$$

From (25.28) it follows that

$$\Big\| T^{ij''}(w_0, \widetilde{\varphi}, \widetilde{\widetilde{\varphi}}) \Big\|_{L_{2\Omega}} \leq m(\nu, \Gamma) D_{s1} D_{f1}^{-1} \mathcal{L} \|\widetilde{\varphi}\|_{H_1} \|\widetilde{\widetilde{\varphi}}\|_{H_1} . \tag{25.29}$$

In the same way it follows from (25.16) that

$$\|\omega'\|_{H_5} \leq m(\nu, \Gamma) D_{s1}^{1/2} D_{f1}^{-1} \mathcal{L} \|w_0\|_{H_1} \|\widetilde{\varphi}\|_{H_1} \tag{25.30}$$

and

$$\Big\| T^{ij'}(w_0, \widetilde{\varphi}) \Big\|_{L_{2\Omega}} \leq m(\nu, \Gamma) D_{s1} D_{f1}^{-1} \mathcal{L} \|w_0\|_{H_1} \|\widetilde{\varphi}\|_{H_1} , \tag{25.31}$$

as well as

$$\Big\| T^{ij'}(w_0, \widetilde{\widetilde{\varphi}}) \Big\|_{L_{2\Omega}} \leq m(\nu, \Gamma) D_{s1} D_{f1}^{-1} \mathcal{L} \|w_0\|_{H_1} \|\widetilde{\widetilde{\varphi}}\|_{H_1} . \tag{25.32}$$

From (25.27), (25.29), (25.31), (25.32) we have

$$(\mathcal{P}''(w_0, \widetilde{\varphi}, \widetilde{\widetilde{\varphi}}) \cdot \varphi)_{H_1} \leq m(\nu, \Gamma) D_{s1} D_{f1}^{-1} \mathcal{L} \Big(\|\widetilde{\varphi}\|_{H_1} \|\widetilde{\widetilde{\varphi}}\|_{H_1} \sum_{i,j=1}^{2} \|w_{0\alpha^i\alpha^j}\|_{L_{2\Omega}} \\ + \|w_0\|_{H_1} \|\widetilde{\varphi}\|_{H_1} \|\widetilde{\widetilde{\varphi}}\|_{H_1} + \|w_0\|_{H_1} \|\widetilde{\widetilde{\varphi}}\|_{H_1} \sum_{i,j=1}^{2} \|\widetilde{\varphi}_{\alpha^i\alpha^j}\|_{H_1} \Big) \|\varphi\|_{C_\Omega} . \tag{25.33}$$

Let us now take into account the relations

$$\sum_{i,j=1}^{2} \|w_{0\alpha^i\alpha^j}\|_{L_{2\Omega}} \leq m(\nu, \Gamma) D_{f1}^{-1/2} \|w_0\|_{H_1} , \\ \sum_{i,j=1}^{2} \|\widetilde{\varphi}_{\alpha^i\alpha^j}\|_{L_{2\Omega}} \leq m(\nu, \Gamma) D_{f1}^{-1/2} \|\widetilde{\varphi}\|_{H_1} , \\ \sum_{i,j=1}^{2} \|\widetilde{\widetilde{\varphi}}_{\alpha^i\alpha^j}\|_{L_{2\Omega}} \leq m(\nu, \Gamma) D_{f1}^{-1/2} \|\widetilde{\widetilde{\varphi}}\|_{H_1} , \tag{25.34}$$

$$\|\varphi\|_{C_\Omega} \leq m(\nu, \Gamma) D_{f1}^{-1/2} \mathcal{L} \|\varphi\|_{H_1} . \tag{25.35}$$

From (25.33)–(25.35) we obtain

$$(\mathcal{P}''(w_0, \widetilde{\varphi}, \widetilde{\widetilde{\varphi}}) \cdot \varphi)_{H_1} \leq m(\nu, \Gamma) D_{s1} D_{f1}^{-2} \mathcal{L}^2 \, \|w_0\|_{H_1} \, \|\widetilde{\varphi}\|_{H_1} \, \|\widetilde{\widetilde{\varphi}}\|_{H_1} \, \|\varphi\|_{H_1}, \tag{25.36}$$

whence

$$\|\mathcal{P}''\|_{H_1} \leq m(\nu, \Gamma) D_{s1} D_{f1}^{-2} \mathcal{L}^2 \, \|w_0\|_{H_1}, \tag{25.37}$$

and from (25.26), (25.27) we have for η_2 the inequality

$$\eta_2 \leq m(\nu, \Gamma) D_{s1} D_{f1}^{-2} \mathcal{L}^2 \frac{\|w_0\|_{H_1}}{1 - \lambda_1(w_0, \nu)}. \tag{25.38}$$

Finally, it follows from (25.25), (25.38) that

$$\eta_1 \eta_2 \leq m(\nu, \Gamma) \eta_3 \eta_4 \eta_5 \eta_6, \tag{25.39}$$

where the dimensionless parameters η_i are given by

$$\eta_3 = \left(\frac{\mathcal{L}}{h}\right)^5, \quad \eta_4 = D_{f1}^{-1/2} \|w_0\|_{H_1} \quad \eta_5 = \mathcal{L}^{-2/p} E^{-1} \left\| R_0^3 - R^3 \right\|_{L_{q\Omega}},$$
$$\eta_6 = \frac{1}{[1 - \lambda_1(w_0, \nu)]^2}. \tag{25.40}$$

Relations (25.39), (25.40) demonstrate the influence of various factors on the convergence of the NK method. A good choice of w_0, its closeness to the desired solution w, determines the size of η_5. The parameter η_3 shows that the NK method is quite sensitive to the thin-walledness parameter of the construction, while η_6 takes into consideration the closeness of a solution to a singular one. In practical applications of the NK method, dependence of the solution w on the load is constructed step by step, and at each step the value of w_0 is taken to be the value of w at the previous step.

In case 2, to estimate $\widetilde{\varphi}$ from (25.21), we take into account the fact that by assumption w_0 is a nondegenerate solution and therefore

$$\widetilde{\varphi} = T_1 w_{**}, \tag{25.41}$$

where T_1 is a bounded operator in H_1. It is easily seen that T_1 depends only on w_0, ν, and therefore

$$\eta_1 \leq \|T_1\| \, \|w_{**}\|_{H_1} \leq m(\nu, \Gamma) \, \|T_1(w_0, \nu)\| \, \mathcal{L}^{3-2/p} D_f^{-1/2} \left\| R_0^3 - R^3 \right\|_{L_{p\Omega}}. \tag{25.42}$$

Similarly, for η_2 from (25.21) with $(\mathcal{P}'' \cdot \varphi)_{H_1}$ on the right-hand side, we have

$$\eta_2 \leq \|T_1(w_0, \nu)\| \cdot \|\mathcal{P}''\|_{H_1} \leq m(\nu, \Gamma) \, \|T_1(w_0, \nu)\| \, D_{s1} D_f^{-2} \mathcal{L}^2 \, \|w_0\|_{H_1}. \tag{25.43}$$

Finally, from (25.42), (25.43) it follows that

$$\eta_1 \eta_2 \leq m(\nu, \Gamma) \, \|T_1(w_0, \nu)\|^2 \, \eta_3 \eta_4 \eta_5. \tag{25.44}$$

In case 2 as well the relation (25.44) reveals the influence of different factors on the convergence rate of the NK method. The constants $m(\nu, \Gamma)$, $\|T_1(w_0, \nu)\|$, $\lambda_1(w_0, \nu)$ can be computed in each particular case numerically.

As a result we have obtained the following result.

Theorem 25.2. *Suppose a homogeneous isotropic plate satisfies Conditions 2, 3 of Section 13 and that moreover, the homogeneous boundary conditions* (20.1)–(20.3) *hold on the entire boundary curve. Assume furthermore that the plate is subjected only to a transverse load* R^3. *Then, if conditions* (25.2) *hold with* $\eta_1\eta_2$ *given by* (25.39) *or by* (25.44), *the boundary value Problem 51 has a unique solution in the ball* (25.4). *If* $\eta_1\eta_2 = \frac{1}{2}$, *then the boundary value Problem 51 has a unique solution in the closed ball* (25.4). *Here, if* $\eta_1\eta_2 < \frac{1}{2}$, *this solution can be found both by the NK method and the modified NK method. Thus, the rate of convergence is estimated using* (25.5), (25.6).

25.3. Let us consider Problem 91 with homogeneous boundary conditions for a homogeneous isotropic plate. Let us also assume that the plate is only subjected to the load R^3, in other words, we set

$$\widetilde{R}_t \equiv \widetilde{T}^{ij}_{\mathrm{p}} \equiv 0.$$

Under these assumptions relations (17.2), (17.3) take the form

$$(w \cdot \varphi)_{H_1} = \int_\Omega [1,\ \Psi,\ w]\varphi \, d\Omega + \int_\Omega R^3 \varphi \, d\Omega, \tag{25.45}$$

$$(\Psi \cdot \Theta)_{H_9} = \frac{1}{2}\int_\Omega [1,\ w,\ w]\Theta \, d\Omega, \tag{25.46}$$

where the symbol $[a,\ b,\ c]$ was introduced in (7.31).

In this case the operator $\mathcal{P}(w)$ is defined by the formula

$$(\mathcal{P}(w)\cdot\varphi)_{H_1} = (w\cdot\varphi)_{H_1} - \int_\Omega ([1,\ \Psi,\ w] + R^3)\varphi \, d\Omega, \tag{25.47}$$

where Ψ is expressed in terms of w by (25.46). In this case as well the operator $\mathcal{P}(w_0)$, where w_0 is some initial approximation, can be given a more explicit form. For that we substitute w_0 into the right-hand side of (25.46), and, having found $\Psi(w_0)$, substitute the result in (25.47). Thus we have

$$(\mathcal{P}(w_0)\cdot\varphi)_{H_1} = (w_0\cdot\varphi)_{H_1} - \int_\Omega ([1,\ \Psi(w_0),\ w] + R_0^3)\varphi \, d\Omega, \tag{25.48}$$

from which we find the corresponding load R_0^3.

From (25.47), (25.48) we obtain (25.13), and using (25.14) we find the corresponding deflection w_{**}. From (25.41) it follows that

$$(\mathcal{P}'(w_0, \widetilde{\varphi})\cdot\varphi)_{H_1}$$

$$= \int_\Omega \{[1,\ \Psi'(w_0,\ \widetilde{\varphi}),\ w_0] + [1,\ \Psi(w_0), \widetilde{\varphi}]\}\varphi \, d\Omega, \tag{25.49}$$

$$(\Psi'(w_0,\ \widetilde{\varphi})\cdot\Theta)_{H_9} = \int_\Omega [1,\ w_0,\ \widetilde{\varphi}]\Theta \, d\Omega, \tag{25.50}$$

$$(\mathcal{P}''(w_0, \widetilde{\varphi},\ \widetilde{\widetilde{\varphi}})\cdot\varphi)_{H_1}$$

$$= \int_\Omega \{[1,\ \Psi''(w_0,\ \widetilde{\varphi},\ \widetilde{\widetilde{\varphi}}),\ w_0] \\ +[1,\ \Psi'(w_0,\ \widetilde{\varphi}),\ \widetilde{\widetilde{\varphi}}] + [1,\ \Psi'(w_0,\ \widetilde{\widetilde{\varphi}}),\ \widetilde{\varphi}]\}\varphi\, d\Omega, \tag{25.51}$$

$$(\Psi''(w_0,\ \widetilde{\varphi},\ \widetilde{\widetilde{\varphi}}) \cdot \Theta)_{H_9} = \int_\Omega [1,\ \widetilde{\widetilde{\varphi}},\ \widetilde{\varphi}]\Theta\, d\Omega. \tag{25.52}$$

To estimate η_1 in this case we need to consider

$$(\widetilde{\varphi}\cdot\varphi)_{H_1} - \int_\Omega \{[1,\ \Psi'(w_0,\ \widetilde{\varphi}),\ w_0]+[1,\ \Psi(w_0), \widetilde{\varphi}]\}\varphi\, d\Omega = (w_{**}\cdot\varphi)_{H_1}. \tag{25.53}$$

In (25.45), $\widetilde{\varphi}$ is the required function, with respect to which (25.45) is linear, while the second term is symmetric with respect to $\widetilde{\varphi}$, φ and completely continuous in $\widetilde{\varphi}$. Therefore, we can introduce the functional

$$\mathcal{E}(\widetilde{\varphi}) = \int_\Omega \{[1,\ \Psi'(w_0,\ \widetilde{\varphi}),\ w_0] + [1,\ \Psi(w_0), \widetilde{\varphi}]\}\widetilde{\varphi}\, d\Omega,$$

which turns out to be weakly continuous in H_1. Furthermore, let λ_1 be the maximum of $\mathcal{E}(\widetilde{\varphi})$ on the sphere $\|\widetilde{\varphi}\|^2_{H_1} = 1$. It is easy to see that in the case under consideration λ_1 will depend only on w_0: $\lambda_1 = \lambda_1(w_0)$. Furthermore, we shall distinguish the two cases of (25.22). In case 1, we obtain the estimate (25.23), and thus (25.25) for η_1 from (25.53). Under these conditions we shall also have the inequality (25.26), and we have to specify an estimate of $\left\|\mathcal{P}''\right\|_{H_1}$. For that we note that from (25.51) it follows that

$$\left|(\mathcal{P}''(w_0,\ \widetilde{\varphi},\ \widetilde{\widetilde{\varphi}}) \cdot \varphi)_{H_1}\right| \\ \le \sum_{i,j,k,l=1}^{2} \left\{ \left\|\Psi''_{\alpha^i\alpha^j}\right\|_{L_{2\Omega}} \|w_{0\alpha^k\alpha^l}\|_{L_{2\Omega}} + \|\Psi_{\alpha^i\alpha^j}(w_0,\ \widetilde{\varphi})\|_{L_{2\Omega}} \left\|\widetilde{\widetilde{\varphi}}_{\alpha^k\alpha^l}\right\|_{L_{2\Omega}} \right. \\ \left. + \left\|\Psi_{\alpha^i\alpha^j}(w_0,\ \widetilde{\widetilde{\varphi}})\right\|_{L_{2\Omega}} \|\widetilde{\varphi}_{\alpha^k\alpha^l}\|_{L_{2\Omega}} \right\} \|\varphi\|_{C_\Omega}. \tag{25.54}$$

Furthermore, let us take into account the following relation, which is deduced from (25.52):

$$\left|(\Psi''(w_0,\ \widetilde{\varphi},\ \widetilde{\widetilde{\varphi}}) \cdot \Theta)_{H_9}\right| \le \left| \int_\Omega [1,\ \widetilde{\varphi},\ \widetilde{\widetilde{\varphi}}]\Theta\, d\Omega \right| \\ \le \sum_{i,j,k,l=1}^{2} \|\widetilde{\varphi}_{\alpha^i\alpha^j}\|_{L_{2\Omega}} \left\|\widetilde{\widetilde{\varphi}}_{\alpha^k\alpha^l}\right\|_{L_{2\Omega}} \|\Theta\|_{C_\Omega}, \tag{25.55}$$

and moreover,

$$\|\Theta\|_{C_\Omega} \le m(\Gamma)C^{-1/2}\mathcal{L}\,\|\Theta\|_{H_9},\ C = \frac{1}{2Eh}. \tag{25.56}$$

Let us observe that the constant $m(\Gamma)$ in (25.56) can be taken to depend only on the shape of the domain Ω, which is symbolized by introducing the dependence on the boundary curve Γ in its notation. Furthermore, we have

$$\|\widetilde{\varphi}_{\alpha^i\alpha^j}\|_{L_{2\Omega}} \leq m(\nu,\,\Gamma)D_{\text{f1}}^{-1/2}\,\|\widetilde{\varphi}\|_{H_1}\,, \tag{25.57}$$

$$\left\|\widetilde{\widetilde{\varphi}}_{\alpha^k\alpha^l}\right\|_{L_{2\Omega}} \leq m(\nu,\,\Gamma)D_{\text{f1}}^{-1/2}\,\left\|\widetilde{\widetilde{\varphi}}\right\|_{H_1}. \tag{25.58}$$

From (25.55)–(25.57) we obtain

$$\left\|\Psi''(w_0,\,\widetilde{\varphi},\,\widetilde{\widetilde{\varphi}})\right\|_{H_9} \leq m(\nu,\,\Gamma)C^{-1/2}\mathcal{L}\sum_{i,j,k,l=1}^{2}\|\widetilde{\varphi}_{\alpha^i\alpha^j}\|_{L_{2\Omega}}\left\|\widetilde{\widetilde{\varphi}}_{\alpha^k\alpha^l}\right\|_{L_{2\Omega}}. \tag{25.59}$$

Next, we take into consideration the inequalities

$$\begin{aligned}
&\left\|\Psi''_{\alpha^r\alpha^t}\right\|_{L_{2\Omega}} \leq m(\Gamma)C^{-1/2}\,\|\Psi\|_{H_9}\,,\\
&\|\widetilde{\varphi}_{\alpha^i\alpha^j}\|_{L_{2\Omega}} \leq m(\nu,\,\Gamma)\,\|\widetilde{\varphi}\|_{H_1}\,D_{\text{f}}^{-1/2},\\
&\left\|\widetilde{\widetilde{\varphi}}_{\alpha^k\alpha^l}\right\|_{L_{2\Omega}} \leq m(\nu,\,\Gamma)\,\left\|\widetilde{\widetilde{\varphi}}\right\|_{H_1}\,D_{\text{f}}^{-1/2}.
\end{aligned} \tag{25.60}$$

Finally, we derive from (25.59), (25.60),

$$\left\|\Psi''_{\alpha^r\alpha^t}\right\|_{L_{2\Omega}} \leq m(\nu,\,\Gamma)C^{-1}D_{\text{f1}}^{-1}\mathcal{L}\,\|w_0\|_{H_1}\,\|\widetilde{\varphi}\|_{H_1}\,. \tag{25.61}$$

Let us now estimate $\left\|\Psi'_{\alpha^r\alpha^t}(w_0,\,\widetilde{\varphi})\right\|_{L_{2\Omega}}$. Using the same arguments as in the derivation of (25.55)–(25.61), we obtain from (25.50),

$$\left\|\Psi'_{\alpha^r\alpha^t}\right\|_{L_{2\Omega}} \leq m(\nu,\,\Gamma)C^{-1}D_{\text{f1}}^{-1}\mathcal{L}\,\|w_0\|_{H_1}\,\|\widetilde{\varphi}\|_{H_1}\,. \tag{25.62}$$

Finally, taking into account the inequality (25.35), we derive from (25.54), (25.61), (25.62) the relation (25.36), from which follow (25.37)–(25.40). In case 2 of (25.22) in problem 91 we can repeat all the arguments of (25.41)–(25.44) almost verbatim and thus obtain the estimates (25.44). Therefore, all the conclusions concerning the influence of various factors on the rate of convergence of the NK method are also valid in the case of problem 91.

Theorem 25.3. *Suppose a homogeneous isotropic plate satisfies Conditions 2, 3 of Section 13;* $\widetilde{T}^{ij}_{\text{p}} \equiv 0$, *and that it is subjected only to a transverse load* $R^3 \in \overline{H}_1$. *Furthermore, assume that homogeneous boundary conditions are satisfied on the entire boundary curve* Γ. *Then all the claims made in theorem 25.2 hold true for the Problem 91.*

25.4. The Newton–Kantorovich method has been quite widely used in nonlinear shell theory [261, 309, 205]. The closely related method of "successive loading" has also been extensively developed. Initial work in direction was done by Petrov [229, 230]. Consequently, this method was developed and applied to problems of nonlinear shell theory in [132, 304]. It is also connected to the method of partial linearization.

CHAPTER VII

Direct Methods in the Nonlinear Theory of Shallow Shells

26. Variational Methods for Approximate Solutions of Problems $t\kappa$ ($\kappa = 1, 2, 3, 4$; $t = 5, 6, 7, 8$). The Version of Papkovich

26.1. The foundations of the Bubnov–Galerkin method were laid in the classical papers of Bubnov [45] and Galerkin [87] at the beginning of the twentieth century. At the same time Ritz suggested his well-known method of approximate computation of an extremum of a functional [262, 263]. These methods were destined to play an exceptionally important role in mathematical physics, and in particular, in mechanics.

Let us present the general scheme of the Bubnov–Galerkin method. Let us consider the operator equation

$$P(x) = 0,$$

where the operator P acts on a separable Hilbert space X. Furthermore, let $\{e_k\}_1^\infty$ be a complete family of elements in X, each finite subset of which is linearly independent (a family is called complete if for each $\epsilon > 0$ and each vector x we can find a finite linear combination of elements of the family such that $\|x - \sum_{i=1}^N c_i e_i\| < \epsilon$). An orthonormal basis of the Hilbert space is an example of such a family.

Equations of the nth approximation of the Bubnov–Galerkin method have the form

$$(P(\sum_{i=1}^n c_i e_i) \cdot e_j) = 0, \quad j = 1, \ldots, n.$$

This is a system consisting of n equations for n real coefficients c_i. In the general case, the system is nonlinear. There are different modifications of this method, in particular the Petrov–Galerkin method.

A number of versions of these methods are in use in the nonlinear theory of shallow shells. A frequently used scheme is the one developed by Papkovich [228]. This scheme was widely used in the work of Panov, Feodos'ev, and other researchers to solve many important engineering problems. Numerous examples of the use of this version can be found in the monographs of Vol'mir, Mushtari and Galimov, Feodos'ev [336, 212, 78]. The Papkovich version [228] is as follows. We seek an approximate solution w_n in the form

$$w_n = \sum_{k=1}^{n} D_{nk}\varphi_k(\mathcal{P}), \tag{26.1}$$

where the D_{nk} are unknown constants and $\varphi_k(\mathcal{P})$ is an orthonormal basis in H_κ. Substituting (26.1) into (13.6) or into (13.33), we have

$$\omega_n = \mathbf{K}(w_n,\, w^*,\, \overset{0}{\omega}). \tag{26.2}$$

Next, substituting (26.1), (26.2) into (13.15) and consecutively setting there $\varphi = \varphi_1, \ldots, \varphi_n$, we obtain

$$D_{nk} = -(\overset{0}{w} \cdot \varphi_k)_{H_\kappa} + \int_\Omega T^{ij}(\mathbf{a}_n + \mathbf{a}^*)[B_{ij}\varphi_k - (w_n + w^*)_{\alpha^i}\varphi_{k\alpha^j}]d\Omega. \tag{26.3}$$

In (26.3) it is assumed that the components of $\mathbf{a}_n$ are ω_n from (26.2) and w_n. Therefore, in view of the relations (4.15), which define T^{ij} in terms of $\overset{0}{\epsilon}_{ij}$, and relations (3.16), which define ϵ_{ij} in terms of w_i, and finally, in view of (26.2), we have that (26.3) is a system of cubic polynomial equations with respect to D_{nk}. Solving it, we obtain an approximate solution of the boundary value problem $t\kappa$. Of course, explicit computation of ω in terms of w_n from (26.2) is not a simple problem even though it is linear. However, assuming it to be solved, we can consider the problem of writing down the system (26.3) to be solved as well. There naturally arise questions of the behavior of w_n as $n \to \infty$ and the relation of this limit to the solutions w of the boundary value problem.

Using the Ritz method [263, 262] to determine D_{nk}, we substitute (26.2) and w_n into the relations for $\overset{0}{\epsilon}_{ij}, \overset{1}{\epsilon}_{ij}$, and then into (21.1), and we set

$$\frac{\partial \mathcal{J}_{\kappa\kappa}}{\partial D_{nk}} = 0, \quad k = 1, \ldots, n. \tag{26.4}$$

The system of equations (26.4) is cubic with respect to D_{nk} as well.

Lemma 26.1. *The systems of equations* (26.3) *and* (26.4) *are the same.*

To prove this, let us note that it is easy to show that

$$\frac{\partial \mathcal{J}_{\kappa\kappa}}{\partial D_{nk}} = \left(\operatorname{grad}_{H_\kappa} \mathcal{J}_{\kappa\kappa} \cdot \frac{\partial w_n}{\partial D_{nk}}\right)_{H_\kappa} = (\operatorname{grad}_{H_\kappa} \mathcal{J}_{\kappa\kappa} \cdot \varphi_k)_{H_\kappa}. \tag{26.5}$$

From Lemma 21.3 (relation (21.20)) and (26.5) we obtain

$$\frac{\partial \mathcal{J}_{\kappa\kappa}}{\partial D_{nk}} = ((w_n - \mathbf{G}_{\kappa\kappa}(w_n)) \cdot \varphi_k)_{H_k}. \tag{26.6}$$

Finally, using the definition of the operator $\mathbf{G}_{\kappa\kappa}$ (given by (13.34)), we obtain (26.3) from (26.6). Lemma 26.1 is proved.

Consider the vector field

$$\begin{aligned} \mathcal{N}_{nk}(D_{n1}, \ldots, D_{nn}) = D_{nk} + (\overset{0}{w} \cdot \varphi_k)_{H_\kappa} \\ - \int_\Omega T^{ij}(\mathbf{a}_n + \mathbf{a}^*)[B_{ij}\varphi_k - (w_n + w^*)_{\alpha^i}\varphi_{k\alpha^j}]d\Omega. \end{aligned} \tag{26.7}$$

in the n-dimensional real space R^n of coefficients D_{nk}. Since the right-hand side of (26.7) is a third-degree polynomial, the field $\mathcal{N}_n = (\mathcal{N}_{n1}, \ldots, \mathcal{N}_{nn})$ is continuous.

We shall assume that the conditions of Theorems 16.1–16.3 are satisfied.

Lemma 26.2. *The following statements hold* [339, 340, 342, 372, 349]*:*

(1) *On spheres of large radius in R^n the vector field $\mathbf{N}_n$ is homotopic to the identity field* **I**.

(2) *for each n, the system* (26.7) *has at least one real solution;*

(3) *all the solutions of* (26.7) *are such that the corresponding w_n defined by* (26.1) *lie inside a sphere*

$$\|w_n\|_{H_\kappa} = R,$$

where R is independent of n.

To prove the lemma, let us consider the vector field $\widetilde{\mathcal{N}}_n(D_{nk}, t)$ defined by

$$\begin{aligned} \widetilde{\mathcal{N}}_n(D_{nk}, t) = D_{nk} + t(\overset{0}{w} \cdot \varphi_k)_{H_\kappa} \\ - t\int_\Omega T^{ij}(\mathbf{a}_n + \mathbf{a}^*)[B_{ij}\varphi_k - (w_n + w^*)_{\alpha^i}\varphi_{k\alpha^j}]d\Omega, \end{aligned}$$

$$\widetilde{\mathcal{N}}_{nk}(D_{nk}, 0) = D_{nk}, \ \widetilde{\mathcal{N}}_{nk}(D_{nk}, 1) = \mathcal{N}_{nk},$$

and let us prove that on spheres of sufficiently large radius in R^n,

$$\widetilde{\mathbf{N}}_n(D_{nk}, t) \neq 0 \text{ for } 0 \leq t \leq 1.$$

Assuming the contrary means that for some $0 \leq t_0 \leq 1$ we have

$$\widetilde{\mathcal{N}}_n(D_{nk}, t_0) = 0. \tag{26.8}$$

Multiplying kth equation (26.8) by D_{nk} and then summing over k from 1 to n, we have

$$\begin{aligned} \sum_{k=1}^{n} D_{nk}^2 + t_0 \sum_{k=1}^{n} D_{nk}\Big((\overset{0}{w} \cdot \varphi_k)_{H_\kappa} - \int_\Omega T^{ij}(\mathbf{a}_n + \mathbf{a}^*)[B_{ij}\varphi_k \\ - (w_n + w^*)_{\alpha^i}\varphi_{k\alpha^j}]d\Omega\Big) = 0. \end{aligned} \tag{26.9}$$

Equation (26.9) can be written in the form

$$\|w_n\|^2_{H_\kappa} + t_0\Big((\overset{0}{w} \cdot w_n)_{H_\kappa} - \int_\Omega T^{ij}(\mathbf{a}_n + \mathbf{a}^*)[B_{ij}w_n - (w_n + w^*)_{\alpha^i} w_{n\alpha^j}]\,d\Omega\Big) = 0, \tag{26.10}$$

or, by (13.34), (19.1),

$$\|w_n\|^2_{H_\kappa} - t_0(\mathbf{G}_{\kappa\kappa}(w_n) \cdot w_n)_{H_\kappa} = \Phi(w_n,\, t_0) = 0,$$

which is impossible by Theorem 16.1 if

$$\|w_n\|_{H_\kappa} \geq R_0,$$

where R_0 is independent of n by the content of that theorem. Therefore, part 1 of Lemma 26.2 has been proved. From that follow parts 2 and 3. Lemma 26.2 has been proved.

Below we shall denote the whole set of approximate solutions of equations (26.3), (26.4) by $\{w_n\}$. By Lemma 26.2 the set $\{w_n\}$ is weakly compact.

Remark 26.1. In connection with Lemma 26.2, let us note that once we have established that $\mathcal{J}_{\kappa\kappa}$ is an increasing functional in H_κ (Theorem 21.3), the existence of real solutions of (26.7) can be deduced from that fact [319, 162]. However, the above method of proof gives us more information: It gives us existence of a solution with nonzero index, as well as a computation of the total index of all the solutions. On the other hand, the variational method [339, 342] demonstrates the existence of a solution that minimizes $\mathcal{J}_{\kappa\kappa}$ on the hyperplane (26.1).

Lemma 26.3. *Every weak limit w_0 of elements in $\{w_n\}$ is a generalized solution of the Problem $t\kappa$.*

To prove the lemma, we need to show that for an arbitrary $\varphi \in H_\kappa$ we have the relation

$$(w_0 \cdot \varphi)_{H_\kappa} = -(\overset{0}{w} \cdot \varphi)_{H_\kappa} + \int_\Omega T^{ij}(\mathbf{a}_0 + \mathbf{a}^*)[B_{ij}\varphi - (w_0 + w^*)_{\alpha^i}\varphi_{\alpha^j}]d\Omega. \tag{26.11}$$

Let w_n be a subsequence of $\{w_n\}$ such that w_n weakly converges to w_0. Then from (14.31) it follows that

$$T^{ij}(\mathbf{a}_n + \mathbf{a}^*) \to T^{ij}(\mathbf{a}_0 + \mathbf{a}^*) \text{ in } L_{2\Omega}, \tag{26.12}$$

and from Theorem 12.3 we have that

$$w_n + w^* \to w_0 + w^* \text{ in } W^{(1)}_{q\Omega} \text{ for any } q \geq 1. \tag{26.13}$$

Next, the system of equations (26.3) can be written as

$$\begin{aligned}&((w_n + \overset{0}{w}) \cdot \varphi_k)_{H_\kappa}\\ &\quad - \int_\Omega T^{ij}(\mathbf{a}_n + \mathbf{a}^*)[B_{ij}\varphi_k - (w_n + w^*)_{\alpha^i}\varphi_{k\alpha^j}]\,d\Omega = 0,\ k \leq n.\end{aligned} \tag{26.14}$$

In (26.14) we can pass to the limit as $n \to \infty$, and since the first term on the left-hand side of (26.14) is a linear functional in H_κ, and (26.12), (26.13) hold, we immediately obtain (26.11) for each $\varphi = \varphi_k$, $k = 1, \ldots, \infty$. In the general case,

$$\varphi = \sum_{k=1}^{\infty} \alpha_k \varphi_k, \ \varphi^n = \sum_{k=1}^{n} \alpha_k \varphi_k, \text{ and } R_n(\varphi) = \sum_{k=n+1}^{\infty} \alpha_k \varphi_k \to 0, \ n \to \infty \tag{26.15}$$

as the remainder of the Fourier series (expansion in an orthonormal basis). Furthermore,

$$((w_0 + \overset{0}{w}) \cdot \varphi)_{H_\kappa} - \int_\Omega T^{ij}(\mathbf{a}_0 + \mathbf{a}^*)[B_{ij}\varphi - (w_0 + w^*)_{\alpha^i}\varphi_{\alpha^j}]\, d\Omega = 0$$

$$((w_0 + \overset{0}{w}) \cdot R_n(\varphi))_{H_\kappa} - \int_\Omega T^{ij}(\mathbf{a}_0 + \mathbf{a}^*)[B_{ij} R_n(\varphi) - (w_0 + w^*)_{\alpha^i}(R_n(\varphi))_{\alpha^j}]\, d\Omega = 0$$

for any n, since (26.9) is already proved for $\varphi = \varphi_k$, for arbitrary k. Furthermore,

$$\left| ((w_0 + \overset{0}{w}) \cdot R_n(\varphi))_{H_\kappa} \right| \le \left\| w_0 + \overset{0}{w} \right\|_{H_\kappa} \| R_n(\varphi) \|_{H_\kappa}, \tag{26.16}$$

and by the embedding Theorem 12.3 (see (12.26), (12.28)),

$$\left| \int_\Omega T^{ij}(\mathbf{a}_0 + \mathbf{a}^*)[B_{ij} R_n(\varphi) - (w_0 + w^*)_{\alpha^i}(R_n(\varphi))_{\alpha^j}]\, d\Omega \right|$$
$$\le m\left(\left\| T^{ij}(\mathbf{a}_0 + \mathbf{a}^*) B_{ij} \right\|_{L_{2\Omega}} + \left\| T^{ij}(\mathbf{a}_0 + \mathbf{a}^*)(w_0 + w^*)_{\alpha^i} \right\|_{L_{2-\epsilon,\Omega}} \right) \| R_n(\varphi) \|_{H_\kappa}. \tag{26.17}$$

From (26.16), (26.17), the relation (26.11) follows for any $\varphi \in H_\kappa$. Lemma 26.3 is proved.

Lemma 26.4. *Every weakly convergent sequence of elements $\{w_n\}$ converges strongly in $H_{\kappa n}$, and the whole set of approximate solutions $\{w_n\}$ is strongly compact in H_κ.*

It suffices to prove the first assertion of the lemma. To prove it, we note that for $t_0 = 1$ we obtain from (26.3) and (26.10)

$$\|w_n\|^2_{H_\kappa} = -(\overset{0}{w} \cdot w_n)_{H_\kappa} + \int_\Omega T^{ij}(\mathbf{a}_n + \mathbf{a}^*)[B_{ij} w_n - (w_n + w^*)_{\alpha^i} w_{n\alpha^j}]\, d\Omega. \tag{26.18}$$

At the same time, setting in (13.15) $\varphi = w_0$, we have

$$\|w_0\|^2_{H_\kappa} = -(\overset{0}{w} \cdot w_0)_{H_\kappa} + \int_\Omega T^{ij}(\mathbf{a}_0 + \mathbf{a}^*)[B_{ij} w_0 - (w_0 + w^*)_{\alpha^i} w_{0\alpha^j}]\, d\Omega. \tag{26.19}$$

From (26.18), (26.19) it follows that

$$\begin{aligned}\|w_0\|^2_{H_\kappa} &- \|w_n\|^2_{H_\kappa}\\ &= -(\overset{0}{w}\cdot(w_0 - w_n))_{H_\kappa} + \int_\Omega \{T^{ij}(\mathbf{a}_0 + \mathbf{a}^*)[B_{ij}w_0 - (w_0 + w^*)_{\alpha^i} w_{0\alpha^j}]\\ &\quad - T^{ij}(\mathbf{a}_n + \mathbf{a}^*)[B_{ij}w_n - (w_n + w^*)_{\alpha^i} w_{n\alpha^j}]\}\, d\Omega. \end{aligned} \tag{26.20}$$

Since $w_n \rightharpoonup w_0$ in H_κ, passing to the limit, we see that the first term on the right-hand side of (26.20) vanishes. Moreover, by (26.12) and (12.29) we conclude that the second term on the right-hand side of (26.20) vanishes as well. Therefore, from (26.20) it follows that

$$\|w_n\|_{H_\kappa} \to \|w_0\|_{H_\kappa},$$

and therefore by Theorem 9.3,

$$w_n \to w_0.$$

Lemma 26.4 is proved.

Lemma 26.5. *The set* $\{w_n\}$ *contains at least one absolutely minimizing sequence for the functional* $\mathcal{I}_{\kappa\kappa}(w)$.

To prove this lemma, we observe that if w_0 is an absolute minimum of the functional $\mathcal{I}_{\kappa\kappa}$ and $w_0 = \sum_{k=1}^{\infty} D_k\varphi_k$, then the sequence $w_{0n} = \sum_{k=1}^{n} D_k\varphi_k$ will be absolutely minimizing for the functional $\mathcal{I}_{\kappa\kappa}$. This claim follows directly from complete continuity of $\mathcal{I}_{\kappa\kappa}$ in H_κ, which in turn follows from Lemma 21.4. Here we have the following obvious string of inequalities:

$$\mathcal{I}_{\kappa\kappa}(w_{0n}) \geq \mathcal{I}_{\kappa\kappa}(w_n) \geq d. \tag{26.21}$$

On the other hand, the sequence of approximate solutions w_n is constructed at each stage using (26.4) so that it contains w_n that minimize $\mathcal{I}_{\kappa\kappa}$ on the linear subspace (26.1) and for which

$$\mathcal{I}_{\kappa\kappa}(w_{0n})|_{n\to\infty} = d. \tag{26.22}$$

Relations (26.21), (26.22) prove Lemma 26.5.

Thus the set $\{w_n\}$ necessarily contains a sequence that converges to an absolute minimum of $\mathcal{I}_{\kappa\kappa}(w)$. If $\mathcal{I}_{\kappa\kappa}(w)$ has a unique minimum, then the sequence of approximations obtained using the Bubnov-Galerkin-Ritz (BGR) method in the Papkovich version will converge to it.

Lemma 26.6. *Let* w_0 *be an isolated solution of Problem* $t\kappa$ *of nonzero index. Then* w_0 *belongs to the set of limit points of* $\{w_n\}$, *and for any such solution* w_0 *and any arbitrarily small* r *we can therefore find a number* N *such that for all* $n \geq N$ *the system* (26.3) *(respectively,* (26.4)*) has for* D_{nk} *real solutions such that* $w_n \in B_{H_\kappa}(r, w_0)$.

To prove it, we note that the defining system (26.3) can be written in the form

$$D_{nk} = (\mathbf{G}_{\kappa\kappa}(w_n) \cdot \varphi_k)_{H_\kappa}$$

or

$$w_n = P_n \mathbf{G}_{\kappa\kappa}(w_n),$$

where P_n is the projection operator onto the finite-dimensional space spanned by $(\varphi_1, \varphi_2, \ldots, \varphi_n)$. In other words, the version of the Bubnov–Galerkin–Ritz method used here is equivalent to the version considered in [162]. Therefore, Lemma 26.6 follows from the results of [162].

Lemma 26.7. *Let w_0 be a generalized solution of Problem $t\kappa$, which furnishes a strict relative minimum of the functional (maximum) of the functional $\mathcal{I}_{\kappa\kappa}(w)$ in H_κ.*

Let w_0 belong to the set of limit points of $\{w_n\}$. Then for any such solution w_0 and any arbitrarily small number r we can find a number N such that for all $n \geq N$ the system *(26.3)* (respectively, *(26.4)*) has for D_{nk} real solutions such that $w_n \in B_{H_\kappa}(r, w_0)$,

$$\mathcal{I}_{\kappa\kappa}(w_n) \geq \mathcal{I}_{\kappa\kappa}(w_0) + 3\delta, \tag{26.23}$$

where δ is some fixed constant.

Indeed, if (26.23) does not hold, then there is a sequence $w_k \in \Sigma_{H_\kappa}(r, w_0)$ such that

$$\lim_{k\to\infty} \mathcal{I}_{\kappa\kappa}(w_k) = \mathcal{I}_{\kappa\kappa}(w_0). \tag{26.24}$$

But then w_k will be a relatively minimizing sequence, and by Theorem 21.9, w_k will be strongly compact. By complete continuity of $\mathcal{I}_{\kappa\kappa}(w)$ we have

$$\mathcal{I}_{\kappa\kappa}(w_0) = \mathcal{I}_{\kappa\kappa}(w_{00}),$$

where w_{00} is some limit point of w_k. But w_{00} necessarily belongs to $\Sigma_{H_\kappa}(r, w_0)$, which is impossible by Definition 21.2. Therefore (26.23) is established.

Moreover, it is easily seen that by complete continuity of $\mathcal{I}_{\kappa\kappa}(w)$ in H_κ we can find a number $\Delta r < r/2$ such that in the closed ball $\overline{B}_{H_\kappa}(r, w_0)$ we shall have the inequality

$$|\mathcal{I}_{\kappa\kappa}(w_1) - \mathcal{I}_{\kappa\kappa}(w_2)| \leq \delta \tag{26.25}$$

if $\|w_1 - w_2\|_{H_\kappa} \leq \Delta r$ for any $w_1, w_2 \in \overline{B}_{H_\kappa}(r, w_0)$. Let us construct an annulus in H_κ contained between the spheres $\Sigma_{H_\kappa}(r, w_0)$ and $\Sigma_{H_\kappa}(r - 2\Delta r, w_0)$. Clearly, in this annulus by (26.23), (26.25) we have the relation

$$\mathcal{I}_{\kappa\kappa}(w) \geq \mathcal{I}_{\kappa\kappa}(w_0) + 2\delta. \tag{26.26}$$

Now let us consider the expansion (26.15) for w_0 and let us choose n such that we have the inequality

$$\|w - w_{0n}\|_{H_\kappa} \leq \Delta r, \quad w_{0n} = \sum_{k=1}^{n} D^*_{nk}\varphi_k.$$

Let us now introduce in the space R^n of coefficients D_{nk} the sphere $\Sigma_{R^n}(r - \Delta r,\ w_{0n})$ defined by the relation

$$\left\| \sum_{k=1}^{n} D_{nk}\varphi_k - w_{0n} \right\|_{H_\kappa}^2 = \sum_{k=1}^{n} (D_{nk} - D_{nk}^*)^2 = (r - \Delta r)^2, \tag{26.27}$$

and let us estimate the value of the functional $\mathcal{I}_{\kappa\kappa}$ on this sphere. We have

$$\begin{aligned} \left\| w_0 - \sum_{k=1}^{n} D_{nk}\varphi_k \right\|_{H_\kappa} &\le \|w_0 - w_{0n}\|_{H_\kappa} + \left\| w_{0n} - \sum_{k=1}^{n} D_{nk}\varphi_k \right\|_{H_\kappa} \\ &\le \Delta r + r - \Delta r = r. \end{aligned} \tag{26.28}$$

At the same time we write down the inequality

$$\begin{aligned} \left\| w_0 - \sum_{k=1}^{n} D_{nk}\varphi_k \right\|_{H_\kappa} &= \left\| w_0 - w_{0n} + w_{0n} - \sum_{k=1}^{n} D_{nk}\varphi_k \right\|_{H_\kappa} \\ &\ge \left\| w_{0n} - \sum_{k=1}^{n} D_{nk}\varphi_k \right\|_{H_\kappa} - \|w_0 - w_{0n}\|_{H_\kappa} \\ &\ge r - \Delta r - \Delta r = r - 2\Delta r. \end{aligned} \tag{26.29}$$

From (26.27)–(26.29) it follows that when $D_{n1}, \ldots, D_{nn} \in \Sigma_{R^n}(r - \Delta r,\ w_{0n})$, the function $\sum_{k=1}^{n} D_{nk}\varphi_k$ is contained in the annulus between the spheres $\Sigma_{H_\kappa}(r,\ w_0)$ and $\Sigma_{H_\kappa}(r - 2\Delta r,\ w_0)$, and hence by (26.26),

$$\mathcal{I}_{\kappa\kappa}\left(\sum_{k=1}^{n} D_{nk}\varphi_k \right) \ge \mathcal{I}_{\kappa\kappa}(w_0) + 2\delta. \tag{26.30}$$

Furthermore, by (26.24), (26.25) we have

$$\mathcal{I}_{\kappa\kappa}(w_{0n}) \ge \mathcal{I}_{\kappa\kappa}(w_0) \ge \mathcal{I}_{\kappa\kappa}(w_{0n}) - \delta.$$

Finally, from (26.26), (26.30) we have

$$\mathcal{I}_{\kappa\kappa}\left(\sum_{k=1}^{n} D_{nk}\varphi_k \right) \ge \mathcal{I}_{\kappa\kappa}(w_{0n}) + \delta.$$

Therefore, on the sphere (26.27) the value of $\mathcal{I}_{\kappa\kappa}(w)$ is everywhere greater than in the center. Therefore, at an interior point of $\Sigma_{R^n}(r - \Delta r,\ w_{0n})$, $\mathcal{I}_{\kappa\kappa}(w)$ attains a minimum on the linear subspace (26.1). Relations (26.3), (26.4) will hold at that point.

The case of a strict relative maximum is considered in a similar way. Lemma 26.7 is proved.

26.2. Experience with numerical computations shows that it is frequently beneficial to consider the iterated sequence $\widetilde{w}_n$ defined by

$$\widetilde{w}_n = \mathbf{G}_{\kappa\kappa}(w_n).$$

It turns out that $\widetilde{w}_n$ always converges faster than w_n. Moreover, in a number of cases the sequences $\widetilde{w}_n$ are appropriate for computation of stresses in a shell, while direct computation of stresses using w_n encounters difficulties. We shall now justify this algorithm.

Lemma 26.8. *The set of iterated solutions* $\{\widetilde{w}_n\}$ *is strongly compact in* H_κ, *and every strong limit of* $\widetilde{w}_n$ *in* H_κ *is a generalized solution of Problem* $t\kappa$. *Furthermore, we have*

$$\|\widetilde{w}_n - w_n\|_{H_\kappa} \to 0$$

for every strongly convergent sequence w_n.

To prove this lemma we consider the definition of $\widetilde{w}_n$. From (13.15) we have (see the notation of (26.15))

$$\begin{aligned}(\widetilde{w}_n \cdot \varphi)_{H_\kappa} &= -(\overset{0}{w} \cdot \varphi)_{H_\kappa} + \int_\Omega T^{ij}(\mathbf{a}_n + \mathbf{a}^*)[B_{ij}\varphi - (w_n + w^*)_{\alpha^i}\varphi_{\alpha^j}]\, d\Omega \\ &= -(\overset{0}{w} \cdot \varphi^n)_{H_\kappa} + \int_\Omega T^{ij}(\mathbf{a}_n + \mathbf{a}^*)[B_{ij}\varphi^n - (w_n + w^*)_{\alpha^i}\varphi^n_{\alpha^j}]\, d\Omega \\ &\quad - (\overset{0}{w} \cdot R_n(\varphi))_{H_\kappa} \\ &\quad + \int_\Omega T^{ij}(\mathbf{a}_n + \mathbf{a}^*)[B_{ij}R_n(\varphi) - (w_n + w^*)_{\alpha^i}(R_n(\varphi))_{\alpha^j}]\, d\Omega.\end{aligned} \tag{26.31}$$

Taking now into account (26.3), which defines the approximate solutions w_n, we obtain from (26.31),

$$\begin{aligned}((\widetilde{w}_n - w_n) \cdot \varphi)_{H_\kappa} &= -(\overset{0}{w} \cdot R_n(\varphi))_{H_\kappa} + \int_\Omega T^{ij}(\mathbf{a}_n + \mathbf{a}^*) \\ &\quad \times [B_{ij}R_n(\varphi) - (w_n + w^*)_{\alpha^i}(R_n(\varphi))_{\alpha^j}]\, d\Omega.\end{aligned}$$

Next, it is easily seen that the sequence of elements $R_n(\varphi)$, where $\varphi \in \Sigma_{H_\kappa}(1, 0)$, weakly converges to zero uniformly in φ. The last statement means that for any element w^{**} the scalar product $(w^{**} \cdot R_n(\varphi))_{H_\kappa}$ can be made arbitrarily small for sufficiently large n for all $\varphi \in \Sigma_{H_\kappa}(1, 0)$ simultaneously. Indeed,

$$|(w^{**} \cdot R_n\varphi)_{H_\kappa}| \le \left|\sum_{n=k}^{\infty} w_k^{**}\alpha_k\right| \le \left(\sum_{n=k}^{\infty} |w_k^{**}|^2\right)^{\frac{1}{2}} \to 0, \; \alpha_k = (\varphi \cdot \varphi_k)_{H_\kappa}, \tag{26.32}$$

where $w_k^{**} = (w^{**}, \varphi_k)_{H_\kappa}$, independently of φ. From (26.32) we immediately obtain the relations

$$(\overset{0}{w} \cdot R_n(\varphi))_{H_\kappa} \le \epsilon_n \|\varphi\|_{H_\kappa}, \tag{26.33}$$

$$\left|\int_\Omega T^{ij}(\mathbf{a}_n + \mathbf{a}^*)B_{ij}R_n(\varphi)\, d\Omega\right| \le \frac{\|R_n(\varphi)\|_{C_\Omega}}{\|\varphi\|_{H_\kappa}} \|\varphi\|_{H_\kappa} \int_\Omega |T^{ij}(\mathbf{a}_n + \mathbf{a}^*)B_{ij}|\, d\Omega$$

$$\leq m \left\| \frac{R_n(\varphi)}{\|\varphi\|_{H_\kappa}} \right\|_{C\Omega} \|\varphi\|_{H_\kappa} \leq \epsilon_n \|\varphi\|_{H_\kappa}, \qquad (26.34)$$

$$\left| \int_\Omega T^{ij}(\mathbf{a}_n + \mathbf{a}^*)(w_n + w^*)_{\alpha^i} (R_n(\varphi))_{\alpha^j}\, d\Omega \right|$$

$$\leq \frac{\|(R_n(\varphi))_{\alpha^j}\|_{L_{p\Omega}} \|\varphi\|_{H_\kappa}}{\|\varphi\|_{H_\kappa}} \int_\Omega \left| T^{ij}(\mathbf{a}_n + \mathbf{a}^*)(w_n + w^*)_{\alpha^i} \right| d\Omega \qquad (26.35)$$

$$\leq m \left| \frac{\|(R_n(\varphi))_{\alpha^j}\|_{L_{p\Omega}}}{\|\varphi\|_{H_\kappa}} \right| \|\varphi\|_{H_\kappa} \leq \epsilon_n \|\varphi\|_{H_\kappa}, \quad p > 2.$$

Here in (26.33)–(26.35), $\epsilon_n \to 0$ as $n \to \infty$. The inequalities (26.34), (26.35) follow from complete continuity of the operator of embedding of H_κ into C_Ω and $L_{p\Omega}$ for any $p > 1$ (Theorem 12.3, (12.29)). From (26.31), (26.32)–(26.35) we have

$$\left| ((\widetilde{w}_n - w_n) \cdot \varphi)_{H_\kappa} \right| \leq 3\epsilon_n \|\varphi\|_{H_\kappa}, \qquad (26.36)$$

and from (26.36),

$$\|\widetilde{w}_n - w_n\|_{H_\kappa} \to 0, \qquad (26.37)$$

which is what had to be shown.

26.3. Let us collect together all the results obtained above.

Theorem 26.1. *Assume that all the conditions of Theorems 16.1–16.3 are satisfied and that the φ_k form an orthonormal basis in H_κ. Assume that in addition, an approximate representation of w_n is sought in the form* (26.1) *using the Papkovich method, where ω is determined in terms of w_n through* (26.2), *while D_{nk} is determined from* (26.36). *Then:*

(1) *Equations of the Bubnov–Galerkin method,* (26.3), *and those of the Ritz method,* (26.4), *lead to the same system of equations.*
(2) *For each n, the finite-dimensional vector field $\mathcal{N}_n(D_{n1}, \ldots, D_{nn})$* (26.7) *has winding number +1 on spheres of large radius in R^n, so that the system* (26.3), (26.4) *has at least one real solution for each n, and all the solutions are contained in a sphere of radius R, which does not depend on n.*
(3) *The set of approximate solutions $\{w_n\}$ is strongly compact in H_κ. In addition, every weakly convergent sequence in $\{w_n\}$ converges strongly, and every limit point of $\{w_n\}$ is a generalized solution of problem $t\kappa$.*
(4) *The set of approximate solutions $\{w_n\}$ contains at least one absolutely minimizing sequence. Any limit point of any absolutely minimizing sequence is a point of absolute minimum of the functional $\mathcal{I}_{\kappa\kappa}$. If $\mathcal{I}_{\kappa\kappa}$ has a unique point of absolute minimum, then any absolutely minimizing sequence for $\mathcal{I}_{\kappa\kappa}$ is convergent.*
(5) *Every relatively minimizing (respectively, relatively maximizing) sequence for the functional $\mathcal{I}_{\kappa\kappa}$ is strongly compact in H_κ, and every strong limit of*

elements of these sequences generates a generalized solution of problem $t\kappa$ *that furnishes a relative minimum (respectively, maximum) of the functional* $\mathcal{I}_{\kappa\kappa}$.

(6) *Every isolated generalized solution* w_0 *of Problem* $t\kappa$ *of nonzero index in* H_κ *belongs to the set of limit points of* $\{w_n\}$ *in* H_κ, *and therefore for any such solution and any arbitrarily small* r *we can find* N *such that for all* $n \geq N$ *the systems of equations* (26.3), (26.4) *will have real solutions* D_{nk} *contained in the ball* $B_{H_\kappa}(r,\ w_0)$.

(7) *Every generalized solution* w_0 *of Problem* $t\kappa$ *furnishing a strict relative minimum (maximum) of the functional* $\mathcal{I}_{\kappa\kappa}(w)$ *belongs to the set of limit points of* $\{w_n\}$ *in* H_κ, *and therefore for any such solution and any arbitrarily small* r *we can find* N *such that for all* $n \geq N$ *the systems of equations* (26.3), (26.4) *will have real solutions* D_{nk} *contained in the ball* $B_{H_\kappa}(r,\ w_0)$.

(8) *The set of iterated solutions* $\widetilde{w}_n$ *is strongly compact in* H_κ, *and every strong limit of* $\{\widetilde{w}_n\}$ *is a generalized solution of Problem* $t\kappa$. *Relation* (26.37) *holds.*

26.4. All the arguments of Sections 26.1–26.3 that justify the use of the BGR method in the version of Papkovich in Problems $t\kappa$ in displacements can be extended, almost verbatim, to cover Problem 9κ with an Airy stress function. An approximate solution of the problem is sought in the form (26.1). To determine D_{nk} we substitute w_n into one of the relations (17.15), (17.39), from which the function $\Psi(w_n)$ is expressed in terms of w_n. Here we naturally assume that such a definition in an explicit or numerically computable form is possible and indeed achievable in practice. Let us now substitute (26.1) and $\Psi(w_n)$ expressed in terms in D_{nk} into one of the expressions (17.14), (17.38), where subsequently we put $\varphi = \varphi_k$. Then we have, for example, using (17.14):

$$\begin{aligned}
&\mathcal{N}_{nk}(D_{n1},\ \ldots,\ D_{nn}) \\
&\quad = D_{nk} + (\overset{0}{w} \cdot \varphi_k)_{H_\kappa} \\
&\qquad - \int_\Omega C^{ik}C^{jl}\big[B_{ij}\varphi_k - ((w_n + w^*)_{\alpha^j} \cdot \varphi_{k\alpha^i})\big]\, \nabla_{kl}\Psi(w_n)\, d\Omega \\
&\qquad - \int_\Omega \widetilde{T}_{\mathrm{p}}^{ij}(w_n + w^*)_{\alpha^i}\varphi_{k\alpha^j}\, d\Omega = 0.
\end{aligned} \tag{26.38}$$

We could construct the system of equations (26.38) in a different way as well. We could substitute the expression (26.1) for w_n and $\Psi(w_n)$ into the functional (7.25) and write down the condition for an extremum in terms of D_{nk}.

Theorem 26.2. *Suppose that Conditions 1–8 of Section 17 are satisfied and that* φ_k *form a basis in* H_κ. *Furthermore, let an approximate solution be sought in the form* (26.1) *using the Papkovich method, where we express* Ψ *in terms of* w_n *using* (17.17), *while the* D_{nk} *are determined from* (26.38) *or from*

$$\frac{\partial \mathcal{I}_\kappa(w_n)}{\partial D_{nk}} = 0, \tag{26.39}$$

where $\mathcal{I}_\kappa(w)$ is the functional $\mathcal{I}_{9\kappa}$ and Ψ has been expressed in terms of w using (17.17). *Then the equations* (26.38) *of the Bubnov–Galerkin method and* (26.39) *of the Ritz method are the same, and all the other statements of Theorem 26.1 hold.*

26.5. This method was further developed in a number of papers dealing with particular problems. In [347, 349, 379, 380, 352, 381, 373, 371, 372, 375, 378, 366, 388, 391] this scheme was used for high-order approximations.

27. The Bubnov–Galerkin–Ritz Method for Approximate Solution of Problems $t\kappa$ ($\kappa = 1, 2, 3, 4; t = 5, 6, 7, 8$). The Versions of Mushtari and Vlasov

27.1. The version of the Bubnov–Galerkin–Ritz method considered in Section 26 was based on exact integration of equations of the tangentially stressed state. Even though in the theory we are considering, these equations are linear, solving them is not always simple. Therefore, methods based on the requirement that all the equations of the system (13.13)–(13.15) be satisfied approximately can prove very useful. Such a method was first suggested by Mushtari [209]; see also [212].

We shall seek an approximate solution of Problem $t\kappa$ in the form

$$\omega_n = \sum_{k=1}^{n} C_{nk}\chi_k, \quad w_n = \sum_{k=1}^{n} D_{nk}\varphi_k, \quad \chi_k = (\varphi_{1k}, \varphi_{2k}), \tag{27.1}$$

where χ_k is an orthonormal basis in H_t and φ_k is an orthonormal basis in H_κ. The indices k and n refer to the number of elements in a sequence φ_n or φ_{nk}, whereas i, j, coming first in the index, refer to components (φ_1, φ_2). It is easily seen that χ_k, φ_k combine to form an orthonormal basis in $H_{t\kappa}$. We should observe that orthonormality of χ_k, φ_k is not necessary. However, in practical numerical computations it is convenient always to be working with an orthonormal basis.

To determine C_{nk} and D_{nk} we use (13.14), (13.15), into which we substitute (27.1) and put consecutively $\chi = \chi_k$, $\varphi = \varphi_k$, $k = 1, 2, \ldots, n$. Then we have

$$\begin{aligned} C_{nk} &= -(\overset{0}{\omega}\cdot\chi_k)_{H_t} + \int_\Omega \Big[B_{sl}(w_n + w^*) - \frac{1}{2}(w_n + w^*)_{\alpha^s}(w_n + w^*)_{\alpha^l}\Big] \\ &\quad \times D_{\mathrm{s}}^{ijsl}\left(\varphi_{ik\alpha^j} - G_{ij}^{\lambda}\varphi_{\lambda k}\right) d\Omega, \end{aligned} \tag{27.2}$$

$$D_{nk} = -(\overset{0}{w}\cdot\varphi_k)_{H_\kappa} + \int_\Omega T^{ij}(\mathbf{a}_n + \mathbf{a}^*)[B_{ij}\varphi_k - (w_n + w^*)_{\alpha^i}\varphi_{k\alpha^j}]\, d\Omega. \tag{27.3}$$

In (27.3) the $T^{ij}(\mathbf{a}_n + \mathbf{a}^*)$ are computed using (4.15), in which the $\overset{0}{\epsilon}_{nij}$ have been expressed using (3.16), where ω_n in w_n are given by (27.1).

A different method of obtaining systems of equations for C_{nk}, D_{nk} consists in substituting (27.1) into $\mathcal{I}_{t\kappa}(\mathbf{a})$ and demanding that it reaches an extremum on the linear space (27.1):

$$\frac{\partial \mathcal{I}_{t\kappa}(\mathbf{a}_n)}{\partial C_{nk}} = 0, \tag{27.4}$$

$$\frac{\partial \mathcal{I}_{t\kappa}(\mathbf{a}_n)}{\partial D_{nk}} = 0. \tag{27.5}$$

Lemma 27.1. *Equations* (27.2), (27.3) *and* (27.4), (27.5) *are the same.*

To prove the claim, we note that

$$\frac{\partial \mathcal{I}_{t\kappa}(\mathbf{a}_n)}{\partial C_{nk}} = \left(\operatorname{grad}_{H_t} \mathcal{I}_{t\kappa}(\mathbf{a}_n) \cdot \frac{\partial \omega_n}{\partial C_{nk}}\right)_{H_t} = ([\omega_n - \mathbf{K}_{t\kappa}(\mathbf{a}_n + \mathbf{a}^*)] \cdot \chi_k)_{H_t}. \tag{27.6}$$

The last equality in (27.6) uses (13.14).

Finally, from the definition of $\mathbf{G}_{t\kappa}$ (see (13.30)) we have

$$\begin{aligned}([\omega_n &- \mathbf{K}_{t\kappa}(\mathbf{a}_n + \mathbf{a}^*)] \cdot \chi_k)_{H_t} \\ &= C_{nk} + (\overset{0}{\omega} \cdot \chi_k)_{H_t} + \int_\Omega \left[B_{sl}(w_n + w^*) - \frac{1}{2}(w_n + w^*)_{\alpha^s}(w_n + w^*)_{\alpha^l}\right] \\ &\quad \times D_{\mathrm{s}}^{ijsl}\left(\varphi_{ik\alpha^j} - G_{ij}^{\lambda}\varphi_{\lambda k}\right) d\Omega.\end{aligned} \tag{27.7}$$

Using (27.7), we conclude that (27.2) and (27.4) are the same. Furthermore, we have

$$\begin{aligned}\frac{\partial \mathcal{I}_{t\kappa}(\mathbf{a}_n)}{\partial D_{nk}} &= \left(\operatorname{grad}_{H_\kappa} \mathcal{I}_{t\kappa}(\mathbf{a}_n) \cdot \frac{\partial w_n}{\partial D_{nk}}\right)_{H_\kappa} = (\operatorname{grad}_{H_\kappa} \mathcal{I}_{t\kappa}(\mathbf{a}_n) \cdot \varphi_k)_{H_\kappa} \\ &= D_{nk} + (\overset{0}{w} \cdot \varphi_k)_{H_\kappa} - \int_\Omega T^{ij}(\mathbf{a}_n + \mathbf{a}^*)[B_{ij}\varphi_k - (w_n + w^*)_{\alpha^i}\varphi_{k\alpha^j}]\, d\Omega.\end{aligned} \tag{27.8}$$

The last computation on (27.8) uses (13.15). From (27.8) we see that (27.3) and (27.5) are equivalent. Lemma 27.1 is proved.

27.2. We study first the system of equations (27.2), (27.3). Substituting (27.2) into (27.3) and introducing a parameter t, we obtain a system of equations that contains only D_{nk}, which can be written in the form

$$\begin{aligned}\widetilde{\mathcal{N}}_{nk}(D_{nk}, t) &= D_{nk} + t(\overset{0}{w} \cdot \varphi_k)_{H_\kappa} \\ &\quad - t\int_\Omega T^{ij}(\mathbf{a}_n + \mathbf{a}^*)[B_{ij}\varphi_k - (w_n + w^*)_{\alpha^i}\varphi_{k\alpha^j}]\, d\Omega = 0, \\ k &= 1, \ldots, n.\end{aligned} \tag{27.9}$$

We can regard $\widetilde{\mathcal{N}}_{nk}(D_{nk}, t)$ as a vector field in R^n. From (27.9) for $t = 0$ we have the identity vector field, while for $t = 1$ we obtain a vector field that corresponds to equation (27.3). Thus a possible argument for solvability of (27.3) could proceed by proving that the fields $\widetilde{\mathcal{N}}_{nk}(D_{nk}, 0)$ and $\widetilde{\mathcal{N}}_{nk}(D_{nk}, 1)$ are homotopic. To prove that this is indeed the case, we consider a functional $\Phi_n(w_n, t)$ defined by the relation

$$
\begin{aligned}
\Phi_n(w_n, t) &= \sum_{k=1}^{n} D_{nk}\widetilde{N}_{nk}(D_{nk}, t) \\
&= \sum_{k=1}^{n} D_{nk}^2 + t\Big(\overset{0}{w} \cdot \sum_{k=1}^{n} D_{nk}\varphi_k\Big)_{H_\kappa} \\
&\quad - t\int_\Omega T^{ij}(\mathbf{a}_n + \mathbf{a}^*) \\
&\quad \times \Big[B_{ij}\sum_{k=1}^{n} D_{nk}\varphi_k - (w_n + w^*)_{\alpha^i}\sum_{k=1}^{n} D_{nk}\varphi_{k\alpha^j}\Big]\,d\Omega \\
&= \|w_n\|_{H_\kappa}^2 + t(\overset{0}{w} \cdot w_n)_{H_\kappa} \\
&\quad - t\int_\Omega T^{ij}(\mathbf{a}_n + \mathbf{a}^*)[B_{ij}w_n - (w_n + w^*)_{\alpha^i} w_{n\alpha^j}]\,d\Omega.
\end{aligned}
\tag{27.10}
$$

Lemma 27.2. *The functional $\Phi_n(w_n, t)$ can be written in the form*

$$
\begin{aligned}
\Phi_n(w_n, t) &= \|w_n\|_{H_\kappa}^2 + t(\overset{0}{w} \cdot w_n)_{H_\kappa} + 2t\int_\Omega Q_{\mathrm{s}}\left(T^{ij}(\mathbf{a}_n + \mathbf{a}^*)\right)\,d\Omega \\
&\quad + t\int_\Omega T^{ij}(\mathbf{a}_n + \mathbf{a}^*)\big[B_{ij}(w_n + 2w^*) - (w_n + w^*)_{\alpha^i} w^*_{\alpha^j} \\
&\quad - 2\big(w^*_{i\alpha^j} - G^\lambda_{ij} w^*_\lambda\big)\big]\,d\Omega - 2t(\omega_{\mathrm{p}} \cdot \omega)_{H_t} \\
&\quad + 2t\Big(\int_{\Gamma_6} k_{\mathrm{s}}^{\tau\tau}(w_{n\tau} + w^*_\tau)w_{n\tau}\,ds + \int_{\Gamma_7} k_{\mathrm{s}}^{mm}(w_{nm} + w^*_m)w_{nm}\,ds \\
&\qquad + \int_{\Gamma_8} k_{\mathrm{s}}^{ij}(w_{ni} + w^*_{ni})w_{nj}\,ds\Big).
\end{aligned}
$$

The proof of Lemma 27.2 basically follows the lines of proof of Lemma 16.1, and therefore will not be presented here.

Let us introduce a countable sequence of operators $\mathbf{K}^n$ from H_κ into the n-dimensional space R^n of coefficients C_{nk} and of the correspondingly finite-dimensional spaces $H^n_{t\kappa}$ defined by the relations (27.2). Let

$$
\begin{aligned}
(\mathbf{K}^n(w) \cdot \chi_k)_{H_t} &= -(\omega^* \cdot \chi_k)_{H_t} + \int_\Omega \Big[B_{sl}(w + w^*) - \frac{1}{2}(w + w^*)_{\alpha^s}(w + w^*)_{\alpha^l}\Big] \\
&\quad \times D_{\mathrm{s}}^{ijsl}\left(\varphi_{ik\alpha^j} - G^\lambda_{ij}\varphi_{\lambda k}\right)\,d\Omega, \ \chi_k \in H^n_{t\kappa}.
\end{aligned}
\tag{27.11}
$$

Relation (27.11) defines an operator $\mathbf{K}^n$ that maps H_κ into C_{nk}. The operator $\mathbf{K}^n$ can be represented in the form

$$
\mathbf{K}^n = \mathbf{K}^n_0 + \mathbf{K}^n_1 + \mathbf{K}^n_2, \tag{27.12}
$$

where $\mathbf{K}^n_\mu$ is a homogeneous operator of order μ, $\mu = 0, 1, 2$, defined by the relations

$$(\mathbf{K}^n_0 \cdot \chi_k)_{H_t} = -(\omega^* \cdot \chi_k)_{H_t} + \int_\Omega \left(B_{sl}w^* - \frac{1}{2}w^*_{\alpha^s}w^*_{\alpha^l}\right)D^{ijsl}_{\mathrm{s}} \times \left(\varphi_{ik\alpha^j} - G^\lambda_{ij}\varphi_{\lambda k}\right) d\Omega, \tag{27.13}$$

$$(\mathbf{K}^n_1(w) \cdot \chi_k)_{H_t} = \int_\Omega \left[B_{sl}w - \frac{1}{2}(w_{\alpha^s}w^*_{\alpha^l} + w_{\alpha^l}w^*_{\alpha^s}\right]D^{ijsl}_{\mathrm{s}} \times \left(\varphi_{ik\alpha^j} - G^\lambda_{ij}\varphi_{\lambda k}\right) d\Omega, \tag{27.14}$$

$$(\mathbf{K}^n_2(w) \cdot \chi_k)_{H_t} = -\frac{1}{2}\int_\Omega w_{\alpha^s}w_{\alpha^l}D^{ijsl}_{\mathrm{s}}\left(\varphi_{ik\alpha^j} - G^\lambda_{ij}\varphi_{\lambda k}\right) d\Omega. \tag{27.15}$$

Lemma 27.3. *Let $\mathcal{I}(f)$ be a linear bounded functional in a separable Hilbert space of functions f. Then we have the inequality*

$$|\mathcal{I}(R_n(f))| \le \epsilon_n \|f\|, \ \epsilon_n \to 0 \text{ as } n \to \infty. \tag{27.16}$$

In (27.16), $R_n(f)$ is the infinite tail of the Fourier expansion of f in a basis φ_k:

$$R_n(f) = \sum_{k=n}^{\infty} f_k\varphi_k,$$

and ϵ_n does not depend on f.

To prove the lemma, we note that by the Riesz theorem.

$$\mathcal{I}(f) = (f \cdot f^*),\ \mathcal{I}(R_n(f)) = (R_n(f) \cdot f^*)$$

and

$$|\mathcal{I}(R_n(f))| = \left|\sum_{k=n}^{\infty} f_k f^*_k\right| \le \left(\sum_{k=n}^{\infty}|f_k|^2\right)^{\frac{1}{2}}\left(\sum_{k=n}^{\infty}|f^*_k|^2\right)^{\frac{1}{2}} \le \|f\|\,\epsilon_n.$$

Clearly, all the claims of Lemma 27.3 are proved.

Lemma 27.4. *We have the relations*

$$\left\|\mathbf{K}^n_\mu(w) - \mathbf{K}_\mu(w)\right\|_{H_t} \le \epsilon_n \|w\|^\mu_{H_\kappa}, \ \epsilon_n \to 0,\ n \to \infty. \tag{27.17}$$

Let us consider, for example the functional

$$((\mathbf{K}^n_2(w) - \mathbf{K}_2(w)) \cdot f)_{H_\kappa} = (\mathbf{K}_2(w) \cdot R_{n+1}(f))_{H_\kappa}. \tag{27.18}$$

By (27.16) we have

$$\left|((\mathbf{K}^n_2(w) - \mathbf{K}_2(w)) \cdot f)_{H_\kappa}\right| \le \epsilon_{n+1}\|\mathbf{K}_2(w)\|_{H_\kappa},$$

and from (14.3) and (27.18) we obtain (27.17) for $\mu = 2$. The cases $\mu = 0, 1$ are proved in a similar fashion. Lemma 27.4 is proved.

Furthermore, let

$$\Pi_n(w_n) = \int_\Omega Q_{\mathrm{s}}\left(T^{ij}(\mathbf{a}_n + \mathbf{a}^*)\right) d\Omega = \int_\Omega D^{ijkl}_{\mathrm{s}}\overset{0}{\epsilon}_{ijn}(w_n)\overset{0}{\epsilon}_{kln}(w_n)\, d\Omega. \tag{27.19}$$

In (27.19) the vector function ω_n is expressed in terms of w_n using (27.2).

Lemma 27.5. *We have the representation*

$$\Pi_n(w_n) = \sum_{\mu=0}^{4} \Pi_{n\mu}(w_n),$$

where $\Pi_{n\mu}(w_n)$ *is a homogeneous functional of order* μ *and we have the inequalities*

$$\left|\Pi_{n\mu}(w_n)\right| \leq m \left\|w_n\right\|_{H_\kappa}^{\mu}, \tag{27.20}$$

where m *does not depend on either* n *or* w.

Lemma 27.5 is proved by using the arguments of Lemma 16.2. There is, however, one difference, which lies in the fact that in general the constant m on the right-hand side of (27.20) will depend on n. However, its value will be determined by an estimate of the form

$$\left\|\mathbf{K}_\mu^n(w_n)\right\|_{H_\kappa} \leq m_n \left\|w_n\right\|_{H_\kappa}^{\mu}. \tag{27.21}$$

From (14.3), (27.17) we have

$$\begin{aligned}\left\|\mathbf{K}_\mu^n(w_n)\right\|_{H_\kappa} &\leq \left\|\mathbf{K}_\mu^n(w_n) - \mathbf{K}_\mu(w_n) + \mathbf{K}_\mu(w_n)\right\|_{H_\kappa} \\ &\leq \left\|\mathbf{K}_\mu^n(w_n) - \mathbf{K}_\mu(w_n)\right\|_{H_\kappa} + \left\|\mathbf{K}_\mu(w_n)\right\|_{H_\kappa} \\ &\leq (\epsilon_n + m) \left\|w_n\right\|_{H_\kappa}^{\mu},\end{aligned}$$

as $\epsilon_n \to 0$, which indeed proves the fact that the constant m_n in (27.21) can be taken to be independent of n. Lemma 27.5 is proved.

27.3. Let us consider a sphere $\Sigma_n(1, 0)$ in the space of D_{nk} and let us define on it the set $\Sigma_n'(1, 0)$ of elements v that satisfy the relation

$$\|v\|_{H_\kappa}^2 - \frac{1}{\epsilon_1} \left\|B_{ij}v - v_{\alpha^i} w_{\alpha^j}^*\right\|_{L_{2\Omega}}^2 - c \left\|\omega_2(v)\right\|_{L_{2\Omega}} \leq \frac{1}{2},$$

where ϵ_1, c are some fixed positive constants, and the vector function $\omega_2(v) = \mathbf{K}_2^n(v)$ is defined by relations (27.12)–(27.15). Next, let $\Sigma_{H_\kappa}'(1, 0) = \cup_{n=1}^{\infty} \Sigma_n'(1, 0)$ and let $\overline{\Sigma}_{H_\kappa}'(1, 0)$ be the weak closure of $\Sigma_{H_\kappa}'(1, 0)$ in H_κ. We shall denote the complement of $\Sigma_{H_\kappa}'(1, 0)$ in the entire sphere $\Sigma_{H_\kappa}(1, 0)$ by $\Sigma_{H_\kappa}''(1, 0)$.

Lemma 27.6. *The set* $\overline{\Sigma}_{H_\kappa}'(1, 0)$ *does not contain zero.*

Indeed, were that not the case, there would exist a sequence $w^m \in \Sigma_{H_\kappa}'(1, 0)$ such that $w^m \rightharpoonup 0$ as $m \to \infty$. There are two possibilities. The first is that all w^m, starting with some m sufficiently large, belong to the same set $\Sigma_n'(1, 0)$. However, this possibility is excluded immediately, since in a finite-dimensional space weak and strong convergence coincide, and then w^m would go to zero in norm, which is impossible, since $\|w^m\|_{H_\kappa} = 1$. The second alternative would occur if the w^m belong to different sets $\Sigma_n'(1, 0)$. However, it has to be excluded as well, since in that case we can use verbatim the arguments of Lemma 15.1. Thus Lemma 27.6 has been proved.

Lemma 27.7. *Assume that Conditions 2–8 of Section 13 are satisfied, $S \in H_{\Omega}^{2,\lambda}$, and the supports on Γ_6, Γ_7, Γ_8 are essentially elastic. Then on $\Sigma'_{H_\kappa}(R, 0)$ we have the inequality*

$$\Pi_{n4}(w_n) \geq m_0 R^4, \tag{27.22}$$

where the constant m on the right-hand side of (27.22) is independent of n.

The proof of Lemma 27.7 follows in essence the proof of Lemma 16.3. We only have to keep in mind that the constant m in (27.22) is independent of n due to the way we constructed the set $\Sigma'_{H_\kappa}(1, 0)$, as a result of which this set itself is independent of n.

Lemma 27.8. *Assume that Conditions 1–8 of Section 13 are satisfied, supports on Γ_6, Γ_7, Γ_8 are essentially elastic, and the shell is geometrically shallow. Then (27.22) holds on $\Sigma'_{H_\kappa}(1, 0)$.*

Lemma 27.9. *Assume that Conditions 1–4, 7 of Section 13 are satisfied and (16.24) holds for a geometrically shallow shell. Then (27.22) is satisfied.*

Lemmas 27.8, 27.9 are proved like Lemma 27.7.

27.4. Let us move on now to estimate $\Phi_n(w_n)$.

Lemma 27.10. *Assume that the conditions of Theorems 16.1–16.3 are satisfied. Then we have the inequality*

$$\Phi_n(w_n, t) \geq m^0 R^2, \ m^0 > 0 \text{ is independent of } n; \ 0 \leq t \leq 1. \tag{27.23}$$

It is possible to give a proof of Lemma 27.10 following the arguments used in proving Theorems 16.1–16.3. Independence of m from n follows from the independence from n of constants m_k in (27.21), (27.22).

Lemma 27.11. *Assume that the conditions of Lemma 27.10 are satisfied. Then the vector fields $\widetilde{N}_{nk}(D_{nk}, 0)$ and $\widetilde{N}_{nk}(D_{nk}, 1)$ are homotopic on spheres $\Sigma_n(R, 0)$ of sufficiently large radius R, for any n.*

Lemma 27.11 follows immediately from (27.23).

Lemma 27.12. *If the conditions of Lemma 27.10 hold, the system of equations (27.10) has at least one real solution.*

The claim of Lemma 27.12 follows from Lemma 27.11.

Lemma 27.13. *Under the conditions of Lemma 27.10 all solutions of equations (27.2), (27.3) are such that*

$$\|\mathbf{a}_n\|_{H_{1\kappa}} \leq R, \ \mathbf{a}_n = (w_{1n}, w_{2n}, w_n),$$

where the constant R is independent of n.

27.5.

Lemma 27.14. *Every weak limit* $\mathbf{a}_0 = (w_{10}, w_{20}, w_0)$ *of the sequence of approximations* $\{\mathbf{a}_n\}$, $\mathbf{a}_n = (w_{1n}, w_{2n}, w_n)$, *is a generalized solution of problem* $t\kappa$.

To prove the lemma, let us consider the integral identities (13.14), (13.15), which define a generalized solution, and let $\mathbf{a}_0 = (w_{10}, w_{20}, w_0)$ be a weak limit of a subsequence $\{\mathbf{a}_{n_j}\}$ in $\{\mathbf{a}_n\}$. Let us consider the relation

$$(\omega_{n_j} \cdot \chi_k)_{H_t} + (\overset{0}{\omega} \cdot \chi_k)_{H_t} - \int_\Omega \big[B_{sl}(w_{n_j} + w^*) - \frac{1}{2}(w_{n_j} + w^*)_{\alpha^s}(w_{n_j} + w^*)_{\alpha^l}\big] D_{\mathrm{s}}^{ijsl} \left(\varphi_{ik\alpha^j} - G_{ij}^{\lambda}\varphi_{\lambda k}\right) d\Omega = 0. \tag{27.24}$$

This will always hold for a fixed k and for $n_j \geq k$. But in this relation we could pass to the limit, $n \to \infty$ keeping k fixed. Then we clearly have

$$\left(\omega_0 + \overset{0}{\omega}\right) \cdot \chi_k)_{H_t} - \int_\Omega \left[B_{sl}(w_0 + w^*) - \frac{1}{2}(w_0 + w^*)_{\alpha^s}(w_0 + w^*)_{\alpha^l}\right] \times D_{\mathrm{s}}^{ijsl} \left(\varphi_{ik\alpha^j} - G_{ij}^{\lambda}\varphi_{\lambda k}\right) d\Omega = 0. \tag{27.25}$$

Indeed, in the first term of the right-hand side of (27.25) we can pass to the limit, since it is a linear functional in H_t. In the third term the passage to the limit is justified by the relations

$$\left\| w_{n_j} - w_0 \right\|_{C_\Omega} \to 0, \ \left\| w_{n_j} - w_0 \right\|_{W_{q\Omega}^{(1)}} \to 0, \ n \to \infty, \tag{27.26}$$

which follow from the embedding Theorem 12.3 (see (12.26)). Let χ be an arbitrary element of H_t and

$$\chi = \sum_{p=1}^{\infty} \alpha_p \chi_p = \sum_{p=1}^{n} \alpha_p \chi_p + R_n(\chi), \ \chi_p = (\varphi_{1p}, \varphi_{2p}), \tag{27.27}$$

$$\begin{aligned}\left((\omega_0 + \overset{0}{\omega}) \cdot \chi\right)_{H_t} &- \int_\Omega \left[B_{kl}(w_0 + w^*) - \frac{1}{2}(w_0 + w^*)_{\alpha^k}(w_0 + w^*)_{\alpha^l}\right] \\ &\times \sum_{s=1}^{\infty} D_{\mathrm{s}}^{ijkl} \left(\varphi_{ik\alpha^j} - G_{ij}^{\lambda}\varphi_{\lambda k}\right) d\Omega \\ &= \left(\left(\omega_0 + \overset{0}{\omega}\right) \cdot R_n(\chi)\right)_{H_t} \\ &- \int_\Omega \left[B_{kl}(w_0 + w^*) - \frac{1}{2}(w_0 + w^*)_{\alpha^k}(w_0 + w^*)_{\alpha^l}\right] D_{\mathrm{s}}^{ijkl} \\ &\times \sum_{s=n+1}^{\infty} \left(\varphi_{is\alpha^j} - G_{ij}^{\lambda}\varphi_{\lambda s}\right) d\Omega.\end{aligned} \tag{27.28}$$

By (27.24), (27.28) holds for any n_j, from which we obtain (13.14) for any element $\chi \in H_t$. Moreover, relation (27.3), which defines w_n, can be written in the form

$$\left(\left(w_{n_j} + \overset{0}{w}\right) \cdot \varphi_k\right)_{H_\kappa} - \int_\Omega T^{ij}(\mathbf{a}_{n_j} + \mathbf{a}^*)\left[B_{ij}\varphi_k - (w_{n_j} + w^*)_{\alpha^i}\varphi_{k\alpha^j}\right] d\Omega = 0. \tag{27.29}$$

It is easily seen that in (27.29) we can pass to the limit as $n_j \to \infty$. Indeed, as far as the first term on the left-hand side of (27.29) is concerned, this is obvious. Next, it is easy to see that we have the relation

$$T^{ij}(\mathbf{a}_{n_j} + \mathbf{a}^*) \to T^{ij}(\mathbf{a}_0 + \mathbf{a}^*) \text{ in } L_{2\Omega}.$$

For a fixed φ_k the second term of the left-hand side of (27.29) is a linear bounded functional with respect to T^{ij} in $L_{2\Omega}$. Therefore, we can pass to the limit in (27.29), as a result of which we obtain (13.15) for $\varphi = \varphi_k$. Obtaining (13.15) for an arbitrary φ is not hard either. Lemma 27.14 is proved.

Lemma 27.15. *Every weakly convergent subsequence $\mathbf{a}_{n_k}$ in $\{\mathbf{a}_n\}$ converges strongly, and therefore the entire set $\{\mathbf{a}_n\}$ is strongly compact.*

To prove the lemma, we set $\chi = \omega$ in (13.14), as a result of which we have

$$\|\omega\|^2_{H_t} = -\left(\overset{0}{\omega} \cdot \omega\right)_{H_t} + \int_\Omega \left[B_{kl}(w + w^*) - \frac{1}{2}(w + w^*)_{\alpha^k}(w + w^*)_{\alpha^l}\right]$$
$$D^{ijkl}_{\mathbf{s}}\left(w_{i\alpha^j} - G^\lambda_{ij} w_\lambda\right) d\Omega, \tag{27.30}$$

and multiplying (27.2) by C_{nk} and summing over k, we have

$$\|\omega_n\|^2_{H_t} = -\left(\overset{0}{\omega} \cdot \omega_n\right)_{H_t} + \int_\Omega \left[B_{kl}(w_n + w^*) - \frac{1}{2}(w_n + w^*)_{\alpha^k}(w_n + w^*)_{\alpha^l}\right]$$
$$\times D^{ijkl}_{\mathbf{s}}\left(w_{in\alpha^j} - G^\lambda_{ij} w_{\lambda n}\right) d\Omega. \tag{27.31}$$

Subtracting (27.31) from (27.30) and taking into consideration (27.26), (27.27), we easily obtain

$$\left\|\omega_{n_k}\right\|_{H_t} \to \|\omega\|_{H_t}. \tag{27.32}$$

In an identical manner we can show that

$$\left\|w_{n_k}\right\|_{H_\kappa} \to \|w_0\|_{H_\kappa}. \tag{27.33}$$

Taking into account weak convergence $w_n \rightharpoonup w_0$, we conclude from (27.32), (27.33) that Lemma 27.15 is true by Theorem 9.3.

Using the arguments of the previous section, we can prove the following four lemmas.

Lemma 27.16. *The set $\{\mathbf{a}_n\}$ contains at least one absolutely minimizing sequence $\mathbf{a}_n$ of $\mathcal{I}_{t\kappa}$.*

Lemma 27.17. *Let* $\mathbf{a}_0$ *be an isolated solution of Problem* $t\kappa$ *of nonzero index in* $H_{t\kappa}$. *Then* $\mathbf{a}_0$ *belongs to set of limit points of* $\{\mathbf{a}_n\}$ *in* $H_{t\kappa}$, *so that for any such solution and any arbitrarily small* r *we can find a number* N *such that for all* $n \geq N$ *the system of equations* (27.2), (27.3) *has real solutions* (C_{nk}, D_{nk}) *and thus the solutions* $\mathbf{a}_n \in B(r, \mathbf{a}_0)$.

Recall that $\mathbf{a}_0$ is a limit point of a sequence $\{\mathbf{a}_n\}$ if there is a subsequence $\{\mathbf{a}_{n_k}\}$ of $\{\mathbf{a}_n\}$ the limit of which is $\mathbf{a}_0$.

Lemma 27.18. *Let* $\mathbf{a}_0$ *be a generalized solution of Problem* $t\kappa$ *that furnishes a strict absolute minimum (maximum) of* $\mathcal{I}_{t\kappa}$ *in* $H_{t\kappa}$. *Then* $\mathbf{a}_0$ *belongs to the set of limit points of* $\{\mathbf{a}_n\}$ *in* $H_{t\kappa}$, *so that for any such solution and any arbitrarily small* r *we can find a number* N *such that for all* $n \geq N$ *the system of equations* (27.2), (27.3) *has real solutions* (C_{nk}, D_{nk}) *and thus the solutions* $\mathbf{a}_n \in B(r, \mathbf{a}_0)$.

Lemma 27.19. *Let* $\{\widetilde{\mathbf{a}}_n\}$ *be the set of iterated approximations*

$$\widetilde{\mathbf{a}}_n = \mathbf{G}_{t\kappa}\mathbf{a}_n.$$

Then $\{\widetilde{\mathbf{a}}_n\}$ *is strongly compact in* $H_{t\kappa}$, *and every strong limit of a subsequence* $\{\widetilde{\mathbf{a}}_{n_k}\}$ *is a generalized solution of Problem* $t\kappa$. *In addition, we have the relation*

$$\left\|\widetilde{\mathbf{a}}_{n_k} - \mathbf{a}_{n_k}\right\|_{H_{t\kappa}} \to 0. \tag{27.34}$$

27.6. Let us collect together all the results obtained above.

Theorem 27.1. *Assume that all the conditions of Theorems 16.1–16.3 are satisfied and that the* (χ_k, φ_k) *form a basis in* $H_{t\kappa}$. *Assume that in addition, an approximate solution is sought in the form* (27.1), *where we are using the Mushtari method to obtain* C_{nk}, D_{nk}, *which leads to* (27.2), (27.3). *Then:*

(1) *The equations of the Bubnov–Galerkin method,* (27.2), (27.3), *and of the Ritz method,* (27.4), (27.5) *are the same.*

(2) *For each* n *the finite-dimensional vector field* $\widetilde{\mathcal{N}}_n(D_{nk}, 1)$, (27.9), *has winding number* $+1$ *on spheres of large radius,* $\Sigma_n(R, 0)$, *so that the system* (27.2), (27.3) *has at least one real solution for each* n.

(3) *The entire set of approximate solutions* $\{\mathbf{a}_n\}$ *is strongly compact in* $H_{t\kappa}$. *In addition, every weakly convergent subsequence in* $\{\mathbf{a}_n\}$ *converges strongly, and every limit point of* $\{\mathbf{a}_n\}$ *is a generalized solution of problem* $t\kappa$.

(4) *The set of approximate solutions* $\{\mathbf{a}_n\}$ *contains at least one absolutely minimizing sequence for the functional* $\mathcal{I}_{t\kappa}$ *in* $H_{t\kappa}$. *If* $\mathcal{I}_{t\kappa}$ *has a unique point of absolute minimum, then any absolutely minimizing sequence in* $\{\mathbf{a}_n\}$ *is convergent.*

(5) *Every relatively minimizing (respectively, relatively maximizing) sequence for the functional* $\mathcal{I}_{t\kappa}(\mathbf{a})$ *is strongly compact in* $H_{t\kappa}$, *and every strong limit of elements of these sequences generates a generalized solution of Problem* $t\kappa$ *that furnishes a relative minimum (respectively, maximum) of the functional* $\mathcal{I}_{t\kappa}$.

(6) *Every isolated generalized solution* $\mathbf{a}_0$ *of Problem* $t\kappa$ *of nonzero index in* $H_{t\kappa}$ *belongs to the set of limit points of* $\{\mathbf{a}_n\}$ *in* $H_{t\kappa}$, *and for any such solution* *and any arbitrarily small r we can find N such that for all $n \geq N$ the systems of equations* (27.2), (27.3) *will have real solutions C_{nk}, D_{nk} such that* $\mathbf{a}_n \in B_{H_{t\kappa}}(r, \mathbf{a}_0)$.

(7) *Every generalized solution $\mathbf{a}_0$ of Problem $t\kappa$ furnishing a strict relative minimum (maximum) of the functional $\mathcal{I}_{t\kappa}$ belongs to the set of limit points of $\{\mathbf{a}_n\}$ in $H_{t\kappa}$, and therefore for any such solution and any arbitrarily small r we can find N such that for all $n \geq N$ the systems of equations* (27.2), (27.3) *will have real solutions C_{nk}, D_{nk} such that $\mathbf{a}_n \in B_{H_{t\kappa}}(r, \mathbf{a}_0)$.*

(8) *The set of iterated approximations $\{\widetilde{\mathbf{a}}_n\}$ is strongly compact in $H_{t\kappa}$, and every strong limit of $\{\widetilde{\mathbf{a}}_n\}$ is a generalized solution of Problem $t\kappa$. Relation* (27.34) *holds.*

Remark 27.1. All the statements of Theorem 27.1 are valid if ω_n is determined not from (27.2) and thus (27.11)–(27.15), but from any sequence of operators $\mathbf{K}^n(w)$ satisfying the conditions

$$\left\|\mathbf{K}^n(w) - \mathbf{K}(w)\right\|_{H_t} \leq \epsilon_n \left(\|w\|_{H_\kappa} + \|w\|_{H_\kappa}^2\right), \quad \epsilon_n \to 0.$$

27.7. Vlasov [333] suggested seeking an approximate solution of the two equations (7.42), (7.49) with an Airy stress function using the Bubnov–Galerkin method. This idea has been extensively used by Koltunov [157, 158, 159]. In combination with other techniques it was developed in the works of Petrov, Krys'ko, and their students [9, 170, 230, 231, 232]. More precisely, let us set

$$w_n = \sum_{k=1}^{n} D_{nk}\varphi_k, \quad \Psi_n = \sum_{k=1}^{n} C_{nk}\theta_k, \tag{27.35}$$

where φ_k, θ_k are orthonormal bases in H_κ, H_9, respectively.

Systems of equations for D_{nk}, C_{nk} can be obtained if (27.35) is substituted into (17.14), where we should put consecutively $\varphi = \varphi_k$, $\theta = \theta_k$, $k = 1, \ldots, n$. then we have

$$\mathcal{N}_{nk1} = D_{nk} + (\overset{0}{w} \cdot \varphi_k)_{H_\kappa} - \int_\Omega C^{is}C^{jl}[B_{ij}\varphi_k - \varphi_{k\alpha^i}(w_n + w^*)_{\alpha^j}]\nabla_{sl}\Psi_n, \, d\Omega$$
$$- \int_\Omega \widetilde{T}^{ij}_{\mathrm{p}}(w_n + w^*)_{\alpha^i}\varphi_{k\alpha^j}\, d\Omega = 0, \tag{27.36}$$

$$\mathcal{N}_{nk2} = C_{nk} + \int_\Omega C^{is}C^{jl}\big[B_{ij}(w_n + w^*) - \frac{1}{2}(w_n + w^*)_{\alpha^i}(w_n + w^*)_{\alpha^j}\big]$$
$$\times \nabla_{sl}\theta\, d\Omega - \int_\Omega C^{\lambda\mu}_{*ij}\widetilde{T}^{ij}\nabla_{\lambda\mu}\theta_k\, d\Omega = 0. \tag{27.37}$$

In the construction of system (27.36), (27.37) we could also use the functional $\mathcal{I}_\kappa$ and write down the conditions for an extremum, that is, the Ritz relations

$$\frac{\partial \mathcal{I}_\kappa}{\partial D_{nk}} = 0, \quad \frac{\partial \mathcal{I}_\kappa}{\partial C_{nk}} = 0. \tag{27.38}$$

Since the analysis of the Vlasov scheme described above largely repeats the analysis of the Mushtari scheme of Sections 27.1–27.6, we present only the final result.

Theorem 27.2. *Assume that all the Conditions 1–8 of Section 17 are satisfied. Assume also that the θ_k form a basis in H_9 and that the φ_κ form an orthonormal basis in H_κ. Then the following facts hold:*

(1) *The systems* (27.35), (27.36) *and* (27.37), (27.38) *are the same.*

(2) *The finite-dimensional vector field $\mathcal{N}_{nk1}$, $\mathcal{N}_{nk2}$ in the $2n$-dimensional space $R^n \times R^n$ (D_{nk}, C_{nk}) has winding number $+1$, so that the systems* (27.35), (27.36) *and* (27.37), (27.38) *have at least one real solution for each n. All the solutions $\mathbf{c}_n = (w_n, \Psi_n)$ lie in a sphere*

$$\|\mathbf{c}_n\|_{H_{9\kappa}} \leq R.$$

(3) *The set of approximate solutions $\{\mathbf{c}_n\}$ is strongly compact in $H_{9\kappa}$. In addition, every weakly convergent subsequence in $\{\mathbf{c}_n\}$ converges strongly, and every limit point of $\{\mathbf{c}_n\}$ is a generalized solution of problem 9κ.*

(4) *The set of approximate solutions $\{\mathbf{c}_n\}$ contains at least one absolutely minimizing sequence for the functional $\mathcal{I}_\kappa$ in H_κ. By Theorem 22.4 every absolutely minimizing sequence for $\mathcal{I}_\kappa$, $\mathbf{c}_n = (w_n, \Psi_n)$ converges in $H_{9\kappa}$ to a point of absolute minimum.*

(5) *Every relatively minimizing (respectively, relatively maximizing) sequence for the functional $\mathcal{I}_\kappa$ is strongly compact in H_κ, and every strong limit of elements of these sequences generates a generalized solution of Problem 9κ that furnishes a relative minimum (respectively, maximum) of the functional $\mathcal{I}_\kappa$.*

(6) *Every isolated generalized solution $\mathbf{c}_0$ of Problem 9κ of nonzero index in $H_{9\kappa}$ belongs to the set of limit points of $\{\mathbf{c}_n\}$ in $H_{9\kappa}$, and for any such solution and any arbitrarily small r we can find N such that for all $n \geq N$ the systems of equations* (27.35), (27.36) *((*27.37), (27.38)*, respectively) will have real solutions C_{nk}, D_{nk} such that $\mathbf{c}_n \in B_{H_{9\kappa}}(r, \mathbf{c}_0)$.*

(7) *Every generalized solution $\mathbf{c}_0$ of Problem 9κ furnishing a strict relative minimum (maximum) of the functional $\mathcal{I}_{9\kappa}$ in $H_{9\kappa}$ belongs to the set of limit points of $\{\mathbf{c}_n\}$ in $H_{9\kappa}$, and therefore for any such solution and any arbitrarily small r we can find N such that for all $n \geq N$ the systems of equations* (27.35), (27.36) *((*27.37), (27.38)*, respectively) will have real solutions C_{nk}, D_{nk} such that $\mathbf{c}_n \in B_{H_{9\kappa}}(r, \mathbf{c}_0)$.*

(8) *The set of iterated approximations $\{\widetilde{\mathbf{c}}_n\}$ is strongly compact in $H_{9\kappa}$, and every strong limit of a subsequence of $\{\widetilde{\mathbf{c}}_n\}$ in $H_{9\kappa}$ is a generalized solution of Problem 9κ. We have*

$$\|\mathbf{c}_n - \widetilde{\mathbf{c}}_n\|_{H_{9\kappa}} \to 0 \text{ as } n \to \infty.$$

Remark 27.2. All the statements of Theorem 27.2 remain valid if Ψ_n is determined not from (27.37) but from a system of operators $\mathbf{K}^n(w)$ that satisfy (27.37).

28. Error Estimates for the Bubnov–Galerkin–Ritz (BGR) Method in Some Problems of the Nonlinear Theory of Shallow Shells

28.1. In this section we shall consider the problem of estimating the rate of convergence of the Bubnov–Galerkin–Ritz (BGR) method for some particular types of approximations, first in the energy norm and then in the $C_\Omega^{(k)}$. For the second of these estimates we have to use Green's function and tensor of the corresponding linear boundary value problems. We start with a brief introduction concerning the construction of Green's function.

These estimates are based on the theorem of Krasnosel'skii [162, 164] on the asymptotic equivalence of the error of the BG approximations and the error of approximating the required solution by a Fourier series. The origins of this idea lie in the work of Krylov and Bogolyubov [169]. Concerning the justification of the Bubnov–Galerkin method in application to linear problems, see the work of Kel'dysh [138] and Mikhlin [197].

We have the following theorem.

Theorem ([162]). *Let the equation*

$$f = \mathbf{G}f, \tag{28.1}$$

where $\mathbf{G}$ *is a completely continuous, in general nonlinear, operator in a Banach space* B *be solved using the BG method,*

$$\begin{aligned} f_n &= \sum_{k=1}^{n} C_{nk} g_k, \\ C_{nk} &= \mathbf{P}_k \mathbf{G} f_k \quad k = 1, \ldots, n, \end{aligned} \tag{28.2}$$

where g_k *is an orthonormal basis in* B *and* $\mathbf{P}_k$ *is an orthogonal projection operator onto* g_k. *Let us assume that* f_0 *is a nonsingular solution of* (28.1). *Then the nonlinear system of equations* (28.2), *starting with some sufficiently large* n, *has a solution* f_n *such that*

$$\|f_0 - f_n\| = (1 + \epsilon_n) \left\| f_0 - \sum_{k=1}^{n} (\mathbf{P}_k f_0) g_k \right\|, \quad \epsilon_n \to 0. \tag{28.3}$$

Thus, by this theorem we have to estimate the error of the Fourier series in g_k of the solution f_0.

28.2. Let us consider the boundary value Problem $t\kappa$ under the boundary conditions

$$w_1|_\Gamma = w_2|_\Gamma = w|_\Gamma = \left.\frac{\partial w}{\partial n}\right|_\Gamma = 0. \tag{28.4}$$

According to our convention (Section 7), this problem can be said to be of type 51, as here $t = 5$, $\kappa = 1$. Let us show how to construct a basis for such a problem.

This clearly must be a basis in H_{51}, which includes as components bases in H_5 and H_1. Let the equation of the boundary curve be given by the relation

$$F(\alpha^1, \alpha^2) = 0.$$

Let us construct a set of functions of the form

$$\{F^2(\alpha^1, \alpha^2)(\alpha^1)^{k_1}(\alpha^2)^{k_2}\} \tag{28.5}$$

and let us orthonormalize (28.5) in the norm of H_1 using the Schmidt method. Let us denote the new family of functions by φ_k. Below it will be convenient to use the following scheme for indexing φ_k: We denote by Q'_N all the functions of the form (28.5) satisfying

$$0 \leq k_1,\ k_2 \leq N. \tag{28.6}$$

It is easy to check that there are $(N+1)^2$ such functions.

Theorem 28.1. *Assume that the following conditions are satisfied:*

(1) $F(\alpha^1, \alpha^2) \neq 0$ *anywhere apart from on* Γ *in an open domain* D *that contains* $\overline{\Omega}$.
(2) $|grad\ F|_\Gamma > 0$.
(3) $F \in H_\Omega^{\rho,1}$.
(4) w *lies in the intersection of* H_1 *and* $H^{\rho,\lambda}$.

Then

$$\left\| w - \sum_{m=0}^{(N+1)^2} (\mathbf{P}_m w)\varphi_m \right\|_{H_1} \leq m'(w)N^{-(\rho+\lambda-2)}. \tag{28.7}$$

In (28.7) $\mathbf{P}_m$ *is the projection operator using the basis* φ_m, *and the constant* $m'(w)$ *depends only on the norms of* w *in* H_1 *and* $H^{\rho,\lambda}$.

The proof of Theorem 28.1 is based on the fact that by the results of Kharrik [140], if conditions 1–4 of Theorem 28.1 are satisfied, there is a collection $Q'_N w$ such that

$$\left\| w - Q'_N w \right\|_{H_1} \leq m'(w)N^{-(\rho+\lambda-2)}. \tag{28.8}$$

However, we also have the relation

$$\left\| w - \sum_{m=0}^{(N+1)^2} (\mathbf{P}_m w)\varphi_m \right\|_{H_1} \leq \left\| w - Q'_N w \right\|_{H_1}. \tag{28.9}$$

Theorem 28.1 follows from (28.8), (28.9).

28.3. Let us consider a set of the functions of the form

$$F(\alpha^1, \alpha^2)(\alpha^1)^{k_1}(\alpha^2)^{k_2}, \tag{28.10}$$

and use it to construct a set of vectors of the form

$$\{F(\alpha^1, \alpha^2)(\alpha^1)^{k_1}(\alpha^2)^{k_2},\ F(\alpha^1, \alpha^2)(\alpha^1)^{m_1}(\alpha^2)^{m_2}\} \tag{28.11}$$

and orthonormalize (28.11) in the metric of H_5. The vector functions obtained in this manner will be denoted by χ_k. As before, we denote all the functions (28.11) in (28.10) of degree N and below that satisfy (28.6), by Q''_N. Clearly, in this case the total number of such vector functions is $(N+1)^4$.

Theorem 28.2. *Assume that conditions 1–4 of Theorem 28.1 are satisfied. Then if $\omega = (w_1, w_2)$ lies in the intersection of H_5 and $H_\Omega^{\rho,\lambda}$,*

$$\left\| \omega - \sum_{m=0}^{(N+1)^4} \mathbf{P}_m(\omega)\omega_m \right\|_{H_5} \leq m''(\omega)N^{-(\rho+\lambda-1)}. \tag{28.12}$$

The proof of Theorem 28.2 is also based on the results of Kharrik [140]. If conditions 1–4 of Theorem 28,1 hold, there is a linear combination $Q''_N\omega$ of vector functions in (28.11) such that

$$\left\| \omega - Q''_N\omega \right\|_{H_5} \leq m''(\omega)N^{-(\rho+\lambda-2)}.$$

Taking into account the relation

$$\left\| \omega - \sum_{m=0}^{(N+1)^4} \mathbf{P}_m(\omega)\omega_m \right\|_{H_5} \leq \left\| \omega - Q''_N\omega \right\|_{H_5},$$

we obtain (28.12), and Theorem 28.2 is established.

28.4. Let us now move on consider directly the error of the BGR method applied to solutions of problems of nonlinear shallow shell theory. Assume that an approximate solution of Problem 51 under conditions (28.4) is sought in the form (26.1), and that the φ_k are constructed using (28.5). For definiteness, we shall use the BGR method in the Papkovich version for the system of equations in displacements. Thus D_{nk} are to be determined from (26.3). We have the following result.

Theorem 28.3. *Assume that Conditions 2–4 of Section 13, and conditions 1–4 of Theorem 28.1 are satisfied and that moreover,*

$$R^3,\ R^s \in W^\rho_{p\Omega};\ \Gamma \in C^{4+\rho}_\Omega,\ S \in C^{4+\rho}_\Omega,\ D^{ijkl}_{\mathrm{f}} \in C^{2+\rho}_\Omega;\ D^{ijkl}_{\mathrm{s}} \in C^{1+\rho}_\Omega. \tag{28.13}$$

Finally, assume that in addition, the following condition holds:

$$F \in H_\Omega^{\rho+2,2(1-1/p)} \text{ if } p < 2;\ F \in H_\Omega^{\rho+3-0} \text{ if } p = 2;$$
$$F \in H_\Omega^{\rho+3,1-2/p} \text{ if } p > 2. \tag{28.14}$$

Then for any nonsingular solution w_0 of the operator equation (13.33) *there is a sequence $w_{(N+1)^2}$ of BGR approximations* (26.3) *for which we have the asymptotic error estimate*

$$\left\| w_0 - w_{(N+1)^2} \right\|_{H_1} \leq m(w_0)N^{-\beta}, \tag{28.15}$$

where

$$\beta = p + 2\left(1 - \frac{1}{p}\right) \text{ if } p \neq 2;\ \beta = p - 0 \text{ for } p = 2. \tag{28.16}$$

The proof of Theorem 28.3 is based on the following facts. If (28.13) holds, the solution w_0 by Theorem 20.3 (see (20.11)) is in $W_{p\Omega}^{(\rho+4)}$, and therefore by the embedding Theorem 10.6,

$$w \in H_\Omega^{\rho+2,2(1-1/p)} \text{ for } p < 2;\ w \in H_\Omega^{\rho+2-0} \text{ for } p = 2;$$
$$w \in H_\Omega^{\rho+3,1-2/p} \text{ for } p > 2. \tag{28.17}$$

But by Theorem 28.1 (relation (28.7)) we have

$$\left\| w_0 - \sum_{m=0}^{(N+1)^2} \mathbf{P}_m(w_0)\varphi_m \right\|_{H_1} \le m'(w_0)N^{-\beta},$$

where β is given by (28.16). But then (28.15) follows from (28.1), (28.3). Theorem 28.3 is proved.

Theorem 28.4. *Assume that Conditions 2–4 of Section 13 and conditions 1–4 of Theorem 28.1 are satisfied, and that moreover,*

$$R^3,\ R^s \in H_\Omega^{\rho,\lambda};\ \Gamma \in H_\Omega^{\rho+4,\lambda};\ S \in H_\Omega^{\rho+4,\lambda};$$
$$D_{\mathrm{f}}^{ijkl} \in H_\Omega^{\rho+2,\lambda};\ D_{\mathrm{s}}^{ijkl} \in H_\Omega^{\rho+1,\lambda}.$$

Finally, assume that in addition,

$$F \in H_\Omega^{\rho+4,\lambda}. \tag{28.18}$$

Then for any nonsingular solution w_0 of the operator equation (13.33) *there is a sequence $w_{(N+1)^2}$ of BGR approximations for which we have the asymptotic error estimate* (28.15), *where*

$$\beta = p + 2 + \lambda \text{ if } \lambda \le 1;\ \beta = \rho + 3 - 0 \text{ if } \lambda = 1. \tag{28.19}$$

Theorem 28.4 is proved exactly as Theorem 28.3, and therefore details are omitted.

Remark 28.1. Theorems 28.1 and 28.2 demonstrate that if the initial data of the problem are analytic, then the theoretical estimate (28.15) claims that the error of the BGR method decays faster than any negative power of N. This will occur, for example, in the case of a spherical or an ellipsoidal shell, as in this case the support curve Γ is a circle or an ellipse, while the load is a polynomial.

However, in this case one should keep in mind that the convergence in practice does not have to be as good as that. For more details on this, see [198].

Remark 28.2. In the case of an axisymmetric deformation of a physically and geometrically axisymmetric shell, its stress-deformed state depends only on the coordinate r, and the exponent in (28.15) has to be doubled.

Remark 28.3. Approximate values of the lateral displacements w_1, w_2 are determined in this form of the BGR method by the relation

$$\omega_{(N+1)^2} = \mathbf{K}_{\iota\kappa}\left(w_{(N+1)^2}\right). \tag{28.20}$$

Therefore, it is easily shown that under the conditions of Theorems 28.3, 28.4 we have the following estimate of the error in lateral and total displacements:

$$\left\|\omega_0-\omega_{(N+1)^2}\right\|_{H_5}\le m(\mathbf{a}_0)N^{-\beta},\quad \left\|\mathbf{a}_0-\mathbf{a}_{(N+1)^2}\right\|_{H_{51}}\le m(\mathbf{a}_0)N^{-\beta}.$$

Remark 28.4. Let us observe that in the problems of nonlinear shallow shell theory we are considering, by Lemmas 26.2, 27.12, solvability of the main finite-dimensional systems of equations of the BGR methods was obtained starting with the first approximation, $N=0$. On the other hand, by the theorem of Krasnosel'skii [162], their solvability is guaranteed only for sufficiently large N.

28.5. Under the boundary conditions (28.4), to find an approximate solution, one can use the BGR method in its Mushtari version (Section 27). Then the approximate solution is sought in the form (27.1), where w_n is constructed using (28.5), while for ω_n one uses (28.10). The coefficients D_{nk} and C_{nk} are determined, respectively, from (27.2), (27.3).

Obviously, the total number of vector functions of degree at most N in (28.5), (28.10) will be $(N+1)^6$.

Theorem 28.5. *Assume that all the conditions of Theorem 28.3 are satisfied. Then for any nonsingular solution* $\mathbf{a}_0$ *of the operator equation* (13.31) *there is a sequence of approximate solutions* $\mathbf{a}_{(N+1)^6}$ *such that*

$$\left\|\mathbf{a}_0-\mathbf{a}_{(N+1)^6}\right\|_{H_{51}}\le m(\mathbf{a}_0)N^{-\beta},\tag{28.21}$$

where β *is given by* (28.16) *with* ρ *replaced by* $\rho-1$.

The proof of this theorem is also based on the relation (28.3); to use it we need to have estimates for the approximation of $\mathbf{a}_0$ by the basis $\mathbf{b}_k=(\chi_k,\varphi_k)$. Then the estimates are given by relations (28.7), (28.12). If the conditions of Theorem 28.3 hold, we have (28.17), and as follows from Theorem 20.3 (relation (20.11)),

$$\omega_0\in H_\Omega^{\rho,2(1-1/p)}\text{ for }p<2;\ \omega_0\in H_\Omega^{\rho-0}\text{ for }p=2;$$

$$\omega_0\in H_\Omega^{\rho+1,1-2/p}\text{ for }p>2;$$

From (28.7), (28.12) we have

$$\left\|\mathbf{a}_0-\sum_{k=1}^{(N+1)^6}(\mathbf{a}_0\cdot\mathbf{b}_k)_{H_{51}}\mathbf{b}_k\right\|_{H_{51}}\le m(\mathbf{a}_0)N^{-\beta},\ \mathbf{b}_k=(\chi_k,\varphi_k).\tag{28.22}$$

From (28.22) we have (28.20) with ρ replaced by $\rho-1$.

Theorem 28.6. *Assume that all the conditions of Theorem 28.4 are satisfied. Then for any nonsingular solution* $\mathbf{a}_0$ *of the operator equation* (13.31) *there is a sequence of approximate solutions* $\mathbf{a}_{(N+1)^6}$ *such that*

$$\left\|\mathbf{a}_0-\mathbf{a}_{(N+1)^6}\right\|_{H_{51}}\le m(\mathbf{a}_0)N^{-\beta},\tag{28.23}$$

where β *is given by* (28.19) *with* ρ *replaced by* $\rho-1$.

We do not present the proof of (28.23), as it is identical to the proof of (28.21).

28.6. The basis φ_k can also be used in the approximate solution of equation (17.38) for problem 91 under the condition $\Gamma_1 = \Gamma$.

For that we use the Papkovich version (Section 26). We seek an approximate solution in the form (26.1), while the D_{nk} are found as solutions of the system (26.38).

Theorem 28.7. *Assume that Conditions 1–4 of Section 17 are satisfied and that moreover,*

$$R^3,\ C^{kl}_{*ij} \cdot \widetilde{T}^{ij}_{\mathrm{p}} \in W^{(\rho)}_{p\Omega};\ \Gamma \in C^{\rho+4}_{\Omega},\ S \in C^{\rho+4}_{\Omega},\ D^{ijkl}_{\mathrm{f}} \in C^{\rho+2}_{\Omega},\ C^{ijkl}_{*} \in C^{\rho+2}_{\Omega}.$$

Assume that in addition, conditions (28.14) *are satisfied. Then for any nonsingular solution of the operator equation* (17.38) *there is a sequence $w_{(N+1)^2}$ of BGR solutions of* (26.38) *for which we have the asymptotic estimate* (28.15), *where β is given by* (28.16).

Theorem 28.8. *Assume that Conditions 1–4 of Section 17 are satisfied and that moreover,*

$$R^3, C^{kl}_{*ij} \cdot \widetilde{T}^{ij}_{\mathrm{p}} \in H^{\rho,\lambda}_{p\Omega};\ \Gamma \in H^{\rho+4,\lambda}_{\Omega},$$

$$S \in H^{\rho+4,\lambda}_{\Omega};\ D^{ijkl}_{\mathrm{f}} \in H^{\rho+2,\lambda}_{\Omega};\ C^{ijkl}_{*} \in H^{\rho+2,\lambda}_{\Omega}.$$

Assume that in addition, condition (28.18) *is satisfied. Then if w_0 is a nonsingular solution of the operator equation* (17.19), *there is a sequence $w_{(N+1)^2}$ of BGR solutions of* (26.38) *for which we have the asymptotic estimate* (28.15), *where β is given by* (28.19). *Furthermore, if $\Psi_{(N+1)^2}$ are defined by*

$$\Psi_{(N+1)^2} = \mathbf{K}_{91} w_{(N+1)^2},$$

then we also have the estimates

$$\left\| \mathbf{c}_0 - \mathbf{c}_{(N+1)^2} \right\|_{H_{91}} \le m(\mathbf{c}_0) N^{-\beta},$$

where β is given by (28.19).

28.7. We shall solve the operator equation (17.39) by the BGR method using the Vlasov version of Section 27. To determine $w_{(N+1)^4}$, $\Psi_{(N+1)^4}$, we shall use the relations (27.35), (27.36). In (27.35), φ_k and θ_k are chosen from the orthonormal family (28.5).

Theorem 28.9. *Let all the conditions of Theorem 28.7 be satisfied. Then for each nonsingular solution w_0, Ψ_0 of the operator equation* (17.39) *there is a sequence of solutions of equations* (27.35), (27.36), *$w_{(N+1)^4}$ $\Psi_{(N+1)^4}$, such that*

$$\left\| \mathbf{c}_0 - \mathbf{c}_{(N+1)^4} \right\|_{H_{91}} \le m(\mathbf{c}_0) N^{-\beta},$$

where β is given by (28.16).

Theorem 28.10. *Let all the conditions of Theorem 28.8 be satisfied. Then for each nonsingular solution of* (17.39) *there is a sequence of solutions of equations* (27.35), (27.36), *$w_{(N+1)^4}$, $\Psi_{(N+1)^4}$ such that*

$$\left\| \mathbf{c}_0 - \mathbf{c}_{(N+1)^4} \right\|_{H_{91}} \le m(\mathbf{c}_0) N^{-\beta},$$

where β is given by (28.19).

Theorems 28.7–28.10 are proved using the arguments of the proof of Theorem 28.3.

We note that remarks 28.1, 28.2 apply also in the conditions of Theorems 28.5–28.10.

28.8. Of considerable interest are the error estimates of the BGR method in the case when f_0 is a singular solution of equations (28.1). When $\mathbf{G}$ is a completely continuous polynomial operator, we can state the following conjecture: Asymptotic error estimates of the BR are given by the formula

$$\|f_0 - f_n\| \leq (1+\epsilon_n) \left\| f_0 - \sum_{k=1}^{n} (\mathbf{P}_k f_0) g_k \right\|^{\nu}, \quad \epsilon_n \to 0, \tag{28.24}$$

where ν is a positive rational number.

In certain cases the estimate (28.24) can be rigorously justified.

28.9. It is important to understand the conditions that guarantee uniform convergence of the sequences $w_{in\alpha^j}$, $w_{n\alpha^i\alpha^j}$ in $\overline{\Omega}$ in the BGR method [342]. This will also ensure uniform convergence of the strain tensor. Different approaches can be used. The first of these is to use the estimates of Il'in [125]. Here we shall describe a different method of constructing approximations with a strain tensor that converges uniformly in $\overline{\Omega}$. For simplicity, let us consider the case of an anisotropic plate, where $B_{ij} \equiv 0$, $A_{ij} = \delta_{ij}$. Then the basic equations of the boundary value problem 51 assume the form

$$\left(D_{\mathrm{s}}^{ijkl} w_{l\alpha^k}\right)_{\alpha^j} = -\frac{1}{2}\left(D_{\mathrm{s}}^{ijkl} w_{\alpha^k} w_{\alpha^l}\right)_{\alpha^j} - R^i = f^i\{w\}, \quad i = 1,\, 2, \tag{28.25}$$

$$\left(D_{\mathrm{f}}^{ijkl} w_{\alpha^k\alpha^l}\right)_{\alpha^i\alpha^j} = T^{ij} w_{\alpha^i\alpha^j} + R^3 - R^s w_{\alpha^s} = f^3\{w\}. \tag{28.26}$$

We also assume the boundary conditions (28.4). Relations (28.25), (28.26) are equivalent to

$$w_i(\mathcal{P}) = \int_\Omega \mathcal{G}_{ik}(\mathcal{P},\, \mathcal{Q}) f^k\{w(\mathcal{Q})\}\, d\mathcal{Q}, \tag{28.27}$$

$$w(\mathcal{P}) = \int_\Omega \mathcal{G}(\mathcal{P},\, \mathcal{Q}) f^3\{w(\mathcal{Q})\}\, d\mathcal{Q}, \tag{28.28}$$

where $\mathcal{G}_{ik}(\mathcal{P},\, \mathcal{Q})$, $\mathcal{G}(\mathcal{P},\, \mathcal{Q})$ are, respectively, Green's matrix and Green's function for the operators on the left-hand side of (28.25), (28.26). Methods for constructing $\mathcal{G}_{ik}(\mathcal{P},\, \mathcal{Q})$, $\mathcal{G}(\mathcal{P},\, \mathcal{Q})$ for a wide range of problems are to be found in [166, 167]. These papers also study in detail the properties of Green's tensor.

Theorem 28.11. *Let* $\Gamma \in C^6_\Gamma$, $S \in C^4_\Gamma$, $D_{\mathrm{s}}^{ijkl} \in C^1_\Omega$, $D_{\mathrm{f}}^{ijkl} \in C^2_\Omega$. *Furthermore, assume that the characteristic polynomials corresponding to the highest-order derivatives on the left-hand side of* (28.25), (28.26) *do not have multiple roots in* *the closed domain* $\widetilde{\Omega}$. *Then we have the estimates* [167]

$$|\mathcal{G}_{ik}(\mathcal{P},\ \mathcal{Q})| \leq M(1 + |\ln r_{\mathcal{P}\mathcal{Q}})|,$$

$$\left|\frac{\partial \mathcal{G}_{ik}(\mathcal{P},\ \mathcal{Q})}{\partial \alpha^i_P}\right| + \left|\frac{\partial \mathcal{G}_{ik}(\mathcal{P},\ \mathcal{Q})}{\partial \alpha^i_Q}\right| \leq M r^{-1}_{\mathcal{P}\mathcal{Q}},$$

$$\mathcal{P} = \mathcal{P}(\alpha^1_P,\ \alpha^2_P),\ \mathcal{Q} = \mathcal{Q}(\alpha^1_Q,\ \alpha^2_Q),$$

$$\left|\frac{\partial \mathcal{G}(\mathcal{P},\ \mathcal{Q})}{\partial \alpha^i_P}\right| + \left|\frac{\partial \mathcal{G}(\mathcal{P},\ \mathcal{Q})}{\partial \alpha^i_Q}\right| \leq M,\ \left|\frac{\partial^2 \mathcal{G}(\mathcal{P},\ \mathcal{Q})}{\partial \alpha^i_{l_1} \partial \alpha^j_{l_2}}\right| \leq M(1 + |\ln r_{\mathcal{P}\mathcal{Q}}|),\ l_1 = P,\ Q,\ l_2 = P,\ Q. \tag{28.29}$$

Probably, the requirements on Γ in this theorem can be reduced. We could take only $\Gamma \in H^{4,\lambda}_{\Gamma}$.

Green's function can also be introduced in a different way. Let us consider an operator equation in H_1 of the form

$$(\widetilde{\mathcal{G}}(\mathcal{P},\ \mathcal{Q}) \cdot \varphi(\mathcal{Q}))_{H_1} = \varphi(\mathcal{P}), \tag{28.30}$$

where $\varphi(\mathcal{Q})$ is an arbitrary function in H_1 and $\widetilde{\mathcal{G}}(\mathcal{P},\ \mathcal{Q})$ is the required function, which is an H_1 function in $\mathcal{Q}$ and fixed. For a fixed $\mathcal{P}$, the right-hand side of (28.30) will be a bounded linear functional in H_1. Indeed, its additivity is obvious, while boundedness follows from Theorem 12.3 (see (12.28)). Therefore, existence of $\widetilde{\mathcal{G}}(\mathcal{P},\ \mathcal{Q})$ as an element of H_1 in $\mathcal{Q}$ follows from Riesz's theorem; clearly, this element will depend on $\mathcal{P}$. Next, let φ_k be an orthonormal basis of H_1. Obviously, we have

$$\widetilde{\mathcal{G}}(\mathcal{P},\ \mathcal{Q}) = \sum_{k=1}^{\infty} g_k \varphi_k(\mathcal{Q}), \tag{28.31}$$

where

$$g_k = (\widetilde{\mathcal{G}}(\mathcal{P},\ \mathcal{Q}) \cdot \varphi_k(\mathcal{Q}))_{H_1} = \varphi_k(\mathcal{P}). \tag{28.32}$$

The second equality of (28.32) follows from (28.30). From (28.31), (28.32) we have

$$\widetilde{\mathcal{G}}(\mathcal{P},\ \mathcal{Q}) = \sum_{k=1}^{\infty} \varphi_k(\mathcal{P}) \varphi_k(\mathcal{Q}).$$

Let us consider now an operator equation in H_1,

$$(w \cdot \varphi)_{H_1} = \int_{\Omega} f^3\{w\} \varphi \, d\Omega. \tag{28.33}$$

Here f^3 is given and belongs to a space $L_{p\Omega}$ for $p > 1$, and w is an arbitrary function in H_1. Clearly, (28.33) is solvable. Let us set $\varphi = \varphi_k$, so that

$$(w \cdot \varphi_k)_{H_1} = \int_\Omega f^3\{w\}\varphi_k \, d\Omega. \tag{28.34}$$

Taking into account (28.30), we have from (28.34),

$$w(\mathcal{P}) = \sum_{k=1}^{\infty}(w \cdot \varphi_k)_{H_1}\varphi_k(\mathcal{P}) = \sum_{k=1}^{\infty}\int_\Omega f^3(\mathcal{Q})\varphi_k\{w(\mathcal{Q})\}\, d\mathcal{Q}\, \varphi_k(\mathcal{P}). \tag{28.35}$$

Comparing (28.28), (28.35), we come to the conclusion that

$$w(\mathcal{P}) = \int_\Omega \widetilde{\mathcal{G}}(\mathcal{P}, \mathcal{Q}) f^3\{w(\mathcal{Q})\}\, d\mathcal{Q} = \int_\Omega \mathcal{G}(\mathcal{P}, \mathcal{Q}) f^3\{w(\mathcal{Q})\}\, d\mathcal{Q},$$

so that

$$\widetilde{\mathcal{G}}(\mathcal{P}, \mathcal{Q}) \equiv \mathcal{G}(\mathcal{P}, \mathcal{Q}).$$

Next, we shall seek an approximate solution of equation (28.28) by replacing the kernel $\mathcal{G}(\mathcal{P}, \mathcal{Q})$ by the degenerate kernel

$$\mathcal{G}_n(\mathcal{P}, \mathcal{Q}) = \sum_{k=1}^{n} \varphi_k(\mathcal{P})\varphi_k(\mathcal{Q}),$$

that is, we set

$$w_n(\mathcal{P}) = \int_\Omega \mathcal{G}_n(\mathcal{P}, \mathcal{Q}) f^3\{w_n(\mathcal{Q})\}\, d\mathcal{Q}.$$

It is easily seen that as a result in this case we obtain for D_{nk} the system (28.2), since it is assumed that ω, T^{ij} are determined using the exact solutions of the boundary value problem for (28.25). In other words, replacing the kernel $\mathcal{G}(\mathcal{P}, \mathcal{Q})$ by a degenerate kernel is equivalent to solving the boundary value 51 by the Papkovich version (Section 26) of the BGR method.

Let us now introduce iterated approximations using

$$\widetilde{w}_n(\mathcal{P}) = \int_\Omega \mathcal{G}(\mathcal{P}, \mathcal{Q}) f^3\{w_n(\mathcal{Q})\}\, d\mathcal{Q}, \tag{28.36}$$

and let w_0 be a solution of the integro-differential equation (28.28) for which there exists a subsequence w_n (we have changed the notation for it) such that

$$\|w_0 - w_n\|_{H_1} \to 0. \tag{28.37}$$

Existence of the subsequence w_n is guaranteed by Theorem 26.1. Then from (28.28), (28.36) we obtain

$$w_0(\mathcal{P}) - \widetilde{w}_n(\mathcal{P}) = \int_\Omega \mathcal{G}(\mathcal{P}, \mathcal{Q})[f^3\{w_0\} - f^3\{w_n\}]\, d\mathcal{Q}, \tag{28.38}$$

and from (28.26) we have

$$w_0(\mathcal{P}) - \widetilde{w}_n(\mathcal{P})$$
$$= \int_\Omega \mathcal{G}(\mathcal{P}, \mathcal{Q})\left[T^{ij}(w_0)\nabla_{ij}w_0 - T^{ij}(w_n)\nabla_{ij}w_n + R^s(w_{0\alpha^s} - w_{n\alpha^s})\right] d\mathcal{Q}.$$

Theorem 28.12. *Assume that*

$$\Gamma \in H^{2,\lambda}_\Gamma;\ R^3,\ R^s \in H^{0,\lambda}_\Omega;\ S \in H^{2,\lambda}_\Gamma;\ D^{ijkl}_{\mathrm{s}} \in H^{2,\lambda}_\Omega;\ D^{ijkl}_{\mathrm{f}} \in H^{1,\lambda}_\Omega.$$

Then

$$\|w_0 - \widetilde{w}_n\|_{C^2_\Omega},\ \left\|\overset{0}{\epsilon}_{ij}(w_0) - \overset{0}{\epsilon}_{ij}(\widetilde{w}_n)\right\|_{C_\Omega},\ \left\|\overset{1}{\epsilon}_{ij}(w_0) - \overset{1}{\epsilon}_{ij}(\widetilde{w}_n)\right\|_{C_\Omega},$$
$$\left\|T^{ij}(w_0) - T^{ij}(\widetilde{w}_n)\right\|_{C_\Omega},\ \left\|M^{ij}(w_0) - M^{ij}(\widetilde{w}_n)\right\|_{C_\Omega} \to 0.$$

To prove Theorem 28.12, we note that from (28.25), (28.37) we immediately obtain

$$\left\|f^j(w_0) - f^j(\widetilde{w}_n)\right\|_{L_{2-0,\Omega}} \to 0 \text{ as } \widetilde{w}_n \to w_n.$$

Next, from (28.27) we obtain

$$w_{i\alpha^k}(w_0) - w_{i\alpha^k}(\widetilde{w}_n) = \int_\Omega \mathcal{G}_{ij\alpha^k}(\mathcal{P}, \mathcal{Q})[f^j\{w_0\} - f^j\{w_n\}]\, d\mathcal{Q}. \qquad (28.39)$$

By Theorem 28.11, the right-hand side of (28.39) can be considered as an integral of potential type [196, 285], due to which we conclude that

$$\|w_{i\alpha^k}(w_0) - w_{i\alpha^k}(\widetilde{w}_n)\|_{L_{q\Omega}} \le m \left\|f^j\{w_0\} - f^j\{\widetilde{w}_n\}\right\|_{L_{2-0,\Omega}}. \qquad (28.40)$$

In (28.40), $q \ge 1$ is arbitrary.

Now let us consider (28.38), embedding Theorem 12.3 (see (12.29)), and relations (4.16) for T^{ij}. As a result, we have

$$\left\|T^{ij}(w_0) - T^{ij}(\widetilde{w}_n)\right\|_{L_{q\Omega}} \to 0,\ \forall q \ge 1.$$

But then from (28.26) we have

$$\left\|f^3\{w_0\} - f^3\{\widetilde{w}_n\}\right\|_{L_{q\Omega}} \to 0,\ \forall q \ge 1.$$

In its turn, due to (28.29), it follows from (28.38) that

$$\|w_0(\mathcal{P}) - \widetilde{w}_n(\mathcal{P})\|_{C_\Omega} \to 0,\ \|w_{0\alpha^i} - \widetilde{w}_{n\alpha^i}\|_{C_\Omega} \to 0,$$
$$\|w_{0\alpha^i\alpha^j} - \widetilde{w}_{n\alpha^i\alpha^j}\|_{C_\Omega} \to 0.$$

Thus Theorem 28.12 has been proved with respect to the sequences $\widetilde{w}_n$, $\widetilde{w}_{n\alpha^i}$, $\widetilde{w}_{n\alpha^i\alpha^j}$. But then from (3.17), (4.14) we obtain

$$\left\|\overset{1}{\epsilon}_{ij}(w_0) - \overset{1}{\epsilon}_{ij}(\widetilde{w}_n)\right\|_{C_\Omega},\ \left\|M^{ij}(w_0) - M^{ij}(\widetilde{w}_n)\right\|_{C_\Omega} \to 0.$$

Furthermore, from (28.27) and the estimates (28.29),

$$\|w_{i\alpha^j}(w_0) - w_{i\alpha^j}(\widetilde{w}_n)\|_{C_\Omega} \to 0,$$

and from (3.20)–(3.22), (4.14) it follows that

$$\left\| \overset{0}{\epsilon}_{ij}(w_0) - \overset{0}{\epsilon}_{ij}(\widetilde{w}_n) \right\|_{C\Omega}, \quad \left\| T^{ij}(w_0) - T^{ij}(\widetilde{w}_n) \right\|_{C\Omega} \to 0.$$

Theorem 28.12 has been completely proved.

Remark 28.5. A significant part in the proof of Theorem 28.12 was played by the estimates (28.29), which hold if the characteristic polynomials of the highest-order differential operators in (28.25), (28.26) have no multiple roots. The mechanical interpretation of this fact is that in the domain Ω the plate has no points of isotropy. If there are such points, the estimates (28.29) change, but both the statement and the general result of Theorem 28.12 remain the same. We quote the more general result without a proof.

Theorem 28.13. *Let* $\Gamma \in H_{\Gamma}^{\rho+4,\lambda}$; $D_{\mathrm{f}}^{ijkl} \in H_{\Omega}^{\rho+2,\lambda}$; $D_{\mathrm{s}}^{ijkl} \in H_{\Omega}^{\rho+1,\lambda}$; $R^3, R^s \in H_{\Omega}^{\rho,\lambda}$, *and we are seeking an approximate solution of Problem 51 using the Papkovich version of the BGR method. Let* w_0 *be a solution for which there exists a sequence of approximations for which*

$$\|w_0 - w_n\|_{H_1} \to 0.$$

Then we have

$$\|w_0 - \widetilde{w}_n\|_{H^{\rho+4,\lambda'}} \to 0, \quad \|\omega_0 - \omega(\widetilde{w}_n)\|_{H^{\rho+2,\lambda'}} \to 0.$$

Here

$$\omega(\widetilde{w}_n) = \mathbf{K}_{51}(\widetilde{w}_n).$$

28.10. The scheme of justifying the BGR method in problems of nonlinear shell theory is due to the author [338, 339, 341, 342, 343, 352, 361]. We also mention [413, 414]. It can be immediately extended to other direct methods: finite element, finite difference methods, approximation by splines [387, 223, 300, 301, 81, 384]. It is important that two essential conditions should hold here: (1) The approximation scheme must allow us to approximate any element arbitrarily close in the norm of the corresponding space as the number of constants used in the approximation grows without bound; (2) equations defining the constants of approximation must be obtained using a variational principle, such as that of Lagrange, Alumyae. It is precisely this method of obtaining the constants of approximation that appears to be the most appropriate. This approach has been implemented in [374] to give a theoretical justification of the method.

Application of the BGR method in nonlinear shell theory has an important advantage: in the process of approximating a solution, it allows us to take into account the intuitive picture concerning singularities of the solution and the results of mathematical or experimental analysis. In this respect see, for example, [242]. A history of the development of BGR methods can be found in the survey [361]. The first applications of these methods in nonlinear shell theory are due to Panov [227] and Feodos'ev [78], while the first rigorous justification of the methods is due to the author [338, 339]. In practical implementations of the BGR methods, systems of nonlinear equations with a large number of unknowns have to be solved. A survey of methods of solving such systems of equations can be found in [224]. An efficient method in this connection is the method of transforming the problem into a system of differential equations due to Davidenko [61, 62, 63, 64]. Numerical implementation of the method encounters a number of obstacles, which were discussed in Section 21. A number of valuable recommendations concerning the practical solution of large systems of nonlinear equations is contained in [302, 303]. Finally, let us note the monograph of Mikhlin [198], where the important concept of a similar operator is introduced and where minimality properties of a system of functions are made essential use of.

Interest in these methods has increased recently due to the fast development of computer symbol-manipulation capabilities. The entire process of deriving and solving systems (26.38), (26.39) can be automated. The advantages are that we can at the same time increase the number of elements χ_k and φ_k being considered and choose them in accordance with the nature of the stress-deformed state of the shell. For more details on symbolic computation see [361, 46, 232, 299].

CHAPTER VIII

Formulation of the Problem of Stability. Global Uniqueness of Solutions. Stiffness of Shells. Well-Posedness Classes

29. Formulation of the Problem of Stability in the Nonlinear Theory of Shallow Shells. Local Uniqueness of Solutions. Conditions for Global Uniqueness

29.1. In the static theory of stability of elastic systems one considers qualitative methods to determine the number of equilibrium configurations of an elastic system for a given load and methods of assigning a probability to each of these configurations being realized. In the modern theory of stability of elastic systems it is the first of these problems, namely, a qualitative study of the number of equilibrium configurations, that has been studied in more detail. The probability of encountering different equilibrium configurations is usually estimated by comparing their values of the potential energy.

When the problem of the number of different equilibrium configurations is being considered, one usually tries to work out the ranges of parameters of the loading under which a given elastic system has a unique equilibrium configuration. It could be assumed that these ranges are defined by the first bifurcation point of the nonlinear equations describing the deformation of the elastic system, while the first bifurcation point is taken to be the smallest eigenvalue of the corresponding linearized boundary value problem. It is precisely by identifying these three concepts, namely, that of a point that defines the domain of existence of a unique equilibrium configuration of an elastic system, of a bifurcation point of solutions of the equations of a deformed state of an elastic system, and of the smallest eigenvalue of the linearized problem, that stability problems have been solved ever since the time of Euler [75]. In certain cases this approach has been made rigorous. These questions are considered in the well-known work of Yasinskii [385]; they have been completely resolved in the case of a hinged rod in [16]. At the same time it is completely clear that the identification of the three above concepts cannot always be valid, and this is the question we shall consider first.

A complete solution of the first stability problem, the determination of the number of equilibrium configurations of the system, can be illustrated using as an example Problem $t\kappa$, as follows [349, 352, 359, 362]. In Section 16 we established necessary and sufficient conditions on the load $\overset{0}{\mathbf{a}} = (\overset{0}{\omega}, \overset{0}{w})$ for the boundary value problem $t\kappa$ to be solvable (Theorems 16.5, 16.8, 16.11). Here it makes sense to introduce the Hilbert space H of load complexes $\{R^1, R^2, R^3, \widetilde{M}^m, \widetilde{Q}, \widetilde{T}^\tau, \widetilde{T}^m, \widetilde{\omega}, \widetilde{w}, \widetilde{w}_4\}$ with the scalar product

$$(\overset{1}{H} \cdot \overset{2}{H})_H = (\overset{1}{\omega}_p \cdot \overset{2}{\omega}_p)_{H_t} + (\overset{1}{w}_p \cdot \overset{2}{w}_p)_{H_\kappa} + (\overset{0}{w}_1 \cdot \overset{0}{w}_2)_{H_\kappa} + (\overset{0}{\omega}_1 \cdot \overset{0}{\omega}_2)_{H_t}, \quad (29.1)$$

where $\overset{0}{w}, \overset{0}{\omega}$ are chosen in a certain way using $\widetilde{\omega}, \widetilde{w}, \widetilde{w}_4 = \partial\widetilde{w}/\partial m$.

Theorems 16.5, 16.8, 16.11 assert the existence of a mapping (not univalent) from H into $H_{t\kappa}$ and, conversely, the existence of a mapping from $H_{t\kappa}$ to H.

To solve completely the first problem of stability, it is necessary to decompose the space H into sets B_r (a U-decomposition) in each of which the Problem $t\kappa$ has precisely r solutions. Transition from B_r to B_{r+1} is clearly via a boundary set separating these two sets. By a theorem of Smale [165], all the boundary sets separating B_r and B_{r+1} ("folds") form a "thin" set. Due to extreme difficulties in constructing B_r, we can simplify the problem somewhat. Let us assume that all the external stresses acting on the shell are described by a finite number of parameters, which we denote by $U_1, \ldots, U_n$. For example, in the problem considered in Section 24, the simultaneous effect of a transverse load and of lateral compression on a circular hinged plate [340], we have $U_1 = q$, $U_2 = T$. Let us assume now that the entire n-dimensional space of the parameters U has been decomposed into sets in each of which the number of solutions of Problem $t\kappa$ is constant. We shall call such a decomposition a U-decomposition (this term is borrowed from control theory). For example, in the problem of Section 24 the U-decomposition is as in Figure 29.1 (at least in a neighborhood of the point $q = 0$, $T = T_0$). However, construction of a U-decomposition is also a sufficiently complicated problem, and below we shall discuss such a decomposition only in the case $n = 1$.

The second problem, namely, to determine the most probable equilibrium configuration, can be solved by using a finer analysis of the behavior of a shell that

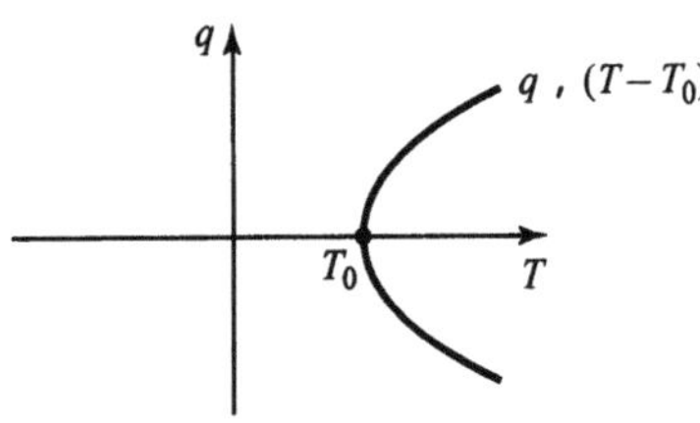

FIGURE 29.1.

takes into consideration random effects. In that case the potential energy of a shell appears as only a very partial characteristic of the degree of stability of an equilibrium configuration. To clarify these points, we conclude the present chapter by presenting a statistical stability theory (Chapter 10) constructed by the author in the mid-fifties. Nevertheless, bearing in mind the terminology of the statistical theory of stability of elastic systems, below we shall call a solution $\mathbf{a}_0$ and thus w_0 of a Problem $t\kappa$ stable if to it corresponds a strict (absolute or relative) minimum of the functional $\mathcal{I}_{t\kappa}(\mathbf{a})$ and therefore $\mathcal{I}_{\kappa\kappa}(w)$. The problem of stability in the Problem 9κ can be formulated along the same lines.

29.2. First of all, let us establish some uniqueness theorems for solutions of the Problem $t\kappa$.

Theorem 29.1. *Assume that the conditions of Theorem 16.5 (respectively, 16.8, 16.11) are satisfied and let the loads satisfy the conditions (of Theorems 16.8 and 16.9)*

$$\|H\|_H \leq \epsilon \tag{29.2}$$

for a sufficiently small ϵ. Then we can find a neighborhood of zero in the space H_κ,

$$\|w\|_{H_\kappa} \leq \delta(\epsilon), \tag{29.3}$$

in which there exists a unique solution of Problem $t\kappa$. Furthermore, $\delta \to 0$ as $\epsilon \to 0$.

To prove the theorem we first show that if (29.2) is satisfied, the vector field $w - \mathbf{G}_{\kappa\kappa}(w)$ will be homotopic to the vector field w on spheres of sufficiently small radius r. Clearly, the two fields can be connected by a deformation $w - t\mathbf{G}_{\kappa\kappa}(w)$, $0 \leq t \leq 1$, and the homotopy of the fields will be demonstrated if we prove that

$$\|\mathbf{G}_{\kappa\kappa}(w)\|_{H_\kappa} < r < 1 \tag{29.4}$$

on spheres of sufficiently small radius. From Lemma 14.1 (relation (14.2)) we have

$$\|\mathbf{G}_{\kappa\kappa}(w)\|_{H_\kappa} \leq \|\mathbf{G}_{\kappa\kappa0}\|_{H_\kappa} + \sum_{\mu=1}^{3} \left\|\mathbf{G}_{\kappa\kappa\mu}(w)\right\|_{H_\kappa} \leq \|\mathbf{G}_{\kappa\kappa0}\|_{H_\kappa} + m \sum_{\mu=1}^{3} \|w\|_{H_\kappa}^{\mu}. \tag{29.5}$$

Let us estimate $\|\mathbf{G}_{\kappa\kappa0}\|_{H_\kappa}$. From (14.18) we have

$$\|\mathbf{G}_{\kappa\kappa0}\|_{H_\kappa} \leq \left\|w^*\right\|_{H_\kappa} + m \sum_{i,j=1}^{2} \left(\left\|T_0^{ij}\right\|_{L_{2\Omega}} + \left\|T^{ij}(\mathbf{a}^*)\right\|_{L_{2\Omega}} \right). \tag{29.6}$$

From (14.16), (14.16) we obtain

$$\left\|T_0^{ij}\right\|_{L_{2\Omega}} \leq m D_{\mathrm{s}}^{ijkl} \left\|\overset{0}{\epsilon}_{kl0}\right\|_{L_{2\Omega}} \leq \|\mathbf{K}_{t\kappa0}\|_{H_t}. \tag{29.7}$$

From (14.5) we immediately have

$$\|\mathbf{K}_{t\kappa0}\|_{H_t} \leq m \left(\left\|\omega^*\right\|_{H_t} + \left\|w^*\right\|_{H_\kappa} + \left\|w^*\right\|_{H_\kappa}^2 \right), \tag{29.8}$$

and from (29.6)–(29.8) we obtain

$$\left\| T_0^{ij} \right\|_{L_{2\Omega}} \leq m \left(\left\| \omega^* \right\|_{H_t} + \left\| w^* \right\|_{H_\kappa} + \left\| w^* \right\|_{H_\kappa}^2 \right).$$

Let us now estimate $\left\| T^{ij}(\mathbf{a}^*) \right\|_{L_{2\Omega}}$. From (4.15) and (3.16) we have

$$\left\| T^{ij}(\mathbf{a}^*) \right\|_{L_{2\Omega}} \leq \sum_{k,l=1}^{2} D_{\mathbf{s}}^{ijkl} \left\| \overset{0}{\epsilon}_{ij}(\mathbf{a}^*) \right\|_{L_{2\Omega}} \leq m \left(\left\| \omega^* \right\|_{H_t} + \left\| w^* \right\|_{H_\kappa} + \left\| w^* \right\|_{H_\kappa}^2 \right).$$

From (29.5), (29.8) we obtain

$$\| \mathbf{G}_{\kappa\kappa 0} \|_{H_\kappa} \leq m \left(\left\| \omega^* \right\|_{H_t} + \left\| w^* \right\|_{H_\kappa} + \left\| w^* \right\|_{H_\kappa}^2 \right). \tag{29.9}$$

Finally, from (29.9), (14.4) we have

$$\| \mathbf{G}_{\kappa\kappa} w \|_{H_\kappa} \leq m \left(\left\| \omega^* \right\|_{H_t} + \left\| w^* \right\|_{H_\kappa} + \left\| w^* \right\|_{H_\kappa}^2 + \sum_{\mu=1}^{3} \| w \|_{H_\kappa}^{\mu} \right). \tag{29.10}$$

Clearly, (29.4) follows from (29.10) for sufficiently small $\| \mathbf{a}^* \|_{H_{t\kappa}}$, $\| w \|_{H_\kappa}$. Thus solvability of Problem $t\kappa$ for sufficiently small $\| \mathbf{a}^* \|_{H_{t\kappa}}$, $\| w \|_{H_\kappa}$ has been established. Let us show now that for sufficiently small ϵ the operator $[\mathbf{I} - \mathbf{G}_{\kappa\kappa 1}(\mathbf{a}^*)]^{-1}$ exists and is bounded. Since we have proved that $\mathbf{G}_{\kappa\kappa 1}(\mathbf{a}^*)$ is a completely continuous operator (Theorem 14.2), our assertion will be established if we show that

$$((w - \mathbf{G}_{\kappa\kappa 1}(\mathbf{a}^*) w) \cdot w)_{H_\kappa} \geq m \, \| w \|_{H_\kappa}^2 . \tag{29.11}$$

However, it is easily seen that the left-hand side of (29.11) is obtained from the expression (16.6) for $\Phi(w, t)$ for $t = 1$ and if we take in that expression only terms of second order in w. If $\mathbf{a}^* = 0$, then

$$((w - \mathbf{G}_{\kappa\kappa 1}(0) w) \cdot w)_{H_\kappa} = \| \omega \|_{H_t}^2 + \| w \|_{H_\kappa}^2 \geq \frac{1}{2} \| w \|_{H_\kappa}^2 . \tag{29.12}$$

Obviously, the inequality (29.12) is preserved if the norm of $\| \mathbf{a}^* \|_{H_{t\kappa}}$ is sufficiently small, which will be ensured by the inequality (29.2). Thus, we have established invertibility of $\mathbf{I} - \mathbf{G}_{\kappa\kappa 1}(\mathbf{a}^*)$. By these arguments, using (14.2) the basic operator equation (13.36) can be written in the form

$$w = [\mathbf{I} - \mathbf{G}_{\kappa\kappa 1}(\mathbf{a}^*)]^{-1} (\mathbf{G}_{\kappa\kappa 0} + \mathbf{G}_{\kappa\kappa 2}(w) + \mathbf{G}_{\kappa\kappa 3}(w)). \tag{29.13}$$

From (29.13) it is seen that due to the structure of (14.19), (14.20), for small $\| \mathbf{a}^* \|_{H_{t\kappa}}$ the operator on the right-hand side of (29.13) is defined in a small sphere in H_κ with center at the origin and will be there a contraction operator. Therefore, for small $\| \mathbf{a}^* \|_{H_{t\kappa}}$ in a ball of H_κ of sufficiently small radius there cannot be more than one solution. On the other hand, existence of a solution was established during the first stage of the proof of Theorem 29.1. Theorem 29.1 is thus completely proved.

It is important to note that under the conditions of Theorem 29.1 uniqueness of solutions is guaranteed only in a neighborhood of zero, and so does not exclude solutions of large norm. These are easily found numerically in a number of cases.

In this respect, see [323, 322]. Such solutions can exist even for zero loads, which will be proved in Section 30.

29.3. Naturally, there arises the question of the conditions that guarantee unique solvability of Problems $t\kappa$ on the whole space (global uniqueness) if the loads are small. In this section we shall establish a global unique solvability theorem under some additional conditions on the physical and geometrical properties of the shell. For that we shall have to obtain a new a priori estimate for the solution. Again, we start with the functional (16.6). If w is a solution of the operator equation (13.36), then clearly, (16.6),

$$\begin{aligned}\Phi(w, 1) = {}& \|w\|^2_{H_\kappa} + (\overset{0}{w} \cdot w)_{H_\kappa} + 2\int_\Omega Q_s \, d\Omega \\ &+ \int_\Omega T^{ij}(\mathbf{a} + \mathbf{a}^*)\big[B_{ij}(w + 2w^*) - (w + w^*)_{\alpha^i} w^*_{\alpha^j} \\ &- 2\nabla_i w^*_j\big] \, d\Omega - 2(\omega_p \cdot \omega)_{H_l} + 2\Big(\int_{\Gamma_6} k_s^{\tau\tau}(w_\tau + w^*_\tau) w^*_\tau \, ds + \\ &+ \int_{\Gamma_7} k_s^{mm}(w_m + w^*_m) w^*_m \, ds + \int_{\Gamma_6} k_s^{ij}(w_i + w^*_i)\, w^*_j \, ds\Big) = 0.\end{aligned} \tag{29.14}$$

Let us assume, in addition, that we have the relations

$$k_s^{\tau\tau} = k_s^{mm} = k_s^{ij} = 0, \tag{29.15}$$

and then from (29.14) it follows that

$$\begin{aligned}\Phi(w, 1) = {}& \|w\|^2_{H_\kappa} + 2\int_\Omega Q_s \, d\Omega \\ &+ \int_\Omega T^{ij}\big[B_{ij}(w + 2w^*) - (w + w^*)_{\alpha^i} w^*_{\alpha^j} - 2\nabla_i w^*_j\big] \, d\Omega \\ &- 2(\omega_p \cdot \omega)_{H_l} + (\overset{0}{w} \cdot w)_{H_\kappa} = 0.\end{aligned} \tag{29.16}$$

Let us consider in detail the second and third terms of the left-hand side of (29.16). These form a second-degree polynomial in T^{ij}. The higher order terms of this polynomial, $2Q_p$, are a positive definite quadratic form with respect to T^{ij}. Therefore, the entire polynomial will have a minimum in the space of T^{ij}; it is determined by the condition

$$\frac{\partial}{\partial T^{ij}}\Big(2Q_s + T^{ij}(\mathbf{a} + \mathbf{a}^*)\big[B_{ij}(w + 2w^*) - (w + w^*)_{\alpha^i} w^*_{\alpha^j} - 2\nabla_i w^*_j\big]\Big) = 0,$$

whence

$$2\frac{\partial Q_s}{\partial T^{ij}} = (w + w^*)_{\alpha^i} w^*_{\alpha^j} - 2\nabla_i w^*_j - B_{ij}(w + 2w^*) = \overset{*}{\epsilon}_{ij}. \tag{29.17}$$

Equation (29.17) is a linear system of equations in T^{ij} having the form

$$\gamma_{11} = C_{11kl,s}T^{kl} = \frac{1}{2}\overset{*}{\epsilon}_{11}; \; \gamma_{12} = C_{12kl,s}T^{kl} = \frac{1}{2}\overset{*}{\epsilon}_{12} \; (1 \to 2 \to 1), \tag{29.18}$$

which can be written in the form

$$\gamma_{ij} = \frac{1}{2}\overset{**}{\epsilon}_{ij}, \quad \overset{**}{\epsilon}_{ii} = \overset{*}{\epsilon}_{ii}, \quad \overset{**}{\epsilon}_{12} = 2\overset{*}{\epsilon}_{12}.$$

Let us now introduce the notation

$$\mathcal{E} = \int_\Omega \left(2Q_s + T^{ij}(\mathbf{a} + \mathbf{a}^*)\overset{*}{\epsilon}_{ij}\right) d\Omega. \tag{29.19}$$

If we substitute (19.15) into (29.19), we obtain

$$\mathcal{E}_{\min} = -\frac{1}{8}\int_\Omega D_s^{ijkl}\overset{**}{\epsilon}_{ij}\overset{**}{\epsilon}_{kl}\, d\Omega. \tag{29.20}$$

Let us study the structure of $\overset{**}{\epsilon}_{ij}$ in more detail. From (29.17) we obtain

$$\overset{**}{\epsilon}_{ij} = \overset{**}{\epsilon}_{ij1} + \overset{**}{\epsilon}_{ij0}, \tag{29.21}$$

where

$$\overset{**}{\epsilon}_{ij1} = -B_{ij}w + w_{\alpha^i}w^*_{\alpha^j}; \; \overset{**}{\epsilon}_{ij0} = w^*_{\alpha^i}w^*_{\alpha^j} + 2\nabla_i w^*_j - 2B_{ij}w^*.$$

Substituting (29.20)–(29.21) into (29.19), we obtain

$$\mathcal{E}_{\min} = \overset{2}{\mathcal{E}}_{\min} + \overset{1}{\mathcal{E}}_{\min} + \overset{0}{\mathcal{E}}_{\min},$$

where $\overset{k}{\mathcal{E}}_{\min}$ is a homogeneous functional in w of order k. We have

$$\overset{2}{\mathcal{E}}_{\min} = -\frac{1}{8}\int_\Omega D_s^{ijkl}\overset{**}{\epsilon}_{ij1}\overset{**}{\epsilon}_{kl1}\, d\Omega, \tag{29.22}$$

$$\overset{1}{\mathcal{E}}_{\min} = -\frac{1}{4}\int_\Omega D_s^{ijkl}\overset{**}{\epsilon}_{ij1}\overset{**}{\epsilon}_{kl0}\, d\Omega, \tag{29.23}$$

$$\overset{0}{\mathcal{E}}_{\min} = -\frac{1}{8}\int_\Omega D_s^{ijkl}\overset{**}{\epsilon}_{ij0}\overset{**}{\epsilon}_{kl0}\, d\Omega. \tag{29.24}$$

Let us estimate (29.22)–(29.24). We have

$$\begin{aligned} |\overset{0}{\mathcal{E}}_{\min}| \le \frac{1}{8}\int_\Omega & \left|D_s^{ijkl}\right| \left|w^*_{\alpha^i}w^*_{\alpha^j} + 2\nabla_i w^*_j - 2B_{ij}w^*\right| \\ & \times \left|w^*_{\alpha^k}w^*_{\alpha^l} + 2\nabla_k w^*_l - 2B_{kl}w^*\right| d\Omega. \end{aligned} \tag{29.25}$$

Using Theorems 12.3 (relation (12.25)) and 11.4 (see (11.40), (11.41)), we obtain

$$|\overset{0}{\mathcal{E}}_{\min}| \le m\left(\|\omega^*\|^2_{H_l} + \|w^*\|_{H_\kappa} + \|w^*\|^2_{H_\kappa}\right).$$

Next we have

$$|\overset{1}{\mathcal{E}}_{\min}| \leq \frac{1}{4}\int_{\Omega} |D_{\mathrm{s}}^{ijkl}|\,|-B_{ij}w + w_{\alpha^i} w^*_{\alpha^j}| \times \left|w^*_{\alpha^k} w^*_{\alpha^l} + 2\nabla_k w^*_l - 2B_{kl} w^*\right| d\Omega.$$

By the same Theorems 12.3 and 11.4 we have

$$|\overset{1}{\mathcal{E}}_{\min}| \leq m\,\|w\|_{H_\kappa}\,(1+\|w\|_{H_\kappa})\left(\left\|\omega^*\right\|^2_{H_t} + \left\|w^*\right\|_{H_\kappa} + \left\|w^*\right\|^2_{H_\kappa}\right).$$

Substituting inequalities (29.25) into (29.15), we obtain

$$\begin{aligned}\|w\|^2_{H_\kappa} &+ \overset{2}{\mathcal{E}}_{\min} - |\overset{1}{\mathcal{E}}_{\min}| - |\overset{0}{\mathcal{E}}_{\min}| - 2(\omega_{\mathrm{p}}\cdot\omega_2) - \\ &- 2|(\omega_{\mathrm{p}}\cdot\omega_1)| - 2|(\omega_{\mathrm{p}}\cdot\omega_0)| - \left|\left(\overset{0}{w}\cdot w\right)_{H_\kappa}\right| \leq 0.\end{aligned} \tag{29.26}$$

Here we have taken into account (14.10). Moreover, we have

$$2|(\omega_{\mathrm{p}}\cdot\omega_1)| \leq m\left\|\omega_{\mathrm{p}}\right\|_{H_t}\|\omega_1\|_{H_t} \leq m\left\|\omega_{\mathrm{p}}\right\|_{H_t}\|w\|_{H_\kappa},$$

$$2|(\omega_{\mathrm{p}}\cdot\omega_0)_{H_t}| \leq m\left\|\omega_{\mathrm{p}}\right\|_{H_t}, \quad \left|\left(\overset{0}{w}\cdot w\right)_{H_\kappa}\right| \leq \left\|\overset{0}{w}\right\|_{H_\kappa}\|w\|_{H_\kappa}.$$

Then from (29.26) we obtain

$$\|w\|^2_{H_\kappa} + \overset{2}{\mathcal{E}}_{\min} - 2(\omega_{\mathrm{p}}\cdot\omega_2) \leq A\,\|w\|_{H_\kappa} + B, \tag{29.27}$$

where

$$\begin{aligned}&A \leq m\left[1 + \left\|w^*\right\|_{H_\kappa}\left(\left\|\omega^*\right\|_{H_t} + \left\|w^*\right\|_{H_\kappa} + \left\|w^*\right\|^2_{H_\kappa} + \left\|\omega_{\mathrm{p}}\right\|_{H_t} + \left\|\overset{0}{w}\right\|_{H_\kappa}\right)\right],\\ &B \leq m\left(\left\|\omega^*\right\|_{H_t} + \left\|w^*\right\|_{H_\kappa} + \left\|w^*\right\|^2_{H_\kappa} + \left\|\omega_{\mathrm{p}}\right\|_{H_t}\right).\end{aligned} \tag{29.28}$$

Let us introduce a parameter μ defined by the relation

$$\mu = \max\left(\overset{2}{\mathcal{E}}_{\min} - 2(\omega_{\mathrm{p}}\cdot\omega_2)_{H_t}\right) \text{ on the sphere } \|w\|^2_{H_\kappa} = 1;$$

μ exists by weak continuity of $\overset{2}{\mathcal{E}}_{\min} - 2(\omega_{\mathrm{p}}\cdot\omega_2)_{H_t}$. Then from (29.27) we obtain

$$\|w\|^2_{H_\kappa}(1-\mu) \leq A\,\|w\|_{H_\kappa} + B. \tag{29.29}$$

Assume now that the condition

$$\mu < 1 \tag{29.30}$$

is satisfied. Then from (29.29) we obtain

$$\|w\|_{H_\kappa} \leq \frac{A + \sqrt{A^2 + 4(1-\mu)B}}{2(1-\mu)}. \tag{29.31}$$

Using the above, we can state the following result.

Theorem 29.2. *Assume that the conditions of Theorem 29.1 and inequality* (29.30) *are satisfied. Then for ϵ sufficiently small in* (29.2), *Problem $t\kappa$ has a unique solution in the whole of H_κ.*

To prove this theorem, we note that if (29.30) holds, all the possible solutions are contained in a sphere of H_κ defined by (29.31). However, if (29.2) holds, both A and B will be small; this follows from (29.28). Therefore, we shall have the inequality

$$\|w\|_{H_\kappa} \leq \epsilon. \tag{29.32}$$

But for small ϵ the operator $\mathbf{G}_{\kappa\kappa}(w)$ is a contraction operator. Thus, Theorem 29.2 is proved.

Let us consider the particular case of $\omega_{\rm p} \equiv 0$. Then

$$\mu = \max \overset{2}{\mathcal{E}}_{\min} \text{ on the sphere } \|w\|^2_{H_\kappa} = 1.$$

At the same time,

$$\overset{2}{\mathcal{E}}_{\min} = -\frac{1}{8}\int_\Omega D_{\rm s}^{ijkl}\widetilde{B}_{ij}\widetilde{B}_{kl}w^2\,d\Omega, \tag{29.33}$$

where $\widetilde{B}_{ii} = B_{ii}$, $\widetilde{B}_{12} = 2B_{12}$. Equation (29.33) shows that the physical and geometrical properties of the shell largely depend on the invariant $D_{\rm s}^{ijkl}\widetilde{B}_{ij}\widetilde{B}_{kl}$. The smaller the curvature parameters B_{ij} are here, the more obvious is its stiffness. For more on this, see Section 30.

Thus by the theorem proved above, under condition (29.32) we have uniqueness in problem (29.32). If this condition is violated, as we already mentioned, in the general case, boundary value problems will not be uniquely solvable, even if the loads are arbitrarily small in the norm (29.1).

The result obtained in Theorem 29.2 can be interpreted in the sense that if (29.32) holds, we can identify a small neighborhood of zero in the Hilbert space H of loads that is contained in B_1.

29.4. In this section we shall study the boundary value problems 9κ. By Theorem 19.3, for a fixed tensor $\widetilde{T}_{\rm p}^{ij} \in L_{2\Omega}$ to each solution w, Ψ of the operator equation (17.20) there corresponds a unique complex $[\widetilde{T}_{\rm p}^{ij}, R^3, \widetilde{M}^m] \in \overline{H}_\kappa$ and boundary displacements $\widetilde{w}$, $\widetilde{w}_4$. At the same time, for a given admissible complex $[\widetilde{T}_{\rm p}^{ij}, R^3, \widetilde{M}^m, \widetilde{w}, \widetilde{w}_4]$ we have one or several generalized solutions of problem 9κ.

For the analysis below, we introduce the Hilbert space H of loads with the scalar product

$$\begin{aligned}(H_1 \cdot H_2)_H &= ([\overset{1}{\widetilde{T}}{}_{\rm p}^{ij}, \overset{1}{R}{}^3, \overset{1}{\widetilde{M}}{}^m, \overset{1}{\widetilde{w}}, \overset{1}{\widetilde{w}}_4] \cdot [\overset{2}{\widetilde{T}}{}_{\rm p}^{ij}, \overset{2}{R}{}^3, \overset{2}{\widetilde{M}}{}^m, \overset{2}{\widetilde{w}}, \overset{2}{\widetilde{w}}_4]) \\ &= (\overset{1}{w}_{\rm p} \cdot \overset{2}{w}_{\rm p})_{H_\kappa} + \int_\Omega \overset{1}{\widetilde{T}}{}_{\rm p}^{ij}\overset{1}{\widetilde{T}}_{{\rm p}ij}\,d\Omega + (\overset{0}{w}_1 \cdot \overset{0}{w}_2)_{H_\kappa},\end{aligned} \tag{29.34}$$

where $\overset{0}{w}_1$, $\overset{0}{w}_2$ are chosen in a particular way using $\widetilde{w}$, $\widetilde{w}_4$.

Clearly, for a full solution of the first problem we need to perform a U-decomposition of the space of loads H. The simplest properties of U-decompositions are provided by uniqueness theorems. We present two such theorems, similar to Theorems 29.1, 29.2. Since the ideas used in the proof are common to all these theorems, we confine ourselves to statements of the theorems.

Theorem 29.3. *Assume that all the conditions of Theorem 19.3 are satisfied, and in addition,* (29.2) *holds. Then for sufficiently small* ϵ, *Problem* 9κ *has a solution in a sphere* (29.3) *that is unique.*

Note that Theorem 29.3 does not preclude the existence of solutions of large norm.

To formulate uniqueness conditions in the whole space H_κ, let us introduce a number μ by the relation

$$\mu = \max \int_\Omega N_{9\kappa} w^2 \, d\Omega \text{ for } \|w\|_{H_\kappa} = 1; \qquad N_{9\kappa} = C_{sijkl} C^{iq} C^{jr} C^{ks} C^{lt} \widetilde{B}_{qr} \widetilde{B}_{st}. \tag{29.35}$$

Clearly, μ exists by weak continuity of the functional on the right-hand side of (29.35) in the space H_κ.

Theorem 29.4. *Assume that the conditions of Theorem 29.3 hold and that the inequality* (29.32) *holds. Then for sufficiently small* ϵ *in* (29.2), *Problem* 9κ *will have a unique solution in the whole space* H_κ *and thus in* $H_{9\kappa}$.

Thus unique solvability of problem 9κ in the entire space H_κ is proved here only under the additional condition (29.30).

Below we shall present an example in which the violation of (29.30) leads to nonunique solvability of Problem 9κ.

The result of Theorem 29.4, just like the result of Theorem 29.2, can be interpreted to mean that if (29.30) holds, we can identify a small neighborhood of zero in the Hilbert space H of loads that is contained in B_1. Here B_i is the domain in the space H of loads to which correspond exactly i solutions of the problem.

In a number of particular cases we can obtain results of global uniqueness. Let us note two facts that follow from Theorems 29.2, 29.4.

Theorem 29.5. *Under the conditions of Theorem 29.1 for sufficiently small* ϵ *we always have global uniqueness for plates.*

Theorem 29.6. *Under the conditions of Theorem 29.3 for sufficiently small* ϵ *we always have global uniqueness for plates.*

29.5. Of significant interest are uniqueness theorems in cases when smallness of loads is not required. Let us formulate a certain principle to do with this case.

Theorem 29.7. *Assume that the conditions of Theorem 16.5 (respectively, 16.8, 16.11) are satisfied. In addition assume that for a given load at each fixed point*

w_0 of the operator $\mathbf{G}_{\kappa\kappa}$ we have the relation

$$((w - \mathbf{G}_{\kappa\kappa 1}(w_0)w) \cdot w)_{H_\kappa} > 0, \quad \text{if } \|w\|_{H_\kappa} \neq 0. \tag{29.36}$$

Then Problem $t\kappa$ has a unique solution in the whole space H_κ and thus in $H_{t\kappa}$.

To prove Theorem 29.7, we shall show that if (29.36) holds, the index of any fixed point w_0 of the operator $\mathbf{G}_{\kappa\kappa}$ is $+1$. To show this, let us make the substitution $w \mapsto w_0 + w$ and obtain for w the equation

$$w = \mathbf{G}_{\kappa\kappa 1}(w_0)w + \mathbf{G}_{\kappa\kappa 2}(w_0, w) + \mathbf{G}_{\kappa\kappa 3}(w_0, w),$$

where the $\mathbf{G}_{\kappa\kappa\mu}(w_0, w)$ are homogeneous operators in w of order μ. Relation (29.36) shows that w_0 is a nonsingular solution, since otherwise the equation

$$w - \mathbf{G}_{\kappa\kappa 1}(w_0)w = 0$$

would have nontrivial solutions, for which the left-hand side of (29.36) would vanish. But then from Lemma 23.1 it follows that w_0 is an isolated solution. Let us now show that the vector field

$$w - \mathbf{G}_{\kappa\kappa 1}(w_0)w$$

on spheres with their center at the origin is homotopic to the identity vector field w, so that its winding number is $+1$. To establish this fact, we connect the fields w and $w - \mathbf{G}_{\kappa\kappa 1}(w_0)w$ by a deformation $w - t\mathbf{G}_{\kappa\kappa 1}(w_0)w$. It is easy to see that for no $t_0 \in [0, 1]$, $w \in H_\kappa$ could we have that

$$w - t_0\mathbf{G}_{\kappa\kappa 1}(w_0)w = 0,$$

since from this would follow that

$$t_0 = \frac{\|w\|^2_{H_\kappa}}{(\mathbf{G}_{\kappa\kappa 1}(w_0)w \cdot w)_{H_\kappa}},$$

which would contradict (29.36), since $t_0 \in [0, 1]$. Thus homotopy of the vector fields w and $w - \mathbf{G}_{\kappa\kappa 1}(w_0)w$ is established, so that the winding number of $w - \mathbf{G}_{\kappa\kappa 1}(w_0)w$ is $+1$. Since we have the inequalities (Lemmas 23.2, 23.3)

$$\left\|\mathbf{G}_{\kappa\kappa\mu}(w_0, w)\right\|_{H_\kappa} \leq m\, \|w\|^{\mu}_{H_\kappa}, \quad \mu = 2, 3,$$

the field $w - \mathbf{G}_{\kappa\kappa 1}(w_0)w - \sum_{\mu=2}^{3} \mathbf{G}_{\kappa\kappa\mu}(w_0, w)$ is homotopic to the identity field w, and therefore its winding number and the index of the fixed point w_0 is $+1$. But since the total index of all the fixed points of the field $w - \mathbf{G}_{\kappa\kappa}(w_0)w$ is also $+1$, the uniqueness theorem is proved.

A similar result can be obtained for the boundary value problem with a stress function. We quote it without proof.

Theorem 29.8. *Assume that in the conditions of Theorem 19.3 at each point w_0 we have the inequality*

$$((w - \mathbf{G}_{\kappa 1}(w_0)w) \cdot w) > 0, \quad \text{if } w \neq 0. \tag{29.37}$$

Then Problem 9κ has a unique solution in the whole space H_κ and thus in $H_{9\kappa}$.

Let us present some specific results.

Theorem 29.9 (Morozov [202, 205]). *Let an isotropic circular plate be subjected to the action of an axisymmetric normal load.*

Then under the boundary conditions

$$w|_\Gamma = \left.\frac{\partial w}{\partial n}\right|_\Gamma = \left.\widetilde{T}^\tau\right|_\Gamma = 0, \quad \left.\widetilde{T}^m\right|_\Gamma \geq 0, \tag{29.38}$$

$$w|_\Gamma = \left.\frac{\partial w}{\partial n}\right|_\Gamma = w_1|_\Gamma = w_2|_\Gamma = 0, \tag{29.39}$$

$$w|_\Gamma = \left.\widetilde{M}^m\right|_\Gamma = \left.\widetilde{T}^\tau\right|_\Gamma = 0, \quad \left.\widetilde{T}^m\right|_\Gamma \geq 0, \tag{29.40}$$

$$w|_\Gamma = \left.\widetilde{M}^m\right|_\Gamma = w_1|_\Gamma = w_2|_\Gamma = 0, \tag{29.41}$$

the boundary value problem of the equilibrium for the system (7.51), (7.60) *has a unique solution in the class of axisymmetric states for any load* $R^3 \in \overline{H}_\kappa$.

Morozov proved relation (29.36) for the boundary conditions (29.38), (29.40); relation (29.37) for the boundary conditions (29.39), (29.41); and then used the Hildebrandt–Graves theorem to complete the proof of Theorem 29.7. Once Theorems 16.4, 16.7, 16.10 showing that the winding number of the vector field $w - \mathbf{G}_{\kappa\kappa} w$ on spheres of large radius is $+1$ and a theorem like Theorem 19.2 for the field $w - \mathbf{G}_\kappa w$ are proved, Theorem 29.9 follows immediately from Theorems 29.7, 29.8. However, it has to be said that the proof of relations (29.36), (29.36), due first to Morozov, is of great interest. We should also note that Morozov [202, 205] requires $R^3 \in L_{2\Omega}$, while our analysis includes a much wider class of R^3 (discontinuities of δ-function type are permitted).

Subsequently, a simpler proof of uniqueness for axisymmetric problems was devised by Srubshchik and Yudovich [298].

Theorem 29.9 completely solves the problem of construction of B-sets in the cases under consideration; namely, the whole space $\overline{H}_\kappa$, to which R^3 belongs, is in B_1. Here, of course, we have axisymmetric solutions in mind. In the class of solutions not having axial symmetry, uniqueness need not hold: On this, see [49].

30. Physical Stiffness of Shells. Connection with Geometrical Stiffness of the Middle Surface

30.1. The results of Section 29 (Theorems 29.2, 29.4) lead us to introduce a class of shells that can naturally be called stiff.

Definition 30.1. We shall call a shell *physically stiff* (or simply *stiff*) if for sufficiently small loads (in relation (29.2) ϵ must be sufficiently small) it has a unique equilibrium state in the entire space H_κ.

Naturally, this equilibrium state will have a correspondingly small norm. On the other hand, if the shell is nonstiff, it will have solutions of finite nonsmall norm for small loads. More precisely, a shell is not stiff if for arbitrarily small ϵ (which can be zero) it has solutions for which

$$\|w\|_{H_\kappa} \geq m_0 > 0,$$

where m_0 does not depend on ϵ. Obviously, a nonstiff shell has for zero exterior loads a stressed nontrivial equilibrium state maintained only by internal elastic forces.

Physical stiffness is an important property of a shell. Obviously, in different practical cases of applications of shells, a shell might be required to be either stiff or nonstiff. Thus for example, in construction shells, stiffness is a necessary condition for function. Buckling safety valves, on the other hand, must be nonstiff. In each particular case the question of stiffness of a shell can be solved by numerically constructing the loading curves "load/deflection." Numerous examples of construction of such characteristics can be found in [323, 322].

Here we shall present some sufficient conditions for stiffness of shells, based on (29.30). First of all let us consider the case of problem $t\kappa$. Stiffness condition (29.30) assumes the form

$$\mu < 1, \text{ where } \mu = \max \frac{1}{8}\int_\Omega D_{\mathrm{s}}^{ijkl}\widetilde{B}_{ij}\widetilde{B}_{kl}\, w^2\, d\Omega, \ \|w\|^2_{H_\kappa} = 1. \tag{30.1}$$

Obviously, μ can be defined as being the maximal value of the parameter for which the eigenvalue problem

$$m(w \cdot \varphi)_{H_\kappa} = \frac{1}{8}\int_\Omega D_{\mathrm{s}}^{ijkl}\widetilde{B}_{ij}\widetilde{B}_{kl}\, w\varphi\, d\Omega, \ (\varphi \in H_\kappa \text{ arbitrary}) \tag{30.2}$$

has a nontrivial solution.

Relation (30.2) can be written in differential form:

$$\mu\nabla_{ij} D D_{\mathrm{f}}^{ijkl}\nabla_{kl} w = \frac{1}{8} D_{\mathrm{s}}^{ijkl} D \widetilde{B}_{ij}\widetilde{B}_{kl} w, \tag{30.3}$$

$$w|_{\Gamma_1+\Gamma_2} = 0, \tag{30.4}$$

$$\left.\frac{dw}{dm}\right|_{\Gamma_1+\Gamma_3} = 0.$$

$$(D_{\mathrm{f}}^{ijkl}\nabla_{kl} w m_i m_j + k_{\mathrm{f}}^{44} w_4)\,|_{\Gamma_2+\Gamma_4} = 0,$$

$$\{D^{-1}(D D_{\mathrm{f}}^{ijkl}\nabla_{kl} w)_s \tau_i m_j + D^{-1}(D D_{\mathrm{f}}^{ijkl}\nabla_{kl} w)_m m_i m_j$$
$$+ (D_{\mathrm{f}}^{ijkl}\nabla_{kl} w m_i \tau_j)_s + D_{\mathrm{f}}^{ijkl}\nabla_{kl} w G_{ij}^k m_k + k^{43} w\}\,|_{\Gamma_3+\Gamma_4} = 0.$$

If we start with the boundary value Problem 9κ, then μ is defined as the maximal value of the parameter for which the eigenvalue problem

$$\mu(w \cdot \varphi)_{H_\kappa} = \frac{1}{8}\int_\Omega C_{\mathrm{s}ijkl} C^{it} C^{jq} C^{kr} C^{ls} \widetilde{B}_{tq}\widetilde{B}_{rs} w\varphi\, d\Omega, \ \varphi \in H_\kappa, \tag{30.5}$$

has a nontrivial solution. In differential form the boundary value problem (30.5) assumes the form

$$\mu \nabla_{ij} D D_{\mathrm{f}}^{ijkl} \nabla_{kl} w = \frac{1}{8} C_{\mathrm{s}ijkl} C^{it} C^{jq} C^{kr} C^{ls} \widetilde{B}_{tq} \widetilde{B}_{rs} w D,$$

$$w|_{\Gamma} = 0, \quad \left.\frac{dw}{dm}\right|_{\Gamma_1} = 0, \quad (D_{\mathrm{f}}^{ijkl} \nabla_{kl} w m_i m_j + k_{\mathrm{f}}^{44} w_4) \, |_{\Gamma_2} = 0.$$

30.2. To clarify the concept of a stiff shell, let us consider the case of an isotropic homogeneous shell under the conditions of problem $t\kappa$ [343, 344]. For an isotropic homogeneous shell, relations (30.3) assume the form

$$\mu D_{\mathrm{f}1} \nabla^4 w = \frac{1}{8} D_{\mathrm{s}}^{ijkl} \widetilde{B}_{ij} \widetilde{B}_{kl} w, \quad D_{\mathrm{f}1} = \frac{2Eh^3}{3(1-\nu^2)}. \tag{30.6}$$

Here we are also assuming that the shell admits the introduction of Cartesian coordinates, that is, it is either developable or essentially shallow. In this case,

$$D_{\mathrm{s}1} = D_{\mathrm{s}}^{1111} = \frac{2Eh}{1-\nu^2} = D_{\mathrm{s}}^{2222}, \quad D_{\mathrm{s}}^{1112} = D_{\mathrm{s}}^{2212} = 0, \tag{30.7}$$

$$D_{\mathrm{s}}^{1122} = \frac{2Eh\nu}{1-\nu^2}, \quad 2D_{\mathrm{s}}^{1212} = \frac{2Eh}{1+\nu} = D_{\mathrm{s}2},$$

$$\widetilde{B}_{11} = \frac{1}{R_1}, \quad \widetilde{B}_{22} = \frac{1}{R_2}, \quad \widetilde{B}_{12} = \frac{1}{R_{12}}, \tag{30.8}$$

and also

$$\frac{1}{8} D_{\mathrm{s}}^{ijkl} \widetilde{B}_{ij} \widetilde{B}_{kl} = \frac{1}{8}\left(D_{\mathrm{s}}^{1111} \frac{1}{R_1^2} + D_{\mathrm{s}}^{2222} \frac{1}{R_2^2} + 2D_{\mathrm{s}}^{1122} \frac{1}{R_1 R_2} + 4D_{\mathrm{s}}^{1212} \frac{1}{R_{12}^2}\right)$$
$$= \frac{Eh}{4(1-\nu^2)}\left(\frac{1}{R_1^2} + \frac{1}{R_2^2} + 2\nu \frac{1}{R_1 R_2} + \frac{2}{R_{12}^2}(1-\nu)\right). \tag{30.9}$$

Substituting (30.7)–(30.9) into (30.6), we obtain

$$\mu \nabla^4 w = \frac{3}{4h^2}(2\mathcal{H}^2 - (1-\nu)K)w, \quad 2\mathcal{H} = \frac{1}{R_1} + \frac{1}{R_2}, \quad K = \frac{1}{R_1 R_2} - \frac{1}{R_{12}^2}. \tag{30.10}$$

From (30.10) it follows immediately that decreasing the thickness of the shell $2h$, all other properties of the shell being kept constant, leads to a violation of the stiffness condition and predisposes the shell to be nonstiff.

Let us consider a doubly curved hinged panel, of length a and width b (see Figure 30.1). Then for (30.10) we have the boundary conditions

$$w = w_{\alpha^1 \alpha^1} = 0 \text{ if } \alpha^1 = 0, \ a,$$

$$w = w_{\alpha^2 \alpha^2} = 0 \text{ if } \alpha^2 = 0, \ b,$$

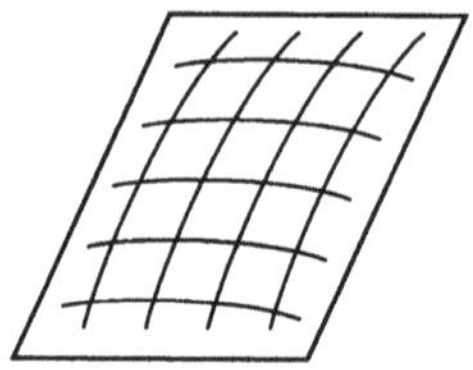

FIGURE 30.1.

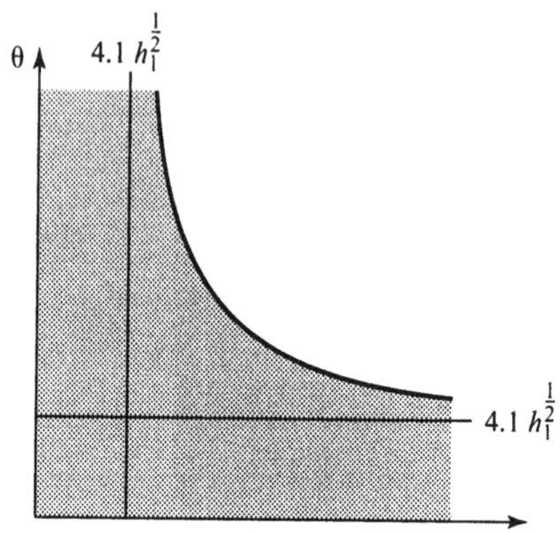

FIGURE 30.2.

and then as is easily seen, condition (29.32) assumes the form

$$\pi^4\left(\frac{1}{a^2}+\frac{1}{b^2}\right)^2 > \frac{3}{4h^2}(2\mathcal{H}^2-(1-\nu)K). \tag{30.11}$$

Let us consider in more detail the case of a cylindrical circular panel of length a, opening angle θ, and radius of curvature R. Then condition (30.11) becomes

$$\frac{1}{a_1^2}+\frac{1}{\theta^2} > \frac{0.0612}{h_1},\quad a_1=\frac{a}{R},\, h_1=\frac{h}{R},\, \theta=\frac{b}{R}. \tag{30.12}$$

In Figure 30.2 we have plotted the domain of stiffness of the shell (30.12). It is immediately seen that increasing either of the parameters a_1, θ will lead to a violation of the stiffness condition. However, for each of these parameters there is a critical value $a_{1\text{cr}}$, θ_{cr} such that if $a_1 < a_{1\text{cr}}$ or $\theta < \theta_{\text{cr}}$, increasing θ or a_1 to infinity already cannot make the shell nonstiff. From (30.12),

$$\theta_{\text{cr}} = a_{1\text{cr}} \approx 4.01 h_1^{\frac{1}{2}}.$$

Thus if $h_1 \sim 10^{-2}$, then θ_{cr} and $a_{1\text{cr}} \sim 0.4$. For a closed shell,

$$a_1 < 4.01 h_1^{\frac{1}{2}}. \tag{30.13}$$

From (30.13) it follows that for a sufficiently long closed shell, condition (30.13) is violated, which indicates that a closed cylindrical shell is nonstiff. This question requires rigorous study.

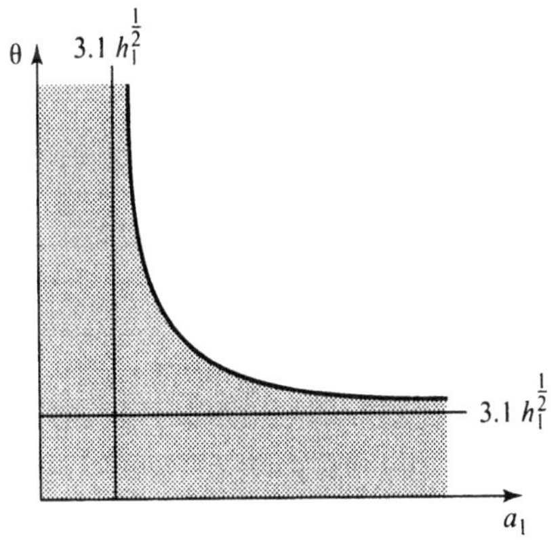

FIGURE 30.3.

Let us consider a shell of a hinged spherical shell of rectangular projection. Then (30.11) becomes

$$\pi^4\left(\frac{1}{a^2}+\frac{1}{b^2}\right)^2 > \frac{3}{4h^2R^2}(1+\nu),$$

or in dimensionless form ($\nu = \frac{1}{3}$),

$$\frac{1}{a_1^2}+\frac{1}{\theta^2} > \frac{0.0867}{h_1^{1/2}}. \tag{30.14}$$

Comparing (30.12) and (30.14), we conclude that the domain of stiffness of a spherical panel is larger than the domain of stiffness of a cylindrical one (see Figure 30.3).

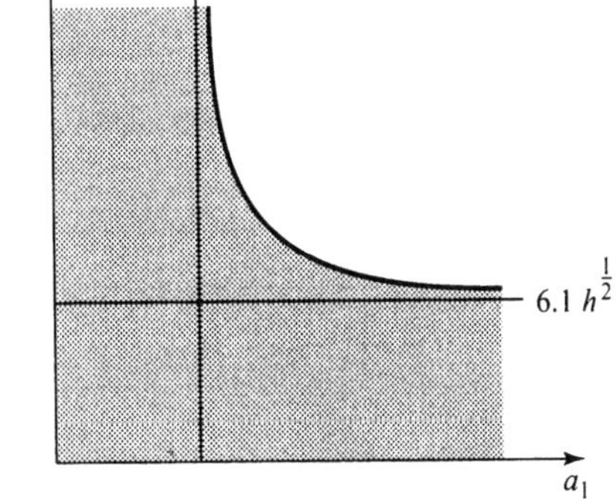

FIGURE 30.4.

Let us consider the case of a fixed shell of rectangular projection that corresponds to the conditions

$$w = w_{\alpha^1} = 0 \text{ on } \alpha^1 = 0 \text{ and } \alpha^1 = a; \; w = w_{\alpha^2} = 0 \text{ on } \alpha^2 = 0 \text{ and } \alpha^2 = b. \tag{30.15}$$

In this case the largest eigenvalue of the boundary value problem (30.10) can be found by the variational method, and condition (29.32) becomes

$$520\left(\frac{1}{a^4} + \frac{1}{b^4} + \frac{2}{3}\frac{1}{a^2b^2}\right) > \frac{3}{4h^2}(2\mathcal{H}^2 - (1-\nu)K). \tag{30.16}$$

Let us study first the case of a cylindrical circular shell. Condition (30.16) becomes $(\nu = \frac{1}{3})$

$$\frac{1}{a_1^4} + \frac{1}{b_1^4} + \frac{2}{3}\frac{1}{a_1^2b_1^2} > \frac{1}{1387h_1^2}. \tag{30.17}$$

Comparing (30.12) and (30.17), where we should put $b_1 = \theta$, it is easy to notice that with the boundary conditions (30.15) the domain of stiffness of the shell is significantly increased. For a closed fixed shell the stiffness condition has the form

$$a_1 < 6.103h_1^{\frac{1}{2}}.$$

In the case of a fixed shell we also have critical values $a_{1\text{cr}}$, $b_{1\text{cr}}$,

$$a_{1\text{cr}},\; b_{1\text{cr}} \approx 6.103h_1^{\frac{1}{2}}.$$

If either a_1 or b_1 is less than the critical value, we see that further increase of the other parameter cannot lead to nonstiffness of the shell. The stiffness condition (30.17) is illustrated in Figure 30.4.

Let us consider a fixed spherical shell of rectangular projection onto the plane. From (30.10) the stiffness condition can be shown to have the following form:

$$\frac{1}{a_1^4} + \frac{1}{b_1^4} + \frac{2}{3}\frac{1}{a_1^2b_1^2} > \frac{1+\nu}{693.3h_1^2}. \tag{30.18}$$

Comparing (30.16) and (30.18), we see an increase in the stiffness domain when we pass from a cylindrical to a spherical panel.

Let us consider now a spherical shell with a circular support of radius a. First we assume that it is fixed (Figure 30.5). Here the stiffness condition has the form

$$\frac{a^2}{hR} < 11.9\sqrt{1+\nu}, \quad a_1 < 3.45h_1^{1/2}\sqrt{1+\nu}, \tag{30.19}$$

FIGURE 30.5.

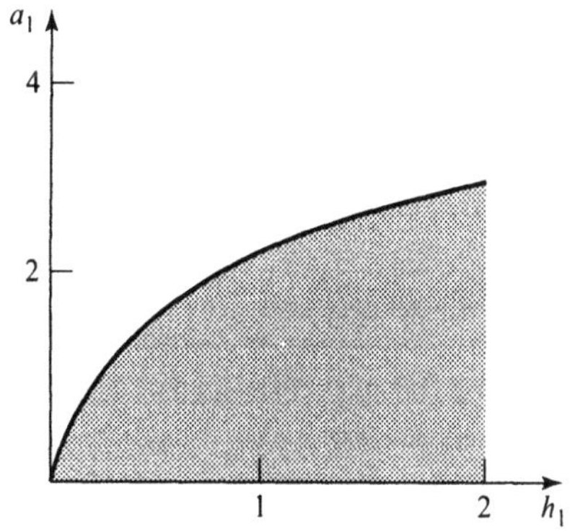

FIGURE 30.6.

(see Figure 30.6). Since $a^2/(2R) = H$, where H is the height of the dome, the first of the relations (30.19) assumes the form ($\nu = \frac{1}{3}$)

$$\frac{H}{h} < 5.95.$$

Thus, stiff shells are the ones with small elevation. In the case of a hinged spherical shell with a circular support, we have a stiffness condition in the form

$$\frac{a^2}{hR} < 5.02,$$

so that

$$\frac{H}{h} < 2.01. \tag{30.20}$$

30.3. Let us consider a cylindrical circular shell, rigidly connected by two elastic circular bottoms. The stiffness condition, approximate in this case, since the corresponding eigenvalues were computed by the variational method, has the form

$$\frac{a_1^2 + 2.05a_1 + 0.84}{0.254a_1^2 + 1.106a_1 + 0.337} > 0.234\frac{a_1^4}{h_1^2}, \quad a_1 = \frac{a}{R}, \quad h_1 = \frac{h}{R}.$$

30.4. From (30.1), (30.2), (30.5) we can state certain general assertions concerning physical stiffness of shells.

Theorem 30.1. *Assume that Conditions 2–6 of Section 13 are satisfied. Then a plate ($B_{ij} \equiv 0$) is a physically stiff shell for all the Problems $t\kappa$.*

The proof follows immediately from the definition of μ in (30.1), (30.2), which is 0 in this case, so that relation (30.1) is automatically satisfied.

Theorem 30.2. *Assume that Conditions 2–5 of Section 17 are satisfied. Then a plate ($B_{ij} \equiv 0$) is a physically stiff shell for all Problems 9κ.*

The proof follows immediately from the definition of μ in (30.5), which is 0 in this case, so that relation (30.5) is again automatically satisfied.

Theorem 30.3. *Assume that the conditions of Theorems 30.1, 30.2 are satisfied. Then if the shell is stiff and either condition* (30.2) *or* (30.9) *holds, then imposition of additional boundary conditions on w will not cause it to become nonstiff.*

The proof follows from the Courant principle [60].

30.5. A well-known important geometrical property of surfaces is the uniqueness of their form under certain conditions; see, for example, [240]. This can be regarded as a physical characteristic of the surface as well. The question of the uniqueness of the form of a surface under given pinning conditions can be interpreted as the problem on the number of equilibrium states of a two-dimensional continuum having zero bending stiffness and infinite extension stiffness. For a uniquely defined surface, the number of equilibrium states will be one. We could generalize the problem by assuming that the continuum has zero bending stiffness and a finite extension stiffness, and that the energy stored in the continuum is proportional to the increase in its area. Thus we arrive at the problem of Plateau [225]. In this case, instead of the question of uniqueness of solutions we are interested in the number of solutions of the Plateau problem. The property of stiffness considered here can thus be interpreted as an extension of the geometrical concept of uniqueness for a surface.

Since analysis of uniqueness of surfaces encounters severe mathematical difficulties, a purely geometrical concept of stiffness has been introduced (see, for example, [325, 73, 241]).

Namely, we say that a surface is stiff if linearized bending equations admit only trivial solutions, which correspond to small motions of the surface considered as a solid body.

It is natural to compare the physical and geometrical concepts of stiffness. Infinitely small bending is described by second-order partial differential equations. Therefore, one boundary condition is usually imposed on the edge. Thus, in the well-known theorem of Liebman [325, 73, 241] it is assumed that edges of a sector of a sphere slide on a plane, that is, it is assumed that displacements of points of the bounding circle are in the direction of the meridian. On the other hand, the concept of physical stiffness introduced above is dependent on four geometrical factors at the edge: three displacements and the angle of rotation. As far as results are concerned, a comparison with Liebman's theorem needs to be considered. By this theorem, any spherical segment smaller than a half-sphere is stiff. As we found in this section, a segment will be physically stiff if its elevation is sufficiently small. The greatest elevation of a segment for which it will be stiff depends strongly on the boundary conditions. It also depends on the elastic properties of the shell via the constant ν.

Geometrical considerations in the problem of stability of thin shells have been developed and widely used by Pogorelov [242]. Analytical methods for such a passage to the limit were developed by Yudovich and Srubshchik [297].

30.6. In Section 30.2 we obtained explicit conditions that guarantee stiffness of a shell. Thus, for example, relation (30.20) tells us that stiffness of a hinged spherical dome is ensured if its height H is small enough. A natural question that arises here concerns the consequences of violating condition (30.20). To clarify it, let us consider an isotropic homogeneous elliptical shell, the middle surface of which, S, is given by

$$\rho = \alpha^1 \mathbf{i}_1 + \alpha^2 \mathbf{i}_2 + (a - b(\alpha^1)^2 - c(\alpha^2)^2)\mathbf{i}_3, \quad a,\ b,\ c > 0.$$

Furthermore, let the shell be hinged on the boundary Γ, which is defined by the equation

$$(b(\alpha^1)^2 + c(\alpha^2)^2) = a;$$

besides, there are no tangential stresses.

Theorem 30.4. *Under the above conditions for sufficiently small thickness h, under no external loads, the shell will have in addition to the trivial state* $w \equiv 0$ *an equilibrium state* $w(\alpha^1, \alpha^2)$*, which is close to the symmetrically reflected one,*

$$\rho = \alpha^1 \mathbf{i}_1 + \alpha^2 \mathbf{i}_2 - (a - b(\alpha^1)^2 - c(\alpha^2)^2)\mathbf{i}_3,$$

and

$$\|w\|_{H_1} \leq m \cdot h.$$

In other words, the shell is nonstiff.

This theorem was proved by Srubshchik [292]. See also [291]. The proof of the theorem is based on the so-called asymptotic method of integrating nonlinear equations of the theory of shallow shells [297]. Using this method, one can construct the approximate solution, based on the symmetrically reflected state of the shell, and then one uses the well-known Kantorovich theorem [130].

31. Well-Posedness of Problems of the Nonlinear Theory of Shallow Shells: Its Relation to Physical Stability

31.1. To formulate the problem of well-posedness, we introduce a class of shells O such that each shell o of class O is parametrized by α^1, α^2 in a domain Ω, which is the same for all the shells in the class O, and furthermore, the decomposition of the boundary $\Gamma = \Gamma_1 + \Gamma_2 + \Gamma_3 + \Gamma_4 = \Gamma_5 + \Gamma_6 + \Gamma_7 + \Gamma_8$ is the same for all $o \in O$. Each shell $o \in O$ has a set of characteristics that completely determine it. For problems $t\kappa$ these are, in the first place, the loading complexes H introduced in Section 29, which form a Hilbert space with a scalar product (29.1). To characterize a shell completely, we need to add to this complex the middle surface S, the elastic and geometrical characteristics D_{s}^{ijkl}, D_{f}^{ijkl}, as well as the quantities that characterize the elastic supports, $k_{\mathrm{s}}^{\tau\tau}$, k_{s}^{mm}, k_{s}^{ij}, k_{f}^{33}, k_{f}^{44}, k_{f}.

As before, we are assuming that $S \in C_{\Omega}^2$, D_{s}^{ijkl}, D_{f}^{ijkl} are piecewise continuous in the respective closed domains of their definition. Let us now introduce on O the metric

$$\rho(o_1, o_2) = \|\Delta\rho\|_{C_\Omega^{(2)}} + \sum_{i,j,k,l=1}^{2} \left(\left\|\Delta D_{\mathrm{s}}^{ijkl}\right\|_{C_\Omega} + \left\|\Delta D_{\mathrm{f}}^{ijkl}\right\|_{C_\Omega} \right)$$

$$+ \left\|\Delta k_{\mathrm{s}}^{\tau\tau}\right\|_{C_{\Gamma_6}} + \left\|\Delta k_{\mathrm{s}}^{mm}\right\|_{C_{\Gamma_7}} + \sum_{i,j=1}^{2} \left\|\Delta k_{\mathrm{s}}^{ij}\right\|_{C_{\Gamma_8}} + \left\|\Delta k_{\mathrm{f}}^{44}\right\|_{C_{\Gamma_2}}$$

$$+ \left\|\Delta k_{\mathrm{f}}^{33}\right\|_{C_{\Gamma_3}} + \sum_{i,j=1}^{2} \left\|\Delta k_{\mathrm{f}}^{ij}\right\|_{C_{\Gamma_4}} + \|\Delta H\|_H . \tag{31.1}$$

For the definition of $\|\Delta H\|_H$ see (29.34), while

$$\Delta\rho(\alpha^1, \alpha^2) = \overset{2}{\rho}(\alpha^1, \alpha^2) - \overset{2}{\rho}(\alpha^1, \alpha^2),$$

$$\Delta D_{\mathrm{s,f}}^{ijkl} = \overset{2}{D}{}_{\mathrm{s,f}}^{ijkl} - \overset{1}{D}{}_{\mathrm{s,f}}^{ijkl}, \quad \Delta k_{\mathrm{s}}^{\tau\tau} = \overset{2}{k}{}_{\mathrm{s}}^{\tau\tau} - \overset{1}{k}{}_{\mathrm{s}}^{\tau\tau},$$

$$\Delta k_{\mathrm{s}}^{mm} = \overset{2}{k}{}_{\mathrm{s}}^{mm} - \overset{1}{k}{}_{\mathrm{s}}^{mm}, \quad \Delta k_{\mathrm{s}}^{ij} = \overset{2}{k}{}_{\mathrm{s}}^{ij} - \overset{1}{k}{}_{\mathrm{s}}^{ij}, \; i, \; j = 1, 2,$$

$$\Delta k_{\mathrm{f}}^{33} = \overset{2}{k}{}_{\mathrm{f}}^{33} - \overset{1}{k}{}_{\mathrm{f}}^{33}, \quad \Delta k_{\mathrm{f}}^{44} = \overset{2}{k}{}_{\mathrm{f}}^{44} - \overset{1}{k}{}_{\mathrm{f}}^{44},$$

$$\Delta k_{\mathrm{f}}^{ij} = \overset{2}{k}{}_{\mathrm{f}}^{ij} - \overset{1}{k}{}_{\mathrm{f}}^{ij}, \; i, \; j = 3, 4.$$

The metric (31.1) formally turns O into a metric space. As can be seen from conditions 1–8, we proved solvability theorems for problems $t\kappa$ under the assumption that the shell is in O. Therefore, it is natural to consider the well-posedness problem in the class O of shells.

31.2.

Lemma 31.1. *Let the shells o_1, o_2 be such that*

$$\|\Delta\rho\|_{C_\Omega^2} \le \epsilon. \tag{31.2}$$

Then for sufficiently small ϵ we have

$$\left\|\Delta A_{ij}\right\|_{C_\Omega^1}, \; \left\|\Delta A^{ij}\right\|_{C_\Omega^1}, \; \left\|\Delta B_{ij}\right\|_{C_\Omega}, \; \left\|\Delta B^{ij}\right\|_{C_\Omega}, \; \left\|\Delta G_{ij}^k\right\|_{C_\Omega} \le m\epsilon,$$

$$\overset{1}{D}(1 - m\epsilon) \le \overset{2}{D} \le \overset{1}{D}(1 + m\epsilon). \tag{31.3}$$

Here m is fixed for a given surface S_1, $\rho = \overset{1}{\rho}(\alpha^1, \alpha^2)$.

Lemma 31.1 is easily obtained from the corresponding relations (1.2), (1.3), (1.5), (1.21), which define the characteristics of (31.3).

Next, suppose that $\overset{1}{H}_t$, $\overset{1}{H}_\kappa$, $\overset{1}{H}_{t\kappa}$ are the spaces constructed for o_1 and that $\overset{2}{H}_t$, $\overset{2}{H}_\kappa$, $\overset{2}{H}_{t\kappa}$ are the spaces constructed for o_2. It is easily seen that the spaces $\overset{1}{H}_t$ and $\overset{2}{H}_t$ are equivalent, since each of them by Theorem 11.3 is equivalent to $\overset{0}{W}{}^{(1)}_{2t\Omega}$.

Therefore, every vector function $\omega \in \overset{1}{H}_t$ can be regarded as an element of $\overset{2}{H}_t$ and vice versa. Similarly, the spaces $\overset{1}{H}_\kappa$ and $\overset{2}{H}_\kappa$ are equivalent, since each of them by Theorem 12.2 is equivalent to $\overset{0}{W}{}^{(2)}_{2\kappa\Omega}$. Therefore, every function $w \in \overset{1}{H}_\kappa$ can be regarded as an element of $\overset{2}{H}_\kappa$ and vice versa. By the above, every element $\mathbf{a} = (\omega, w) \in \overset{1}{H}_{t\kappa}$ belongs to $\overset{2}{H}_{t\kappa}$ and vice versa.

Lemma 31.2. *Assume that the lateral supports of* o_1 *are essentially elastic (Section 11), that condition* (31.2) *holds, and that in addition,*

$$\|\Delta D_{\mathrm{s}}^{ijkl}\|_{C\Omega}, \ \|\Delta k_{\mathrm{s}}^{\tau\tau}\|_{\Gamma_6}, \ \|\Delta k_{\mathrm{s}}^{mm}\|_{\Gamma_7}, \ \|\Delta k_{\mathrm{s}}^{ij}\|_{\Gamma_8} \le \epsilon. \tag{31.4}$$

Then for sufficiently small ϵ *we have the relations*

$$(\omega\cdot\chi)_{\overset{1}{H}_t} - m\epsilon\,\|\omega\|_{\overset{1}{H}_t}\,\|\chi\|_{\overset{1}{H}_t} \le (\omega\cdot\chi)_{\overset{2}{H}_t} \le (\omega\cdot\chi)_{\overset{1}{H}_t} + m\epsilon\,\|\omega\|_{\overset{1}{H}_t}\,\|\chi\|_{\overset{1}{H}_t}. \tag{31.5}$$

To prove the lemma, we note that if the supports are essentially elastic for o_1, they remain so for o_2 if ϵ is sufficiently small. Next, from relation (11.4), which defines the scalar product in H_t, we have

$$(\omega\cdot\chi)_{\overset{2}{H}_t} = (\omega\cdot\chi)_{\overset{1}{H}_t} + \int_\Omega \Delta D_{\mathrm{s}}^{ijkl}\overset{\tilde{0}}{\gamma}_{ij}(\omega)\overset{\tilde{0}}{\gamma}_{kl}(\chi)\,d\Omega$$
$$+ \int_{\Gamma_6}\Delta k_{\mathrm{s}}^{\tau\tau}w_\tau\varphi_\tau\,ds + \int_{\Gamma_7}\Delta k_{\mathrm{s}}^{mm}w_m\varphi_m\,ds + \int_{\Gamma_8}\Delta k_{\mathrm{s}}^{ij}\varphi_i w_j\,ds.$$

Next, since $\overset{\tilde{0}}{\gamma}_{ij} = \nabla_i w_j$, taking into account Theorem 11.4 (relations (11.40), (11.41)), in which $d \sim \Gamma_6$, Γ_7, Γ_8, and (30.4), we have

$$\Big|\int_\Omega \Delta D_{\mathrm{s}}^{ijkl}\overset{\tilde{0}}{\gamma}_{ij}(\omega)\overset{\tilde{0}}{\gamma}_{kl}(\chi)\,d\Omega + \int_{\Gamma_6}\Delta k_{\mathrm{s}}^{\tau\tau}w_\tau\varphi_\tau\,ds$$
$$+ \int_{\Gamma_7}\Delta k_{\mathrm{s}}^{mm}w_m\varphi_m\,ds + \int_{\Gamma_8}\Delta k_{\mathrm{s}}^{ij}w_i\varphi_j\,ds\Big|$$
$$\le \epsilon\Big(\sum_{i,j,k,l=1}^{2}\Big\|\overset{\tilde{0}}{\gamma}_{ij}(\omega)\Big\|_{L_{2\Omega}}\Big\|\overset{\tilde{0}}{\gamma}_{kl}(\chi)\Big\|_{L_{2\Omega}} + \|w_\tau\|_{L_{2\Gamma_6}}\|\varphi_\tau\|_{L_{2\Gamma_6}} \tag{31.6}$$
$$+ \|w_m\|_{L_{2\Gamma_7}}\|\varphi_m\|_{L_{2\Gamma_7}} + \sum_{i,j=1}^{2}\|w_i\|_{L_{2\Gamma_8}}\|\varphi_j\|_{L_{2\Gamma_8}}\Big)$$
$$\le m\epsilon\,\|\omega\|_{H_t}\,\|\chi\|_{H_t}.$$

From (31.6) for small ϵ we obtain (31.5). We remind the reader that by m we decided to denote various constants if it is only the fact of their existence that is of importance, and not their particular values. Let us also note that in (31.5) m can be taken to depend only on the parameters of o_1.

Lemma 31.3. *Assume that the transverse supports of o_1 are essentially elastic, that conditions* (31.2) *hold, and that in addition,*

$$\left\|\Delta D_{\mathrm{f}}^{ijkl}\right\|_{C_\Omega}, \ \left\|\Delta k_{\mathrm{f}}^{33}\right\|_{C_\Omega}, \ \left\|\Delta k_{\mathrm{f}}^{44}\right\|_{C_\Omega}, \ \left\|\Delta k_{\mathrm{f}}^{ij}\right\|_{C_\Omega} \le \epsilon. \tag{31.7}$$

Then for sufficiently small ϵ we have the relations

$$(w\cdot\varphi)_{\overset{1}{H}_\kappa} - m\epsilon\,\|w\|_{\overset{1}{H}_\kappa}\|\varphi\|_{\overset{1}{H}_\kappa} \le (w\cdot\varphi)_{\overset{2}{H}_\kappa} \le (w\cdot\varphi)_{\overset{1}{H}_\kappa} + m\epsilon\,\|w\|_{\overset{1}{H}_\kappa}\|\varphi\|_{\overset{1}{H}_\kappa}. \tag{31.8}$$

The proof of Lemma 31.3 uses the same arguments as that of Lemma 31.2. However, instead of Theorem 11.4 we apply Theorem 12.3 (relations (12.25), (12.27)). Here as well we can take m to depend only on the parameters of o_1.

Lemma 31.4. *Assume that the transverse and lateral supports of o_1 are essentially elastic and that conditions* (31.3), (31.5), *and* (31.7) *hold. Then*

$$\begin{aligned}(\mathbf{a}\cdot\mathbf{b})_{\overset{1}{H}_{\iota\kappa}} - m\epsilon\,\|\mathbf{a}\|_{\overset{1}{H}_{\iota\kappa}}\|\mathbf{b}\|_{\overset{1}{H}_{\iota\kappa}} &\le (\mathbf{a}\cdot\mathbf{b})_{\overset{2}{H}_{\iota\kappa}} \\ &\le (\mathbf{a}\cdot\mathbf{b})_{\overset{1}{H}_{\iota\kappa}} + m\epsilon\,\|\mathbf{a}\|_{\overset{1}{H}_{\iota\kappa}}\|\mathbf{b}\|_{\overset{1}{H}_{\iota\kappa}}.\end{aligned} \tag{31.9}$$

Lemma 31.4 is a direct corollary of Lemmas 31.2, 31.3.

Let us introduce the operator $\mathbf{A}_\tau$, defined by the relation

$$(\mathbf{A}_\tau\omega\cdot\chi)_{\overset{1}{H}_\iota} = (\omega\cdot\chi)_{\overset{2}{H}_\iota}, \tag{31.10}$$

where $\omega \in \overset{2}{H}_\iota$ is fixed and $\chi \in \overset{2}{H}_\iota$ is arbitrary.

Lemma 31.5. *The operator $\mathbf{A}_\tau$ exists, satisfies the inequalities*

$$1 - m\epsilon \le \|\mathbf{A}_\tau\| \le 1 + m\epsilon, \tag{31.11}$$

and has an inverse $\mathbf{A}_\tau^{-1}$ that satisfies the relations

$$1 - m\epsilon \le \left\|\mathbf{A}_\tau^{-1}\right\| \le 1 + m\epsilon; \tag{31.12}$$

m depends on o_1, and ϵ is the same as in (31.4), (31.5).

To prove the lemma, we quote the relation [310]

$$\|\mathbf{A}_\tau\| = \sup \frac{\left|(\mathbf{A}_\tau\omega\cdot\chi)_{\overset{1}{H}_\iota}\right|}{\|\omega\|_{\overset{1}{H}_\iota}\|\chi\|_{\overset{1}{H}_\iota}} = \frac{\left|(\omega\cdot\chi)_{\overset{2}{H}_\iota}\right|}{\|\omega\|_{\overset{1}{H}_\iota}\|\chi\|_{\overset{1}{H}_\iota}}. \tag{31.13}$$

The relation (31.11) is then obtained from (31.5), (31.13).

Next, the operator $\mathbf{A}_\tau^{-1}$ is defined by the relation

$$(\omega\cdot\chi)_{\overset{1}{H}_\iota} = (\mathbf{A}_\tau^{-1}\omega\cdot\chi)_{\overset{2}{H}_\iota}. \tag{31.14}$$

Next, using the complete symmetry of the spaces $\overset{1}{H}_t$ and $\overset{2}{H}_t$, we note that in addition to (31.5) we have the relation

$$(\omega \cdot \chi)_{\overset{2}{H}_t} - m\epsilon \|\omega\|_{\overset{2}{H}_t} \|\chi\|_{\overset{2}{H}_t} \le (\omega \cdot \chi)_{\overset{1}{H}_t} \le (\omega \cdot \chi)_{\overset{2}{H}_t} + m\epsilon \|\omega\|_{\overset{2}{H}_t} \|\chi\|_{\overset{2}{H}_t}, \tag{31.15}$$

and (31.12) follows from (31.14), (31.15).

Let us now introduce the operator $\mathbf{B}_\tau$ defined by

$$(\mathbf{B}_\tau w \cdot \varphi)_{\overset{1}{H}_\kappa} = (w \cdot \varphi)_{\overset{2}{H}_\kappa}.$$

Lemma 31.6. *The operator* $\mathbf{B}_\tau$ *exists and satisfies the inequalities*

$$1 - m\epsilon \le \|\mathbf{B}_\tau\| \le 1 + m\epsilon$$

and has an inverse that satisfies the relations

$$1 - m\epsilon \le \left\|\mathbf{B}_\tau^{-1}\right\| \le 1 + m\epsilon.$$

The proof of Lemma 31.6 uses the same arguments as that of Lemma 31.5, and we omit the details.

Lemma 31.7. *From the relation*

$$\left\|\overset{1}{\omega} - \mathbf{A}_\tau \overset{2}{\omega}\right\|_{\overset{1}{H}_t} \le \epsilon \tag{31.16}$$

follows the inequality

$$\left\|\overset{1}{\omega} - \overset{2}{\omega}\right\|_{\overset{1}{H}_t} \le m\epsilon \left(\left\|\overset{1}{\omega}\right\|_{\overset{1}{H}_t} + 1\right); \tag{31.17}$$

conversely, from

$$\left\|\overset{1}{\omega} - \overset{2}{\omega}\right\|_{\overset{1}{H}_t} \le \epsilon \tag{31.18}$$

follows the inequality

$$\left\|\overset{1}{\omega} - \mathbf{A}_\tau \overset{2}{\omega}\right\|_{\overset{1}{H}_t} \le m\epsilon \left(\left\|\overset{1}{\omega}\right\|_{\overset{1}{H}_t} + 1\right). \tag{31.19}$$

In Lemmas 31.4–31.7, ϵ is assumed to be sufficiently small.

To prove (31.17) let us consider the identity

$$\overset{1}{\omega} - \overset{2}{\omega} = \mathbf{A}_\tau^{-1} \left(\overset{1}{\omega} - \mathbf{A}_\tau \overset{2}{\omega}\right) - (\mathbf{A}_\tau^{-1} - \mathbf{I})\overset{1}{\omega},$$

from which we have

$$\left\|\overset{1}{\omega} - \overset{2}{\omega}\right\|_{\overset{1}{H}_t} \le \|\mathbf{A}_\tau^{-1}\| \left\|\left(\overset{1}{\omega} - \mathbf{A}_\tau \overset{2}{\omega}\right)\right\|_{\overset{1}{H}_t} + \|\mathbf{A}_\tau^{-1} - \mathbf{I}\| \left\|\overset{1}{\omega}\right\|_{\overset{1}{H}_t}. \tag{31.20}$$

By (31.12) we have

$$\left\|\mathbf{A}_\tau^{-1} - \mathbf{I}\right\| \le m\epsilon, \tag{31.21}$$

and from (31.20), (31.21), (31.12) we obtain (31.17). To derive (31.19) from (31.18) we use the identity

$$\overset{1}{\omega} - \mathbf{A}_\tau\overset{2}{\omega} = \mathbf{A}_\tau\left(\overset{1}{\omega} - \overset{2}{\omega}\right) + (\mathbf{I} - \mathbf{A}_\tau)\overset{1}{\omega},$$

whence

$$\left\|\overset{1}{\omega} - \mathbf{A}_\tau\overset{2}{\omega}\right\|_{\overset{1}{H}_\iota} \le \|\mathbf{A}_\tau\| \left\|\overset{1}{\omega} - \overset{2}{\omega}\right\|_{\overset{1}{H}_\iota} + \|\mathbf{I} - \mathbf{A}_\tau\| \left\|\overset{1}{\omega}\right\|_{\overset{1}{H}_\iota}. \tag{31.22}$$

By (31.11) we have

$$\|\mathbf{A}_\tau - \mathbf{I}\| \le m\epsilon. \tag{31.23}$$

Using (31.22), (31.23), and (31.11), we obtain (31.19). Lemma 31.7 is proved.

Lemma 31.8. *From the relation*

$$\left\|\overset{1}{w} - \mathbf{B}_\tau\overset{2}{w}\right\|_{\overset{1}{H}_\kappa} \le \epsilon$$

follows the inequality

$$\left\|\overset{1}{w} - \overset{2}{w}\right\|_{\overset{1}{H}_\kappa} \le m\epsilon\left(\left\|\overset{1}{w}\right\|_{\overset{1}{H}_\kappa} + 1\right); \tag{31.24}$$

conversely, from

$$\left\|\overset{1}{w} - \overset{2}{w}\right\|_{\overset{1}{H}_\kappa} \le \epsilon$$

follows

$$\left\|\overset{1}{w} - \mathbf{B}_\tau\overset{2}{w}\right\|_{\overset{1}{H}_\kappa} \le m\epsilon\left(\left\|\overset{1}{w}\right\|_{\overset{1}{H}_\kappa} + 1\right). \tag{31.25}$$

We do not prove Lemma 31.8, since the proof is identical to the proof of Lemma 31.7. We only note that the constant m in (31.11), (31.12), (31.17), (31.19), (31.24), (31.25) can be taken to depend only on the parameters of the shell o_1.

31.3. Assume that we have two shells, $o_1, o_2 \in S$, and

$$\|o_1 - o_2\|_O \le \epsilon,$$

which is equivalent to (31.1). Let us consider for each of the shells operator equations (13.32), (13.33), which define the operator $\mathbf{K}_{t\kappa}(w)$. We have

$$(\overset{1}{\mathbf{K}}_{t\kappa} \cdot \chi)_{\overset{1}{H}_\iota} = -(\overset{1}{\omega}{}^* \cdot \chi)_{\overset{1}{H}_\iota} + \int_\Omega \left[\overset{1}{B}_{kl}(w + \overset{1}{w}{}^*)\right.$$

$$-\frac{1}{2}(w+\overset{1}{w}{}^*)_{\alpha^k}(w+\overset{1}{w}{}^*)_{\alpha^l}\Big]\overset{1}{D}{}_{\mathrm{s}}^{ijkl}\nabla_i\varphi_j\,d\Omega, \tag{31.26}$$

$$(\overset{2}{\mathbf{K}}_{t\kappa}\cdot\chi)_{\overset{2}{H}_t} = -(\overset{2}{\omega}{}^*\cdot\chi)_{\overset{2}{H}_t} + \int_\Omega\Big[\overset{2}{B}_{kl}(w+\overset{2}{w}{}^*)$$
$$-\frac{1}{2}(w+\overset{2}{w}{}^*)_{\alpha^k}(w+\overset{2}{w}{}^*)_{\alpha^l}\Big]\overset{2}{D}{}_{\mathrm{s}}^{ijkl}\nabla_i\varphi_j\,d\Omega. \tag{31.27}$$

Let w_0 be some element of H_κ and let us set in (31.17), (31.18), $w = w_0 + v$. From (31.17), (31.18) we have

$$\overset{1}{\mathbf{K}}_{t\kappa}(w_0,\ v) = \overset{1}{\mathbf{K}}_{t\kappa 0}(w_0) + \overset{1}{\mathbf{K}}_{t\kappa 1}(w_0,\ v) + \overset{1}{\mathbf{K}}_{t\kappa 2}(w_0,\ v), \tag{31.28}$$

$$\overset{2}{\mathbf{K}}_{t\kappa}(w_0,\ v) = \overset{2}{\mathbf{K}}_{t\kappa 0}(w_0) + \overset{2}{\mathbf{K}}_{t\kappa 1}(w_0,\ v) + \overset{2}{\mathbf{K}}_{t\kappa 2}(w_0,\ v), \tag{31.29}$$

where the $\overset{i}{\mathbf{K}}_{t\kappa\mu}$ are homogeneous operators in v, $i = 1,\ 2$ of order μ, $\mu = 0,\ 1,\ 2$. To define them, we have from (31.26)–(31.29) the integral identities

$$(\overset{s}{\mathbf{K}}_{t\kappa 0}(w_0)\cdot\chi)_{\overset{s}{H}_t} = -(\overset{s}{\omega}{}^*\cdot\chi)_{\overset{s}{H}_t} + \int_\Omega\Big[\overset{s}{B}_{kl}(w_0+\overset{s}{w}{}^*)$$
$$-\frac{1}{2}(w_0+\overset{s}{w}{}^*)_{\alpha^k}(w_0+\overset{s}{w}{}^*)_{\alpha^l}\Big]\overset{s}{D}{}_{\mathrm{s}}^{ijkl}\nabla_i\varphi_j\,d\Omega, \tag{31.30}$$

$$(\overset{s}{\mathbf{K}}_{t\kappa 1}(w_0,\ v)\cdot\chi)_{\overset{s}{H}_t} = \int_\Omega\Big[\overset{s}{B}_{kl}v - \frac{1}{2}v_{\alpha^k}(w_0+\overset{s}{w}{}^*)_{\alpha^l}$$
$$-\frac{1}{2}v_{\alpha^l}(w_0+\overset{s}{w}{}^*)_{\alpha^k}\Big]\overset{s}{D}{}_{\mathrm{s}}^{ijkl}(\varphi_{i\alpha^j} - \overset{s}{G}{}^\lambda_{ij}\varphi_\lambda)\,d\Omega \tag{31.31}$$

$$(\overset{s}{\mathbf{K}}_{t\kappa 2}(w_0,\ v)\cdot\chi)_{\overset{s}{H}_t} = -\frac{1}{2}\int_\Omega v_{\alpha^k}v_{\alpha^l}\cdot\overset{s}{D}{}_{\mathrm{s}}^{ijkl}\nabla_i\varphi_j\,d\Omega. \tag{31.32}$$

In relations (31.30)–(31.32), $s = 1,\ 2$. Relations (31.28), (31.29) are expansions of the operators $\overset{s}{\mathbf{K}}_{t\kappa}$ in a neighborhood of the point w_0.

Lemma 31.9. *We have the relations*

$$\Big\|\overset{s}{\mathbf{K}}_{t\kappa 0}\Big\|_{\overset{s}{H}_t} \le m\bigg(\Big\|\overset{s}{\omega}{}^*\Big\|_{\overset{s}{H}_t} + \sum_{i,j=1}^{2}\Big\|\Big[\overset{s}{B}_{kl}(\overset{s}{w}_0+\overset{s}{w}{}^*)$$
$$-\frac{1}{2}(\overset{s}{w}_0+\overset{s}{w}{}^*)_{\alpha^k}(\overset{s}{w}_0+\overset{s}{w}{}^*)_{\alpha^l}\Big]\overset{s}{D}{}_{\mathrm{s}}^{ijkl}\Big\|_{L_{2\Omega}}\bigg),$$

$$\Big\|\overset{s}{\mathbf{K}}_{t\kappa 1}\Big\|_{\overset{s}{H}_t} \le m\bigg(\sum_{i,j=1}^{2}\Big\|\overset{s}{B}_{kl}\overset{s}{D}{}_{\mathrm{s}}^{ijkl}\Big\|_{C_\Omega} + \sum_{i,j,l=1}^{2}\Big\|(\overset{s}{w}_0+\overset{s}{w}{}^*)_{\alpha^k}\overset{s}{D}{}_{\mathrm{s}}^{ijkl}\Big\|_{L_{2\Omega}}\bigg)\|v\|_{\overset{1}{H}_\kappa},$$

$$\Big\|\overset{s}{\mathbf{K}}_{t\kappa 2}\Big\|_{\overset{s}{H}_t} \le m\sum_{i,j,k,l=1}^{2}\Big\|\overset{s}{D}{}_{\mathrm{s}}^{ijkl}\Big\|_{C_\Omega}\|v\|^2_{H_\kappa}.$$

The lemma follows immediately from (31.30)–(31.32) if we take into account the embedding Theorem 11.4 (relations (11.40)) and the embedding Theorem 12.3 (relations (12.26), (12.28)).

Theorem 31.1. *Assume that* (31.2) *is satisfied and that in addition,*

$$\left\| \overset{1}{\omega}{}^* - \overset{2}{\omega}{}^* \right\|_{\overset{1}{H}_\iota}, \quad \left\| \overset{1}{w}{}^* - \overset{2}{w}{}^* \right\|_{\overset{1}{H}_\kappa}, \quad \left\| \overset{1}{D}{}^{ijkl}_{\mathbf{s}} - \overset{2}{D}{}^{ijkl}_{\mathbf{s}} \right\|_{C_\Omega} \leq \epsilon. \tag{31.33}$$

Then for sufficiently small ϵ *we have*

$$\left\| \overset{1}{\mathbf{K}}_{\iota\kappa} - \overset{2}{\mathbf{K}}_{\iota\kappa} \right\|_{\overset{1}{H}_\iota} \leq m\epsilon \left(1 + \|v\|_{H_\kappa} + \|v\|^2_{H_\kappa}\right). \tag{31.34}$$

To prove the theorem, let us consider first the difference $\overset{1}{\mathbf{K}}_{\iota\kappa 0} - \overset{2}{\mathbf{K}}_{\iota\kappa 0}$. From (31.30) we have

$$(\overset{1}{\mathbf{K}}_{\iota\kappa 0} \cdot \chi)_{\overset{1}{H}_\iota} - (\overset{2}{\mathbf{K}}_{\iota\kappa 0} \cdot \chi)_{\overset{2}{H}_\iota} = (\overset{2}{\omega}{}^* \cdot \chi)_{\overset{2}{H}_\iota} - (\overset{1}{\omega}{}^* \cdot \chi)_{\overset{1}{H}_\iota} + \Delta\mathcal{D}_0, \tag{31.35}$$

where

$$\Delta\mathcal{D}_0 = \int_\Omega \Big\{ \Big[\overset{1}{B}_{kl}(w_0 + \overset{1}{w}{}^*) - \frac{1}{2}(w_0 + \overset{1}{w}{}^*)_{\alpha^k}(w_0 + \overset{1}{w}{}^*)_{\alpha^l} \Big] \overset{1}{D}{}^{ijkl}_{\mathbf{s}}$$
$$- \Big[\overset{2}{B}_{kl}(w_0 + \overset{2}{w}{}^*) - \frac{1}{2}(w_0 + \overset{2}{w}{}^*)_{\alpha^k}(w_0 + \overset{2}{w}{}^*)_{\alpha^l} \Big] \overset{2}{D}{}^{ijkl}_{\mathbf{s}} \Big\} \nabla_i \varphi_j \, d\Omega.$$

By (31.10), relation (31.35) can be written in the form

$$((\overset{1}{\mathbf{K}}_{\iota\kappa 0} - \mathbf{A}_\tau \overset{2}{\mathbf{K}}_{\iota\kappa 0}) \cdot \chi)_{\overset{1}{H}_\iota} = ((-\overset{1}{\omega}{}^* + \mathbf{A}_\tau \overset{2}{\omega}{}^*) \cdot \chi)_{\overset{1}{H}_\iota} + \Delta\mathcal{D}_0. \tag{31.36}$$

From (31.36), by taking into account the last two relations in (31.33), as well as the embedding Theorems 12.3 (see (12.26), (12.28)) and 11.4 (relation (11.40)) and the inequality

$$\left\| \overset{1}{B}_{kl} - \overset{1}{B}_{kl} \right\|_{C_\Omega} \leq m\epsilon, \tag{31.37}$$

which follows from (31.2), we have

$$|\Delta\mathcal{D}_0| \leq m\epsilon \, \|\chi\|_{\overset{1}{H}_\iota},$$

whence

$$\left\| \overset{1}{\mathbf{K}}_{\iota\kappa 0} - \mathbf{A}_\tau \overset{2}{\mathbf{K}}_{\iota\kappa 0} \right\|_{\overset{1}{H}_\iota} \leq m_1\epsilon + \left\| \overset{1}{\omega}{}^* - \mathbf{A}_\tau \overset{2}{\omega}{}^* \right\|_{\overset{1}{H}_\iota} \leq m\epsilon. \tag{31.38}$$

Here we used the first assertion of Lemma 31.7 (relations (31.16), (31.17)), as well as our convention concerning the constants m. Finally, again from Lemma 31.7

and (31.38), we obtain the relation

$$\left\| \overset{1}{\mathbf{K}}_{\iota\kappa 0} - \overset{2}{\mathbf{K}}_{\iota\kappa 0} \right\|_{\overset{1}{H}_{\iota}} \leq m\epsilon. \tag{31.39}$$

Next, let us consider the difference $\overset{1}{\mathbf{K}}_{\iota\kappa 1} - \overset{2}{\mathbf{K}}_{\iota\kappa 1}$, for which from (31.31) it follows that

$$(\overset{1}{\mathbf{K}}_{\iota\kappa 1} \cdot \chi)_{\overset{1}{H}_{\iota}} - (\overset{2}{\mathbf{K}}_{\iota\kappa 1} \cdot \chi)_{\overset{2}{H}_{\iota}} = \Delta\mathcal{D}_1,$$

and

$$\begin{aligned} \Delta\mathcal{D}_1 = & \int_\Omega \left(\overset{1}{B}_{kl} \overset{1}{D}{}^{ijkl}_{s} - \overset{2}{B}_{kl} \overset{2}{D}{}^{ijkl}_{s}\right) v \nabla_i \varphi_j \, d\Omega \\ & - \frac{1}{2} \int_\Omega v_{\alpha^k} \left[(w_0 + \overset{1}{w}{}^*)_{\alpha^l} \overset{1}{D}{}^{ijkl}_{s} - (w_0 + \overset{2}{w}{}^*)_{\alpha^l} \overset{2}{D}{}^{ijkl}_{s} \right] \nabla_i \varphi_j \, d\Omega \\ & - \frac{1}{2} \int_\Omega v_{\alpha^l} \left[(w_0 + \overset{1}{w}{}^*)_{\alpha^k} \overset{1}{D}{}^{ijkl}_{s} - (w_0 + \overset{2}{w}{}^*)_{\alpha^k} \overset{2}{D}{}^{ijkl}_{s} \right] \nabla_i \varphi_j \, d\Omega. \end{aligned} \tag{31.40}$$

From the last two relations in (31.33), as well as the embedding Theorems 12.3 (see (12.26), (12.28)) and 11.4 (relation (11.40)), as well as (31.37), we have

$$|\Delta\mathcal{D}_1| \leq m\epsilon \, \|\chi\|_{\overset{1}{H}_{\iota}} \, \|v\|_{\overset{1}{H}_{\kappa}} . \tag{31.41}$$

From (31.40), (31.41), repeating the arguments of the step dealing with the estimate of $\overset{1}{\mathbf{K}}_{\iota\kappa 0} - \overset{2}{\mathbf{K}}_{\iota\kappa 0}$, we have

$$\left\| \overset{1}{\mathbf{K}}_{\iota\kappa 1} - \overset{2}{\mathbf{K}}_{\iota\kappa 1} \right\|_{\overset{1}{H}_{\iota}} \leq m\epsilon \, \|v\|_{\overset{1}{H}_{\kappa}} . \tag{31.42}$$

Finally, using completely identical arguments, from (31.32) we obtain

$$\left\| \overset{1}{\mathbf{K}}_{\iota\kappa 2} - \overset{2}{\mathbf{K}}_{\iota\kappa 2} \right\|_{\overset{1}{H}_{\iota}} \leq m\epsilon \, \|v\|^2_{\overset{1}{H}_{\kappa}} . \tag{31.43}$$

From (31.39), (31.42), we obtain (31.34). Theorem 31.1. has been established.

31.4. Let us now consider a decomposition of the operator $\overset{s}{\mathbf{G}}_{\iota\kappa}$ introduced in (13.34) in a neighborhood of w_0. First of all let us note that from (31.28), (31.39),

$$\overset{s}{T}{}^{ij} = \overset{s}{T}{}^{ij}_{0} + \overset{s}{T}{}^{ij}_{1} v + \overset{s}{T}{}^{ij}_{2} v, \tag{31.44}$$

where by Theorem 31.1 and Lemma 31.9 we have

$$\begin{aligned} & \left\| \overset{1}{T}{}^{ij}_{0} - \overset{2}{T}{}^{ij}_{0} \right\|_{L_{2\Omega}} \leq m\epsilon, \quad \left\| \overset{1}{T}{}^{ij}_{1} v - \overset{2}{T}{}^{ij}_{1} v \right\|_{L_{2\Omega}} \leq m\epsilon \, \|v\|_{\overset{1}{H}_{\kappa}} , \\ & \left\| \overset{1}{T}{}^{ij}_{2} v - \overset{2}{T}{}^{ij}_{2} v \right\|_{L_{2\Omega}} \leq m\epsilon \, \|v\|^2_{\overset{1}{H}_{\kappa}} , \quad \left\| \overset{1}{T}{}^{ij}_{\mu} \right\|_{L_{2\Omega}} \leq m \, \|v\|^{\mu}_{\overset{1}{H}_{\kappa}} , \ \mu = 0, 1, 2. \end{aligned} \tag{31.45}$$

Next, for each shell o we substitute relations (31.44) and $w = w_0 + v$ into (13.34), which defines $\mathbf{G}_{\kappa\kappa}$. We obtain

$$\overset{s}{\mathbf{G}}_{\kappa\kappa}(w) = \overset{s}{\mathbf{G}}_{\kappa\kappa 0}(w_0) + \overset{s}{\mathbf{G}}_{\kappa\kappa 1}(v) + \overset{s}{\mathbf{G}}_{\kappa\kappa 2}(v) + \overset{s}{\mathbf{G}}_{\kappa\kappa 3}(v),$$

and the $\mathbf{G}_{\kappa\kappa\mu}$ are defined by the relations

$$(\overset{s}{\mathbf{G}}_{\kappa\kappa 0}(w_0)\cdot\varphi)_{\overset{s}{H}_\kappa} = -(\overset{s}{\overset{0}{w}}\cdot\varphi)_{\overset{s}{H}_\kappa} + \int_\Omega \overset{s}{T}{}^{ij}_0\Big[\overset{s}{B}_{ij}\varphi - (\overset{s}{w}_0 + \overset{s}{w}{}^*)_{\alpha^i}\varphi_j\Big]\, d\Omega \tag{31.46}$$

$$(\overset{s}{\mathbf{G}}_{\kappa\kappa 1}(v)\cdot\varphi)_{\overset{s}{H}_\kappa} = -\int_\Omega \Big\{\overset{s}{T}{}^{ij}_0 v_{\alpha^i}\varphi_{\alpha^j} + \overset{s}{T}{}^{ij}_1\Big[\overset{s}{B}_{ij}\varphi - (\overset{s}{w}_0 + \overset{s}{w}{}^*)_{\alpha^i}\varphi_j\Big]\Big\}\, d\Omega \tag{31.47}$$

$$(\overset{s}{\mathbf{G}}_{\kappa\kappa 2}(v)\cdot\varphi)_{\overset{s}{H}_\kappa} = \int_\Omega \Big\{\overset{s}{T}{}^{ij}_2\Big[\overset{s}{B}_{ij}\varphi - (\overset{s}{w}_0 + \overset{s}{w}{}^*)_{\alpha^i}\varphi_j\Big]\Big\}\, d\Omega, \tag{31.48}$$

$$(\overset{s}{\mathbf{G}}_{\kappa\kappa 3}(v)\cdot\varphi)_{\overset{s}{H}_\kappa} = -\int_\Omega \overset{s}{T}{}^{ij}_2 v_{\alpha^i}\varphi_{\alpha^j}\, d\Omega. \tag{31.49}$$

Theorem 31.2. *Assume the condition*

$$\|o_1 - o_2\|_O \le \epsilon. \tag{31.50}$$

Then for a sufficiently small ϵ we have the inequalities

$$\Big\|\overset{s}{\mathbf{G}}_{\kappa\kappa\mu}(v)\Big\| \le m\,\|v\|^\mu_{H_\kappa},\ \ \mu = 0,\ 1,\ 2,\ 3, \tag{31.51}$$

and furthermore,

$$\Big\|\overset{1}{\mathbf{G}}_{\kappa\kappa}(w) - \overset{2}{\mathbf{G}}_{\kappa\kappa}(w)\Big\|_{\overset{1}{H}_\kappa} \le m\epsilon\left(1 + \|v\|_{\overset{1}{H}_\kappa} + \|v\|^2_{\overset{1}{H}_\kappa} + \|v\|^3_{\overset{1}{H}_\kappa}\right), \tag{31.52}$$

where m can be taken to depend only on the parameters of o_1.

The inequalities (31.51) follow immediately from (31.46)–(31.49) by the last of the relations (31.45) and the embedding Theorem 12.3 (relations (12.26), (12.28)). To prove (31.52) we shall demonstrate that

$$\Big\|\overset{1}{\mathbf{G}}_{\kappa\kappa\mu}(v) - \overset{2}{\mathbf{G}}_{\kappa\kappa\mu}(v)\Big\| \le m\epsilon\,\|v\|^\mu_{\overset{1}{H}_\kappa},\ \ \mu = 0,\ 1,\ 2,\ 3, \tag{31.53}$$

where m can again be taken to depend only on o_1. Let us consider, for example, the case of $\mu = 1$. From (31.47) it follows that

$$\begin{aligned}
&(\overset{1}{\mathbf{G}}_{\kappa\kappa 1}(v)\cdot\varphi)_{\overset{1}{H}_\kappa} - (\overset{2}{\mathbf{G}}_{\kappa\kappa 1}(v)\cdot\varphi)_{\overset{1}{H}_\kappa}\\
&= \int_\Omega (\overset{2}{T}{}^{ij}_0 - \overset{1}{T}{}^{ij}_0)v_{\alpha^i}\varphi_{\alpha^j}\, d\Omega + \int_\Omega (\overset{2}{T}{}^{ij}_1 \overset{2}{B}_{ij} - \overset{1}{T}{}^{ij}_1 \overset{1}{B}_{ij})\varphi\, d\Omega\\
&\quad + \int_\Omega \big[\overset{1}{T}{}^{ij}_1(\overset{1}{w}_0 + \overset{1}{w}{}^*)_{\alpha^i} - \overset{2}{T}{}^{ij}_1(\overset{2}{w}_0 + \overset{2}{w}{}^*)_{\alpha^i}\big]\varphi_{\alpha^j}\, d\Omega.
\end{aligned} \tag{31.54}$$

From (31.45) and (12.26), (12.31), (31.37) we easily obtain

$$\left|\int_\Omega (\overset{2}{T}{}_0^{ij} - \overset{1}{T}{}_0^{ij}) v_{\alpha^i} \varphi_{\alpha^j}\, d\Omega\right| \le m\epsilon \, \|v\|_{\overset{1}{H}_\kappa} \|\varphi\|_{\overset{1}{H}_\kappa}, \tag{31.55}$$

$$\left|\int_\Omega (\overset{2}{T}{}_1^{ij} \overset{2}{B}_{ij} - \overset{1}{T}{}_1^{ij} \overset{1}{B}_{ij}) \varphi\, d\Omega\right| \le m\epsilon \, \|v\|_{\overset{1}{H}_\kappa} \|\varphi\|_{\overset{1}{H}_\kappa}. \tag{31.56}$$

In addition, taking into consideration (31.50) we derive

$$\left|\int_\Omega [\overset{1}{T}{}_1^{ij} (\overset{1}{w}_0 + \overset{1}{w}{}^*)_{\alpha^i} - \overset{2}{T}{}_1^{ij} (\overset{2}{w}_0 + \overset{2}{w}{}^*)_{\alpha^i}] \varphi_{\alpha^j}\, d\Omega\right| \le m\epsilon \, \|v\|_{\overset{1}{H}_\kappa} \|\varphi\|_{\overset{1}{H}_\kappa}. \tag{31.57}$$

From (31.54)–(31.57) we have

$$\left\|(\overset{1}{\mathbf{G}}_{\kappa\kappa 1}(v) - \mathbf{B}_\tau \overset{2}{\mathbf{G}}_{\kappa\kappa 1})(v) \cdot \varphi\right\|_{\overset{1}{H}_\kappa} \le m\epsilon \, \|v\|_{\overset{1}{H}_\kappa} \|\varphi\|_{\overset{1}{H}_\kappa},$$

so that

$$\left\|\overset{1}{\mathbf{G}}_{\kappa\kappa 1}(v) - \mathbf{B}_\tau \overset{2}{\mathbf{G}}_{\kappa\kappa 1}(v)\right\|_{\overset{1}{H}_\kappa} \le m\epsilon \, \|v\|_{\overset{1}{H}_\kappa}. \tag{31.58}$$

From (31.58) and Lemma 31.8 we obtain (31.53) for $\mu = 1$. The remaining relations (31.53) are demonstrated in the same way. Theorem 31.2 has been proved.

Theorem 31.3. *Assume that Conditions 1–8 of Section 13 hold for a shell $o_1 \in O$ and that there exists a nonsingular solution w_0 of the operator equation* (13.39). *Let us also be given a shell $o_2 \in O$ such that condition* (31.50) *holds. Then for sufficiently small ϵ, the operator equation* (13.36) *for o_2 will have a nonsingular solution $w_0 + \Delta w_0$ for which*

$$\|\Delta w_0\|_{\overset{1}{H}_\kappa} \le \delta(\epsilon), \;\; \delta(\epsilon) \to 0 \text{ as } \epsilon \to 0. \tag{31.59}$$

Furthermore, the ball $B(\delta(\epsilon), w_0)$ will contain precisely one solution of equation (13.39) *for o_1, o_2.*

To prove the theorem we note that if w_0 is a nonsingular solution of (13.36), then it necessarily must be isolated (Lemma 23.1), and a ball $B(w_0, \delta)$ will not contain any other solution. Let us consider on the sphere $\Sigma_{\overset{1}{H}_\kappa}(w_0, \delta)$ the vector fields $\Pi_1 w = w - \overset{1}{\mathbf{G}}_{\kappa\kappa}(w)$ and $\Pi_2 w = w - \overset{2}{\mathbf{G}}_{\kappa\kappa}(w)$, and let us show that they are homotopic if δ and ϵ are sufficiently small. Let us attempt to relate them by a vector field $\Pi(t)$ depending on a parameter t,

$$\Pi(t)(w) = w - \overset{1}{\mathbf{G}}_{\kappa\kappa}(w) + t(\overset{1}{\mathbf{G}}_{\kappa\kappa}(w) - \overset{2}{\mathbf{G}}_{\kappa\kappa}(w)).$$

Obviously, $\Pi(0)(w) = \Pi_1(w)$, $\Pi(1)(w) = \Pi_2(w)$. Next, we have

$$\begin{aligned}
\|\Pi(t)(w)\|_{\overset{1}{H}_\kappa} &= \left\| w - \overset{1}{\mathbf{G}}_{\kappa\kappa}(w) + t(\overset{1}{\mathbf{G}}_{\kappa\kappa}(w) - \overset{2}{\mathbf{G}}_{\kappa\kappa}(w)) \right\|_{\overset{1}{H}_\kappa} \\
&= \Big\| v - \overset{1}{\mathbf{G}}_{\kappa\kappa 1}(w_0, v) - \overset{1}{\mathbf{G}}_{\kappa\kappa 2}(w_0, v) - \overset{1}{\mathbf{G}}_{\kappa\kappa 3}(w_0, v) + t\big(\overset{1}{\mathbf{G}}_{\kappa\kappa}(w) \\
&\qquad - \overset{2}{\mathbf{G}}_{\kappa\kappa}(w)\big) \Big\|_{\overset{1}{H}_\kappa} \\
&\geq \left\| v - \overset{1}{\mathbf{G}}_{\kappa\kappa 1}(w_0, v) \right\|_{\overset{1}{H}_\kappa} - \left\| \overset{1}{\mathbf{G}}_{\kappa\kappa 2}(w_0, v) \right\|_{\overset{1}{H}_\kappa} \\
&\qquad - \left\| \overset{1}{\mathbf{G}}_{\kappa\kappa 3}(w_0, v) \right\|_{\overset{1}{H}_\kappa} \\
&quad - \left\| \overset{1}{\mathbf{G}}_{\kappa\kappa}(w) - \overset{2}{\mathbf{G}}_{\kappa\kappa}(w) \right\|_{\overset{1}{H}_\kappa}, \quad w = w_0 + v.
\end{aligned} \tag{31.60}$$

Now, since w_0 is a nonsingular solution, $v - \overset{1}{\mathbf{G}}_{\kappa\kappa 1}(w_0, v)$ has no eigenvectors, and therefore

$$\left\| v - \overset{1}{\mathbf{G}}_{\kappa\kappa 1}(v) \right\|_{\overset{1}{H}_\kappa} \geq m \, \|v\|_{\overset{1}{H}_\kappa}, \quad m > 0. \tag{31.61}$$

Using (31.61), (31.52), from (31.60) we obtain

$$\begin{aligned}
\|\Pi(t)(w)\|_{\overset{1}{H}_\kappa} &\geq m \, \|v\|_{\overset{1}{H}_\kappa} - m \left(\|v\|^2_{\overset{1}{H}_\kappa} + \|v\|^3_{\overset{1}{H}_\kappa} \right) \\
&\qquad - m \cdot \epsilon \left(1 + \|v\|_{\overset{1}{H}_\kappa} + \|v\|^2_{\overset{1}{H}_\kappa} + \|v\|^3_{\overset{1}{H}_\kappa} \right) \\
&= m[\delta(1-\epsilon) - \epsilon] - m(1+\epsilon)(\delta^2 + \delta^3).
\end{aligned} \tag{31.62}$$

It is easy to see that if in the right-hand side we set $\delta = 2\epsilon$, then for small ϵ we have

$$\|\Pi(t)(w)\|_{\overset{1}{H}_\kappa} \geq m(\epsilon) > 0.$$

Therefore, the homotopy of $\Pi_1(w)$ and $\Pi_2(w)$ on spheres of small radius $\Sigma_{\overset{1}{H}_\kappa}(w_0, \delta)$ has been proved. Now we note that on these spheres the winding number of $\Pi_1(w)$ is ± 1, since w_0 is a nonsingular solution. Therefore, the winding number of $\Pi_2(w)$ on $\Sigma_{\overset{1}{H}_\kappa}(w_0, \delta)$ is also ± 1, so that equation (13.36) has a solution $w_0 + \Delta w_0$ inside $\Sigma_{\overset{1}{H}_\kappa}(w_0, \delta)$, (31.59) holds and by $\delta = 2\epsilon$, and we have

$\delta \to 0$ as $\epsilon \to 0$. It is also easy to see that $w_0 + \Delta w_0$ is a nonsingular solution of equation (13.36) for o_2, and therefore it is an isolated solution of (13.36). Theorem 31.3 is proved.

We observe that from (14.1) we immediately obtain

$$\left\| \mathbf{K}_{t\kappa}(w_0) - \mathbf{K}_{t\kappa}(w_0 + \Delta w_0) \right\|_{H_t} \le \delta(\epsilon).$$

Remark 31.1. The point of Theorem 31.3 is that it establishes well-posedness classes for Problems $t\kappa$. These classes are determined by (31.50) in view of (31.1) and (29.1). Conditions on ρ, D_{s}^{ijkl}, and D_{f}^{ijkl} are easy to understand. The condition on the elastic coefficients of the supports says that the change of the solution in energy norm is small if the change in the coefficients of the supports is small uniformly on closed sets in their domain of definition. This condition can be weakened. Namely, using arguments similar to the above we can show that the change of the solution in energy norm will be small if the change in the elastic coefficients of the support is small in $L_{p\Gamma_i}$ for some $p > 1$ in the region Γ_i of definition of the elastic coefficient of the support. A small change of the shell in the loading norm $\|H\|$ implies by (29.1) a small change of ω_{p} in H_t and w_{p} in H_κ, as well a small change in $\widetilde{w}_3$, w_4, $\widetilde{w}_m$, $\widetilde{w}_\tau$ in norms corresponding to (29.1). A small change in ω_{p} will be ensured if, for example, the changes in R^s in $L_{p\Omega}$ for any $p > 1$ and in $\widetilde{T}^\tau$, $\widetilde{T}^m$ in some $L_{p,\Gamma_6+\Gamma_7}$, $L_{p,\Gamma_6+\Gamma_8}$ will be small. A small change in w_{p} is guaranteed if the change in R^3 is small in Ω even in classes of functions that include δ-functions. The change in $\widetilde{Q}$ is small in classes that include δ-functions on $\Gamma_3 + \Gamma_4$, while the change in $\widetilde{M}^m$ is small in any $L_{p,\Gamma_2+\Gamma_4}$ for $p > 1$.

31.5. Theorem 31.3 establishes well-posedness of the generalized formulation of problems $t\kappa$ in spaces H_κ. It is important to establish similar well-posedness results but in stronger spaces that allow us, for example, to derive a uniform estimate of the change in stresses. We present one such result without proof.

Consider two shells, $o_1, o_2 \in O$. We shall assume that the complexes of parameters that define them have the following properties:

$$\overset{s}{\rho} \in C_\Omega^{4+k}(H_\Omega^{4+k,\lambda}),\ \overset{s}{D}{}_{\mathrm{s}}^{ijrl} \in C_\Omega^{1+k}(H_\Omega^{1+k,\lambda}),$$

$$\overset{s}{D}{}_{\mathrm{f}}^{ijrl} \in C_\Omega^{2+k}(H_\Omega^{2+k,\lambda}),\ \overset{s}{\overset{0}{w}} \in W_{p\Omega}^{(4+k)}(H_\Omega^{4+k,\lambda}),$$

$$\overset{s}{\overset{0}{\omega}} \in W_{p\Omega}^{(2+k)}(H_\Omega^{2+k,\lambda}),\ \overset{s}{R}{}^3,\ \overset{s}{R}{}^t \in W_{p\Omega}^{(k)}(H_\Omega^{k,\lambda}).$$

Theorem 31.4. *Assume that under the boundary conditions* (21.1)–(21.3) *the shell* o_1 *has a nonsingular solution* w_0, ω_0 *of the operator equation* (13.36). *In in addition*

$$\Gamma \in C_\Gamma^{4+k}(H_\Gamma^{4+k,\lambda})$$

and for sufficiently small ϵ,

$$\left\| \overset{1}{\rho} - \overset{2}{\rho} \right\|_{C_\Omega^{4+k}(H_\Omega^{4+k,\lambda})}, \quad \left\| \overset{1}{D}{}_{\mathrm{s}}^{ijrl} - \overset{2}{D}{}_{\mathrm{s}}^{ijrl} \right\|_{C_\Omega^{1+k}(H_\Omega^{1+k,\lambda})},$$

$$\left\| \overset{1}{D}{}_{\mathrm{f}}^{ijrl} - \overset{1}{D}{}_{\mathrm{f}}^{ijrl} \right\|_{C_\Omega^{2+k}(H_\Omega^{2+k,\lambda})}, \quad \left\| \overset{\overset{1}{0}}{w} - \overset{\overset{2}{0}}{w} \right\|_{W_{p\Omega}^{(4+k)}(H_\Omega^{4+k,\lambda})} \tag{31.63}$$

$$\left\| \overset{1}{R}{}^3 - \overset{2}{R}{}^3 \right\|_{W_{p\Omega}^{(k)}(H_\Omega^{k,\lambda})} \leq \epsilon,$$

$$\left\| \overset{\overset{1}{0}}{\omega} - \overset{\overset{2}{0}}{\omega} \right\|_{W_{p\Omega}^{(2+k)}(H_\Omega^{2+k,\lambda})}, \quad \left\| \overset{1}{R}{}^3 - \overset{2}{R}{}^3 \right\|_{W_{p\Omega}^{(k)}(H_\Omega^{k,\lambda})} \leq \epsilon,$$

then the operator equation (13.36) *for* o_2 *also has a nonsingular solution* $w_0 + \Delta w_0$, $\omega_0 + \Delta\omega_0$, *and*

$$\|\Delta w_0\|_{W_{p\Omega}^{(4+k)}(H_\Omega^{4+k,\lambda'})}, \; \|\Delta\omega_0\|_{W_{p\Omega}^{(2+k)}(H_\Omega^{2+k,\lambda'})} \leq \delta(\epsilon),$$

where $\delta(\epsilon) \to 0$ *as* $\epsilon \to 0$.

Clearly, if $k = 0$, we necessarily have small changes in stresses in the shell.

31.6. Let us present without proof some results dealing with well-posedness of problems 9κ. The class O will contain shells with a common domain of definition Ω, and the boundary curve Γ has for all shells the same decomposition $\Gamma = \Gamma_1 + \Gamma_2$. Clearly, for problems 9κ the complexes defining the shell include $\rho(\alpha^1, \alpha^2)$, D_{f}^{ijkl}, C_{s}^{ijkl}, $\widetilde{T}_{\mathrm{p}}^{ij}$, $\widetilde{M}^m$, R^3, $\overset{0}{w}$. We introduce a norm on the set of these complexes,

$$\|o\|_O = \|\rho\|_{C_\Omega^2} + \sum_{i,j,k,l=1}^{2} \left\| D_{\mathrm{f}}^{ijkl} \right\|_{C_\Omega} + \sum_{i,j,k,l=1}^{2} \left\| C_{\mathrm{s}}^{ijkl} \right\|_{C_\Omega}$$

$$+ \sum_{i,j=1}^{2} \left\| \widetilde{T}_{\mathrm{p}}^{ij} \right\|_{L_{2\Omega}} + \left\| \overset{0}{w} \right\|_{W_{2\Omega}^{(2)}} + \left\| w_{\mathrm{p}}^* \right\|_{H_\kappa}.$$

Theorem 31.5. *Assume that each of the shells* o_1, o_2 *satisfies Conditions 1–8 of Section 17, and that in addition, the operator equation* (17.19) *for* o_1 *has a nonsingular solution* w_0, Ψ_0.

Then if

$$\|o_1 - o_2\|_O \leq \epsilon,$$

then for sufficiently small ϵ, *equation* (17.19) *for* o_2 *will also have a nonsingular solution* $w_0 + \Delta w_0$, $\Psi_0 + \Delta\Psi_0$ *such that*

$$\|\Delta w_0\|_{H_\kappa}, \; \|\Delta\Psi_0\|_{H_9} \leq \delta(\epsilon), \tag{31.64}$$

where $\delta(\epsilon) \to 0$ *as* $\epsilon \to 0$.

In (31.64), $\Delta\Psi_0 = \mathbf{K}_{9\kappa}(w_0 + \Delta w_0) - \mathbf{K}_{9\kappa}(w_0)$.

Next, we assume that the complexes of parameters for o_1, o_2 under the conditions of Problem 9κ have the following properties:

$$\overset{s}{\rho} \in C_{\Omega}^{4+k}(H_{\Omega}^{4+k,\lambda}), \ \overset{s}{D}{}_{\mathrm{f}}^{ijrl} \in C_{\Omega}^{2+k}(H_{\Omega}^{2+k,\lambda}),$$

$$\overset{s}{R}{}^{3} \in W_{p\Omega}^{(k)}(H_{\Omega}^{k,\lambda}), \ \overset{s}{\overset{0}{w}} \in W_{p\Omega}^{(4+k)}(H_{\Omega}^{4+k,\lambda}),$$

$$\overset{s}{C}{}_{*ij}^{\lambda\mu}\overset{s}{\widetilde{T}}{}_{\mathrm{p}}^{ij} \in W_{p\Omega}^{(k)}(H_{\Omega}^{k,\lambda}).$$

Theorem 31.6. *Assume that there exists a nonsingular solution w_0, Ψ_0 of the operator equation* (17.55) *for o_1 under the boundary conditions* (21.42). *If for sufficiently small ϵ we also have that* (31.84) *holds and*

$$\Gamma \in C_{\Gamma}^{4+k}(H_{\Gamma}^{4+k,\lambda}), \ \left\| \overset{1}{C}{}_{*ij}^{\lambda\mu}\overset{1}{\widetilde{T}}{}_{\mathrm{p}}^{ij} - \overset{2}{C}{}_{*ij}^{\lambda\mu}\overset{2}{\widetilde{T}}{}_{\mathrm{p}}^{ij} \right\|_{W_{p\Omega}^{(k)}(H_{\Omega}^{k,\lambda})} \le \epsilon,$$

then the operator equation (17.19) *for o_2 will also have a nonsingular solution $w_0 + \Delta w_0$, $\Psi_0 + \Delta\Psi_0$ such that*

$$\|\Delta w_0\|_{W_{p\Omega}^{(4+k)},(H_{\Omega}^{4+k,\lambda'})} \ \|\Delta\Psi_0\|_{W_{p\Omega}^{(4+k)},(H_{\Omega}^{4+k,\lambda'})} \le \delta(\epsilon),$$

where $\delta(\epsilon) \to 0$ as $\epsilon \to 0$.

CHAPTER IX

Stability in the Large of the Membrane State of a Shallow Shell. Existence of the Lower Critical Value

32. Momentless State of Shells. Passage to the Linearized Problem. Spectral Properties of the Linearized Problem

32.1. We shall say that a shell is in the momentless (membrane) state (MlS) if all the moments M^{ij} vanish. Under our boundary conditions this corresponds to the condition

$$w \equiv 0. \tag{32.1}$$

In this case from (13.6) we obtain

$$(\omega_{\text{Ml}} \cdot \chi)_{H_t} = (\omega_{\text{p}} \cdot \chi)_{H_t}, \ \omega_{\text{Ml}} = (w_{1\text{Ml}}, w_{2\text{Ml}}), \ \chi = (\varphi_1, \varphi_2), \tag{32.2}$$

where the element $\omega_{\text{p}} \in H_t$ has been defined by the relation (11.50).

From (13.7) it follows that

$$\int_\Omega T^{ij}_{\text{Ml}} B_{ij} \varphi d\Omega + (w_{\text{p}} \cdot \varphi)_{H_\kappa} = 0, \tag{32.3}$$

where $w_{\text{p}} \in H_\kappa$ has been defined by the relation (12.40). In the sequel we shall denote all the quantities relating to an MlS by the subscript "Ml," as we already did in (32.2), (32.3). It should be noted that it does not follow from (32.2) that $\omega_{\text{Ml}} \equiv \omega_{\text{p}}$, since it is not being assumed that ω_{Ml} belongs to H_t in view of the possibility of geometric boundary conditions on Γ_5, Γ_6. However, the scalar product $(\omega_{\text{Ml}} \cdot \chi)_{H_t}$ makes sense, since $\omega_{\text{Ml}} \in W^{(1)}_{2\Omega}$.

In this section we shall always assume that the conditions of Theorem 16.1 (or, respectively, 16.2, 16.3) hold.

Theorem 32.1. *The conditions*

$$\widetilde{M}^m\big|_{\Gamma_2+\Gamma_4} \equiv \widetilde{Q}\big|_{\Gamma_3+\Gamma_4} \equiv 0, \tag{32.4}$$

$$\widetilde{w}|_{\Gamma_1+\Gamma_2} \equiv \widetilde{w}|_{\Gamma_1+\Gamma_3} \equiv 0, \tag{32.5}$$

$$T^{ij}_{\mathrm{Ml}}B_{ij} + R^3 = 0 \tag{32.6}$$

are both necessary and sufficient for the existence of an MlS of the equilibrium of a shell.

To prove necessity, let us note that (32.5) follows directly from the definition of an MlS. The relation (32.6) follows from (32.3), since here we can choose φ to be a function with compact support in H_κ. Once (32.6) is established, the relations (32.4) follow from (32.3), where we already take $\forall\varphi \in H_\kappa$. Thus, necessity of (32.4)–(32.6) has been established. Moving on to the proof of sufficiency, we observe that if (32.4)–(32.6) hold, the integral relations (13.6)–(13.7) will be satisfied if we substitute there $w \equiv 0$ and ω as defined by (32.2).

Remark 32.1. In relation (32.6) it is assumed that

$$T^{ij}_{\mathrm{Ml}} = D^{ijkl}_{\mathrm{s}}\nabla_k w_{l\mathrm{Ml}},$$

while the $w_{i\mathrm{Ml}}$ are defined by (32.2). Thus in essence, (32.6) is a condition on the loads R^i on the surface S^0 tangential to the midsurface, $\widetilde{T}^\tau\big|_{\Gamma_6+\Gamma_8}$, $\widetilde{T}^m\big|_{\Gamma_7+\Gamma_8}$, the displacements $\widetilde{w}_m|_{\Gamma_5+\Gamma_6}$, $\widetilde{w}_\tau|_{\Gamma_5+\Gamma_7}$, as well as on R^3.

Let us assume now that the load terms R^i, R^3, $\widetilde{T}^\tau$, $\widetilde{T}^m$, $\widetilde{w}_m$, $\widetilde{w}_\tau$, $\widetilde{w}$, $\widetilde{w}_4$ satisfy the conditions of Theorem 32.1. In this case, clearly, for every λ loads of the form $-\lambda R^i$, $-\lambda R^3$, $-\lambda\widetilde{T}^\tau$, $\lambda\widetilde{T}^m$, $\lambda\widetilde{w}_m$, $\lambda\widetilde{w}_\tau$ induce the MlS

$$-\lambda\omega_{\mathrm{Ml}} = -\lambda(w_{1\mathrm{Ml}},\, w_{2\mathrm{Ml}}),\ w \equiv 0,\ -\lambda T^{ij}_{\mathrm{Ml}}.$$

Let us make in (13.6), (13.7) the change of variable

$$\omega \sim \omega_M - \lambda\omega_{\mathrm{Ml}},\ T^{ij} \sim T^{ij}_M - \lambda T^{ij}_{\mathrm{Ml}}. \tag{32.7}$$

As a result, we obtain

$$(\omega_M \cdot \chi)_{H_t} = \int_\Omega \left(B_{kl}w - \frac{1}{2}w_{\alpha^k}w_{\alpha^l}\right) D^{ijkl}_{\mathrm{s}}\nabla_i\varphi_j\, d\Omega, \tag{32.8}$$

$$(w \cdot \varphi)_{H_\kappa} = \int_\Omega T^{ij}_M\left(B_{ij}\varphi - w_{\alpha^i}\varphi_{\alpha^j}\right) d\Omega + \lambda\int_\Omega T^{ij}_{\mathrm{Ml}} w_{\alpha^i}\varphi_{\alpha^j}\, d\Omega. \tag{32.9}$$

In (32.7)–(32.9), T^{ij}_M is determined from (4.14)–(4.15), in which $\omega = (w_{1M},\, w_{2M})$ is found from (32.8), and this determines all the characteristics of a moment state (ms).

32.2. For each λ, (32.8), (32.9) have the obvious solution

$$\omega_M \equiv w \equiv T^{ij}_M \equiv 0,$$

to which corresponds an MlS. Using the stability framework of Section 29, we shall endeavor to construct a U-decomposition on the real line $-\infty < \lambda < +\infty$. Here an important part will be played by the linearized system of operator equations (32.8), (32.9) constructed by neglecting terms that are nonlinear in w. As is well known, the passage to linearized systems was the crucial step in the development of a general theory of stability; it constitutes the greatest achievement of Leonard Euler in mathematical physics [76]. For more details on this see [74].

In accordance with (14.6), we have from (32.8),

$$\omega_M = \mathbf{K}_{t\kappa 1}(w) + \mathbf{K}_{t\kappa 2}(w) = \omega_{1M} + \omega_{2M}, \tag{32.10}$$

where

$$(\mathbf{K}_{t\kappa 1}(w)\cdot\chi)_{H_t} = \int_\Omega B_{kl} w D_{\mathrm{s}}^{ijkl} \nabla_i \varphi_j d\Omega, \tag{32.11}$$

$$(\mathbf{K}_{t\kappa 2}(w)\cdot\chi)_{H_t} = -\frac{1}{2}\int_\Omega w_{\alpha^k} w_{\alpha^l} D_{\mathrm{s}}^{ijkl} \nabla_i \varphi_j d\Omega. \tag{32.12}$$

By (32.10)–(32.12), linearizing (32.8), (32.9), we obtain

$$(w\cdot\varphi)_{H_\kappa} = \int_\Omega T_{1M}^{ij} B_{ij}\varphi d\Omega + \lambda\int_\Omega T_{\mathrm{MI}}^{ij} w_{\alpha^i}\varphi_{\alpha^j} d\Omega. \tag{32.13}$$

In (32.13) the tensor T_{1M}^{ij} is given by (14.16), where the ϵ_{ij1M} are computed using the second formula of (14.15) for $w^* \equiv 0$.

The linear operator equation (LOE) (32.13) has the trivial solution $w \equiv 0$ for all λ. Therefore, below we shall consider (32.13) as an eigenvalue problem.

32.3. Let us now consider the MIS for the boundary value problems 9κ. In this section we shall assume that the conditions of Theorem 19.3 and (32.1) are satisfied. But in this case we obtain from (17.2), (17.3),

$$\int_\Omega \{[(C^{ik}C^{jl}\nabla_{kl}\Psi_{\mathrm{MI}} + \widetilde{T}_{\mathrm{p}}^{ij})]B_{ij} + R^3\}\varphi d\Omega + \int_{\Gamma_2} \widetilde{M}^m \frac{\partial\varphi}{\partial m} ds = 0, \tag{32.14}$$

$$(\Psi_{\mathrm{MI}}\cdot\theta)_{H_9} = -\int_\Omega C_{*ij}^{\lambda\mu}\widetilde{T}_{\mathrm{p}}^{ij}\nabla_{\lambda\mu}\theta\, d\Omega. \tag{32.15}$$

Relations (32.14), (32.15) must hold for all $\varphi \in H_\kappa$, $\theta \in H_9$. Since φ is arbitrary, it then follows from (32.14) that

$$T_{\mathrm{MI}}^{ij} B_{ij} + R^3 = 0, \tag{32.16}$$

$$T_{\mathrm{MI}}^{ij} = C^{ik}C^{jl}\nabla_{kl}\Psi_{\mathrm{MI}} + \widetilde{T}_{\mathrm{p}}^{ij}, \quad \widetilde{M}^m\big|_{\Gamma_2} \equiv 0. \tag{32.17}$$

Here it is assumed that Ψ_{MI} has been computed from (32.15) and substituted into the left-hand side of (32.16). Thus the entire left-hand side has been expressed in terms of the loads R^3 and T_{p}^{ij}.

Theorem 32.2. *For the existence of an MIS in the Problem 9κ it is both necessary and sufficient to satisfy conditions* (32.16), (32.17), *in which we perform the substitution* $\Psi_{\mathrm{MI}} \mapsto \widetilde{T}_{\mathrm{p}}^{ij}$ *from* (32.15).

Necessity of (32.16), (32.17) has in essence been established in the process of derivation. Sufficiency follows from the fact that the operator equations (17.2), (17.3) have a solution for which $w \equiv 0$ and Ψ are determined from (32.13).

It is easy to see that to loads of the form $-\lambda \widetilde{T}_{\mathrm{p}}^{ij}$, $-\lambda R^3$ correspond MIS defined by the Airy stress function $-\lambda \Psi_{\mathrm{MI}}$. Let us make in (17.2), (17.3) a substitution of the form

$$\Psi \sim \Psi_M - \lambda \Psi_{\mathrm{MI}}. \tag{32.18}$$

Then we obtain

$$(w \cdot \varphi)_{H_\kappa} = \int_\Omega C^{ik} C^{jl} (B_{ij}\varphi - \varphi_{\alpha^i} w_{\alpha^j}) \nabla_{kl} \Psi_M d\Omega + \lambda \int_\Omega T_{\mathrm{MI}}^{ij} w_{\alpha^i} \varphi_{\alpha^j} d\Omega, \tag{32.19}$$

$$(\Psi_M \cdot \theta)_{H_9} = \int \Omega \left(\frac{1}{2} w_{\alpha^i} w_{\alpha^j} - w B_{ij} \right) C^{ik} C^{jl} \nabla_{kl} \theta d\Omega. \tag{32.20}$$

The system of equations (32.19), (32.20) has the obvious solution

$$w \equiv 0, \ \Psi_M \equiv 0.$$

We shall try to elucidate properties of the U-decomposition of the λ-axis. In this question as well, an important part is played by the linearized problem obtained from (32.19), (32.20) by retaining only terms of first order in w. To derive it, it is necessary to recall that from (32.20) we have that

$$\Psi_M = \Psi_{1M} + \Psi_{2M}; \ \Psi_i M = \mathbf{K}_{9\kappa i} w,$$

$$(\Psi_{1M} \cdot \Theta)_{H_9} = -\int_\Omega C^{ik} C^{jl} w B_{ij} \nabla_{kl} \Theta d\Omega, \tag{32.21}$$

$$(\Psi_{2M} \cdot \Theta)_{H_9} = \frac{1}{2} \int_\Omega C^{ik} C^{jl} w_{\alpha^i} w_{\alpha^j} \nabla_{kl} \Theta d\Omega. \tag{32.22}$$

Then, linearizing, we have

$$(w \cdot \varphi)_{H_\kappa} = \int_\Omega C^{ik} C^{jl} B_{ij} \varphi \nabla_{kl} \Psi_{1M} d\Omega + \lambda \int_\Omega T_{\mathrm{MI}}^{ij} w_{\alpha^i} \varphi_{\alpha^j} d\Omega. \tag{32.23}$$

In (32.23) we take Ψ_{1M} to be expressed in terms of w by (32.21). Then we obtain a linear operator equation, the eigenvalue problem for which we are going to consider.

32.4. Let us consider an abstract linear operator equation that contains as particular cases both (32.13) and (32.23). Let **A**, **B**, **C** be additive self-adjoint operators acting in some Hilbert space $\overset{0}{H}$. Furthermore, assume that

(1) **A** is a positive definite self-adjoint operator on $\overset{0}{H}$, and $H_{\mathbf{A}}$ is a Hilbert space with the scalar product [197]

$$(w \cdot \varphi)_{H_{\mathbf{A}}} = (\mathbf{A} w \cdot \varphi)_{\overset{0}{H}};$$

H_A is the closure of $\overset{0}{H}$ in the corresponding norm.

(2) **B** is a self-adjoint completely continuous operator from H_A into $\overset{0}{H}$ such that

$$(\mathbf{B}w \cdot w)_{\overset{0}{H}} \leq 0. \tag{32.24}$$

(3) **C** is a self-adjoint completely continuous operator from H_A into $\overset{0}{H}$ such that

$$(\mathbf{C}w \cdot w)_{\overset{0}{H}} \geq 0 \text{ and from } (\mathbf{C}w \cdot w)_{\overset{0}{H}} = 0 \text{ it follows that } w \equiv 0.$$

(4) The space $\overset{0}{H}$ is infinite-dimensional.

Let us be given an LOE

$$\mathbf{A} = \mathbf{B} + \lambda \mathbf{C}. \tag{32.25}$$

Theorem 32.3. *Let conditions (1)–(4) hold. Then*

1) *The LOE* (32.25) *has a countable set of eigenelements* λ_k, w_k.
2) $\lambda_1 > 0,\ \lambda_{k+1} \geq \lambda_k$ *and* $\lambda_k \to \infty$ *as* $k \to \infty$. (32.26)
3) *The* w_k *form an orthonormal basis in the space* H *defined by the relation*
$$(w \cdot \varphi)_H = (w \cdot \varphi)_{H_A} - (\mathbf{B}w \cdot \varphi)_{\overset{0}{H}} \tag{32.27}$$
4) $(\mathbf{C}w_k \cdot w_r)_{\overset{0}{H}} = \delta_{kr} \cdot \lambda_k^{-1} = \delta_{kr}\lambda_r^{-1}.$ (32.28)

Below it will make sense to use the Hilbert space H_C with the scalar product

$$(w \cdot \varphi)_{H_C} = (\mathbf{C}w \cdot \varphi)_{\overset{0}{H}}, \tag{32.29}$$

so that the relation (32.28) can be written in the form

$$(w_k \cdot w_r)_{H_C} = \delta_{kr}\lambda_k^{-1} = \delta_{kr}\lambda_r^{-1}.$$

Turning now to the proof of Theorem 32.3, we note certain inequalities to be used below. By positive definiteness of **A**,

$$\|w\|_{H_A} \geq m\,\|w\|_{\overset{0}{H}}.$$

Lemma 32.1. *We have the inequality*

$$M\,\|w\|_{H_A} \geq \|w\|_{H_C}. \tag{32.30}$$

Since **C** is a completely continuous operator from H_A into $\overset{0}{H}$, the functional $(\mathbf{C}w \cdot w)_{\overset{0}{H}} = \|w\|_{H_C}^2$ is weakly continuous in H_A. On the sphere $\|w\|_{H_A}^2 = 1$, and $(\mathbf{C}w \cdot w)_{\overset{0}{H}}$ has a maximum, so that the inequality (32.40) holds.

Lemma 32.2. *The spaces* H_A *and* H *are equivalent.*

To prove this claim, we note that from (32.24), (32.25) it follows that

$$\|w\|_H \geq \|w\|_{H_A}. \tag{32.31}$$

Furthermore, since $\mathbf{B}$ is a completely continuous operator, the functional $(\mathbf{B}w \cdot w)_{\overset{0}{H}}$ is weakly continuous, and therefore by Theorem 9.4, on the unit sphere in H we have that

$$\left|(\mathbf{B}w \cdot w)_{\overset{0}{H}}\right| \leq M \tag{32.32}$$

and in the whole of $\overset{0}{H}$,

$$\left|(\mathbf{B}w \cdot w)_{\overset{0}{H}}\right| \leq M\|w\|_H^2.$$

From (32.27), (32.32) we obtain

$$\|w\|_H^2 \leq (M+1)\|w\|_{H_A}^2. \tag{32.33}$$

Lemma 32.2 now follows from (32.31), (32.33).

Lemma 32.3. *The linear operator equation* (32.25) *can be written in the following equivalent form:*

$$\operatorname{grad} \|w\|_H^2 = \lambda \operatorname{grad} \|w\|_{H_C}^2. \tag{32.34}$$

The proof of Lemma 32.3 is obvious.

Thus the search for eigenelements of the linear operator equation (32.25) has been reduced, by Lyusternik's theorem [188], [310], to the search for the extrema of $\|w\|_H^2$ on unit spheres $\Sigma_{H_C}(1, 0)$ with center at the origin.

Lemma 32.4. *On* $\Sigma_{H_C}(1, 0)$ *the functional* $\|w\|_H^2$ *has an absolute minimum at* w_1 *and* $\lambda_1 = \|w_1\|_H^2$.

To prove the lemma, we observe that $\|w\|_H^2$ is bounded from below on $\Sigma_{H_C}(1, 0)$, so that there is a minimizing sequence w_k that can be taken to be weakly convergent in H and therefore also in H_A. Let us consider the numbers α_{kl} defined by the relation

$$\left\|\frac{1}{2}\alpha_{kl}(w_k' + w_l')\right\|_{H_C}^2 = 1. \tag{32.35}$$

(Here and below there is no summation over k, l.) The numbers α_{kl} of (32.35) are well-defined, since the sequence w_k' is weakly convergent in H, while the functional $\|w\|_{H_C}$ is weakly continuous in H. Furthermore, if at the same time

$$k, l \to \infty, \text{ we have } \alpha_{kl} \to 1. \tag{32.36}$$

Next, we have

$$\left\|\frac{1}{2}\alpha_{kl}(w_k' - w_l')\right\|_H^2 = \alpha_{kl}^2\left(\frac{1}{2}\|w_k'\|_H^2 + \frac{1}{2}\|w_l'\|_H^2\right) - \left\|\frac{(w_k' + w_l')}{2}\alpha_{kl}\right\|_H^2, \tag{32.37}$$

and for sufficiently large k, l we obtain

$$\begin{gathered}\frac{1}{2}\alpha_{kl}^2\|w_k'\|_H^2 \le \frac{1}{2}\lambda_1 + \epsilon_{kl}, \quad \frac{1}{2}\alpha_{kl}^2\|w_l'\|_H^2 \le \frac{1}{2}\lambda_1 + \epsilon_{kl}, \\ \left\|\alpha_{kl}\frac{(w_k' + w_l')}{2}\right\|_H^2 \ge \lambda_1.\end{gathered} \tag{32.38}$$

From (32.37), (32.38) it follows that

$$\|w_k' - w_l'\|_H^2 \le \frac{4}{\alpha_{kl}^2}\left(\frac{1}{2}\lambda_1 + \epsilon_{kl} + \frac{\lambda_1}{2} + \epsilon_{kl} - \lambda_1\right) \le \frac{8\epsilon_{kl}}{\alpha_{kl}^2}.$$

From (32.33) we deduce strong convergence of w_k'. Let w_1 be the strong limit of w_k' in H. Clearly, w_1 satisfies (32.34). Then also $\lambda_1 \neq 0$, since if $\lambda_1 = 0$, then $w_1 \equiv 0$ as well and $w_1 \notin \Sigma_{H\mathrm{c}}(1, 0)$. Also clearly, $\lambda_1 > 0$.

To construct the subsequent eigenelements λ_k, w_k, let us assume that $k - 1$ eigenelements have already been constructed. Let us consider the minimum of $\|w\|_H^2$ on the set of elements

$$\|w\|_{H\mathrm{c}}^2 = 1, \ (w \cdot w_r)_{H\mathrm{c}} = 0, \ r = 1, \ldots, k - 1. \tag{32.39}$$

Lemma 32.5. *On the set* (32.39) *the functional* $\|w\|_H^2$ *has a minimum point* w_k *that is a solution of* (32.8), (32.34). *Furthermore,* $\lambda_k = \|w_k\|_H^2$.

The lemma is proved using arguments similar to those of Lemma 32.3. By construction, we obviously have $\lambda_{k+1} \ge \lambda_k$.

Lemma 32.6. *The eigenfunctions* w_k, w_l *can be picked in such a way that for any* k, l *we have that*

$$(w_k \cdot w_l)_H = (w_k \cdot w_l)_{H\mathrm{c}} = 0, \ k \neq l. \tag{32.40}$$

To prove this lemma, let us consider first the case when w_k, w_l belong to different eigenvalues λ_k, λ_l. Then

$$(w_k \cdot \varphi)_H = \lambda_k (w_k \cdot \varphi)_{H\mathrm{c}}, \tag{32.41}$$

$$(w_l \cdot \varphi)_H = \lambda_l (w_l \cdot \varphi)_{H\mathrm{c}}. \tag{32.42}$$

Setting $\varphi = w_l$ in (32.41) and $\varphi = w_k$ in (32.42), we obtain the system of equations

$$(w_k \cdot w_l)_H = \lambda_k (w_k \cdot w_l)_{H\mathrm{c}}, \quad (w_k \cdot w_l)_H = \lambda_l (w_k \cdot w_l)_{H\mathrm{c}},$$

for the two unknowns $(w_k \cdot w_l)_H$ and $(w_k \cdot w_l)_{H\mathrm{c}}$. The determinant of this system is $\lambda_k - \lambda_l$, from which (32.40) follows.

If $w_k, w_{k+1}, \ldots, w_{k+p}$ correspond to the same eigenvalue, they can be orthonormalized by the Schmidt process [310] in the space H. But if they are orthogonal in H, they will also be orthogonal in H_c. Therefore, (32.40) will be satisfied.

It is not impossible that some k different minimizing sequences would lead to different solutions w_k. It could also happen that as we are constructing w_{k+1} we obtain the same value λ_k. This situation is controlled by the following lemma.

Lemma 32.7. *To each value λ_k corresponds a finite number of independent eigenfunctions $w_k, w_{k+1}, \ldots, w_{k+p}$.*

The lemma will be proved by arguing from the contrary.

Let us assume the existence of an infinite number of linearly independent w_k, $w_{k+1}, \ldots$, which is possible by condition 4 eigenfunctions. Let us orthonormalize them in the norm of H, but then by (32.40) they will be orthogonal in H_C, and furthermore,

$$\|w_r\|_{H_C} = \frac{1}{\sqrt{\lambda_k}}, \quad r = k;\ k+1,\ \ldots,\ \infty. \tag{32.43}$$

On the other hand,

$$w_k \rightharpoonup 0 \text{ in } H,$$

and by complete continuity,

$$w_k \Rightarrow 0 \text{ in } H_C,$$

and (32.43) is impossible. The resulting contradiction proves the lemma.

Lemma 32.8. *The process of constructing w_k, λ_k cannot terminate at any k.*

Indeed, suppose that for some k we obtain a complete set of linearly independent solutions $w_k, w_{k+1}, \ldots, w_{k+p}$, the number of which must be finite by Lemma 32.7. Let us move on now to the construction of eigenelements by minimizing $\|w\|_H^2$ on the set of elements that satisfy (32.39). We cannot obtain the same eigenvalue λ_k, since then the corresponding eigenfunction $\widetilde{w}_{k+1}$ must be linearly dependent on $w_k, w_{k+1}, \ldots, w_{k+p}$. However, this cannot happen, since we are looking for $\widetilde{w}_{k+1}$ in the set (32.39), which is orthogonal in H_C to all $w_{k+1}, \ldots, w_{k+p}$, which is nonempty by condition 4.

Lemma 32.9. *The set of the points λ_k cannot have an accumulation point at a finite distance from the origin.*

Let us assume the opposite. Furthermore, let w_k be an eigenelement corresponding to λ_k. By Lemma 32.6 all the w_k have been orthonormalized in H_C. But then by boundedness of λ_k they will be bounded in H and therefore weakly compact. Without loss of generality, we can take w_k to be weakly convergent in H. Let us consider now the quantity

$$\|w_{r_1} - w_{r_2}\|_{H_C}, \tag{32.44}$$

where w_{r_1}, w_{r_2} are some eigenfunctions corresponding to large numbers r_1 and r_2. Clearly, by strong continuity of $\mathbf{C}$ the quantity (32.44) can be made arbitrarily small for sufficiently large r_1, r_2. On the other hand, by orthonormality of w_{r_1}, w_{r_2}

in H_C we have

$$\|w_{r_1} - w_{r_2}\|^2_{HC} = 2. \tag{32.45}$$

This contradiction proves Lemma 32.9.

Lemma 32.10. *The eigenfunctions $w_k/\sqrt{\lambda_k}$ of the linear operator equation* (32.29), (32.34) *form an orthonormal basis in H.*

Let $\widetilde{w} \in H$ and moreover,

$$(\widetilde{w} \cdot w_k)_H = 0, \; k = 1, \ldots, \infty.$$

By (32.40),

$$\|\widetilde{w}\|^2_H \geq \lambda_k \|\widetilde{w}\|^2_{HC}$$

for any k, and thus by Lemmas 32.8 and 32.9,

$$\|\widetilde{w}\|_{HC} = 0, \quad \tilde{w} \equiv 0.$$

Lemma 32.11. *The linear operator equation* (32.36), (32.34) *has no eigenelements apart from the ones constructed in this section.*

Indeed, assume that $\widetilde{\lambda}$, $\widetilde{w}$ is a new eigenelement. But then by Lemma 32.6, the equality (32.45) holds, while by Lemma 32.10, $\widetilde{w} \equiv 0$, so that $\widetilde{w}$ is not an eigenfunction.

Theorem 32.3 follows from Lemmas 32.1–32.11 if we renorm $w_k \sim w_k \lambda_k{}^{-\frac{1}{2}}$. We note that the proofs of Lemmas 32.1–32.11 are quite standard [197, 310]; however, we decided to reproduce them here for completeness and due to the specific properties of equation (32.25), (32.34).

32.5. Let us go back to the linear operator equation (32.13). We shall show that it satisfies all the conditions of Theorem 32.3 if we set

$$H_A \sim H_\kappa, \; (\mathbf{B}_w \cdot \varphi)_{\overset{0}{H}} = \int_\Omega T_1^{ij}(w) B_{ij} \varphi \, d\Omega,$$

$$(w \cdot \varphi)_{HC} = \int_\Omega T^{ij}_{M1} w_{\alpha^i} \varphi_{\alpha^j} \, d\Omega. \tag{32.46}$$

The space $\overset{0}{H}$ contains elements of $L_{2\Omega}$, and moreover,

$$(\varphi_1 \cdot \varphi_2)_{\overset{0}{H}} = \int_\Omega \varphi_1 \varphi_2 \, d\Omega.$$

Theorem 32.4. *Assume that the conditions*

$$\int_\Omega T^{ij}_{M1} w_{\alpha^i} w_{\alpha^j} \, d\Omega \geq 0 \textit{ and } \int_\Omega T^{ij}_{M1} w_{\alpha^i} w_{\alpha^j} \, d\Omega \equiv 0 \textit{ implies } w \equiv 0 \tag{32.47}$$

hold. Then:

(1) *The linear operator equation* (32.13) *has a countable number of eigenelements λ_k, w_k, where to each λ_k corresponds a finite number of eigenfunctions w_k.*

(2) $\lambda_1 > 0$, $\lambda_{k+1} \geq \lambda_k$ *and* $\lambda_k \to \infty$ *as* $k \to \infty$.
(3) w_k *form an orthonormal basis in* H *(the space* H *will be defined below).*
(4) $\int_\Omega T_{\text{M1}}^{ij} w_{k\alpha^i} w_{r\alpha^j} = \delta_{kr}$.

Let us verify that all the conditions of Theorem 32.3 are satisfied. Let us start with condition 3. We have

$$(\mathbf{B}w \cdot \varphi)_{\overset{0}{H}} = \int_\Omega T_1^{ij}(w) B_{ij} \varphi d\Omega.$$

From (32.11), which defines $\mathbf{K}_{t\kappa 1}(w)$ and thus T_1^{ij}, it follows that

$$\int_\Omega T_1^{ij}(w) \nabla_i \varphi_j d\Omega + \int_{\Gamma_6} k_{\text{s}}^{\tau\tau} w_\tau \varphi_\tau ds + \int_{\Gamma_7} k_{\text{s}}^{mm} w_m \varphi_m ds + \int_{\Gamma_8} k_{\text{s}}^{ij} w_i \varphi_j ds = 0. \tag{32.48}$$

Here the w_j are solutions of (32.11) if on the right-hand side we have w, while the φ_j are the solutions of (32.11) if φ replaces w on the right-hand side. From (32.48) we have

$$\int_\Omega T_1^{ij}(w) \overset{0}{\epsilon}_{ij1}(\varphi) d\Omega + \int_{\Gamma_6} k_{\text{s}}^{\tau\tau} w_\tau \varphi_\tau ds + \int_{\Gamma_7} k_{\text{s}}^{mm} w_m \varphi_m ds + \int_{\Gamma_8} k_{\text{s}}^{ij} w_i \varphi_j ds$$
$$= -\int_\Omega T_1^{ij}(w) B_{ij} \varphi d\Omega = -(\mathbf{B}w \cdot \varphi)_{\overset{0}{H}}. \tag{32.49}$$

The left-hand side of (32.49) is easily seen to be symmetric in w, φ, which proves self-adjointness of the operator $\mathbf{B}$. From (32.49) we have

$$(\mathbf{B}w \cdot w)_{\overset{0}{H}} = -\int_\Omega T_1^{ij}(w) \overset{0}{\epsilon}_{ij1}(w)\, d\Omega - \int_{\Gamma_6} k_{\text{s}}^{\tau\tau} w_\tau^2 ds - \int_{\Gamma_7} k_{\text{s}}^{mm} w_m^2 ds - \int_{\Gamma_8} k_{\text{s}}^{ij} w_i w_j ds < 0. \tag{32.50}$$

Thus (32.24) holds as well.

Self-adjointness can be seen from (32.46). Complete continuity of $\mathbf{B}$ follows from Theorem 12.3 (relation (12.29)). Thus, condition 4 holds. Finally, let us define the space H explicitly in this case using (32.27). We have

$$(w \cdot \varphi)_H = (w \cdot \varphi)_{H_\kappa} - (\mathbf{B}w \cdot \varphi)_{\overset{0}{H}} = (w \cdot \varphi)_{H_\kappa} - \int_\Omega T_1^{ij}(w) B_{ij} \varphi d\Omega. \tag{32.51}$$

That H as introduced above is well-defined follows from the properties of the operator $\mathbf{B}w$ established above (see (32.49), (32.50)).

In view of the great importance of the first eigenvalue λ of linear operator equation (32.13) in the theory of stability of shells, it will be denoted below by $\lambda_{t\kappa\text{E}}$. The number $\lambda_{t\kappa\text{E}}$ determines the critical load if we use the method of linearization of Euler. We note the inequality

$$\|w\|^2_{HC} \le \frac{1}{\lambda_{t\kappa E}} \|w\|^2_H .$$

32.6. Let us consider equation (32.23). We shall show that it also satisfies all the requirements of Theorem 32.3 if we take

$$H_A = H_\kappa, \ (\mathbf{B}w \cdot \varphi)_{\overset{0}{H}} = \int_\Omega C^{ik}C^{jr}\nabla_{kr}\Psi_1(w)B_{ij}\varphi \, d\Omega. \tag{32.52}$$

For **C** we have the same definition (32.46), and therefore in relation to **C** the conditions of Theorem 32.3 will be satisfied if (32.47) holds, which we shall assume henceforth. To prove self-adjointness of **B**, we note that from (32.21), (32.52) we have

$$(\mathbf{B}w \cdot \varphi)_{\overset{0}{H}} = \int_\Omega C^{ik}C^{jl}\nabla_{kl}\Psi_1(w)B_{ij}\varphi \, d\Omega = -(\Psi_1(w) \cdot \Psi_1(\varphi))_{H_{9\kappa}}. \tag{32.53}$$

From symmetry of the right-hand side of (32.53) with respect to w and φ self-adjointness of **B** follows. From (32.53) it is also seen that condition 2 of Theorem 32.3 holds, that is, (32.24) is satisfied. The space H in this case is defined by the relation

$$(w \cdot \varphi)_H = (w \cdot \varphi)_{H_\kappa} - \int_\Omega C^{ik}C^{jl}\nabla_{kl}\Psi_1(w)B_{ij}\varphi \, d\Omega. \tag{32.54}$$

Theorem 32.5. *Assume that condition* (32.47) *is satisfied. Then the linear operator equation* (32.23) *satisfies all the parts 1–4 of Theorem 32.4.*

Due the great importance of the first eigenvalue λ_1 of LOE (32.23) in the theory of stability of shells, it will be denoted below by $\lambda_{9\kappa E}$. The number $\lambda_{9\kappa E}$ determines the critical load if we use the method of linearization of Euler around an MIS.

33. Global Stability of Shells in Problems $t\kappa$. Existence of Lower Critical Numbers. Some Estimates for U-Decompositions

33.1. Let us assume that in Problem $t\kappa$ the loads P acting on the shell are such that conditions (32.4)–(32.6) for the existence of MIS are satisfied, and we shall consider the shell under the load $-\lambda P$. It is described by the nonlinear operator equation (32.8), (32.9). We shall say that λ belongs to the spectrum of the nonlinear operator equation (SNOE) (32.8), (32.9) if for that λ there is a solution with w different from zero. Here, clearly, if $\lambda \in$ SNOE, in addition to MIS there is a nonmomentless state (ms), and transition to an ms is possible, that is, MIS can lose stability. All the points of the λ-axis where only zero solutions of the equation

(32.8), (32.9) exist will be called regular, and the set of such points will be called RNOE.

Below it will be useful to transform the total energy functional $\mathcal{I}_{t\kappa}$ of the system shell/elastic supports/external forces, defined by (12.44), taking into consideration (32.4)–(32.6). Taking into account (32.7), we have

$$\begin{aligned}
\mathcal{I}^{\lambda}_{\kappa\kappa} = \frac{1}{2}\Big(\|w\|^2_{H_\kappa} &+ \int_\Omega D^{ijkl}_{\mathrm{s}}(\overset{0}{\epsilon}_{ijM} - \lambda\overset{0}{\epsilon}_{ij\mathrm{MI}})(\overset{0}{\epsilon}_{klM} - \lambda\overset{0}{\epsilon}_{kl\mathrm{MI}})\, d\Omega \\
&+ \int_{\Gamma_6} k^{\tau\tau}_{\mathrm{s}}(w_{\tau M} - \lambda w_{\tau\mathrm{MI}})^2\, ds + \int_{\Gamma_7} k^{mm}_{\mathrm{s}}(w_{mM} - \lambda w_{m\mathrm{MI}})^2\, ds \\
&+ \int_{\Gamma_8} k^{ij}_{\mathrm{s}}(w_{iM} - \lambda w_{i\mathrm{MI}})(w_{jM} - \lambda w_{j\mathrm{MI}})\, ds\Big) \\
&+ \lambda(w_{\mathrm{p}}\cdot w)_{H_\kappa} + \lambda(\omega_{\mathrm{p}}\cdot\omega_M)_{H_t}.
\end{aligned} \tag{33.1}$$

Multiplying out the first integrand of (33.1), we obtain

$$\begin{aligned}
\mathcal{I}^{\lambda}_{\kappa\kappa} = \frac{1}{2}\Big(\|w\|^2_{H_\kappa} &+ \int_\Omega D^{ijkl}_{\mathrm{s}}\overset{0}{\epsilon}_{ijM}\overset{0}{\epsilon}_{klM}\, d\Omega + \int_{\Gamma_6} k^{\tau\tau}_{\mathrm{s}} w^2_{\tau M}\, ds \\
&+ \int_{\Gamma_7} k^{mm}_{\mathrm{s}} w^2_{mM}\, ds + \int_{\Gamma_8} k^{ij}_{\mathrm{s}} w_{iM} w_{jM}\, ds\Big) \\
&- \lambda\Big[\int_\Omega D^{ijkl}_{\mathrm{s}}\overset{0}{\epsilon}_{ijM}\overset{0}{\epsilon}_{kl\mathrm{MI}}\, d\Omega + \int_{\Gamma_6} k^{\tau\tau}_{\mathrm{s}} w_{\tau M} w_{\tau\mathrm{MI}}\, ds \\
&+ \int_{\Gamma_7} k^{mm}_{\mathrm{s}} w_{mM} w_{m\mathrm{MI}}\, ds + \int_{\Gamma_8} k^{ij}_{\mathrm{s}} w_{iM} w_{j\mathrm{MI}}\, ds \\
&- (w_{\mathrm{p}}\cdot w)_{H_\kappa} - (\omega_{\mathrm{p}}\cdot\omega_M)_{H_t}\Big].
\end{aligned} \tag{33.2}$$

Here we have neglected the terms that depend only on the MIS, since they do not enter the considerations below. In (33.1), $\overset{0}{\epsilon}_{ij\mathrm{MI}}$ is determined from the first of the relations (14.15), where we should set $w^* \equiv 0$ and $w_{i0} \sim w_{i\mathrm{MI}}$, which are determined from (32.2). Knowing $w_{i\mathrm{MI}}$, we compute $w_{\tau\mathrm{MI}}$, $w_{m\mathrm{MI}}$, which are the tangential and normal components to Γ of the vector of tangential displacements $\omega_{\mathrm{MI}} = (w_{1\mathrm{MI}}, w_{2\mathrm{MI}})$ of the MIS. The $\overset{0}{\epsilon}_{ijM}$ are found from (3.20)–(3.23), where we put $w_i \sim w_{iM}$. Then in accordance with (32.11)–(32.12), we have

$$\overset{0}{\epsilon}_{ijM} = \overset{0}{\epsilon}_{ijM1} + \overset{0}{\epsilon}_{ijM2}, \tag{33.3}$$

where the $\overset{0}{\epsilon}_{ijM1}$ are computed, as we already mentioned in Section 32, using the second formula (14.15), where we put $w^* \equiv 0$, while w_{i1} is determined using ω_{1M}, which in turn are determined from (32.11). The coefficients $\overset{0}{\epsilon}_{ijM2}$ are computed using the third formula (14.15), where w_{i2} is determined using ω_{2M}, which are found in turn from (32.12).

Let us consider the expression

$$\begin{aligned}\int_\Omega D_{\mathrm{s}}^{ijkl}\overset{0}{\epsilon}_{ijM}\overset{0}{\epsilon}_{klM1}\,d\Omega &= \int_\Omega T_{\mathrm{M1}}^{ij}\overset{0}{\epsilon}_{ijM}\,d\Omega \\ &= \int_\Omega T_{\mathrm{M1}}^{ij}\Big(\nabla_i w_{jM} - B_{ij}w + \frac{1}{2}w_{\alpha^i}w_{\alpha^j}\Big)\,d\Omega \\ &= \int_\Omega T_{\mathrm{M1}}^{ij}\nabla_i w_{jM}\,d\Omega - \int_\Omega T_{\mathrm{M1}}^{ij}B_{ij}w\,d\Omega \\ &\quad + \frac{1}{2}\int_\Omega T_{\mathrm{M1}}^{ij}w_{\alpha^i}w_{\alpha^j}\,d\Omega \\ &= \int_\Omega D_{\mathrm{s}}^{ijkl}\nabla_i w_{jM}\nabla_k w_{lM}\,d\Omega \\ &\quad - \int_\Omega T_{\mathrm{M1}}^{ij}B_{ij}w\,d\Omega + \frac{1}{2}\int_\Omega T_{\mathrm{M1}}^{ij}w_{\alpha^i}w_{\alpha^j}\,d\Omega.\end{aligned} \tag{33.4}$$

Setting $\chi = \omega_M$ in (32.2) and in (32.3) $\varphi = w$, we obtain

$$\begin{aligned}\int_\Omega D_{\mathrm{s}}^{ijkl}\nabla_i w_{j\mathrm{M1}}\nabla_k w_l\,d\Omega = (\omega_{\mathrm{p}}\cdot\omega_M)_{H_t} &- \int_{\Gamma_6} k_{\mathrm{s}}^{\tau\tau}w_{\tau M}w_{\tau\mathrm{M1}}\,ds \\ &- \int_{\Gamma_7} k_{\mathrm{s}}^{mm}w_{mM}w_{m\mathrm{M1}}ds \\ &- \int_{\Gamma_8} k_{\mathrm{s}}^{ij}w_{iM}w_{j\mathrm{M1}}ds,\end{aligned} \tag{33.5}$$

$$\int_\Omega T_{\mathrm{M1}}^{ij}B_{ij}w\,d\Omega = -(w_{\mathrm{p}}\cdot w)_{H_\kappa}. \tag{33.6}$$

Substituting (33.4)–(33.6) into (33.2), we obtain

$$\begin{aligned}\mathcal{I}_{\kappa\kappa}^{\lambda} = \frac{1}{2}\Big(\|w\|_{H_\kappa}^2 &+ \int_\Omega D_{\mathrm{s}}^{ijkl}\overset{0}{\epsilon}_{ijM}\overset{0}{\epsilon}_{klM}\,d\Omega + \int_{\Gamma_6} k_{\mathrm{s}}^{\tau\tau}w_{\tau M}^2\,ds \\ &+ \int_{\Gamma_7} k_{\mathrm{s}}^{mm}w_{mM}^2 ds + \int_{\Gamma_8} k_{\mathrm{s}}^{ij}w_{iM}w_{jM}ds - \lambda\|w\|_{HC}^2\Big).\end{aligned} \tag{33.7}$$

Recall that $\|w\|_{HC}^2$ is given by (32.46). Next, taking into account (33.3), we have from (33.7),

$$\mathcal{I}_{\kappa\kappa}^{\lambda} = \mathcal{I}_{\kappa\kappa 2}^{\lambda} + \mathcal{I}_{\kappa\kappa 3} + \mathcal{I}_{\kappa\kappa 4}, \tag{33.8}$$

where

$$\begin{aligned}\mathcal{I}_{\kappa\kappa 2}^{\lambda} = \frac{1}{2}\Big(\|w\|_{H_\kappa}^2 &+ \int_\Omega D_{\mathrm{s}}^{ijkl}\overset{0}{\epsilon}_{ijM1}\overset{0}{\epsilon}_{klM1}\,d\Omega + \int_{\Gamma_6} k_{\mathrm{s}}^{\tau\tau}w_{\tau M1}^2\,ds \\ &+ \int_{\Gamma_7} k_{\mathrm{s}}^{mm}w_{mM1}^2\,ds + \int_{\Gamma_8} k_{\mathrm{s}}^{ij}w_{jM1}w_{iM1}\,ds - \lambda\,\|w\|_{HC}^2\Big),\end{aligned} \tag{33.9}$$

$$\mathcal{I}_{\kappa\kappa 3} = \int_\Omega D_{\mathrm{s}}^{ijkl} \overset{0}{\epsilon}_{ijM2} \overset{0}{\epsilon}_{klM1} \, d\Omega + \int_{\Gamma_6} k_{\mathrm{s}}^{\tau\tau} w_{\tau M2} w_{\tau M1} \, ds + \int_{\Gamma_7} k_{\mathrm{s}}^{mm} w_{mM1} \cdot w_{mM2} \, ds + \int_{\Gamma_8} k_{\mathrm{s}}^{ij} w_{iM1} w_{jM2} \, ds, \tag{33.10}$$

$$\mathcal{I}_{\kappa\kappa 4} = \frac{1}{2}\Big(\int_\Omega D_{\mathrm{s}}^{ijkl} \overset{0}{\epsilon}_{ijM2} \overset{0}{\epsilon}_{klM2} \, d\Omega + \int_{\Gamma_6} k_{\mathrm{s}}^{\tau\tau} w_{\tau M2}^2 \, ds + \int_{\Gamma_7} k_{\mathrm{s}}^{mm} w_{mM2}^2 \, ds + \int_{\Gamma_8} k_{\mathrm{s}}^{ij} w_{iM1} w_{jM2} \Big). \tag{33.11}$$

Finally, taking into account (32.50), (32.51), we have from (33.9) for $\mathcal{I}^\lambda_{\kappa\kappa 2}$,

$$\mathcal{I}^\lambda_{\kappa\kappa 2} = \frac{1}{2}\left(\|w\|^2_{H_\kappa} - \lambda \, \|w\|^2_{HC} \right).$$

For the considerations below we shall need the relation

$$\Phi(w, \ 1) = 0,$$

which is constructed using the MlS conditions (32.4)–(32.6). It can be obtained from (16.6), which defines the functional Φ if (32.7) is substituted. However, it can also be obtained in a more elementary fashion if we set $\varphi \equiv w$ in (32.9) and use (32.8), in which we take $\chi = \omega_M$. We have the sequence of transformations

$$\begin{aligned} 0 &= \|w\|^2_{H_\kappa} - \int_\Omega T_M^{ij}(B_{ij} w - w_{\alpha^i} w_{\alpha^j}) \, d\Omega - \lambda \, \|w\|^2_{HC} \\ &= \|w\|^2_{H_\kappa} + \int_\Omega T_M^{ij}(-2B_{ij} w + w_{\alpha^i} w_{\alpha^j}) \, d\Omega - \lambda \, \|w\|^2_{HC} + \int_\Omega T_M^{ij} B_{ij} w \, d\Omega \\ &= \|w\|^2_{H_\kappa} + \int_\Omega T_M^{ij}(\nabla_i w_{jM} + \nabla_j w_{iM} - 2B_{ij} w + w_{\alpha^i} w_{\alpha^j}) \, d\Omega - \lambda \, \|w\|^2_{HC} \\ &\quad + \int_\Omega T_M^{ij} B_{ij} w \, d\Omega - \int_\Omega T_M^{ij}(\nabla_i w_{jM} + \nabla_j w_{iM}) \, d\Omega. \end{aligned} \tag{33.12}$$

From (32.8) we obtain

$$\int_\Omega T_M^{ij}(\nabla_i w_{jM} + \nabla_j w_{iM}) \, d\Omega = -2\Big(\int_{\Gamma_6} k_{\mathrm{s}}^{\tau\tau} w_{\tau M}^2 \, ds + \int_{\Gamma_7} k_{\mathrm{s}}^{mm} w_{mM}^2 \, ds + \int_{\Gamma_8} k_{\mathrm{s}}^{ij} w_{iM} w_{jM} \, ds \Big), \tag{33.13}$$

and from (33.12), (33.13) it follows that

$$\|w\|^2_{H_\kappa} + \int_\Omega T_M^{ij} \overset{0}{\epsilon}_{ijM} \, d\Omega + 2\Big(\int_{\Gamma_6} k_{\mathrm{s}}^{\tau\tau} w_{\tau M}^2 \, ds + \int_{\Gamma_7} k_{\mathrm{s}}^{mm} w_{mM}^2 \, ds + \int_{\Gamma_8} k_{\mathrm{s}}^{ij} w_{iM} w_{jM} \, ds \Big) + \int_\Omega T_M^{ij} B_{ij} w \, d\Omega - \lambda \, \|w\|^2_{HC} = 0. \tag{33.14}$$

Lemma 33.1. *Assume that condition* (32.47) *is satisfied. Then the functional*

$$\psi(w) = \|w\|^2_{H_\kappa} - \int_\Omega \widetilde{N} w^2 \, d\Omega, \tag{33.15}$$

where $\widetilde{N} \in C_\Omega$, *has a finite minimum on the surface*

$$\|w\|^2_{HC} = \int_\Omega T^{ij}_{\mathrm{MI}} w_{\alpha^i} w_{\alpha^j} \, d\Omega = 1. \tag{33.16}$$

To prove the lemma, let us assume the opposite. For a sequence w_k let

$$\psi(w_k) \to -\infty \;\; k \to \infty.$$

Clearly, we must necessarily have

$$\|w_k\|_{H_\kappa} \to \infty,$$

and moreover, the sequence $w_k^* = w_k / \|w_k\|_{H_\kappa}$ can be taken to be weakly convergent in H_κ. It is easy to see that in H_κ,

$$w_k^* \rightharpoonup 0,$$

for otherwise (33.16) cannot hold. Indeed, if $w_k^* \rightharpoonup w^{**} \neq 0$, then

$$\|w_k\|^2_{HC} = \|w_k\|^2_{H_\kappa} \left\|w_k^*\right\|^2_{HC} \sim \|w_k\|^2_{H_\kappa} \left\|w^{**}\right\|^2_{HC} \to \infty$$

by weak continuity of $\|w\|^2_{HC}$ in H_κ and condition (32.47). Thus $w_k^* \rightharpoonup 0$. But from (33.15) we have

$$\psi(w_k) = \|w_k\|^2_{H_\kappa} \left(1 - \int_\Omega \widetilde{N} w_k^{*2} \, d\Omega\right). \tag{33.17}$$

By Theorem 12.3 the second term on the right-hand side of (33.17) vanishes and $\psi(w_k) \to \infty$. This contradiction proves Lemma 33.1.

Theorem 33.1. *Assume that all the conditions of Theorem 16.1 (respectively, 16.2, 16.3) are satisfied and that in addition, condition* (32.47) *holds. Then the SNOE* (32.8), (32.9) *is to the right of some point* $\lambda^*_{t\kappa}$. *All the points* $\lambda < \lambda^*_{t\kappa}$ *belong to RNOE of* (32.8), (32.9).

To prove this theorem let us consider in (33.14) the quantity $\mathcal{E}(T^{ij}_M)$ defined by the relation

$$\mathcal{E}(T^{ij}_M) = \int_\Omega \left(T^{ij}_M \overset{0}{\epsilon}_{ijM} + T^{ij}_M B_{ij} w \right) d\Omega.$$

In this expression for $\mathcal{E}$ the integrand is a second-degree polynomial with respect to T^{ij}_M. Its quadratic form is a positive definite form of T^{ij}_M. Therefore, the integrand has a minimum in the three-dimensional space of $T^{11}_M, T^{12}_M, T^{22}_M$. It can be found by the usual methods. In fact, we presented such arguments in Section 29. Therefore, without repeating the computations, we show the final result:

$$\mathcal{E}(T_M^{ij}) = -\frac{1}{8}\int_\Omega \widetilde{N}_{\kappa\kappa} w^2\, d\Omega, \ \widetilde{N}_{\kappa\kappa} = D_s^{ijkl}\widetilde{B}_{ij}\widetilde{B}_{kl}, \ \widetilde{B}_{ii} = B_{ii}, \qquad (33.18)$$
$$\widetilde{B}_{ij} = 2B_{ij}, \ i \neq j.$$

Here $\widetilde{N}_{\kappa\kappa}$ is a new elastic and geometric complex, a convolution of fourth rank of tensor D_s^{ijkl} and second-rank tensors B_{ij}, B_{kl}, which we introduced in Section 29 (see (29.33)).

From (33.14), (33.18) we obtain

$$\|w\|^2_{H_\kappa} - \frac{1}{8}\int_\Omega \widetilde{N}_{\kappa\kappa} w^2\, d\Omega - \lambda\, \|w\|^2_{Hc} \leq 0. \qquad (33.19)$$

By Lemma 33.1 we have from (33.19),

$$(\lambda^*_{t\kappa} - \lambda)\, \|w\|^2_{Hc} \leq 0,$$

where $\lambda^*_{t\kappa}$ is the minimum of the functional (33.15) on (33.16) for

$$\widetilde{N} = \frac{1}{8}N_{\kappa\kappa} = \frac{1}{8}D_s^{ijkl}\widetilde{B}_{ij}\widetilde{B}_{kl},$$

whence by (32.47), we have that $w \equiv 0$ for $\lambda < \lambda^*_{t\kappa}$. This establishes Theorem 33.1.

Corollary 33.2. *The entire semi-infinite straight line $\lambda < \lambda^*_{t\kappa}$ is in the set U_1; that is, here we have only the MlS of the shell. Clearly, by definition, $\lambda^*_{t\kappa}$ is the smallest eigenvalue of the linear operator equation*

$$(w\cdot\varphi)_{H_\kappa} = -\frac{1}{4}\int_\Omega \widetilde{N}_{\kappa\kappa} w\varphi\, d\Omega - \lambda_{t\kappa}(w\cdot\varphi)_{Hc} = 0,$$

which can be written in the form

$$\widetilde{\nabla}_{ij} D D_f^{ijkl}\nabla_{kl} w - \frac{1}{4}\widetilde{N}_{\kappa\kappa} w\cdot D + \lambda^*_{t\kappa}(T_{\mathrm{Ml}}^{ij} w_{\alpha^i} D)_{\alpha^j} = 0,$$
$$w|_{\Gamma_1+\Gamma_2} = 0, \qquad (33.20)$$
$$\left.\frac{dw}{dm}\right|_{\Gamma_1+\Gamma_3} = 0, \qquad (33.21)$$
$$D_f^{ijkl}\nabla_{ij} w m_k\, m_l|_{\Gamma_2+\Gamma_4} = -k_f^{44} w_4,$$
$$\big\{D^{-1}(DD_f^{ijkl}\nabla_{kl}w)_s m_i\tau_j + D^{-1}(DD_f^{ijkl}\nabla_{kl}w)_m m_i\tau_j + (D_f^{ijkl}\nabla_{kl} w m_i\tau_j)_s + D_f^{ijkl}\nabla_{kl} w G^s_{ij} m_s\big\}_{\Gamma_3+\Gamma_4} = -k_f^{43} w_3.$$

33.2.

Theorem 33.3. *Assume that the conditions of Theorem 33.1 are satisfied. Then the semi-infinite line $\lambda > \lambda^*_{t\kappa\mathrm{E}}$ belongs to SNOE* (33.11). *Furthermore, at points of the spectrum in addition to the MlS the shell will have a moment state having lower total energy of the system shell/elastic supports/external loads.*

To prove this assertion, let us consider the formula (33.8) for $\mathcal{I}_{\kappa\kappa}^{\lambda}$, the potential energy of this system. Furthermore, let $\lambda = \lambda_1 = \lambda_{t\kappa E}$, and w_1 the first eigenelement of the LOE (32.13) (or respectively (32.25) or (32.34)). Let us substitute tw_1 for w into (33.8); after normalizing, we have

$$\mathcal{I}_{\kappa\kappa}^{\lambda} = \frac{t^2}{2}\left(1 - \frac{\lambda}{\lambda_{t\kappa e}}\right) + t^3\mathcal{I}_{\kappa\kappa 3}(w_1) + t^4\mathcal{I}_{\kappa\kappa 4}(w_1). \tag{33.22}$$

For sufficiently small t the sign of $\mathcal{I}_{\kappa\kappa}^{\lambda}$ will be determined only by the sign of the first term on the right-hand side of (33.22), which will be negative for $\lambda > \lambda_{t\kappa e}$.

But by Theorem 21.4, $\mathcal{I}_{\kappa\kappa}^{\lambda}$ will assume its absolute minimum at a point $w_{\min}$, and furthermore, at that point

$$\mathcal{I}_{\kappa\kappa}^{\lambda}(w_{\min}) < 0.$$

Clearly, $w_{\min}$ cannot be zero, and therefore there exists a state with lower energy than that of MIS.

Theorem 33.2 establishes an important property of U-decompositions of the λ axis. Under its conditions on the line $\lambda > \lambda_{t\kappa e}$, the emergent moment configuration is more stable than the existent momentless one. Finally, let us note that in some particular cases local methods allow us to establish for $\lambda > \lambda_{t\kappa e}$ in a small neighborhood the existence of three equilibrium states, a momentless one and two moment ones.

33.3. Theorems 33.1 and 33.2 allow us to introduce a new important indicator $\lambda_{t\kappa l}$ in the problem of stability. Let $\lambda_{t\kappa l}$ be a value such that for $\lambda < \lambda_{t\kappa l}$ the shell has a unique MIS, while in every interval $\lambda_{t\kappa l} \le \lambda < \lambda_{t\kappa l} + \epsilon$ there exists at least one moment state. In other words, the semi-infinite interval $\lambda < \lambda_{t\kappa l}$ belongs to RNOE (31.8), (31.9), while in any interval $\lambda_{t\kappa l} \le \lambda < \lambda_{t\kappa l} + \epsilon$ there is at least one point of SNOE (31.8). In the terminology of U-sets, the semi-infinite interval $\lambda < \lambda_{t\kappa l}$ is a U_1-set, while any interval $\lambda_{t\kappa l} \le \lambda < \lambda_{t\kappa l} + \epsilon$ contains points of U_k for $k \ge 2$. Below we shall call $\lambda_{t\kappa l}$ the *lower critical value.*

Theorem 33.4. *Assume that the conditions of Theorem 33.1 are satisfied. Then $\lambda_{t\kappa l}$ exists, and we have the inequalities*

$$\lambda_{t\kappa}^{*} \le \lambda_{t\kappa l} \le \lambda_{t\kappa e}.$$

Theorem 33.3 follows immediately from Theorems 33.1 and 33.2.

In connection with Theorem 33.3, we observe that determining the lower critical value $\lambda_{t\kappa l}$ is the subject of an enormous number of important studies, in which numerical methods were used to that end; a list of such works, which does not claim to be exhaustive, is to be found in Section 25–26. The importance of Theorem 33.3 is that the existence of the lower critical value is established in a sufficiently general situation in a rigorous manner.

In view of the importance of $\lambda_{t\kappa l}$ as an indicator of the carrying capacity of a structure, we present some estimates of that quantity.

Lemma 33.2. *Assume that there is an element $w \in H_\kappa$ such that*

$$\mathcal{I}_{\kappa\kappa}^{\lambda_{t\kappa e}}(w) < 0. \tag{33.23}$$

Then

$$\lambda_{t\kappa 1} < \lambda_{t\kappa e}. \tag{33.24}$$

Indeed, if (33.23) holds, then for sufficiently small δ,

$$\mathcal{I}_{\kappa\kappa}^{\lambda_{t\kappa e}-\delta}(w) < 0,$$

and then by Theorem 21.2 the absolute minimum of $\mathcal{I}_{\kappa\kappa}^{\lambda_{t\kappa e}-\delta}$ in H_κ exists and is also negative.

Lemma 33.3. *Let w_1 be an eigenelement of the linear operator equation* (31.14) *belonging to $\lambda_{t\kappa e}$, and assume*

$$\mathcal{I}_{\kappa\kappa 3}(w_1) \neq 0.$$

Then (33.24) *holds.*

To prove this lemma, we use the following relation, which follows from (33.22):

$$\mathcal{I}_{\kappa\kappa}^{\lambda_{t\kappa e}}(tw_1) = t^3\mathcal{I}_{\kappa\kappa 3}(w_1) + t^4\mathcal{I}_{\kappa\kappa 4}(w_1).$$

It is easily seen that for small t the sign of $\mathcal{I}_{\kappa\kappa}^{\lambda_{t\kappa e}}(tw_1)$ is defined by the sign of the first term, and we can choose the sign of t such that $\mathcal{I}_{\kappa\kappa}^{\lambda_{t\kappa e}}(tw_1)$ is a negative quantity. Then the assertion of Lemma 33.3 follows from Lemma 33.2. Furthermore, a simple computation shows that if

$$\delta < \frac{\mathcal{I}_{\kappa\kappa 3}^2(w_1)}{4\mathcal{I}_{\kappa\kappa 4}(w_1)}, \tag{33.25}$$

then

$$\mathcal{I}_{\kappa\kappa}^{\lambda_{t\kappa e}-\delta} < 0, \tag{33.26}$$

so that the set $\lambda > \lambda_{t\kappa e} - \delta$ belongs to SNOE, more precisely, to U_k, $k \geq 2$. Thus

$$\lambda_{t\kappa 1} \leq \lambda_{t\kappa e}\left(1 - \frac{\mathcal{I}_{\kappa\kappa 3}^2(w_1)}{4\mathcal{I}_{\kappa\kappa 4}(w_1)\lambda_{t\kappa e}}\right). \tag{33.27}$$

In (33.25)–(33.27), w_1 is normalized by the condition $\|w_1\|_{Hc}^2 = 1$.

The importance of the assertions of Lemmas 33.2, 33.3 lies in the fact that they provide us with conditions under which the solution of the problem of stability of MIS on the basis of Euler's linearization principle in a neighborhood of the MIS is impossible, as ms appear for $\lambda < \lambda_{t\kappa e}$, and transition to those states of stress is possible.

Lemma 33.4. *Assume that the inequality*

$$\mathcal{I}_{\kappa\kappa 2}^{\lambda_{t\kappa e}}(w) + \frac{3}{2}\mathcal{I}_{\kappa\kappa 3}(w) + 2\mathcal{I}_{\kappa\kappa 4}(w) \neq 0 \tag{33.28}$$

holds for any $w \in H_\kappa$. *Then*

$$\lambda_{t\kappa l} = \lambda_{t\kappa e}. \tag{33.29}$$

To prove the lemma, we note that from (33.28) it follows that

$$\mathcal{I}_{\kappa\kappa 2}^{\lambda_{t\kappa e}}(w) + \frac{3}{2}\mathcal{I}_{\kappa\kappa 3}(w) + 2\mathcal{I}_{\kappa\kappa 4}(w) > 0. \tag{33.30}$$

Indeed, if we assume that (33.30) does not hold, then for some w^{**} we should have

$$\mathcal{I}_{\kappa\kappa 2}^{\lambda_{t\kappa e}}(w^{**}) + \frac{3}{2}\mathcal{I}_{\kappa\kappa 3}(w^{**}) + 2\mathcal{I}_{\kappa\kappa 4}(w^{**}) < 0. \tag{33.31}$$

But (33.31) can hold only if

$$\mathcal{I}_{\kappa\kappa 3}(w^{**}) < 0, \tag{33.32}$$

since the remaining terms in (33.29) are positive, as is easily seen. But then

$$\mathcal{I}_{\kappa\kappa 2}^{\lambda_{t\kappa e}}(w^{**}) - \frac{3}{2}\mathcal{I}_{\kappa\kappa 3}(w^{**}) + 2\mathcal{I}_{\kappa\kappa 4}(w^{**}) > 0. \tag{33.33}$$

Let us now consider a functional of the form

$$\begin{aligned} F(t) = \mathcal{I}_{\kappa\kappa 2}^{\lambda_{t\kappa e}}((1-2t)w^{**}) &+ \frac{3}{2}\mathcal{I}_{\kappa\kappa 3}((1-2t)w^{**}) \\ &+ 2\mathcal{I}_{\kappa\kappa 4}((1-2t)w^{**}). \end{aligned} \tag{33.34}$$

Obviously, $F(t)$ is a fourth degree polynomial in t. Furthermore, it is easy to see that

$$F(0) < 0, \;\; F(1) > 1,$$

which is a consequence of (33.31), (33.33). Therefore, between 0 and 1 there is a t^* for which the left-hand side of (33.34) is zero, which contradicts (33.28). Thus we have shown that (33.28) implies (33.30).

Moving on now to prove the main claim of Lemma 33.4, let us assume that for some $\lambda < \lambda_{t\kappa e}$ the nonlinear operator equation (32.8), (32.9) has a nontrivial solution w. Then (33.14) will hold. But then we shall have the inequality

$$\begin{aligned} \|w\|_{H_\kappa}^2 + \int_\Omega \mathcal{E}(T_M^{ij})\,d\Omega + 2\Big(\int_{\Gamma_6} k_s^{\tau\tau} w_{\tau M}^2\,ds + \int_{\Gamma_7} k_s^{mm} w_{mM}^2\,ds + \\ + \int_{\Gamma_8} k_s^{ij} w_{iM} w_{jM}\,ds\Big) - \lambda_{t\kappa e}\,\|w\|_{H_C}^2 < 0, \end{aligned} \tag{33.35}$$

where $\mathcal{E}$ is given in Theorem 33.1. If we now take into account (32.12), (33.2) and homogeneity of the functionals $\mathcal{I}_{\kappa\kappa 2}^{\lambda}$, $\mathcal{I}_{\kappa\kappa 3}$, $\mathcal{I}_{\kappa\kappa 4}$, then (33.35) assumes the form

$$\mathcal{I}_{\kappa\kappa 2}^{\lambda_{t\kappa e}}(w) + \frac{3}{2}\mathcal{I}_{\kappa\kappa 3}(w) + 2\mathcal{I}_{\kappa\kappa 4}(w) < 0,$$

which contradicts (33.30), which follows from (33.29). Therefore, $\lambda_{t\kappa l} < \lambda_{t\kappa e}$ cannot happen, and thus (33.29) holds. Lemma 33.4 is proved.

Its importance is in the fact that it provides us with a sufficient criterion to be able to use Euler's linearization criterion around an MIS in the problem of stability.

33.4. Since the lower critical value is an important characteristic of a shell, we shall consider here some additional estimates of $\lambda_{t\kappa 1}$. From (33.14) we obtain

$$\lambda(w) = \frac{E(w)}{\|w\|^2_{Hc}}, \quad E(w) = \|w\|^2_{H_\kappa} + \widetilde{E}(w),$$

$$\widetilde{E}(w) = \int_\Omega T^{ij}_M(\overset{0}{\epsilon}_{ijM} + B_{ij}w)\,d\Omega$$
$$+2\Big(\int_{\Gamma_6} k^{\tau\tau}_s w^2_{\tau M}\,ds + \int_{\Gamma_7} k^{mm}_s w^2_{mM}\,ds + \int_{\Gamma_8} k^{ij}_s w_{iM} w_{jM}\,ds\Big). \quad (33.36)$$

In (33.36), $w_{\tau M}$, w_{mM}, w_{iM} are determined in terms of w from (32.8), and $\widetilde{E}(w)$ can be regarded as a quadratic functional in H_κ.

Theorem 33.5. *Let the conditions of Theorem 16.1 (Theorems 16.2, 16.3, respectively) be satisfied. Assume that the tensor T^{ij}_{MI} is continuous in $\overline{\Omega}$ and positive definite; that is, we have the inequalities*

$$m\left(w^2_{\alpha^1} + w^2_{\alpha^2}\right) \le \left|T^{ij}_{\mathrm{MI}} w_{\alpha^i} w_{\alpha^j}\right| \le M\left(w^2_{\alpha^1} + w^2_{\alpha^2}\right), \; m,\; M > 0. \quad (33.37)$$

Then the functional $\lambda(w)$ given by (33.36) *has a minimum $\lambda_{t\kappa\min}$ at a point $w_{\min}$ (there can be several such points).*

To prove the theorem, we note that $\widetilde{E}$ is essentially the same as the functional $\Phi(w, 1)$ if there we take

$$\overset{0}{w} \equiv 0, \; \mathbf{a}^* \equiv \omega_{\mathrm{p}} \equiv 0.$$

By Theorem 16.1 (or Theorems 16.2, 16.3, respectively), (16.25), and (32.30), for sufficiently large norm $\|w\|_{H_\kappa} = R$ we have

$$\widetilde{E}(w) \ge m\,\|w\|^2_{H_\kappa},$$

and then for $\|w\|_{H_\kappa} \ge R$ we obtain for λ,

$$\lambda(w) \ge m.$$

Thus we have to consider $\lambda(w)$ for $\|w\|_{H_\kappa} \le R$. Due to the structure of λ (33.36) we have

$$\lambda(w) \ge \|w\|^{-2}_{Hc}\Big(\|w\|^2_{H_\kappa} + \int_\Omega T^{ij}_M(\overset{0}{\epsilon}_{ijM} + B_{ij}w)\,d\Omega\Big).$$

Here we have neglected positive line integrals. Next, we have

$$\lambda(w) \ge \widetilde{\lambda} + \|w\|^{-2}_{Hc}\Big(\int_\Omega C_{ijkl}T^{ij}_M T^{kl}_M\,d\Omega - \int_\Omega \left|T^{ij}_M B_{ij}\right| |w|\,d\Omega\Big),$$
$$\widetilde{\lambda} = \min \|w\|^2_{H_\kappa}\|w\|^{-2}_{Hc}.$$

On the right-hand side, λ_1 is given by relation (32.26) of Theorem 32.3, and C_{ijkr} is a positive definite tensor introduced in (4.16). Furthermore, for any $\epsilon > 0$ we have the inequality

$$\int_\Omega \left|T_M^{ij} B_{ij}\right| |w| \, d\Omega \le \frac{1}{2}\left(\epsilon \int_\Omega \left|T_M^{ij} B_{ij}\right|^2 d\Omega + \frac{1}{\epsilon}\int_\Omega |w|^2 \, d\Omega\right),$$

and then

$$\lambda(w) \ge \widetilde{\lambda} + \|w\|_{Hc}^{-2}\left(\int_\Omega \left[C_{ijkl} T_M^{ij} T_M^{kl} - \frac{\epsilon}{2}\left|T_M^{ij} B_{ij}\right|^2\right] d\Omega - \frac{1}{2\epsilon}\int_\Omega |w|^2 \, d\Omega\right).$$

For sufficiently small ϵ we have

$$\lambda(w) \ge \widetilde{\lambda} - \frac{\|w\|_{L_{2\Omega}}^2}{2\epsilon \, \|w\|_{Hc}^2} \ge \lambda_1 - \frac{m}{2\epsilon}\frac{\|w\|_{L_{2\Omega}}^2}{\|w\|_{W_{2\Omega}^{(1)}}^2}.$$

Here we have taken into account the inequality

$$m \, \|w\|_{Hc}^2 \ge \|w\|_{W_{2\Omega}^{(1)}}^2,$$

which follows from Theorem 10.8 (relation (10.12)) if we take $R(w) = \|w\|_{Hc}^2$. Then condition 2 of Theorem 10.8 (relation (10.11)) will hold due to (33.37). Finally, we note that on a sphere $\|w\|_{W_{2\Omega}^{(1)}} = 1$, the functional $\|w\|_{L_{2\Omega}}^2$ has a maximum m_1, and then we obtain

$$\lambda(w) > \widetilde{\lambda} - \frac{m m_1}{2\epsilon}.$$

Therefore, the functional $\lambda(w)$ is bounded from below on the entire space H_κ, from which we infer the existence of the lower bound of $\lambda(w)$ in H_κ, which we denote by $\lambda_{t\kappa\min}$.

Let us show that there exists at least one element $w_{\min}$ such that

$$\lambda_{t\kappa\min} = \frac{E(w_{\min})}{\|w_{\min}\|_{Hc}^2}.$$

The arguments one uses here largely repeat the proof of Theorem 21.4 and will be only briefly sketched. Let w_k be a minimizing sequence for $\lambda(w)$. We have already shown that w_k is bounded and weakly compact. We shall take it to be weakly convergent, and $w_k \rightharpoonup w_0$; initially, we assume that $w_0 \equiv 0$. Then we have the identity

$$\begin{aligned}\left\|\frac{w_m - w_n}{2}\right\|_{H_\kappa}^2 &= \|w_m\|_{Hc}^2 \frac{1}{2}\frac{\|w_m\|_{H_\kappa}^2 + \widetilde{E}(w_m)}{\|w_m\|_{Hc}^2} + \|w_n\|_{Hc}^2 \frac{1}{2}\frac{\|w_n\|_{H_\kappa}^2 + \widetilde{E}(w_n)}{\|w_n\|_{Hc}^2} \\ &\quad - \left\|\frac{w_m + w_n}{2}\right\|_{Hc}^2 \frac{\left\|\frac{w_m+w_n}{2}\right\|_{H_\kappa}^2 + \widetilde{E}\left(\frac{w_m+w_n}{2}\right)}{\left\|\frac{w_m+w_n}{2}\right\|_{Hc}^2} \\ &\quad - \frac{1}{2}\widetilde{E}(w_m) - \frac{1}{2}\widetilde{E}(w_n) + \widetilde{E}\left(\frac{w_m + w_n}{2}\right).\end{aligned} \tag{33.38}$$

We recall now that as can be seen from (33.35), $\widetilde{E}$, $\|w\|^2_{HC}$ are weakly continuous functionals in H_κ, which follows from (14.26) and Theorem 12.3 (see (12.29)). Next, for sufficiently large m and n we have

$$\begin{aligned} \lambda_{t\kappa\min} &\le \frac{\|w_m\|^2_{H_\kappa} + \widetilde{E}(w_m)}{\|w_m\|^2_{HC}} \le \lambda_{t\kappa\min} + \epsilon_m, \\ \lambda_{t\kappa\min} &\le \frac{\|w_n\|^2_{H_\kappa} + \widetilde{E}(w_n)}{\|w_n\|^2_{HC}} \le \lambda_{t\kappa\min} + \epsilon_n, \end{aligned} \tag{33.39}$$

where ϵ_m, $\epsilon_n \to 0$ as m, $n \to \infty$; finally,

$$\lambda_{t\kappa\min} \le \frac{\left\|\frac{w_m+w_n}{2}\right\|^2_{H_\kappa} + \widetilde{E}\left(\frac{w_m+w_n}{2}\right)}{\left\|\frac{w_m+w_n}{2}\right\|^2_{HC}}. \tag{33.40}$$

Relations (33.39), (33.40) follow from the definition of $w_{\min}$ as giving the lower bound of $E(w)\,\|w\|^{-2}_{HC}$. From (33.39), (33.40) it follows that the right-hand side of (33.38) vanishes for sufficiently large m, n, which in turn implies strong convergence of w_m. This concludes the proof of Theorem 33.4.

Theorem 33.6. *Assume that the conditions of Theorem 33.1 hold. Then*

$$\lambda_{t\kappa\min} \le \lambda_{t\kappa 1}. \tag{33.41}$$

Indeed, by definition $\lambda_{t\kappa 1}$ can be interpreted as the lower bound of the values of $\lambda(w)$ on the set of solutions w of the nonlinear operator equation (32.9); we have defined $\lambda_{t\kappa\min}$ as the lower bound of λ on the entire space H_κ, that is, on a larger set, so that (33.41) follows.

33.5. The equation (32.8), (33.11) can be can be reduced to an operator equation of the form of (13.30). Since the method of reduction was presented in detail in Section 13, we quote only the final result:

$$w = \overset{0}{\mathbf{G}}_{\kappa\kappa}(w) + \lambda \overset{1}{\mathbf{G}}_{\kappa\kappa}(w), \tag{33.42}$$

where

$$(\overset{0}{\mathbf{G}}_{\kappa\kappa}(w)\cdot\varphi)_{H_\kappa} = \int_\Omega T^{ij}\left(B_{ij}\varphi - w_{\alpha^i}\varphi_{\alpha^j}\right)\,d\Omega,$$

$$(\overset{1}{\mathbf{G}}_{\kappa\kappa}(w)\cdot\varphi)_{H_\kappa} = (w\cdot\varphi)_{HC}.$$

Definition 33.1. A solution $w \neq 0$ of the nonlinear operator equation (32.8), (33.11) corresponding to λ_0 will be called *nonsingular* if $\sigma = 1$ is not an eigenvalue of the equation

$$w = \sigma\left(\operatorname{grad}_{H_\kappa} \overset{0}{\mathbf{G}}_{\kappa\kappa}(w) + \lambda_0 \operatorname{grad}_{H_\kappa} \overset{1}{\mathbf{G}}_{\kappa\kappa}(w)\right) \text{ at the point } w_0.$$

Lemma 33.5. *Let w_0, λ_0 be a nonsingular solution of (33.42). Then the interval*

$$\lambda_0 - \delta \le \lambda \le \lambda_0 + \delta$$

belongs to SNOE (33.42), *so that* (32.8), (33.11) *hold.*

To prove Lemma 33.5, we write (33.42) for $\lambda = \lambda_0 + \delta$. We have

$$w = \overset{0}{\mathbf{G}}_{\kappa\kappa}(w) + \lambda_0 \overset{1}{\mathbf{G}}_{\kappa\kappa}(w) + \delta \overset{1}{\mathbf{G}}_{\kappa\kappa}(w). \tag{33.43}$$

Equation (33.43) is of the type we considered in Section 23. Lemma 33.5 is proved almost precisely in the same way as Theorem 23.1; we omit the details.

Lemma 33.6. *Let w_{l} be an eigenfunction corresponding to $\lambda_{t\kappa\mathrm{l}}$. Then w_{l} is a singular solution of equation* (33.42) *for $\lambda = \lambda_{t\kappa\mathrm{l}}$.*

Indeed, if we assume that w_{l} is a nonsingular solution, then by Lemma 33.5 the interval $\lambda_{t\kappa\mathrm{l}} - \delta \le \lambda \le \lambda_{t\kappa\mathrm{l}}$ belongs to the spectrum of the nonlinear operator equation (33.10), (33.11), which is impossible by definition of $\lambda_{t\kappa\mathrm{l}}$.

33.6. Let us present some general facts characterizing the behavior of shells in problem $t\kappa$ after the loss of stability.

Theorem 33.7. *Assume that the conditions of Theorem 33.1 are satisfied. Then each surface in the space H_κ defined by the equation*

$$\mathcal{I}^0_{\kappa\kappa}(w) = \frac{1}{2}\,\|w\|^2_H + \mathcal{I}_{\kappa\kappa 3}(w) + \mathcal{I}_{\kappa\kappa 4}(w) = c, \ \ c \ge \min \mathcal{I}^0_{\kappa\kappa}(w) \ on \ H_\kappa, \tag{33.44}$$

contains at least one solution of the nonlinear operator equation (33.10), (33.11), *and therefore of* (33.42).

To prove the theorem we note that the set

$$\mathcal{I}^0_{\kappa\kappa}(w) \le c \tag{33.45}$$

is weakly closed [319] in H_κ. Indeed, let $w_k \in \{\mathcal{I}^0_{\kappa\kappa}(w) \le c\}$, and $w_k \rightharpoonup w_0$. Let us show that $w_0 \in \{\mathcal{I}^0_{\kappa\kappa}(w) \le c\}$. We have

$$\frac{1}{2}\,\|w_k\|^2_H + \mathcal{I}_{\kappa\kappa 3}(w_k) + \mathcal{I}_{\kappa\kappa 4}(w_k) \le c. \tag{33.46}$$

From (33.46) it follows that

$$\frac{1}{2}\underline{\lim}\,\|w_k\|^2_H + \mathcal{I}_{\kappa\kappa 3}(w_0) + \mathcal{I}_{\kappa\kappa 4}(w_0) \le c. \tag{33.47}$$

But since $w_k \rightharpoonup w_0$,

$$\frac{1}{2}\,\|w_0\|^2_H \le \frac{1}{2}\underline{\lim}\,\|w_k\|^2_H\,. \tag{33.48}$$

Weak closure of (33.45) follows from (33.47), (33.48).

Now we observe that $\|w\|^2_{Hc}$ is a weakly continuous functional in H, so that $\|w\|^2_{Hc}$ reaches a maximum on (33.45). If we assume that this point is interior to (33.45), we will have there

$$\mathrm{grad}_{H_\kappa}\,\|w\|^2_{Hc} = 0. \tag{33.49}$$

But since $\|w\|_{Hc}^2$ is a homogeneous quadratic functional in H, from (33.49) it follows that $w \equiv 0$, which is impossible, since the maximum of $\|w\|_{Hc}^2$ on (33.45) cannot be zero. Therefore, the maximum is achieved at some point of (33.44). But by Lyusternik's theorem [281, 312], at that point

$$\operatorname{grad}_{H_\kappa} \mathcal{I}_{\kappa\kappa}^0(w) = \lambda(c)\operatorname{grad}_{H_\kappa} \|w\|_{Hc}^2, \tag{33.50}$$

where $\lambda(c)$ is a function of the level c.

Theorem 33.8. *Assume that the conditions of Theorem 33.1 are satisfied. Then any surface of H_κ defined by the equation*

$$\|w\|_{Hc}^2 = c^2 > 0 \tag{33.51}$$

contains at least one eigenelement λ, w of the nonlinear operator equation (33.10), (33.11), *and therefore of* (33.42).

The proof of this theorem is based on the arguments used in Theorem 21.3, and therefore will be only briefly sketched. First of all, let us note that we have the representation

$$\mathcal{I}_{\kappa\kappa}^0(w) = \frac{1}{2}\|w\|_H^2 + \widetilde{\mathcal{I}}_{\kappa\kappa}^0(w), \tag{33.52}$$

where $\widetilde{\mathcal{I}}_{\kappa\kappa}^0(w)$ is some weakly continuous functional in H_κ.

Furthermore, $\mathcal{I}_{\kappa\kappa}^0(w)$ is bounded from above in H_κ, since this functional coincides with the functional $\mathcal{I}_{\kappa\kappa}(w)$ defined by (21.1) if $\mathbf{a}^* \equiv \omega_{\mathrm{p}}^* \equiv 0$, $w_{\mathrm{p}}^* \equiv 0$, and its boundedness from below follows from Theorem 21.1 for the conditions of Theorem 16.1. Therefore, if w_k is an absolutely minimizing sequence for $\mathcal{I}_{\kappa\kappa}^0(w)$, it can be taken to be weakly convergent in H_κ; so assume that $w_k \rightharpoonup w_0$. Let us show that in fact,

$$w_k \to w_0.$$

For this let us associate with each pair w_k, w_l of terms of the sequence w_k a number α_{kl} such that

$$\left\|\alpha_{kl}\frac{w_k + w_l}{2}\right\|_{Hc} = c^2.$$

In Section 32 we showed that

$$\alpha_{kl} \to 1 \text{ as } k,\ l \to \infty.$$

Moreover, we have

$$\begin{aligned}\left\|\frac{w_k - w_l}{2}\alpha_{kl}\right\|_H^2 &= \frac{\alpha_{kl}^2}{2}\|w_k\|_H^2 + \frac{\alpha_{kl}^2}{2}\|w_l\|_H^2 - \left\|\frac{w_k + w_l}{2}\alpha_{kl}\right\|_H^2 \\ &+ \widetilde{\mathcal{I}}_{\kappa\kappa}^0(\alpha_{kl}w_k) + \widetilde{\mathcal{I}}_{\kappa\kappa}^0(\alpha_{kl}w_l) - 2\widetilde{\mathcal{I}}_{\kappa\kappa}^0\Big(\frac{w_k + w_l}{2}\alpha_{kl}\Big) \qquad (33.53) \\ &+ 2\widetilde{\mathcal{I}}_{\kappa\kappa}^0\Big(\frac{w_k + w_l}{2}\alpha_{kl}\Big) - \widetilde{\mathcal{I}}_{\kappa\kappa}^0(\alpha_{kl}w_k) - \widetilde{\mathcal{I}}_{\kappa\kappa}^0(\alpha_{kl}w_l).\end{aligned}$$

Let us now take into account the relations

$$\mathcal{I}^0_{\kappa\kappa}(\alpha_{kl} w_k) \to d,\ \mathcal{I}^0_{\kappa\kappa}(\alpha_{kl} w_l) \to d, \quad \text{as } k,\ l \to \infty.$$

But then for large $k,\ l$ we have the inequalities

$$\begin{aligned} d &\le \frac{\alpha^2_{kl}}{2}\, \|w_k\|^2_H + \widetilde{\mathcal{I}}^0_{\kappa\kappa}(\alpha_{kl} w_k) \le d + \epsilon_{kl}, \\ d &\le \frac{\alpha^2_{kl}}{2}\, \|w_l\|^2_H + \widetilde{\mathcal{I}}^0_{\kappa\kappa}(\alpha_{kl} w_l) \le d + \epsilon_{kl}, \\ 2d &\le \left\| \frac{w_k + w_l}{2} \alpha_{kl} \right\|^2_H + 2\widetilde{\mathcal{I}}^0_{\kappa\kappa}\Big(\frac{w_k + w_l}{2} \alpha_{kl}\Big), \quad \epsilon_{kl} \to 0 \text{ as } k,\ l \to \infty. \end{aligned} \tag{33.54}$$

From (33.53) , (33.54) we have

$$\begin{aligned} \|w_k - w_l\|^2_H \le 8d + 8\epsilon_{kl} - 8d + 8\widetilde{\mathcal{I}}^0_{\kappa\kappa}\Big(\frac{w_k + w_l}{2} \alpha_{kl}\Big) \\ - 4\widetilde{\mathcal{I}}^0_{\kappa\kappa}(\alpha_{kl} w_k) - 4\widetilde{\mathcal{I}}^0_{\kappa\kappa}(\alpha_{kl} w_l). \end{aligned}$$

Due to weak continuity of $\mathcal{I}^0_{\kappa\kappa}(w)$ in H as well as the fact that $\epsilon_{kl} \to 0$, we have from (33.53) that

$$\|w_k - w_l\|_H \to 0.$$

Therefore, w_k converges strongly in H to w_0, and then the functional $\mathcal{I}^0_{\kappa\kappa}(w)$ defined by (33.52) has a minimum at the point w_0 on (33.51). Hence by a theorem of [312] we have (33.50) and the existence of an eigenelement of the nonlinear operator equation (32.8), (32.9), and thus (33.42) is proved.

34. Global Stability of Shells in Problems 9κ. Existence of Lower Critical Values. Some Estimates for U-Decompositions

34.1. Let us consider the nonlinear operator equation (32.19), (32.20). Since the method of analyzing these equations is only technically different from the arguments used in Section 33, we restrict ourselves to stating the main results. First of all, let us note that the concepts of the spectrum of a nonlinear operator equation and of the regular set of a nonlinear operator equation are introduced here precisely in the same way as in problems $t\kappa$. The expression for the potential energy $\mathcal{I}^\lambda_\kappa(w)$ has the form

$$\mathcal{I}^\lambda_\kappa(w) = \frac{1}{2}\big(\|w\|^2_H - \lambda\, \|w\|^2_{Hc} \big) + \mathcal{I}_{9\kappa 3}(w) + \mathcal{I}_{9\kappa 4}(w), \tag{34.1}$$

where

$$\|w\|^2_H = \|w\|^2_{H_\kappa} + \int_\Omega C^{ijkl}_{\mathrm{s}} \nabla_{ij} \Psi_1 \nabla_{kl} \Psi_1 \, d\Omega, \tag{34.2}$$

$$\mathcal{I}_{9\kappa 3}(w) = 2\int_\Omega C_{\mathrm{s}}^{ijkl}\nabla_{ij}\Psi_1\nabla_{kl}\Psi_2\, d\Omega, \tag{34.3}$$

$$\mathcal{I}_{9\kappa 4}(w) = \int_\Omega C_{\mathrm{s}}^{ijkl}\nabla_{ij}\Psi_2\nabla_{kl}\Psi_2\, d\Omega, \tag{34.4}$$

In (34.2)–(34.4), Ψ_1, Ψ_2 are expressed in terms of w by (32.21), (32.22). Equation (34.1) is obtained if in (7.25) we perform the substitution (32.18) and take into account conditions (32.16), (32.17), which allow the existence of an MIS.

34.2. For the considerations below we need the relation

$$\Phi(w,\ 1) = 0, \tag{34.5}$$

derived using the MIS conditions (32.16), (32.17).

Equation (34.5) can be obtained if we substitute (32.18) in (19.2). However, it is easier to obtain (34.5) if we set $\varphi = w$ in (32.19) and take (32.20) into account. We have

$$\|w\|^2_{H_\kappa} = \int_\Omega C^{ik}C^{jl}(B_{ij}w - w_{\alpha^i}w_{\alpha^j})\nabla_{kl}\Psi_M\, d\Omega + \lambda\,\|w\|^2_{Hc}\,. \tag{34.6}$$

Substituting $\theta = \Psi_M$ in (32.20), we obtain

$$\|\Psi_M\|^2_{H_9} = \int_\Omega (w_{\alpha^i}w_{\alpha^j} - B_{ij}w)C^{ik}C^{jl}\nabla_{kl}\Psi_M\, d\Omega. \tag{34.7}$$

From (34.6)–(34.7) we obtain

$$\|w\|^2_{H_\kappa} + 2\,\|\Psi_M\|^2_{H_9} + \int_\Omega C^{ik}C^{jl}B_{ij}\nabla_{kl}\Psi_M w\, d\Omega - \lambda\,\|w\|^2_{Hc} = 0. \tag{34.8}$$

Equation (34.8) is the same as (34.5).

Theorem 34.1. *Assume that the conditions of Theorem 19.1, the conditions* (32.19), (32.17) *for the existence of an MIS, and condition* (32.47) *hold. Then SNOE* (32.19), (32.20) *is to the right of a point $\lambda^*_{9\kappa}$. All the points $\lambda < \lambda^*_{9\kappa}$ are in RNOE* (32.19), (32.20).

To prove Theorem 34.1, we consider a collection of terms in (34.8) defined by

$$\mathcal{E}(\Psi_M) = 2\,\|\Psi_M\|^2_{H_9} + \int_\Omega C^{ik}C^{jl}B_{ij}\nabla_{kl}\Psi_M w\, d\Omega.$$

It can be studied by the methods of Section 29 as we did for the functional $\mathcal{E}(T^{ij}_M)$, and it can be shown that it has a minimum defined by the relation

$$\mathcal{E}_{\min}(\Psi_M) = -\frac{1}{8}\int_\Omega N_{9\kappa}w^2\, d\Omega;$$

$$N_{9\kappa} = C_{ijkl,\mathrm{s}}C^{iq}C^{jr}C^{ks}C^{lt}\widetilde{B}_{qr}\widetilde{B}_{st}.$$

Here $N_{9\kappa}$ is a complex describing the elastic and geometric properties of the shell introduced in (29.35). From (34.8) we obtain

$$\|w\|^2_{H_\kappa} - \lambda\,\|w\|^2_{Hc} - \frac{1}{8}\int_\Omega N_{9\kappa}w^2\, d\Omega \le 0. \tag{34.9}$$

Let us now introduce the parameter

$$\lambda_{9\kappa}^{*} = \min\left(\|w\|_{H_\kappa}^{2} - \frac{1}{8}\int_{\Omega} N_{9\kappa} w^{2}\, d\Omega\right) \text{ for } \|w\|_{Hc}^{2} = 1. \tag{34.10}$$

Existence of $\lambda_{9\kappa}^{*}$ is guaranteed by Lemma 33.1. From (34.9), (34.10) it follows that

$$(\lambda_{9\kappa}^{*} - \lambda)\,\|w\|_{Hc}^{2} \leq 0,$$

from which for $\lambda < \lambda_{9\kappa}^{*}$ it follows that

$$\|w\|_{Hc}^{2} = 0,$$

so that $w \equiv 0$. The proof of Theorem 34.1 is complete.

Clearly, by definition, $\lambda_{9\kappa}^{*}$ is the smallest eigenvalue of the LOE

$$(w \cdot \varphi)_{H_\kappa} - \frac{1}{4}\int_{\Omega} N_{9\kappa} w\varphi\, d\Omega - \lambda(w \cdot \varphi)_{Hc} = 0,$$

which can be written in the form

$$\widetilde{\nabla}_{ij} D D_{\mathrm{f}}^{ijkl} \nabla_{kl} w - \frac{1}{4} N_{9\kappa} w D + \lambda_{9\kappa}^{*}\left(T_{\mathrm{MI}}^{ij} w_{\alpha^i} D\right)_{\alpha^j} = 0$$

with the boundary conditions (33.20), (33.21).

Theorem 34.2. *Let the conditions of Theorem 34.1 be satisfied. Then the unbounded interval* $\lambda > \lambda_{9\kappa\mathrm{e}}$ *is in the SNOE* (32.19), (32.20). *In addition to the MIS, the shell will also have a moment strained state with lower total energy of the system shell/elastic supports/external loads.*

The proof of Theorem 34.2 is exactly the same as that of Theorem 33.2, the only difference being that instead of $\mathcal{I}_{t\kappa}^{\lambda}$ we use the functional $\mathcal{I}_{9\kappa}^{\lambda}$ defined by (34.1). So we will not repeat the arguments.

Thus in problem 9κ the unbounded interval $\lambda > \lambda_{9\kappa\mathrm{e}}$ belongs to U_k, $k \geq 2$, and in this problem Theorem 34.2 reveals the important role played by Euler's characteristic $\lambda_{9\kappa\mathrm{e}}$ in the global analysis of the problem of stability, which has to do with the fact that for $\lambda > \lambda_{9\kappa\mathrm{e}}$ the moment strained state that appears is more stable than the momentless one. Here we also note that in fact for $\lambda > \lambda_{9\kappa\mathrm{e}}$ there emerge two moment strained states. In a number of cases this can be shown by analytic methods for λ slightly above $\lambda_{9\kappa\mathrm{e}}$.

34.3. As in Problem $t\kappa$, we introduce $\lambda_{9\kappa\mathrm{l}}$ as the value such that for $\lambda < \lambda_{9\kappa\mathrm{l}}$ the shell has a unique MIS, while in a small interval $\lambda_{9\kappa\mathrm{l}} \leq \lambda < \lambda_{9\kappa\mathrm{l}} + \epsilon$ there exists at least one moment strained state.

Theorem 34.3. *Assume that the conditions of Theorem 34.1 are satisfied. Then* $\lambda_{9\kappa\mathrm{l}}$ *exists, and we have the inequalities*

$$\lambda_{9\kappa}^{*} \leq \lambda_{9\kappa\mathrm{l}} \leq \lambda_{9\kappa\mathrm{e}}. \tag{34.11}$$

Theorem 34.3 is a direct consequence of Theorems 34.1 and 34.2.

Under the conditions of problem 9κ, computation of $\lambda_{9\kappa l}$ is the subject of an enormous number of studies. A list of these papers, which is far from being exhaustive, is given in Sections 25–28; see the references in Chapter VI. Theorem 34.3 establishes the existence of the lower critical value as a rigorous mathematical fact.

Let us present some estimates of $\lambda_{9\kappa l}$.

Lemma 34.1. *Assume that under the conditions of Theorem 34.1 there exists an element $w' \in H_\kappa$ such that*

$$\mathcal{I}_\kappa^{\lambda_{9\kappa e}}(w') < 0.$$

Then we have the strong inequality

$$\lambda_{9\kappa l} < \lambda_{9\kappa e}. \tag{34.12}$$

Lemma 34.2. *Let w_1 be an eigenfunction of the LOE* (33.21) *belonging to $\lambda_{9\kappa e}$ and*

$$\mathcal{I}_{\kappa 3}(w_1) \neq 0.$$

Then (34.12) *holds. We also have the inequality*

$$\lambda_{9\kappa l} \leq \lambda_{9\kappa e}\Big(1 - \frac{\mathcal{I}_{\kappa 3}^2(w_1)}{4\mathcal{I}_{\kappa 4}(w_1)\lambda_{9\kappa e}}\Big). \tag{34.13}$$

In (34.13), *w_1 is normalized by the condition $\|w_1\|_{Hc}^2 = 1$.*

The proofs of Lemmas 34.1, 34.2 are analogous to those of Lemmas 33.2, 33.3, and are thus omitted. As in Problem $t\kappa$ these lemmas assert that it is impossible to use Euler's linearization principle in a neighborhood of an MIS in order to resolve questions of stability, since moment strained states already appear for $\lambda < \lambda_e$ and transition to these states of stress is possible.

Lemma 34.3. *Let the condition*

$$\mathcal{I}_{\kappa 2}^{\lambda_{9\kappa e}}(w) + \frac{3}{2}\mathcal{I}_{\kappa 3}(w) + 2\mathcal{I}_{\kappa 4}(w) \neq 0$$

hold for any $w \in H_\kappa$. Then

$$\lambda_{9\kappa l} = \lambda_{9\kappa e}.$$

34.4. Let us consider some additional estimates for $\lambda_{9\kappa l}$. For this we introduce the functional

$$\begin{aligned} \lambda(w) &= \frac{\mathcal{E}(w)}{\|w\|_{Hc}^2}, \\ \mathcal{E}(w) &= \|w\|_{H_\kappa}^2 - \int_\Omega c^{ik}c^{jl}\left(B_{ij}w - w_{\alpha^i}w_{\alpha^j}\right)\nabla_{kl}\Psi_M\, d\Omega, \end{aligned} \tag{34.14}$$

in which Ψ_M is defined in terms of w by (32.20).

Theorem 34.4. *Assume that the conditions of Theorem 19.3 and in addition, relations* (33.37) *are satisfied. Then the functional* $\lambda(w)$ *defined by* (34.14) *attains an absolute minimum* $\lambda_{9\kappa\min}$ *at a point* $w_{\min} \in H_\kappa$ *(there can be more than one such point).*

Theorem 34.5. *Assume that the conditions of Theorem 34.4 are satisfied. Then we have the inequalities*

$$\lambda_{9\kappa\min} \le \lambda_{9\kappa 1}. \tag{34.15}$$

34.5. The nonlinear operator equation (32.19), (32.20) can be reduced to an operator equation of the form of (17.19) and (17.20):

$$w = \overset{0}{\mathbf{G}}_\kappa(w) + \lambda \overset{1}{\mathbf{G}}_\kappa(w), \tag{34.16}$$

where

$$(\overset{0}{\mathbf{G}}_\kappa(w)\cdot\varphi)_{H_\kappa} = \int_\Omega C^{ik}C^{jl}(B_{ij}\varphi - \varphi_{\alpha^i} w_{\alpha^j})\nabla_{kl}\Psi_M\, d\Omega,$$

$$(\overset{1}{\mathbf{G}}_\kappa(w)\cdot\varphi)_{H_\kappa} = (w\cdot\varphi)_{H_C}.$$

Definition 34.1. A solution $w_0 \not\equiv 0$, λ_0 of (32.19),(32.20) will be called *nonsingular* if the linear operator equation

$$w = \sigma(\operatorname{grad}_{H_\kappa} \overset{0}{\mathbf{G}}_\kappa(w) + \lambda_0 \operatorname{grad}_{H_\kappa} \overset{1}{\mathbf{G}}_\kappa(w))$$

does not have $\sigma = 1$ as an eigenvalue.

Lemma 34.4. *Let* w_0, λ_0 *be a nonsingular solution of* (34.16). *Then the interval*

$$\lambda_0 - \delta \le \lambda \le \lambda_0 + \delta$$

for sufficiently small δ *belongs to SNOE of* (32.19), (32.20) *and therefore of* (34.16).

Lemma 34.5. *Let* w_0 *be any solution of* (32.19), (32.20) *corresponding to* $\lambda_{9\kappa 1}$. *Then* w_0, $\lambda_{9\kappa 1}$ *is a singular solution of* (32.19), (32.20), (34.16).

34.6. Let us present some general results concerning the behavior of shells in Problem 9κ after the loss of stability.

Theorem 34.6. *If the conditions of Theorem 34.1 are satisfied, every surface in the space* H_κ *defined by the equation*

$$\mathcal{I}^0_\kappa(w) = \frac{1}{2}\|w\|^2_H + \mathcal{I}_{\kappa 3}(w) + \mathcal{I}_{\kappa 4}(w) = c,\ c \ge \min \mathcal{I}^0_\kappa(w) \text{ in } H_\kappa, \tag{34.17}$$

where $\|w\|^2_H$ *is given by the relation* (32.54), *contains at least one solution of the equation* (32.19), (32.20) *and therefore of* (34.16).

Theorem 34.7. *Assume that the conditions of Theorem 34.1 are satisfied. Then every surface in the space* H_κ *defined by the equation*

$$\|w\|^2_{H_C} = c^2 \tag{34.18}$$

contains at least one eigenelement w, λ *of the nonlinear operator equation* (32.19), (32.20) *and therefore of* (34.16).

35. Bifurcation of Solutions in a Neighborhood of the Momentless State

35.1.

Definition 35.1. We say that a point $\bar{\lambda} \in (-\infty, \infty)$ is a *bifurcation point* of the nonlinear operator equation (32.8), (32.9) if for any arbitrarily small ϵ there exists a solution $w(\lambda)$ such that

$$\|w(\lambda)\|_{H_\kappa} \leq \delta(\epsilon), \ \left|\lambda - \bar{\lambda}\right| \leq \epsilon, \ \delta(\epsilon) \to 0, \ \epsilon \to 0.$$

The study of bifurcations in the nonlinear theory of elasticity (actually in natural sciences in general) was initiated, as we already mentioned, by Euler [76]. The present section is devoted to establishing a connection between bifurcation of solutions of the nonlinear operator equations (32.8), (32.9), (32.19), (32.20) and eigenelements of the linear operator equation (32.13), (32.23).

Theorem 35.1. *Assume that the conditions of Theorem 16.1 hold. Then* $\bar{\lambda}$ *is a bifurcation point of the nonlinear operator equation* (32.8), (32.9) *if and only if* $\bar{\lambda}$ *belongs to the spectrum of the linear operator equation* (32.13), *that is, if and only if* $\bar{\lambda}$ *is an eigenvalue of the linear operator equation* (32.13).

To prove necessity in Theorem 35.1, we represent (33.42) in the following form:

$$w = \mathbf{G}^0_{\kappa\kappa 1}(w) + \mathbf{G}^0_{\kappa\kappa 2}(w) + \mathbf{G}^0_{\kappa\kappa 3}(w) + \lambda \mathbf{G}^1_{\kappa\kappa 1}(w). \tag{35.1}$$

Here the operators $\mathbf{G}^0_{\kappa\kappa\mu}$, $\mu = 1, 2, 3$, are given by the relations (14.18)–(14.20), in which we should put $a^* \equiv T_0^{ij} \equiv 0$. Let $\bar{\lambda}$ be a bifurcation point; then there must exist a sequence of solutions of (35.1), $w_k \not\equiv 0$, λ_k, such that

$$\|w_k\|_{H_\kappa} \to 0, \ \lambda_k \to \bar{\lambda}, \ k \to \infty. \tag{35.2}$$

Furthermore, assume that $\bar{\lambda}$ does not belong to the spectrum of the LOE (32.13). Then we have

$$\begin{aligned} w_k - \mathbf{G}^0_{\kappa\kappa 1}(w_k) - \bar{\lambda}\mathbf{G}^1_{\kappa\kappa 1}(w_k) = {} & \mathbf{G}^0_{\kappa\kappa 2}(w_k) + \mathbf{G}^0_{\kappa\kappa 3}(w_k) \\ & + (\lambda_k - \bar{\lambda})\mathbf{G}^1_{\kappa\kappa 1}(w_k), \end{aligned}$$

and so

$$\begin{aligned} w_k = (\mathbf{I} - \mathbf{G}^0_{\kappa\kappa 1} - \bar{\lambda}\mathbf{G}^1_{\kappa\kappa 1})^{-1}\{ & \mathbf{G}^0_{\kappa\kappa 2}(w_k) + \mathbf{G}^0_{\kappa\kappa 3}(w_k) \\ & + (\lambda_k - \bar{\lambda})\mathbf{G}^1_{\kappa\kappa 1}(w_k)\}, \end{aligned} \tag{35.3}$$

and, finally, from (35.3) we obtain

$$\|w_k\|_{H_\kappa} \leq m\left(\|w_k\|_{H_\kappa} + \|w_k\|^2_{H_\kappa} + (\lambda_k - \bar{\lambda})\right)\|w_k\|_{H_\kappa}. \tag{35.4}$$

In the derivation of (35.4) we took into account (14.4) and the inequality

$$\|(\mathbf{I} - \mathbf{G}^0_{\kappa\kappa 1} - \overline{\lambda}\mathbf{G}^1_{\kappa\kappa 1})^{-1}\| \leq m,$$

which follows from the definition of this operator. However, it easy to see that (35.2) contradicts (35.4). Therefore, if $\overline{\lambda}$ does not belong to the spectrum of the LOE (32.13), it cannot be a bifurcation point.

In the proof of sufficiency in Theorem 35.1 it would have been natural to refer to the well-known theorem of Krasnosel'skii [162]. However, the nonlinear operator equation (33.42) and thus (35.1) do not fit into the type of equations considered in [162], where equations of the type

$$w = \lambda \mathbf{A}(w) \tag{35.5}$$

are studied, in which **A** is a potential completely continuous operator. The structure of (35.1) is significantly different from that of (35.5). However, the variational considerations that constitute the basis of the analysis of (35.5) can still be applied in our case. For this let us note that solving (35.1) is equivalent to looking for conditional extrema of the functional $\frac{1}{2}\|w\|^2_{Hc}$ on the surface

$$\frac{1}{2}\|w\|^2_H + \mathcal{I}_{\kappa\kappa 3}(w) + \mathcal{I}_{\kappa\kappa 4}(w) = \frac{1}{2}\epsilon^2, \tag{35.6}$$

and to prove sufficiency of the conditions of Theorem 35.1 we have to show that for each $\overline{\lambda}_k$ in the SLOE (32.13) there is a sequence as in (35.2). To this end, let us construct in H_κ an operator mapping of the form

$$w = \epsilon \overline{w}\mu(\overline{w}), \tag{35.7}$$

where

$$\mu^2(\overline{w}) + 2\epsilon\mathcal{I}_{\kappa\kappa 3}(\overline{w})\mu^3(\overline{w}) + 2\epsilon^2\mathcal{I}_{\kappa\kappa 4}(\overline{w})\mu^4(\overline{w}) = 1. \tag{35.8}$$

In (35.7) the functional $\mu(\overline{w})$ is determined from (35.8).

Lemma 35.1. *For small ϵ the mapping* (35.7), (35.8) *is a homeomorphism in H under which the surface* (35.9) *becomes the sphere $\|\overline{w}\|_H = 1$.*

To prove Lemma 35.1, let us note that from (35.7), (35.8) we can for small ϵ express $\overline{w}$ in terms of w. For this we must agree to choose in (35.8) the only positive root for small ϵ. This mapping can be written in the form

$$\overline{w} = \frac{w}{\epsilon} + \cdots, \quad \mu = 1 + \epsilon\mu_1(w) + \cdots. \tag{35.9}$$

From (35.9) we can find $w \mapsto \overline{w}$ for small ϵ.

Thus the task of finding eigenelements of (35.1) has been reduced to that of finding extrema of the functional $\mathcal{I}(\overline{w}) = \mu^2(\overline{w})\,\|\overline{w}\|^2_{Hc}$ on the sphere $\Sigma\ :\ \|\overline{w}\|^2_H = 1$. Further considerations require certain estimates.

Lemma 35.2. *For sufficiently small ϵ we have the estimates*

$$1 - m\epsilon \leq \mu(\overline{w}) \leq 1 + m\omega, \quad \|\overline{w}\|^2_H = 1. \tag{35.10}$$

The estimate (35.10) is obtained from (35.8) if we set $\|\overline{w}\|_H^2 = 1$. Then from weak continuity of $\mathcal{I}_{\kappa\kappa\mu}$ ($\mu = 3, 4$) we have

$$\left|\mathcal{I}_{\kappa\kappa\mu}(\overline{w})\right| \le m, \tag{35.11}$$

and furthermore, if the conditions of Theorem 16.1 are satisfied, we have the inequality (16.18),

$$\mathcal{I}_{\kappa\kappa 4}(\overline{w}) \ge m_0. \tag{35.12}$$

From (35.8), (35.11), (35.12) we deduce (35.10) for small ϵ, from which, by the way, it also follows that

$$\left|\mu^2 - 1\right| \le m\epsilon. \tag{35.13}$$

Lemma 35.3. *We have the relation*

$$\text{grad } \|\overline{w}\|_{Hc}^2 = 2\mathbf{A}\overline{w}, \tag{35.14}$$

where $\mathbf{A}$ *is a bounded operator in* H *and*

$$\text{grad } \mu = -\frac{3\mathbf{G}_{\kappa\kappa 2}(\overline{w}) + 4\epsilon\mu(\overline{w})\mathbf{G}_{\kappa\kappa 3}(\overline{w})}{1 + 3\mu\epsilon\mathcal{I}_{\kappa\kappa 3}(\overline{w}) + 4\mu^2\epsilon^2\mathcal{I}_{\kappa\kappa 4}(\overline{w})}\epsilon\mu^2(\overline{w}). \tag{35.15}$$

In (35.15) the gradient is taken in the space H.

To prove (35.14), let us note that

$$\frac{d}{d\alpha}\|\overline{w} + \alpha\varphi\|_{Hc}^2 \big|_{\alpha=0} = 2(\overline{w}\cdot\varphi)_{Hc} = 2(\mathbf{A}\overline{w}\cdot\varphi)_H. \tag{35.16}$$

The operator $\mathbf{A}$ on the right-hand side of (35.16) exists and is bounded in view of the inequality (32.30). Equation (35.15) is obtained from (35.9) if we take into account the relations

$$\text{grad}\,\mathcal{I}_{\kappa\kappa\mu}(\overline{w}) = \mu\mathbf{G}_{\kappa\kappa\mu-1},$$

which follow from the definition of the gradient operator [319] and the relations (35.1) for the operators $\mathbf{G}_{\kappa\kappa\mu}$ and (33.8) for $\mathcal{I}_{\iota\kappa\mu}$. More precisely, differentiating (35.8), we have

$$\begin{aligned} 2\mu\text{grad } \mu + 6\epsilon\mathbf{G}_{\kappa\kappa 2}(\overline{w})\mu^3(\overline{w}) + 6\epsilon\mu^2(\overline{w})\mathbf{G}_{\kappa\kappa 2}(w)\text{grad}\mu \\ + 8\epsilon^2\mu^4(\overline{w})\mathbf{G}_{\kappa\kappa 3}(\overline{w}) + 8\epsilon^2\mu^3(\overline{w})\text{grad } \mu\mathcal{I}_{\kappa\kappa}(\overline{w}) = 0. \end{aligned} \tag{35.17}$$

From (35.17) we obtain (35.15). Lemma 35.3 is proved.

Lemma 35.4. *The functional* $\mathcal{I}(\overline{w}) = \mu^2(\overline{w})\,\|\overline{w}\|_{Hc}^2$ *is close to the quadratic functional* $\|\overline{w}\|_{Hc}^2$ *for small* ϵ.

To prove this lemma, we need to establish the following two relations on the sphere $\Sigma\ :\ \|\overline{w}\|_H^2 = 1$:

$$\begin{aligned} \mathcal{I}(\overline{w}) - \|\overline{w}\|_{Hc}^2 \to 0, \\ \left\|\text{grad}\,\mathcal{I}(\overline{w}) - \text{grad } \|\overline{w}\|_{Hc}^2\right\|_H \to 0,\ \epsilon \to 0. \end{aligned} \tag{35.18}$$

The first of the relations of (35.18) is established immediately:

$$\mathcal{I}(\overline{w}) - \|\overline{w}\|^2_{Hc} = (\mu^2 - 1)\,\|\overline{w}\|^2_{Hc}, \tag{35.19}$$

and then we have to take into account (35.13) and the inequalities (32.30), (32.31). The second of the relations (35.18) has to be considered more carefully:

$$\begin{aligned}\operatorname{grad}\mathcal{I}(\overline{w}) - \operatorname{grad}\,\|\overline{w}\|^2_{Hc} &= \operatorname{grad}\left[(\mu^2 - 1)\,\|\overline{w}\|^2_{Hc}\right]\\ &= 2\mu\operatorname{grad}\mu\,\|\overline{w}\|^2_{Hc} + (\mu^2 - 1)\operatorname{grad}\,\|\overline{w}\|^2_{Hc}\\ &= -2\epsilon\mu^3\frac{3\mathbf{G}_{\kappa\kappa 2}(\overline{w}) + 4\epsilon\mu(\overline{w})\mathbf{G}_{\kappa\kappa 3}(\overline{w})}{1 + 3\mu\epsilon\mathcal{I}_{\kappa\kappa 3}(\overline{w}) + 4\mu^2\epsilon^2\mathcal{I}_{\kappa\kappa 4}(\overline{w})}\,\|\overline{w}\|^2_{Hc}\\ &\quad + (\mu^2 - 1)2\mathbf{A}\overline{w},\ \ \mu = \mu(\overline{w}).\end{aligned} \tag{35.20}$$

In (35.20) we have used (35.14), (35.15). Next, both the summands on the right-hand side vanish as $\epsilon \to 0$, which follows from the first relation in (35.18), boundedness of the operator **A**, inequality (33.25), and (35.11).

Therefore, all the conditions of the theorem concerning critical points of functionals that are close to quadratic ones are satisfied, and we can conclude that if $\overline{\lambda}_k$ is in SLOE (32.13), that is, if it is a critical point of $\|\overline{w}\|^2_{Hc}$ on the surface

$$\|\overline{w}\|^2_H = 1, \tag{35.21}$$

then for sufficiently small ϵ there are numbers λ_k and element $\overline{w}_k$ such that $\overline{w}_k$ are critical points of $\mu^2(\overline{w})\,\|\overline{w}\|^2_{Hc}$ on (35.21), and in addition,

$$\left|\lambda_k - \overline{\lambda}_k\right| \le m\epsilon.$$

But then by (35.7) we have that $w_k = \epsilon\mu(\overline{w}_k)\overline{w}_k$ supplies an extremum of $\|\overline{w}\|^2_{Hc}$ on the surface (35.6). Here

$$\|w_k\|_H \sim \|w_k\|_{H_\kappa} \le m\epsilon,$$

and the point λ_k is a bifurcation point for the nonlinear operator equation (32.8), (32.9). Thus we have established that each point λ_k of the SLOE (32.13) is a bifurcation point of (32.8), (32.9). Theorem 35.1 is proved.

Its significance lies in that without using analytic tools that are in general typical of bifurcation theory, we have established the following important fact: Each eigenvalue of the LOE is a bifurcation point. This result reflects intrinsic properties of boundary value problems of nonlinear shell theory, uncovered by using topological and variational methods. It is based on the potential nature of the system. It is also known that this situation does not always occur, and there are cases when an eigenvalue of the LOE is not a point of bifurcation of the corresponding nonlinear operator equation [27].

Theorem 35.2. *The branch of solutions coming out of the first eigenvalue* $\lambda_1 = \lambda_{t\kappa e}$ *of* (32.11), (32.12) *can be continued indefinitely; more precisely, to each value of the level* (33.8) *there corresponds at least one eigenelement of the nonlinear operator equation* (32.8), (32.9).

Theorem 35.2 follows immediately from Theorem 33.7. We have only to verify that if

$$\|w\|_{Hc}^2 = c^2 \to 0, \tag{35.22}$$

then

$$\lambda(c) \to \lambda_1.$$

From (33.37), (35.22) it follows that

$$\|w\|_{W_{2\Omega}^{(1)}} \to 0 \text{ if } c \to 0.$$

For the eigenvalue of the nonlinear operator equation we have the relation

$$\lambda = \frac{\|w\|_H^2 + 3\mathcal{I}_{\kappa\kappa 3}(w) + 4\mathcal{I}_{\kappa\kappa 4}(w)}{\|w\|_{Hc}^2}. \tag{35.23}$$

In (35.23), $\|w\|_H$ is given by (32.51). Furthermore, due to the structure of $\mathcal{I}_{\kappa\kappa\mu}(w)$, $\mu = 3,\ 4$, we obtain

$$\frac{3\mathcal{I}_{\kappa\kappa 3}(w) + 4\mathcal{I}_{\kappa\kappa 4}(w)}{\|w\|_{Hc}^2} \to 0 \text{ as } c \to 0.$$

But

$$\frac{\|w\|_H^2}{\|w\|_{Hc}^2} \to \lambda_1 = \lambda_{t\kappa e},$$

since in Theorem 33.7 we were minimizing $\mathcal{I}_{\kappa\kappa}^0$ over the entire set of elements belonging to the sphere (33.51). Theorem 35.2 is proved.

35.2. We state two theorems:

Theorem 35.3. *Assume that Conditions 1–7 of Section 17 hold. Then every eigenvalue λ_k of the linear operator equation* (32.27), (32.23) *is a bifurcation point of the nonlinear operator equation* (32.19), (32.20).

Theorem 35.4. *The branch of solutions coming out of the first eigenvalue $\lambda_1 = \lambda_{9\kappa e}$ of* (32.21), (32.23) *can be continued indefinitely; more precisely, to each value of the level* (34.17) *there corresponds at least one eigenelement of the nonlinear operator equation* (32.19), (32.20).

Since the proofs of Theorems 35.3, 35.4 largely repeat the proofs of Theorems 35.1, 35.2, they will not be presented in detail. Let us note only that instead of the surface (35.6), we are considering the surface

$$\frac{1}{2}\|w\|_H^2 + \mathcal{I}_{\kappa 3}(w) + \mathcal{I}_{\kappa 4}(w) = \frac{1}{2}\epsilon^2, \tag{35.24}$$

where the $\mathcal{I}_{\kappa\mu}$ are defined by by (34.3), (34.4). Instead (35.23) we have the formula

$$\lambda = \frac{\|w\|_H^2 + 3\mathcal{I}_{\kappa 3}(w) + 4\mathcal{I}_{\kappa 4}(w)}{\|w\|_{Hc}^2}.$$

The surfaces (35.24) and (35.6) differ in that in (35.6), $\|w\|_H$ is taken using (32.51), while in (35.24) we are using (32.54).

In conclusion, we note that the main result of the present section can be interpreted as establishing the important equalities

$$\lambda_B = \lambda_{t\kappa 1} = \lambda_{t\kappa e}, \ \lambda_B = \lambda_{9\kappa 1} = \lambda_{9\kappa e},$$

which, in conjunction with (33.22) and (34.11), completely resolve the question of interrelations of these quantities both in the case of Problem $t\kappa$ and of the Problem 9κ [346, 349, 350, 352, 354]. Here λ_B is the smallest bifurcation point on the λ-axis of the boundary value problem (32.19), (32.20).

36. Variational Methods in Global Stability of Shallow Shells

36.1. To solve the question of global stability it is first necessary to determine λ_1, and therefore to make the first step in the construction of U-decompositions for the problem at hand. Here one usually starts with the fact for $\lambda = \lambda_1$, the corresponding solution w_1 of the nonlinear operator equation (13.36), (17.19) is singular. It is precisely this fact that is used in all the numerical approaches. Usually, the solution of the problem is subdivided into two stages. At the first stage the corresponding nonlinear problems are solved for a fixed value of the loading parameter λ, and then values of the parameter that give rise to degeneracy are sought. In some cases the two operations are combined. Let φ_k be a set of functions satisfying the conditions of Section 26. We shall seek an approximate solution of (32.8), (32.9) in the form (26.1). To determine D_{nk} we shall use the system of equations

$$\frac{\partial}{\partial D_{nk}} \left(\mathcal{I}^0_{\kappa\kappa 2}(w) + \mathcal{I}_{\kappa\kappa 3}(w) + \mathcal{I}_{\kappa\kappa 4}(w)\right) = \overset{n}{\lambda} \frac{\partial}{\partial D_{nk}} \frac{1}{2} \|w_k\|^2_{Hc},$$

$$w_n = \sum_{k=1}^{n} D_{nk}\varphi_n \tag{36.1}$$

$$\mathcal{I}^0_{\kappa\kappa}(w) = \mathcal{I}^0_{\kappa\kappa 2}(w) + \mathcal{I}_{\kappa\kappa 3}(w) + \mathcal{I}_{\kappa\kappa 4}(w)$$
$$= c > \min \mathcal{I}^0_{\kappa\kappa}; \ k = 1, \ldots, n$$

$$\mathcal{I}^0_{\kappa\kappa 2}(w) = \frac{1}{2} \|w\|^2_H, \tag{36.2}$$

where $\|w\|_H$ is defined via (32.51). Let us consider the widely used Papkovich version of the method, in which the replacements w_1, w_2 are expressed in terms of w using (13.6) or (13.33). The system (36.1) contains $n+1$ unknowns: D_{nk} and $\overset{n}{\lambda}$. Therefore, it has to be solved together with equation (36.2). Let us present some facts to justify this approach.

Lemma 36.1. *The system of equations* (36.1), (36.2) *has at least one real solution for each n.*

The lemma is easily proved if we note that all the expressions involving D_{nk} in (36.1), (36.2) are polynomials, while the functional $\frac{1}{2}\ \|w_n\|^2_{Hc}$ is bounded from below on (36.2). Next, since w_n belongs to (36.2), all the approximations are contained in a sphere H_κ, and therefore they form a weakly compact set in H_κ.

Lemma 36.2. *Each sequence of $\{w_n\}$ that converges weakly in H_κ either converges weakly to zero or converges strongly. Every strong limit of $\{w_n\}$ in H_κ is an eigenfunction for* (32.8), (32.9).

Let $w_n \rightharpoonup w_0 \neq 0$ in H_κ. First of all, let us show that the sequence of numbers $\overset{n}{\lambda}$ converges. Indeed, for $\overset{n}{\lambda}$ we have the formula (35.23), from which, taking into account (36.2), we obtain

$$\overset{n}{\lambda} = \frac{2c + \mathcal{I}_{\kappa\kappa 3}(w_n) + 2\mathcal{I}_{\kappa\kappa 4}(w_n)}{\|w_n\|^2_{Hc}}, \tag{36.3}$$

and by weak continuity of the right-hand side of (36.3) and the condition that $w_n \neq 0$ for sufficiently large n, we have

$$\overset{n}{\lambda} \to \overset{0}{\lambda} = \frac{2c + \mathcal{I}_{\kappa\kappa 3}(w_0) + 2\mathcal{I}_{\kappa\kappa 4}(w_0)}{\|w_0\|^2_{Hc}}.$$

Let us observe now that equations (36.1) can be brought into the form

$$(w_n \cdot \varphi_k)_{H_\kappa} = \int_\Omega T^{ij}_M(w_n)\left(B_{ij}\varphi_k - w_{n\alpha^i} \cdot \varphi_{k\alpha^j}\right) d\Omega + \overset{n}{\lambda}(w_n \cdot \varphi_k)_{Hc}, \tag{36.4}$$

where the T^{ij}_M are determined from $\omega_M(w_n)$, which in turn have to be found from (32.8), in which we substitute w_n for w. Let us now show that w_0 is a solution of (32.8), (32.9). For this we consider the following sequence of transformations:

$$\begin{aligned} 0 &= (w_0 \cdot \varphi_k)_{H_\kappa} - \int_\Omega T^{ij}_M(w_n)\left(B_{ij}\varphi_k - w_{0\alpha^i}\varphi_{k\alpha^j}\right) d\Omega - \overset{0}{\lambda}(w_0 \cdot \varphi_k)_{Hc} \\ &= ((w_0 - w_n)\cdot \varphi_k)_{H_\kappa} + \int_\Omega \left[T^{ij}_M(w_n) - T^{ij}_M(w_0)\right] B_{ij}\varphi_k\, d\Omega \\ &\quad + \int_\Omega \left[T^{ij}_M(w_0)w_{0\alpha^i} - T^{ij}_M(w_n)w_{n\alpha^i}\right]\varphi_{k\alpha^j}\, d\Omega - ((\overset{0}{\lambda}w_0 - \overset{n}{\lambda}w_n)\cdot\varphi_k)_{Hc}, \\ k &= 1, \ldots, n. \end{aligned} \tag{36.5}$$

The last operation in (36.5) uses (36.4). Furthermore, the first three terms vanish, since $w_n \rightharpoonup w_0$ and since the first term is a linear functional in H_κ. In addition, we have to take into consideration (14.31) and (12.29). Finally, the last term on the right-hand side of (36.5) also vanishes as $n \to \infty$, since $\overset{0}{\lambda}w_0 - \overset{n}{\lambda}w_n \rightharpoonup 0$. Therefore, we have shown that w_0 is a solution of the NOE (32.8), (32.9). Strong convergence of $w_n \to w_0$ is proved using the same arguments as in Lemma 26.4. Lemma 36.2 is proved.

Let us state the results we obtained in Section 36.1.

Theorem 36.1. *Assume that all the conditions of Theorem 16.1 (16.2, 16.3, respectively) are satisfied. Let us seek a solution of problem $\tau\kappa$ on the surface* (36.2) *in the form* (26.1) *using the Papkovich version of the Bubnov–Galerkin method. Then on the surface* (36.2), *the system* (36.1) *has at least one solution w_n for every n. The set $\{w_n\}$ is weakly compact, and every weakly convergent sequence w_n, if it does not converge to zero, converges strongly to a solution of* (32.8), (32.9) *lying on* (36.2).

36.2. Let us indicate a method to find a set of approximations strongly compact in H_κ. For this we observe that for every n the functional $\frac{1}{2}\|w\|^2_{Hc}$ achieves a maximum on the surface c (36.2). The point is that this functional is a quadratic form in D_{nk} that is bounded from above. Furthermore, the sequence $w_n^{(\max)}$ cannot converge weakly to zero. Were that the case, we would have that $\left\|w_n^{(\max)}\right\|^2_{Hc} \to 0$. At the same time we have the obvious relation

$$\left\|w_n^{(\max)}\right\|^2_{Hc} \geq \left\|w_{n-1}^{(\max)}\right\|^2_{Hc},$$

and we cannot have $\left\|w_n^{(\max)}\right\|^2_{Hc} \to 0$. Therefore, we have proved the following result.

Theorem 36.2. *Assume that the conditions of Theorem 36.1 are satisfied. Then the set of approximations $\{w_n\}$ has a strongly compact subset each limit point of which is a solution of the nonlinear operator equation* (32.8), (32.9) *lying on the surface* (36.2).

36.3. Let us consider which eigenelements of the NOE (32.8), (32.9) can be obtained as limits of the set of approximations $\{w_n\}$. To this end, let us substitute λ from (35.23) in (32.9):

$$(w \cdot \varphi)_{H_\kappa} = \int_\Omega T_M^{ij}(B_{ij}\varphi - w_{\alpha^i}\varphi_{\alpha^j})\,d\Omega + \frac{2c + \mathcal{I}_{\kappa\kappa 3}(w) + 2\mathcal{I}_{\kappa\kappa 4}(w)}{\|w\|^2_{Hc}}(w \cdot \varphi)_{Hc}, \tag{36.6}$$

and as usual, we shall assume that in (36.6), T_M^{ij} has been expressed in terms of w by (32.9). Thus, already by itself (36.6) is a nonlinear operator equation with respect to w, but it does not involve the parameter λ. Condition (36.2) is here taken into account automatically.

Let us introduce the operator $\widetilde{\mathbf{G}}_{\kappa\kappa}$ by the relation

$$(\widetilde{\mathbf{G}}_{\kappa\kappa}(w) \cdot \varphi)_{H_\kappa} = \int_\Omega T_M^{ij}\left(B_{ij}\varphi - w_{\alpha^i}\varphi_{\alpha^j}\right)d\Omega + \frac{2c + \mathcal{I}_{\kappa\kappa 3}(w) + 2\mathcal{I}_{\kappa\kappa 4}(w)}{\|w\|^2_{Hc}}(w \cdot \varphi)_{Hc}. \tag{36.7}$$

Lemma 36.3. *The operator $\widetilde{\mathbf{G}}_{\kappa\kappa}$ is completely continuous on any part of H_κ, the weak closure of which does not contain zero.*

The proof of the lemma follows from (14.31), (12.29) and from the fact that the coefficient of $(w \cdot \varphi)_{H_C}$ on the right-hand side of (36.7) is weakly continuous in any part of H_κ, the weak closure of which does not contain zero.

Clearly, (36.6) can now be reduced to an equation of the form

$$w = \widetilde{\mathbf{G}}_{\kappa\kappa}(w). \tag{36.8}$$

Now let w_0 be a solution of (36.8), and let r be any number less than $\|w_0\|_{H_\kappa}$. Clearly, the weak closure of the ball $B(r, w_0)$ does not contain zero, and in that ball $\widetilde{\mathbf{G}}_{\kappa\kappa}$ is completely continuous. Then we haveÿ20the following theorem.

Theorem 36.3. *Let w_0 be a solution of* (36.8) *of nonzero index. Then $w_0 \in \overline{\{w_n\}}$, where $\overline{\{w_n\}}$ is the strong closure of the set of approximations $\{w_n\}$ in H_κ. If furthermore w_0 is a nonsingular solution of* (36.8), *then there is a sequence of approximations w_n, $\overset{n}{\lambda}$ such that*

$$\begin{aligned} \|w_0 - w_n\|_{H_\kappa} &\le (1 + \epsilon_n)\,\|w_0 - P_n w_0\|_{H_\kappa}, \\ \left|\overset{0}{\lambda} - \overset{n}{\lambda}\right| &\le m\,\|w_0 - P_n w_0\|_{H_\kappa}, \end{aligned} \tag{36.9}$$

where P_0 is the orthogonal projection operator in the basis $\{\varphi_k\}$.

Under the conditions of Theorem 36.3, we can give, for certain types of bases, a sharper estimate of the rate of convergence of approximate methods in the problem of global stability. We shall assume that conditions (28.4) are satisfied, and as $\{\varphi_k\}$ we take the set (28.5) orthonormal in H_κ.

Theorem 36.4. *Assume that the conditions of Theorem 28.3 hold. Then we have the following error estimates for the BGR method in the problem of global stability:*

$$\left\| w_0 - \sum_{k=1}^{n} D_{nk}\varphi_k \right\|_{H_1} \le m(w_0) n^{-\beta/2}, \tag{36.10}$$

$$\left|\overset{0}{\lambda} - \overset{n}{\lambda}\right| \le m(w_0) n^{-\beta/2}, \tag{36.11}$$

where β is given by (28.16).

Theorem 36.5. *Assume that the conditions of Theorem 28.4 are satisfied together with the conditions for the existence of an MIS. Then for any nonsingular solution of* (36.9), (36.8) *there is a sequence of BGR approximations of the form* (26.1), *where the φ_k are the functions* (28.7), *orthonormal in H_1, such that* (36.11) *holds, in which β is determined from* (28.19).

Proofs of Theorems 36.4, 36.5 repeat almost verbatim the scheme of proofs of Theorems 28.3, 28.4. The estimate (36.11) for $\overset{0}{\lambda}$ is obtained from the estimate (36.10) of approximations for w_0.

Remark 36.1. We could apply the Mushtari version of the BGR method to find an approximate solution of the NOE (32.8) when the tangential equations (32.8) are also approximately satisfied. Then, using the results of Sections 20, 28 we can obtain estimates similar to (36.9)–(36.11).

In the NOE (36.7), (36.8) we have introduced a new parameter c, determined by the right-hand side of (34.18) instead of the parameter λ. Therefore, solutions of the NOE (36.7), (36.8) can be regarded as functions of the parameter c.

Theorem 36.6. *Assume that for some c_0 the solution $w(c_0)$ of* (36.9)–(36.8) *is nonsingular. In that case we can find an interval*

$$c_0 - \epsilon \leq c \leq c_0 + \epsilon$$

on which

$$w(c) \in H^{0,1}_{c_0-\epsilon\leq c\leq c_0+\epsilon},$$

that is, $w(c)$ is a Lipschitz function of c in a neighborhood of c_0.

Theorem 36.6 demonstrates certain important properties of solutions of (36.7), (36.8). Due to lack of space, we do not present its proof.

36.4. Let us state some results concerning the use of direct methods in problems of global stability in the problem 9κ. Let us consider the nonlinear operator equation (32.19), (32.20) and solve it by the BGR method using (26.1) and assuming that (32.20) is solved exactly for Ψ_M. Then (32.19) becomes a nonlinear operator equation with respect to w_n. Thus, here also we are using the Papkovich version.

According to (34.1) the equations for D_{nk}, $\overset{n}{\lambda}$ assume the form

$$\frac{\partial}{\partial D_{nk}}\left(\mathcal{I}^0_{\kappa 2}(w_n) + \mathcal{I}_{\kappa 3}(w_n) + \mathcal{I}_{\kappa 4}(w_n)\right) = \overset{n}{\lambda}\frac{\partial}{\partial D_{nk}}\frac{1}{2}\,\|w_n\|^2_{Hc}\,, \tag{36.12}$$

where

$$\mathcal{I}^0_{\kappa 2}(w) = \frac{1}{2}\,\|w\|^2_H\,,$$

and $\|w\|_H$ is defined by (32.54). We shall seek w_n on the surface

$$\mathcal{I}^0_{\kappa}(w_n) = \mathcal{I}^0_{\kappa 2}(w_n) + \mathcal{I}_{\kappa 3}(w_n) + \mathcal{I}_{\kappa 4}(w_n) = c > \min \mathcal{I}^0_{\kappa}. \tag{36.13}$$

Theorem 36.7. *Assume that all the conditions of Theorem 19.2 and the conditions for the existence of an MIS* (32.16), (32.17) *are satisfied. Then the system* (36.12), (36.13) *has at least one real solution D_{nk}, $\overset{n}{\lambda}$ for each n. The set of the approximations $\{w_n\}$ is weakly compact. Furthermore, each weakly convergent sequence in $\{w_n\}$ either weakly converges to zero or converges strongly to a solution w_0 of the nonlinear operator equation* (32.24), (32.20). *Then we also have the convergence of the real sequence $\overset{n}{\lambda}$ to an eigenvalue $\overset{0}{\lambda}$ of* (32.19), (32.20) *that corresponds to w_0.*

Theorem 36.8. *Let all the conditions of Theorem 36.6 be satisfied. Then the set of approximations $\{w_n\}$ has a strongly compact subset, each limit point of which is a solution of* (32.19), (32.20) *lying on the surface* (36.13).

36.5. Let us introduce the operator $\widetilde{\mathbf{G}}_\kappa$ by the relation

$$(\widetilde{\mathbf{G}}_\kappa(w) \cdot \varphi)_{H_\kappa} = \int_\Omega C^{ik} C^{jl} \big(B_{ij}\varphi - \varphi_{\alpha^i} w_{\alpha^j}\big) \nabla_{kl} \Psi_M \, d\Omega + \frac{2c + \mathcal{I}_{\kappa\kappa 3}(w) + 2\mathcal{I}_{\kappa\kappa 4}(w)}{\|w\|^2_{HC}} (w \cdot \varphi)_{HC}, \tag{36.14}$$

and then clearly, under the condition (36.13), (32.19) reduces to an NOE of the form

$$w = \widetilde{\mathbf{G}}_\kappa(w). \tag{36.15}$$

In (36.15), Ψ_M is assumed to be expressed in terms of w using (32.20).

Theorem 36.9. *Let w_0 be a solution of (36.15) of nonzero index. Then $w_0 \in \overline{\{w_n\}}$, where $\overline{\{w_n\}}$ is the strong closure of the set of approximations $\{w_n\}$ in H_κ. If furthermore w_0 is a nonsingular solution of (36.15), then there is a sequence of approximations w_n, $\overset{n}{\lambda}$ such that*

$$\|w_0 - w_n\|_{H_\kappa} \leq (1 + \epsilon_n) \|w_0 - P_n w_0\|_{H_\kappa}$$

$$\left|\overset{0}{\lambda} - \overset{n}{\lambda}\right| \leq m \|w_0 - P_n w_0\|_{H_\kappa},$$

where P_0 is the orthogonal projection operator in the basis $\{\varphi_k\}$ (Section 28).

We consider now the case when w satisfies the boundary conditions (28.4), while for Ψ_M we naturally have (7.24). As $\{\varphi_k\}$ we can also take the sets (28.5), orthonormal in H_1.

Theorem 36.10. *Assume that the conditions of Theorem 28.7 are satisfied, as well as (28.13) and the conditions (32.16), (32.17) for the existence of an MIS. Then in $\{w_n\}$ there is a sequence w_n such that we have the estimates (36.10), (36.11), in which β is given by (28.8).*

Theorem 36.11. *Assume that the conditions of Theorem 28.8 are satisfied as well as the condition (32.17) for the existence of an MIS. Then in $\{w_n\}$ there is a sequence w_n such that we have the estimates (36.10), (36.11), in which β is given by (28.19).*

Remark 36.2. We could apply the Vlasov version of the BGR method to find an approximate solution of the NOE (32.19), (32.20); then the equation of tangential deformation (32.20) is also approximately satisfied. Then, using the results of Sections 20, 28 we can obtain estimates similar to (36.9)–(36.11).

Theorem 36.12. *Let the solution $w(c_0)$ of the nonlinear operator equation (36.15) be nonsingular for some c_0. Then Theorem 36.6 holds.*

Remark 36.3. The framework for a rigorous foundation of the BGR methods in problems of global stability of shallow shells presented in this section can be extended to cover the case when the approximation is done by using the method of finite differences of finite elements. Here it is important to satisfy the following two conditions: (1) the approximation scheme must guarantee approximation of any element in H_κ if the Papkovich version is used, or of any element of $H_{t\kappa}$ (respectively, $H_{9\kappa}$) if the Mushtari (respectively, Vlasov) version is used; (2) the computation of the constants of approximation is done using some variational principle of Lagrange or Alumyae.

37. Some Problems of Global Stability of Plates

37.1. Let us consider first the case of Problem $t\kappa$. From (32.6) follows an obvious fact: The necessary and sufficient conditions for existence of an MlS of a plate are given by the relations

$$\widetilde{M}^m\big|_{\Gamma_2+\Gamma_4} \equiv \widetilde{Q}\big|_{\Gamma_3+\Gamma_4} \equiv \widetilde{w}|_{\Gamma_1+\Gamma_3} \equiv \widetilde{w}_4|_{\Gamma_1+\Gamma_3} \equiv R^3 \equiv 0. \tag{37.1}$$

Theorem 37.1. *Assume that Conditions 2–6 of Section 13 as well as condition* (37.1) *of existence of an MlS, are satisfied. Then for a plate we have the exact equalities*

$$\lambda_{t\kappa l} = \lambda_{t\kappa u} = \lambda_{t\kappa 1} = \lambda_{t\kappa e}. \tag{37.2}$$

To prove (37.2) we shall need the following lemma.

Lemma 37.1. *Assume that an eigenfunction of the linear operator equation* (32.13) w *is equal to a constant. Then* $w \equiv 0$*, and the corresponding number* λ *is not an eigenvalue.*

Lemma 37.1 is obvious if the boundary Γ of the shell S contains a segment $\Gamma_1 + \Gamma_2 > 0$. If on the other hand, $\Gamma_1 + \Gamma_2 = 0$, then the elastic supports on Γ_3, Γ_4, due to the conditions required to construct the spaces H_κ, must be essentially elastic, and then by the form of the boundary conditions on Γ_3, Γ_4, $w \equiv$ const implies $w \equiv 0$.

To prove Theorem 37.1, we start with Lemma 33.4. In the case of a plate, as can be seen from (33.8)–(33.10),

$$\mathcal{I}_{\kappa\kappa 2}^{\lambda_{t\kappa e}}(w) = \frac{1}{2}\left(\|w\|_H^2 - \lambda_{t\kappa e}\|w\|_{Hc}^2\right), \tag{37.3}$$

$$\mathcal{I}_{\kappa\kappa 3}(w) \equiv 0, \tag{37.4}$$

$$\mathcal{I}_{\kappa\kappa 4}(w) \geq 0. \tag{37.5}$$

In (37.3), $\|w\|_H$ is given by (32.51), (32.54). Relation (37.4) follows from the fact that for a plate from (32.11) we obtain $\mathbf{K}_{t\kappa 1} \equiv 0$, and thus by (14.15) $\overset{0}{\epsilon}_{ij1} \equiv 0$, since $w^* \equiv 0$ under conditions (37.1). Therefore, for a plate, (33.28) assumes the form

$$\|w\|_H^2 - \lambda_{t\kappa e}\|w\|_{Hc}^2 + 4\mathcal{I}_{\kappa\kappa 4}(w) \neq 0. \tag{37.6}$$

Now let there be an equal sign in (37.6). By definition of $\lambda_{t\kappa e}$ we have

$$\|w\|_H^2 - \lambda_{t\kappa e}\|w\|_{Hc}^2 \geq 0,$$

and thus, taking into account (37.5), we have

$$\|w\|_H^2 - \lambda_{t\kappa e}\,\|w\|_{Hc}^2 = 0,$$

whence

$$w = w_1,$$
$$\mathcal{I}_{\kappa\kappa 4}(w_1) = 0, \tag{37.7}$$

where w_1 is the first eigenfunction of the LOE (32.13).

But (37.7), in view of (33.11), implies that

$$\overset{0}{\epsilon}_{ijM_2}(w_1) \equiv 0,$$
$$w_{\Gamma M_2}\big|_{\Gamma_6} \equiv w_{mM_2}\big|_{\Gamma_7} \equiv w_{iM}|_{\Gamma_8} \equiv 0;\ i = 1,\ 2,$$

and then

$$w_1 = c,$$

and by Lemma 37.1 $w_1 \equiv 0$. Therefore, the conditions of Lemma 33.4 are satisfied, and Theorem 37.1 is proved.

Theorem 37.1 completely resolves the question of the structure of SNOE (32.8), (32.9). Namely, the spectrum fills the entire interval $\lambda > \lambda_{t\kappa e}$. At the same time we have a positive answer concerning the applicability of Euler's linearization method in the study of stability of plates in problems $t\kappa$.

Let us now consider certain facts that govern the behavior of a plate after loss of stability.

Theorem 37.2. *Assume that Conditions 2–6 of Section 13 and the conditions (37.1) for the existence of an MlS are satisfied. Then to each level $c > 0$ of potential energy of the plate after loss of stability corresponds at least a countable number of equilibrium states having that energy level. The corresponding eigenelements have the limiting behavior*

$$w_{kc} \rightharpoonup 0, \quad \lambda_{kc} \to \infty. \tag{37.8}$$

Theorem 37.2 will be proved if we show that the functional $\mathcal{I}(w) = \frac{1}{2}\,\|w\|_{Hc}^2$ has on the surface

$$\mathcal{I}^0_{\kappa\kappa 2}(w) + \mathcal{I}_{\kappa\kappa 4}(w) = c^2 \tag{37.9}$$

a countable number of critical elements w_{kc}, λ_{kc} that satisfy (37.9). To prove this result, we use the well-known theorem of Lyusternik [186] concerning critical points of even functionals on spheres in a Hilbert space. This theorem was later significantly generalized [281]. Of major importance to us are the contributions of Sobolev [284], who removed the assumption of homogeneity of the functional I; of Vainberg [317], who dispensed with some of the smoothness conditions; and of Tsitlanadze [312, 313] who considered the case of multiple critical points. A different approach to the problem of extrema of even functionals was suggested by Krasnosel'skii [162]. We quote the principal result.

Theorem 37.3. *Assume that a functional $I(w)$ satisfies the following conditions:*

(a) *$I(w)$ is odd, that is, $I(-w) = I(w)$.*
(b) *$I(w) \geq 0$ and $I(w) = 0$ if and only if $w \equiv 0$.*
(c) $\text{grad}_{H_\kappa} I(w) \neq 0$ *if* $w \neq 0$; $\text{grad}_{H_\kappa} I(w)|_{w=0} = 0$.
(d) *$I(w)$ is a smooth functional, that is,*

$$I(w+h) - I(w) = (\text{grad}_{H_\kappa} I \cdot h)_{H_\kappa} + \zeta(w, h), \tag{37.10}$$

where

$$\frac{|\zeta(w, h)|}{\|h\|_{H_\kappa}} \to 0 \text{ as } \|h\|_{H_\kappa} \to 0$$

uniformly in w such that $\|w\|_{H_\kappa} \leq R$.

Then each sphere $\|w\|_{H_\kappa} = r \leq R$ contains at least a countable number of eigenelements w_{kc}, λ_{kc} that satisfy (37.9).

To use this fact, we perform the following change of variables:

$$w \mapsto c\mu(w)w, \tag{37.11}$$

where $\mu(w)$ is to be found from the equation

$$\mu^2(w) + c^2\mu^4(w)\mathcal{I}_{\kappa\kappa 4}(w) = 1. \tag{37.12}$$

If we take $\mu(w)$ to be defined by (37.12) (below we shall always take the unique positive root), the surface (37.9) becomes a sphere Σ of radius 1 with center at the origin. Then the functional $\mathcal{I}(w) = \frac{1}{2}\|w\|^2_{Hc}$ becomes

$$\widetilde{\mathcal{I}}(w) = \frac{1}{2}c^2\mu^2(w)\,\|w\|^2_{Hc}\,.$$

Thus we have to study critical points of $\widetilde{\mathcal{I}}(w)$ on Σ.

Let us verify conditions (a)–(d) for the functional $\widetilde{\mathcal{I}}(w)$. Clearly, conditions (a), (b) hold. Let us show now that from the relation

$$\text{grad}_{H_\kappa} \widetilde{\mathcal{I}}(w) = 0 \tag{37.13}$$

it follows that

$$w \equiv 0.$$

We have

$$(\text{grad}_{H_\kappa} \widetilde{\mathcal{I}}(w) \cdot \widetilde{h})_{H_\kappa} = (\text{grad}_{H_\kappa} \mathcal{I}(w) \cdot h)_{H_\kappa}, \tag{37.14}$$

where $\widetilde{h}$ and h are related by

$$h = c\mu(w)\widetilde{h} + c(\text{grad}_{H_\kappa} \mu \cdot \widetilde{h})_{H_\kappa} w. \tag{37.15}$$

Formulae (37.14), (37.15) reflect the invariance of increments of $\mathcal{I}(w)$ under the change of variables (37.11).

Lemma 37.2. *Assume that a constant c and $h,\ w \in H_\kappa$ are given. Then equation* (37.15) *uniquely defines $\widetilde{h}$.*

To prove the lemma, we take the scalar product of (37.15) with $\operatorname{grad}_{H_\kappa} \mu$. Then we have

$$(h \cdot \operatorname{grad}_{H_\kappa} \mu)_{H_\kappa} = c\mu(w)(\widetilde{h} \cdot \operatorname{grad}_{H_\kappa} \mu)_{H_\kappa} + c(\operatorname{grad}_{H_\kappa} \mu \cdot \widetilde{h})_{H_\kappa}(w \cdot \operatorname{grad}_{H_\kappa} \mu)_{H_\kappa}. \tag{37.16}$$

From (37.16) it follows that

$$(\widetilde{h} \cdot \operatorname{grad}_{H_\kappa} \mu)_{H_\kappa} = \frac{(h \cdot \operatorname{grad}_{H_\kappa} \mu(w))_{H_\kappa}}{c[\mu(w) + (w \cdot \operatorname{grad}_{H_\kappa} \mu)_{H_\kappa}]}. \tag{37.17}$$

Substituting (37.17) into (37.15), we obtain the required expression of $\widetilde{h}$ in terms of h:

$$\widetilde{h} = \frac{1}{c\mu(w)}\left[h - w\frac{(h \cdot \operatorname{grad}_{H_\kappa} \mu(w))_{H_\kappa}}{\mu(w) + (w \cdot \operatorname{grad}_{H_\kappa} \mu)_{H_\kappa}}\right]. \tag{37.18}$$

The proof of Lemma 37.2 will be complete if we show that

$$\mu + (\operatorname{grad}_{H_\kappa} \mu \cdot w)_{H_\kappa} \neq 0. \tag{37.19}$$

Applying the grad operator to (37.12), we have

$$\operatorname{grad}_{H_\kappa} \mu + 2c^2\mu^2 \operatorname{grad}_{H_\kappa} \mu \mathcal{I}_{\kappa\kappa 4}(w) + \frac{1}{2}c^2\mu^3 \operatorname{grad}_{H_\kappa} \mathcal{I}_{\kappa\kappa 4}(w) = 0. \tag{37.20}$$

Next we use the relation

$$\operatorname{grad}_{H_\kappa} \mathcal{I}_{\kappa\kappa 4}(w) = \mathbf{G}_{\kappa\kappa 3}(w), \tag{37.21}$$

and by homogeneity of $\mathcal{I}_{\kappa\kappa 4}(w)$, $\mathbf{G}_{\kappa\kappa 3}(w)$ we find from (37.21) that

$$(\operatorname{grad}_{H_\kappa} \mathcal{I}_{\kappa\kappa 4}(w) \cdot w)_{H_\kappa} = (\mathbf{G}_{\kappa\kappa 3}(w) \cdot w)_{H_\kappa} = 4\mathcal{I}_{\kappa\kappa 4}(w). \tag{37.22}$$

From (37.20), (37.22) it follows that if we multiply (37.20) by w, we obtain

$$(\operatorname{grad}_{H_\kappa} \mu \cdot w)_{H_\kappa}(1 + 2c^2\mu^2\mathcal{I}_{\kappa\kappa 4}(w)) = -2c^2\mu^2\mathcal{I}_{\kappa\kappa 4}(w). \tag{37.23}$$

Finally, from (37.23) it follows that

$$\mu + (\operatorname{grad}_{H_\kappa} \mu \cdot w)_{H_\kappa} = \mu - \frac{2c^2\mu^2\mathcal{I}_{\kappa\kappa 4}(w)}{1 + 2c^2\mu^2\mathcal{I}_{\kappa\kappa 4}(w)} = \frac{\mu}{1 + 2c^2\mu^2\mathcal{I}_{\kappa\kappa 4}(w)},$$

and (37.19) has been proved. Thus we have also established Lemma 37.2.

Therefore, (37.15), (37.18), and (37.14) all hold simultaneously. Let us now assume that for some w, (37.13) holds. Then the left-hand side of (37.14) vanishes for any $\widetilde{h}$, and so by (37.18) the right-hand side of (37.14) is zero for any h. This means that

$$(\operatorname{grad}_{H_\kappa} \mathcal{I}(w) \cdot w)_{H_\kappa} = 0. \tag{37.24}$$

By the structure of $\mathcal{I}(w)$ it follows from (37.24) that

$$\mathcal{I}(w) = 0$$

and therefore $w = 0$, so condition (c) is satisfied as well. Smoothness of $\widetilde{\mathcal{I}}(w)$ is established by elementary means.

Therefore, we have demonstrated the existence of at least a countable number of extrema of $\widetilde{I}$ on Σ, and so we have proved the existence of at least a countable number of extrema of $\mathcal{I}(w)$ on the surface (37.9) such that (37.8) holds.

We can estimate the rate of weak convergence of w_{kc} to zero. For this let us consider the relation

$$\lambda_{kc} = 2\frac{c^2 + \mathcal{I}_{\kappa\kappa 4}(w_{kc})}{\|w_{kc}\|^2_{Hc}},$$

from which we have

$$\|w_{kc}\|^2_{Hc} \le \frac{2c^2}{\lambda_{kc}}, \tag{37.25}$$

and finally, since

$$m\,\|w\|_{W^{(1)}_{2\Omega}} \le \|w\|_{Hc}\,,$$

we have from (37.25),

$$\|w_{kc}\|_{W^{(1)}_{2\Omega}} \le \frac{m}{\lambda^{1/2}_{kc}}.$$

Let us clarify the physical content of Theorem 37.2. This can be done by analogy with the phenomenon of stability loss of a rod. It is known that for a given level of energy, after losing stability, the rod has a countable number of equilibrium states. As the number k of the equilibrium configuration grows, its amplitude decreases, but the number of intermediate zeros grows. For example, for a hinged rod this number is $k-1$. This is what weak convergence to zero of equilibria means. Keeping the rod in a state with an increasing number of intermediate zeros requires larger and larger lateral loads, which corresponds to $\lambda_{kc} \to \infty$. We note that physically the weak convergence $w_{kc} \rightharpoonup 0$ means that as k grows, after loss of stability the plate is subdivided into a large number of lobes, the deflection itself is a fast oscillating function, and the phenomenon itself is of an increasingly local nature. Due to the oscillating nature of w_{kc}, the potential energy of bending will dominate the potential energy of stretching. In the case of a plate we also have some additional facts having to do with the application of direct methods.

Theorem 37.4. *Assume that Conditions 2–6 of Section 13 and conditions* (37.1) *for the existence of an MlS hold. Assume that we are seeking a solution of problem* $t\kappa$ *in the form* (26.1) *using the Papkovich version of the BGR method. Then for a plate system,* (36.1) *has on the surface* (37.9) *at least* $n+1$ *real solutions. The family* $\{w_n\}$ *is weakly compact, and each weakly convergent sequence* w_n *(if it does not converge weakly to zero) converges strongly to a solution of the NOE* (32.9) *for a plate; this solution lies on the surface* (37.9).

We remind the reader that the corresponding Theorem 36.1 for shells only claims the existence of a real solution of system (36.1).

Theorem 37.3 follows from the results of [281, 313] if we note that $\mathcal{I}^0_{\kappa\kappa 2}(w_n)$, $\mathcal{I}_{\kappa\kappa 4}(w_n)$, $\|w_n\|^2_{Hc}$ in the case of plates are even functions of D_{nk}, while if we substitute (26.1), (37.9) turns out to be homeomorphic to a sphere. Furthermore, the homeomorphism can be chosen in such a way that the central symmetry of points of the sphere and of (37.9) is preserved.

37.2. In the case of circular plate, we can obtain some additional information on the behavior of the plate after loss of stability.

Theorem 37.5 **([349]).** *Let a circular homogeneous isotropic plate of radius 1 be either fixed at the perimeter or, alternatively, hinged, and let it satisfy one of the following systems of boundary conditions,*

$$w|_\Gamma = \left.\frac{dw}{dr}\right|_\Gamma = 0 \text{ or } w|_\Gamma = \left.\left[\frac{d^2w}{dr^2} + (1+\nu)\frac{dw}{dr}\right]\right|_\Gamma = 0,$$

and be subjected to uniform compression. Furthermore, let λ_n be the eigenvalues of the LOE (32.13) *when $H_\kappa \to H_1$. Then for each λ in the interval*

$$\lambda_n < \lambda < \lambda_{n+1}$$

there exist at least n solutions of the NOE (32.8), (32.9) (w_{nj}, Ψ_{nj}), $(-w_{nj}, \Psi_{nj})$, $j = 1, \ldots, n$; *w_{nj} has $n-1$ zeros in the interval* $0 < r < 1$.

37.3. For Problem 9κ we have assertions analogous to Theorems 37.1–37.3. Since they are proved in a similar fashion, we confine ourselves to formulations.

Theorem 37.6. *Assume that Conditions 2–7 of Section 17 and the conditions for the existence of MlS* (37.1) *are satisfied. Then for a plate we have the exact equalities*

$$\lambda_{9\kappa l} = \lambda_{9\kappa u} = \lambda_{9\kappa 1} = \lambda_{9\kappa e}.$$

Theorem 37.7. *Assume that Conditions 2–7 of Section 17 and the conditions for the existence of MlS* (37.1) *are satisfied. Then to each level $c > 0$ of the potential energy of the shell after loss of stability corresponds at least a countable number of equilibria having that energy level. The corresponding eigenelements satisfy in the limit the relations*

$$w_{kc} \rightharpoonup 0, \ \lambda_{kc} \to \infty.$$

Theorem 37.8. *Assume that Conditions 2–7 of Section 17 and the conditions for the existence of MlS* (37.1) *are satisfied. If a solution of Problem 9κ for a plate is sought in the form* (26.1) *using the Papkovich version of the BGR method, then the system*

$$\frac{\partial}{\partial D_{nk}}(\mathcal{I}^0_{\kappa 2}(w_n) + \mathcal{I}_{\kappa 4}(w_n)) = \lambda \frac{\partial}{\partial D_{nk}} \frac{1}{2} \|w_n\|^2_{Hc}, \ k = 1, \ldots, n,$$

has on the surface

$$\mathcal{I}^0_\kappa(w) = \mathcal{I}^0_{\kappa 2}(w) + \mathcal{I}_{\kappa 4}(w) = c > \min \mathcal{I}^0_\kappa \text{ in } H_\kappa \tag{37.26}$$

at least $n + 1$ *real solutions. The family* $\{w_n\}$ *is weakly compact, and each weakly convergent sequence* w_n *(if it does not converge weakly to zero) converges strongly to a solution of the nonlinear operator equation* (32.19) *for a plate that lies on the surface* (37.26).

37.4. In [343, 344, 346] the author presents a number of results dealing with stability "in the large" in the nonlinear theory of shallow shells. These results were also described in [349, 350, 352, 354, 359, 362]. Let us note first Theorems 33.3 and 34.3 asserting the existence of lower critical values for Problems $t\kappa$ and 9κ. A complete proof of these theorems is presented here for the first time. Theorems 35.1, 35.3 on the bifurcation of solutions from every eigenvalue of the linearized problem are given in [346]. A detailed proof appears here for the first time as well. In [24] Berger applied variational considerations to eigenvalue problems for various types of nonlinear elliptic boundary value problems. Using these results, he considered in [27, 25] what in our terminology is Problem 91 for a plate and confirmed in this particular case all the assertions of Theorem 35.3. He also proved some additional results, such as the fact that if λ_n is an eigenvalue of multiplicity p, then out of it bifurcate at least p branches of solutions of the corresponding nonlinear operator equation. Let us note that this conclusion holds also for all the other Problems 9κ and $t\kappa$ for plates that we considered above. In [28, 26] Berger considered problem 91 for a circular cylindrical shell. Results of these works are a particular case of Theorem 35.3. Reference [346] also contains the main result on the postcritical behavior of plates (Theorems 37.2, 37.4). A detailed proof of Theorem 37.4 is contained in [358].

CHAPTER X

A Probabilistic Approach to the Problem of Stability of Shallow Shells

38. A Probabilistic Model of Operation of a Shallow Shell Under Moderate Bending

38.1. From the results of Sections 29–37 it follows that a relatively typical picture of deformation of a shell is one in which there are several equilibrium configurations under given loading conditions. Moreover, in some cases there are several stable equilibrium configurations. Naturally, there arises the question of choosing the equilibrium configuration that has the best chances of being observed in an experiment. In our terminology (Section 29), this is the second problem of stability theory. It cannot be solved without having more precise information concerning the conditions of operation of the shell and its parameters. We have in mind information about distribution of parameters governing its shape, elastic characteristics, external loading, and thus a probabilistic theory of the operation of the shell. Of course, such a theory must also include the criteria used in the theory of stability of elastic systems, such as an estimate of degree of stability of the system with respect to the level of potential energy. From all the considerations above it follows that we shall cover quite a wide range of problems if we assume that realizations of the stochastic process of deformation of a shell $\mathbf{a} = (w_1, w_2, w)$ belong to $H_{t\kappa}$. Thus, a complete and rigorous analysis of this question involves introducing probability distributions in this function space. Difficulties associated with construction of such a theory are well known. They become even more formidable if we plan to construct numerical algorithms for modern computers.

In this section we present a construction of distributions describing the operation of a shell in finite approximating subspaces W_n, where W_n $(n = 1, 2, \ldots, \infty)$ has a basis $\varphi_1, \varphi_2, \ldots, \varphi_n$.

38.2. For this purpose, let us consider in more detail the structure of random factors that define the behavior of a shell. These can be subdivided into three groups. The first of these groups will include random perturbations in the thickness of the shell, in the shape of the boundary curve, and in elastic characteristics. The second group includes perturbation in parameters determining the clamping of the shell. Finally, the third group includes random loads that act on the shell.

Concerning a probabilistic description of the first two groups, we shall assume that they are described sufficiently precisely by a finite collection of random numbers $a_1, \ldots, a_N$ and a corresponding distribution law $\theta(a_1, \ldots, a_N)$. In the case of random functional parameters, such as the thickness of the shell, we shall do as follows: Let us choose a system of functions $\zeta_k(\alpha^1, \alpha^2)$ that is sufficiently representable for a given parameter and let us approximate the random parameter by linear combinations in $\zeta_k(\alpha^1, \alpha^2)$. The coefficients in these combinations will be the random numbers $a_1, \ldots, a_m$ under consideration, describing the random nature of the functional parameter being considered. For example, to describe random perturbations in the thickness of the shell, we could use the representations

$$h(\alpha^1, \alpha^2) = \sum_{k=1}^{m} a_k \zeta_k(\alpha^1, \alpha^2), \quad \zeta_k(\alpha^1, \alpha^2) = \alpha_1^{n_1} \alpha_2^{n_2},$$

that is, a polynomial approximation. Here the parameters a_k are to be determined for each shell simply by measuring its thickness at various points, and then by a statistical analysis of data to determine $\theta(a_k)$, $k = 1, \ldots, m$. Of course, the family of approximating functions has to be chosen in such a way that all the practically observable range of $h(\alpha^1, \alpha^2)$ can be described by a minimal number of parameters a_k. In the same way we could describe the random field of perturbations of the middle surface or of other functional parameters in the first two groups; in this way we obtain the distribution $\theta_1(a_1, \ldots, a_n)$.

38.3. Let us move on to the probabilistic description of the external load acting on the shell. For simplicity of exposition, let us assume that only $R^3(\alpha^1, \alpha^2, t)$ is involved. Below, starting with natural assumptions on the structure of the random function $R^3(\alpha^1, \alpha^2, t)$, we take

$$R^3(\alpha^1, \alpha^2, t) = R^{31}(\alpha^1, \alpha^2, t) + R^{32}(\alpha^1, \alpha^2, t) + R^{33}(\alpha^1, \alpha^2, t), \quad (38.1)$$

where

$$R^{33}(\alpha^1, \alpha^2, t) = \mathbf{E}\{R^3(\alpha^1, \alpha^2, t)\}, \quad E\{R^{3k}(d^1, d^2, t)\} = 0, \quad k = 1, 2,$$

while R^{31}, R^{32} are random functions, the nature of which will be described below; **E** stands for the operator of mathematical expectation.

We shall assume that $R^{32}(\alpha^1, \alpha^1, t)$ is a process continuous in time, that is, a process each realization of which is a continuous function of time. A probabilistic description of such a process can be obtained as for the factors in the first two groups. Namely, we shall assume that sufficiently precisely for any realization of

$R^{32}(\alpha^1, \alpha^2, t)$ we can take [249]

$$R^{32}(\alpha^1, \alpha^2, t) = \sum_{k_1=1}^{m_1} \sum_{k_2=1}^{m_2} b_{k_1 k_2} \psi_{k_1}(\alpha^1, \alpha^2) \psi_{k_2}(t). \tag{38.2}$$

Here it is assumed that $\psi_{k_1}(\alpha^1, \alpha^2)$ are elements of some basis in $L_{2\Omega}$, while $\psi_{k_2}(t)$ are elements of a basis in L_{2T}, where T is an interval of time during which we are observing the behavior of the shell. Clearly, we shall obtain a sufficiently complete description of $R^{32}(\alpha^1, \alpha^2, t)$ if we define the distribution law $\theta_2(b_{k_1 k_2})$ for the random quantities $b_{k_1 k_2}$.

In general, the random variables a_k and $b_{k_1 k_2}$ may be dependent. For example, if $R^{32}(\alpha^1, \alpha^2, t)$ includes a random component describing the influence of air currents on the shell, then it can depend on random perturbations in the shape of the shell, that is, on parameters entering a_k. Therefore, it makes sense in general to assume that we are given a joint distribution law $\theta_3(a_k, b_{k_1 k_2})$ of all random variables in the first two groups and $R^3(\alpha^1, \alpha^2, t)$. For more details on such a definition of random functions, see [249, 250].

Concerning $R^{31}(\alpha^1, \alpha^2, t)$ we shall assume that this part of the load is the so-called white noise in time [154, 250]. Its statistical properties are described by the relation

$$\mathbf{E}\{R^{31}(\alpha^1, \alpha^2, t_1) R^{31}(\beta^1, \beta^2, t_2)\} = K^{31}(\alpha^1, \alpha^2, \beta^1, \beta^2)\delta(t_1 - t_2). \tag{38.3}$$

Equation (38.3) means that the values of realizations of $R^{31}(\alpha^1, \alpha^2, t)$ at different times are statistically independent. In other words, knowing a realization of $R^{31}(\alpha^1, \alpha^2, t)$ at a moment of time, we cannot say anything about the probability with which any value of it will be taken at other, even arbitrarily close, moments of time. In other words, $R^{31}(\alpha^1, \alpha^2, t)$ is a purely discontinuous process. Of course, this is an idealization of the processes encountered in reality. In a sense, white noise is the opposite of a continuous stochastic process. for the latter, the probability of large values of the quantity $\left\| R^{33}(t + \Delta t) - R^{33}(t) \right\|_{L_{2\Omega}}$ becomes arbitrarily small if Δt is sufficiently small. For processes approximated by white noise, approximations of the form (38.2) are not suitable. To attain the required precision, one must take a large number of terms. On the other hand, the introduction of white noise allows us to use the analytical machinery of the Kolmogorov–Fokker–Planck (KFP) equation and to bring the solution of the problem to a stage where it can conceivably be computed. This approach covers a sufficiently wide variety of problems.

38.4. For simplicity, we restrict ourselves to essentially shallow shells. For this class we can take the metric in the middle surface to be the metric of the plane. To obtain the equations of motion of the shell, we add in (6.37) to the forces $R^3(\alpha^1, \alpha^2, t)$ inertial forces $-\rho w_{tt}$ and friction forces $-2\beta w_t$, which we thus assume to be proportional to the velocity. Here ρ is the area mass density of the shell, 2β is the friction coefficient. As a result, we get

$$\rho w_{tt} + 2\beta w_t + (D_{\mathrm{f}}^{ijkl} w_{\alpha^k\alpha^l})_{\alpha^i\alpha^j} = D_{\mathrm{s}}^{ijkl}\overset{0}{\epsilon}_{ij}(B_{ij} + w_{\alpha^i\alpha^j}) + R^3(\alpha^1, \alpha^2, t),$$
$$w_1|_\Gamma = w_2|_\Gamma = \left.\frac{dw}{dn}\right|_\Gamma = 0,$$
$$w|_{t=0} = w(0), \quad w_t|_{t=0} = v_0. \tag{38.4}$$

Equations (38.4) can be considered as an initial–boundary value problem with respect to w, since w_1, w_2 that enter $\overset{0}{\epsilon}_{kl}$ can be taken to be expressed in terms of w through (14.1), where we should take $R^s \equiv 0$, $s = 1, 2$. Since for $R^3(\alpha^1, \alpha^2, t)$ we are assuming the structure given by (38.1), the solutions $w(\alpha^1, \alpha^2, t)$, $w_t(\alpha^1, \alpha^2, t)$ will now be a Markov process, while w_{tt} will contain white noise.

38.5. For an approximate computation of probabilistic properties of $w(\alpha^1, \alpha^2, t)$, we set

$$w(\alpha^1, \alpha^2, t) \approx \sum_{r=1}^{n} q_r(t)\varphi_r(\alpha^1, \alpha^2),$$

where $\varphi_r(\alpha^1, \alpha^2)$ is an orthonormal basis in H_1; for q_r we obtain by the Bubnov–Galerkin method the system of equations

$$\rho\ddot{q}_r + 2\beta\dot{q}_r = -\frac{\partial}{\partial q_r}\Big[\mathcal{U} - \sum_{s=1}^{n}(R_s^{32} + R_s^{33})q_s\Big] + R_s^{31}(t),$$
$$R_s^{3k}(t) = \int_\Omega R^{3k}(\alpha^1, \alpha^2, t)\varphi_s(\alpha^1, \alpha^2)\, d\alpha^1 d\alpha^2, k = 1, 2, 3. \tag{38.5}$$

In (38.5), $\mathcal{U}(q_1, \ldots, q_n)$ is the internal potential energy of deformation of the shell defined by (4.8)–(4.10). Here w_1, w_2 are given in terms of w by (14.1), so that the entire potential energy of the shell $\mathcal{U}$ turns out to be expressed in terms of $q_1, \ldots, q_n$. In accordance with (38.3), we can determine the characteristics of the white noise R_s^{31}. We have

$$\mathbf{E}\{R_s^{31}(t_1)R_p^{31}(t_2)\} = \mathbf{E}\Big\{\int_\Omega\int_\Omega R^{31}(\alpha^1, \alpha^2, t_1)$$
$$R_p^{31}(\beta^1, \beta^2, t_2)\varphi_s(\alpha^1, \alpha^2, t_1)\varphi_p(\beta^1, \beta^2, t_2) \times d\alpha^1 d\alpha^2 d\beta^1 d\beta^2$$
$$= \int_\Omega\int_\Omega K^{31}(\alpha^1, \alpha^2, \beta^1, \beta^2)$$
$$\times \varphi_s(\alpha^1, \alpha^2)\varphi_p(\beta^1, \beta^2)\, d\alpha^1 d\alpha^2 d\beta^1 d\beta^2\, \delta(t_1 - t_2). \tag{38.6}$$

Thus, (38.5) is a system subject to the action of random factors a_k, $b_{k_1k_2}$ and white noise $R_s^{31}(t)$ with correlation matrix (38.6). Let us assume now that the parameters a_k, $b_{k_1k_2}$ have taken certain values and let us consider the conditional probability distribution $f_n(t, q_r \mid a_k, b_{k_1k_2})$, $r = 1, \dots, n$; $k = 1, \dots, N$; $k_1 = 1, \dots, m_1$, $k_2 = 1, \dots, m_2$. To determine it, let us write (38.5) in the form

$$\dot{q}_r = p_r, \ \dot{p}_r = -\frac{2\beta}{\rho} p_r - \frac{1}{\rho}\frac{\partial}{\partial q_r}\Big[\mathcal{U} - \sum_{s=1}^{n}(R_s^{32} + R_s^{33})q_s\Big] + R_r^{31}(t).$$

Since by definition $R_s^{31}(t)$ is white noise, the distribution $f_n(t, q, p \mid a_k, b_{k_1k_2})$ satisfies the KFP equation [11, 10, 154, 155, 250]

$$\frac{\partial f_n}{\partial t} = -\sum_{r=1}^{n}\frac{\partial}{\partial q_r}(p_r f_n) + \sum_{r=1}^{n}\frac{\partial}{\partial p_r}\Big\langle\Big\{\frac{2\beta}{\rho}p_r + \frac{\partial}{\partial q_r}\Big[\mathcal{U} - \sum_{s=1}^{n}(R_s^{32} + R_s^{33})q_s\Big]\Big\} f_n\Big\rangle + \frac{1}{2\rho^2}\sum_{r,s=1}^{n} K_{rs}^{31}\frac{\partial^2 f_n}{\partial p_r \partial p_s}. \tag{38.7}$$

Equation (38.7) can be rewritten as

$$\frac{\partial f_n}{\partial t} = -\sum_{r=1}^{n} p_r\frac{\partial f_n}{\partial q_r} + \frac{2\beta n}{\rho} f_n + \frac{2\beta}{\rho}\sum_{r=1}^{n} p_r \frac{\partial f_n}{\partial p_r} + \sum_{r=1}^{n}\Big(\frac{\partial \mathcal{U}}{\partial q_r} - R_r^{32} - R_r^{33}\Big)\frac{\partial f_n}{\partial p_r} + \frac{1}{2\rho^2}\sum_{r,s=1}^{n}\mathcal{K}_{rs}^{31}\frac{\partial^2 f_n}{\partial p_r \partial p_s}. \tag{38.8}$$

We shall also assume that we know the initial distribution $f_n(0, q, p \mid a_k, b_{k_1k_2})$. Then under certain conditions that we do not describe here, (38.8) admits a unique solution f_n that satisfies the conditions

(1) $f_n \geq 0$;

(2) $\int_{-\infty}^{\infty}\int_{-\infty}^{\infty} f_n \, dq \, dp = 1;$ (38.9)

(3) $f_n(q, p, t) \to 0$ if $\sum_{r=1}^{n}(q_r^2 + p_r^2) \to \infty$.

If the distribution $f_n(t, q, p, \mid a_k, b_{k_1k_2})$ of (38.8), (38.9) has been determined, the unconditional probability distribution $f^*(t, q, p)$ can be found from

$$f_n^*(t, q, p) = \int_{-\infty}^{\infty}\int_{-\infty}^{\infty} f_n(t, q, p \mid a_k, b_{k_1k_2})\theta_3(a_k, b_{k_1k_2})\, da\, db.$$

Finally, we define the coordinates distribution $f^{**}(t, q)$ by

$$f_n^{**}(t, q) = \int_{-\infty}^{\infty}\int_{-\infty}^{\infty}\int_{-\infty}^{\infty} f_n(t, q, p \mid a_k, b_{k_1k_2})\theta_3(a_k, b_{k_1k_2})\, da\, db\, dp.$$

These functions $f_n^*(t, q, p)$, $f_n^{**}(t, q)$ can be used to estimate the chances of finding in practice a particular configuration of the shell at time t.

Of great interest in the problem of stability are the limiting, as $t \to \infty$, distributions $f_n(\infty, q, p \mid a_k, b_{k_1k_2})$, $f_n^*(\infty, q, p)$, $f^{**}(\infty, q)$. To determine them, let us assume that the limits

$$R^{32}(\alpha^1, \alpha^2, t)\big|_{t\to\infty} = R^{32}_{\infty}(\alpha^1, \alpha^2)$$

exist and that for each realization of $R^{33}(\alpha^1, \alpha^2, t)$ we have

$$R^{33}(\alpha^1, \alpha^2, t)\big|_{t\to\infty} = R^{33}_{\infty}(\alpha^1, \alpha^2).$$

To give a probabilistic description of $R^{32}_{\infty}(\alpha^1, \alpha^2)$, we shall assume that a significant portion of the realizations obey

$$R^{32}_{\infty}(\alpha^1, \alpha^2) + R^{33}_{\infty}(\alpha^1, \alpha^2) = \sum_{p=1}^{m_3} v_p \psi_p(\alpha^1, \alpha^2),$$

where ψ_p is as before an orthonormal basis in $L_{2\Omega}$. Then $R^{32}_{\infty}(\alpha^1, \alpha^2)$ will be described by a distribution $\theta_4(v_p)$. We could introduce a more general probability distribution $\theta_5(a_k, v_p)$ that would take into account a possible dependence between a_k and v_p.

The limiting distribution as $t \to \infty$, $f(\infty, q, p, | a_k, v_p)$ will be found from the stationary equation

$$\begin{aligned} 0 = &-\sum_{r=1}^{n} p_r \frac{\partial f_n}{\partial q_r} + \frac{2\beta n}{\rho} f_n + \frac{2\beta}{\rho} \sum_{r=1}^{n} p_r \frac{\partial f_n}{\partial p_r} \\ &+ \sum_{r=1}^{n} \left(\frac{\partial \mathcal{U}}{\partial q_r} - R^{32}_{r\infty} - R^{33}_{r\infty} \right) \frac{\partial f_n}{\partial p_r} + \frac{1}{2\rho^2} \sum_{r,s=1}^{n} \mathcal{K}^{31}_{rs} \frac{\partial^2 f_n}{\partial p_r \partial p_s}, \end{aligned} \tag{38.10}$$

where

$$R^{3s}_{r\infty} = \int_{\Omega} R^{3s}_{\infty}(\alpha^1, \alpha^2) \varphi_r(\alpha^1, \alpha^2)\, d\alpha^1 d\alpha^2, \quad (s = 2, 3.)$$

Having determined $f_n(\infty, q, p, | a_k, v_p)$, we can find the unconditional probability distributions

$$\begin{aligned} f_n^{***}(\infty, q, p) &= \int_{-\infty}^{\infty} \int_{-\infty}^{\infty} f_n(\infty, q, p, a_k, v_p) \theta_5(a_k, v_p)\, da\, dv, \\ f_n^{*****}(\infty, q) &= \int_{-\infty}^{\infty} \int_{-\infty}^{\infty} \int_{-\infty}^{\infty} \int_{-\infty}^{\infty} f_n(\infty, q, p, a_k, v_p) \theta_5(a_k, v_p)\, da\, dv\, dp. \end{aligned} \tag{38.11}$$

While $f_n^{*}(t, q, p)$, $f_n^{**}(t, q)$ characterize the probabilistic behavior of the shell on a finite interval of time T when the shell is subject to the action of the forces decomposed as in (38.1), then $f_n^{***}(\infty, q, p)$, $f_n^{****}(\infty, q)$ describe asymptotic stability of the shell in the nth approximation. Thus, we have indicated a fundamental method to solve the second problem of elastic stability.

Let us now consider to what extent the existing intuitive ideas concerning criteria of plausibility of a configuration of a shell correspond to the distributions $f_n^{***}(\infty, q, p)$ and $f_n^{****}(\infty, q)$. The most frequently used criterion uses the level of the total energy of the system shell–external forces. To resolve this question, we note that under certain conditions equation (38.10) admits explicit solutions. Indeed, let us assume that

$$\mathcal{K}^{31}_{rs} = \delta_{rs} \cdot \delta, \tag{38.12}$$

where δ_{rs} is the Kronecker delta. We verify directly that

$$\begin{aligned} &f_n(\infty, q, p, a_k, v_p) \\ &= \mathcal{Q}_n \exp\Big(-\frac{2\beta}{\delta}\Big[\frac{\rho}{2}\sum_{r=1}^{n} p_r^2 + \mathcal{U} - \sum_{r=1}^{n}(R^{32}_{r\infty} + R^{33}_{r\infty})q_r\Big]\Big) \end{aligned} \tag{38.13}$$

is a solution of (38.10). Furthermore, it satisfies

$$\begin{aligned} &(1)\ f_n \geq 0; \\ &(2)\ \int_{-\infty}^{\infty}\int_{-\infty}^{\infty} f_n\, dq\, dp = 1; \\ &(3)\ f_n \to 0 \text{ if } \sum_{r=1}^{n}(q_r^2 + p_r^2) \to \infty. \end{aligned} \tag{38.14}$$

In (38.13), a_k enters $\mathcal{U}$ as a parameter, likewise v_p in R^{32}_{∞}. The constant $\mathcal{Q}_n$ is found from the normalization condition 2, (38.14). Equation (38.13) is the Maxwell–Boltzmann distribution.

If we are interested only in the limiting distribution of coordinates, it follows from (38.13) that

$$f_n = D_n \exp\Big(-\frac{2\beta}{\delta}\Big[\mathcal{U} - \sum_{r=1}^{n}(R^{32}_{r\infty} + R^{33}_{r\infty})q_r\Big]\Big), \tag{38.15}$$

and we obtain the Gibbs distribution [348, 350], in which, indeed, the probability of a shell being in a particular limiting equilibrium configuration is determined by its potential energy level. Here the higher the level of potential energy, the less probable is the configuration. Thus in this particular case (38.12), potential energy levels do indeed determine the probability of encountering a particular equilibrium configuration. Let us consider (38.12) in more detail,

$$\mathcal{K}^{31}_{rs} = \int_{\Omega}\int_{\Omega} \mathcal{K}(\alpha^1, \alpha^2, \beta^1, \beta^2)\varphi_r(\alpha^1, \alpha^2)\varphi_s(\beta^1, \beta^2)\, d\alpha^1 d\alpha^2 d\beta^1 d\beta^2 = \delta\delta_{rs}. \tag{38.16}$$

If $\varphi_r(\alpha^1, \alpha^2)$ is, as we assume, a complete orthonormal family in $L_{2\Omega}$, then it follows from (38.16) that

$$\mathcal{K}(\alpha^1, \alpha^2, \beta^1, \beta^2) = \delta\sum_{r=1}^{\infty}\varphi_r(\alpha^1, \alpha^2)\varphi_s(\beta^1, \beta^2) = \delta\delta(r_{PQ}),$$

as was shown in Section 28. Thus (38.12) tells us that $R^{31}(\alpha^1, \alpha^2, t)$ is white noise not only in time, but in space as well. Therefore, in this case the level of potential energy of the system shell–external forces is indeed connected with the probability of encountering a particular equilibrium configuration of the shell (for a particular case of this assertion, see [38]). In other cases, when (38.12) does not hold, we have to use $f_n(\infty, q, \mid a_k, d_p)$ and hence the distribution f_n^{****} defined in (38.11).

We can obtain the Gibbs distribution law if we we assume sufficiently large time t or friction 2β and neglect in (38.5) inertial terms. Then we have

$$2\beta\dot{q}_r = -\frac{\partial}{\partial q_r}\Big[\mathcal{U} - \sum_{s=1}^{n}(R_s^{32} + R_s^{33})q_s\Big] + R_s^{31}(t). \tag{38.17}$$

Therefore, under this assumption, the velocities (38.17) themselves already contain a white noise component. The corresponding KFP equation has the form

$$2\beta\frac{\partial f_n}{\partial t} = -\sum_{r=1}^{n}\frac{\partial}{\partial q_r}\Big\{\Big[\mathcal{U} - \sum_{s=1}^{n}(R_s^{32} + R_s^{33})q_s\Big]f_n\Big\} + \frac{1}{2}\sum_{r,s=1}^{n}\mathcal{K}_{rs}^{31}\frac{\partial^2 f_n}{\partial q_r \partial q_s}. \tag{38.18}$$

If (38.12) holds, then the stationary distribution $f_n(q, \mid a_k, v_p)$ determined from (38.18) will be a Gibbs distribution.

38.6. The probabilistic models of shell operation constructed here were designed to solve the second problem of the theory of stability. At the same time, the theory of Markov processes has been widely used to estimate reliability and to describe the characteristics of fatigue-induced collapse of shells subjected to random loads, acoustic pressure of engines, turbulent atmosphere and so on. In these cases, considering $\{q\}$ or $\{q, \dot{q}\}$ as Markov processes can prove insufficient. We have to use other hypotheses concerning statistical properties of $R^{31}(\alpha^1, \alpha^2, t)$. In particular, it may prove useful to use a representation of the form

$$T\left(\frac{\partial}{\partial \alpha^i}, \frac{\partial}{\partial t}\right) R^{31}(\alpha^1, \alpha^2, t) = I(\alpha^1, \alpha^2, t), \tag{38.19}$$

where T is a rational function of the operators $\partial/\partial\alpha^i$, $\partial/\partial t$, and $I(\alpha^1, \alpha^2, t)$ is white noise in space and time. Under the assumptions of (38.19) we can also use the techniques of the KFP equation, and there are reasons to assume that acceptable numerical algorithms can be thus obtained. Good prospects for the study of stability of elastic systems under the action of random factors are offered by extending the KFP equation to a larger class of stochastic processes. In this respect, let us note that one of the first works in this direction is [248]. Construction of a probabilistic stability theory completes, in a sense, the program of analysis of the problem of stability outlined in the two questions of Section 29.

38.7. The theory of Markov processes in the problem of shell stability was first used in [345]. It was further developed in [348, 350]. In those papers, a classification was made of random factors acting on a shell, and a method for taking all of them into account simultaneously using a total probability theorem. The author restricted himself to assuming that the generalized coordinates are Markovian, which turns out to be sufficient for the analysis of problems of stability in a wide class of problems. Trying to justify the potential energy level criterion as a basis for the construction of a statistical theory, the author [345, 348, 350, 362] considered the case of a load δ-correlated in space and time (see (38.12)). Goncharenko [101, 102, 103] extended this analysis to the general case, in which both the generalized velocities and coordinates are taken to be Markovian. In addition, he studied the case when the external load is not δ-correlated in space. Later Goncharenko considered distributions in Sobolev spaces [104, 105], which is natural for this type of question. A number of problems are discussed in [38, 337]. A large number of studies are in existence dealing with applications of the theory of Markov processes in the accumulation of fatigue faults in shells. Analysis of such problems is beyond the scope of our study.

It would be most interesting to justify the scheme presented here, in particular the question of behavior of the distributions f_n as $n \to \infty$. It is easy to find examples when as $n \to \infty$ they become degenerate. A number of results in this direction have been obtained by Chueshov [56, 57], who constructed probability distributions in infinite-dimensional energy spaces for plates and shells. Let us also note in this regard the work of Goncharenko [104, 105] mentioned above. We observe that in hydrodynamics, in particular, for boundary value problems for the Navier–Stokes equation this problem was already being considered by Hopf and was resolved in the work of Vishik and Fursikov [331].

Some Unsolved Problems of the Mathematical Theory of Shells

i. Formulation of the main boundary value problems of nonlinear shell theory without the assumption of shallowness (3.18) and moderate bending (2.13), that is, for arbitrary rotation.

ii. Construction of a mathematical theory of boundary value problems for the Reissner equations (axisymmetric shell, arbitrary deflection).

iii. Derivation of the main boundary value problems of the nonlinear theory of shallow shells directly from boundary value problems of nonlinear three-dimensional theory of elasticity. Justification of this derivation (see Ciarlet, Rabier), limits of applicability.

iv. Introduction of Airy stress functions in linear and nonlinear theory of non-shallow shells. Reduction of the number of independent variables in the theory.

v. A detailed analysis of the limits of applicability of nonlinear boundary value problems of shallow shell theory. Derivation of theoretical estimates of the influence of nonshallowness on various characteristics of the stress-deformed state.

vi. A mathematical analysis of boundary value problems of shallow shell theory for a wider class of boundary conditions than are being considered in this book. In particular, solvability theorems for free shells that are not subjected to any geometric boundary conditions.

vii. Construction of a mathematical theory of boundary value problems for shells of the type considered by Timoshenko, Reissner, in which in addition to the geometric nonlinearity, shear stresses are taken into account. Justification of approximation methods.

viii. Extension of the main results of the present monograph to the case of nonconservative loads. In this case there are examples where solvability holds only for sufficiently large loads.

ix. Extension of the method for obtaining a priori estimates to the case of linearly viscoelastic shells. Here it is natural to use properties of viscoelastic operators, defined by the stability principle of the natural stressed state of viscoelastic bodies.

x. Isolation of a class of nonlinear boundary value problems of mathematical physics for which a priori estimates of the solution can be found using the methods developed in the monograph.

xi. Analysis of the structure of the energy solution of nonlinear boundary value problems in a neighborhood of a corner point and in a neighbourhood of points and lines of change in the boundary conditions.

xii. Formulation of nonlinear boundary value problems of the theory of shallow shells reinforced with rigid ribbing. Behavior of energy solutions in a neighborhood of a rigid rib. Conjugation with rigid bodies.

xiii. Analysis of singularities caused by singularities in S, $\rho(\alpha^1, \alpha^2)$ of solutions of nonlinear boundary value problems of the theory of shallow shells.

xiv. Development of the theory of a lower critical number for nonshallow shells under arbitrary rotation angles.

xv. Development of the theory of a lower critical number for the states of a shell that are not momentless.

xvi. Estimates of lower critical numbers for shells of various types under different clamping conditions. A criterion of rigidity. Analysis of rigidity of a closed sphere, a closed ellipsoid.

xvii. A rigorous analysis of bifurcation equations in a neighborhood of singular solutions in nonlinear shell theory. Application of theoretical methods. Secondary bifurcation. Analysis of the topological structure of the functionals $\mathcal{I}(w), \mathcal{I}(a), \mathcal{I}(\psi, w)$.

xviii. The use of group-theoretical methods in bifurcation theory.

xix. Study of the possibility of unbounded continuation of branches of solutions of a nonlinear boundary value problem that originate at an eigenelement of the corresponding problem linearized around an MIS.

xx. Extension of Wolkowisky's theorem to the general case of plates of arbitrary form (possibly, under certain symmetry conditions): Under load λ that exceeds the nth eigenvalue of the problem linearized around an MIS of a boundary value problem, there exist at least n distinct solutions of the nonlinear problem.

xxi. Analysis of the number of solutions of nonlinear boundary value problems in the theory of moderate bending under various classes of symmetry.

xxii. Development and justification of asymptotics in thin-wall parameter methods of solutions of nonlinear boundary value problems of shell theory.

xxiii. Analysis of the behavior as $n \to \infty$ of finite-dimensional distributions $f_n(p, q)$ obtained using the Kolmogorov–Fokker–Planck equation for finite-dimensional approximations obtained by the Bubnov–Galerkin or the Ritz method in the main linear boundary value problems of the nonlinear theory of shallow shells.

References

The Russian journals referred to below are translated into English under the following titles:

Dokl. Akad. Nauk SSSR ~ Russian Academy of Sciences. Doklady. Mathematics.
Inzh. Zh. MTT (later, Izv. Acad. Nauk SSSR, MTT) ~ Mechanics of Solids.
Izv. Akad. Nauk SSSR, Ser. mat. ~ Russian Academy of Sciences. Izvestiya. Mathematics.
Izv. Vyssh. Uchebn. Zaved. Mat. ~ Soviet Mathematics.
Mat. Sb. ~ Russian Academy of Sciences. Sbornik. Mathematics.
Prikl. Mat. Mekh. ~ Journal of Applied Mathematics and Mechanics.
Prikl. Mekh. ~ Soviet Applied Mechanics.
Sib. Matem. Zh. ~ Siberian Mathematics Journal.
Trudy Mat. Inst. Akad. Nauk ~ Proceedings of the Steklov Institute of Mathematics.
Trudy Mosk. Mat. Obshch. ~ Transactions of the Moscow Mathematical Society.
Uspekhi Mat. Nauk ~ Russian Mathematical Surveys.
Vestnik MGU, Ser. I Matem. Mekh ~ Moscow University Mathematics bulletin.

References

[1] S. Agmon, A. Douglis, and L. Nirenberg, Estimates near the boundary for solutions of elliptic partial differential equations satisfying general boundary conditions, I. Comm. Pure Appl. Math., 12:623–727, 1959, II, Comm. Pure Appl. Math., 17:35–92, 1964.

[2] O.K. Aksentyan and I.I. Vorovich, Stressed state of a plate of small thickness, Prikl. Mat. Mekh., 27:1057–1074, 1963.

[3] P.S. Aleksandrov, *Combinatorial Topology*, Gostekhizdat, Moscow–Leningrad 1947.

[4] S.A. Alekseev, Post-critical work of flexible elastic plates, Prikl. Matem. Mekh., 20:637–679, 1956.

[5] N.A. Alumyae, Differential equations of equilibria of thin elastic shells in postcritical state, Prikl. Mat. Mekh., 13:95–107, 1949.

[6] N.A. Alumyae, Application of the generalized variational principle of Castigliano to the study of postcritical state of thin-walled elastic shells, Prikl. Mat. Mekh., 14:93–99, 1950.

[7] N.A. Alumyae, A variational formula for the study of thin-walled elastic shells in postcritical state, Prikl. Mat. Mekh., 14:197–203, 1950.

[8] S.A. Ambartsumyan, *General Theory of Anisotropic Shells*, Nauka, Moscow 1974.

[9] V.V. Amel'chenko, I.V. Neverov, and V.V. Petrov, Solution of nonlinear problems of the theory of shallow shells by variational iterations, Izv. Akad. Nauk SSSR, MTT, 3:62–68, 1963.

[10] A.A. Andronov, *Collected Works*, Izd. Akad. Nauk SSSR, Moscow, 1956.

[11] A.A. Andronov, A.A. Vitt, and L.S. Pontryagin, On the statistical point of view of dynamical systems, Zh. Eksp. Teor. Fiz. 33:165–180, 1938.

[12] S.S. Antman, Buckled states of non-linearly elastic plates, Arch. Rational Mech. Anal., 67:111–149, 1978.

[13] V.I. Arnol'd, A.N. Varchenko, and S.M. Gusein-Zade, *Singularities of Differentiable Mappings. Classification of Critical Points, Caustics and Wave Fronts*, Nauka, Moscow 1982.

[14] V.I. Arnol'd, A.N. Varchenko, and S.M. Gusein-Zade, *Singularities of Differentiable Mappings. Monodromy and Asymptotics of Integrals*, Nauka, Moscow 1984.

[15] N. Aronszajn, On coercive integro-differential quadratic forms, Tech. Report., Uni. of Kansas, 14:94–106, 1955.

[16] I.A. Bakhtin and M.A. Krasnosel'skii, On the problem of lateral bending of a rod of variable rigidity, Dokl. Akad. Nauk SSSR, 105:621–624, 1955.

[17] G.A. Baker and P. Graves-Morris, *Pade Approximants. Part I: Basic theory; Part II: Extensions and Applications*, Addison Wesley Publishing Company, London, 1981.

[18] N.A. Bazarenko and I.I. Vorovich, Asymptotics of behavior of solutions of a problem in the theory of elasticity for a hollow cylinder of finite length and small thickness, Prikl. Mat. Mekh., 29:1035–1062, 1965.

[19] N.A. Bazarenko and I.I. Vorovich, Asymptotical behavior of solutions of a problem in theory of elasticity for a cylinder of finite length and small thickness, in: *Proc. V-th All-Union Conference in Shell and Plate Theory (Moscow, 1965)*, pages 13–14, Izd. Akad. Nauk SSSR, Moscow 1965.

[20] N.A. Bazarenko and I.I. Vorovich, Analysis of three-dimensional stressed and deformed state of circular cylindric shells. Construction of applied theories, Prikl. Mat. Mekh., 33:495–510, 1969.

[21] R. Bellman, *Perturbation Techniques in Mathematics, Physics and Engineering*, Holt, Rinehart and Winston, New York 1964.

[22] Yu.M. Berezanskii, *Expansion in Eigenfunctions of Self-Adjoint Operators*, Naukova Dumka, Kiev, 1965.

[23] Yu.M. Berezanskii, S.G. Krein, and Ya.A. Roitberg, A homeomorphism theorem and local improvement of smoothness up to the boundary of solutions of elliptic equations, Dokl. Akad. Nauk SSSR, 148:745–748, 1963.

[24] M.S. Berger, An eigenvalue problem for nonlinear elliptic partial differential equations, Trans. Amer. Math. Soc., 120:145–184, 1965.

[25] M.S. Berger, An application of the calculus of variations in the large to the equations of nonlinear elasticity, Bull. Amer. Math. Soc., 73:520–525, 1967.

[26] M.S. Berger, On von Kármán equations and the buckling of a thin elastic plate, Comm. Pure Appl. Math., 20:687–718, 1968.

[27] M.S. Berger, Bifurcation theory for nonlinear elliptic equations and systems, in [139], pages 71–112.

[28] M.S. Berger, Bifurcation theory for nonlinear elliptic equations and systems (continued), in [139], pages 113–128.

[29] M.S. Berger and P.C. Fife, On von Kármán's equations and the buckling of a thin elastic plate II. Plate with general edge conditions, Comm. Pure Appl. Math., 21:226–241, 1968.

[30] M. Bernadou, J.T. Oden, An existence theorem for a class of non-linear shallow shell problems, J. Math. Pures Appl., 60:285–308, 1981.

[31] O.V. Besov, A study of a family of spaces of functions in connection with embedding and extension theorems, Trudy Mat. Inst. Akad. Nauk SSSR, 60:42–84, 1961.

[32] O.V. Besov, V.P. Il'in, and S.M. Nikol'skii, *Integral Representation of Functions and Embedding Theorems*. Nauka, Moscow 1975.

[33] P.K. Bhattacharyya, On existence of a solution of the clamped shallow non-circular cylindrical shell problem, Int. J. Eng. Sci., 23:359–369, 1985.

[34] L. Bieberbach, *Differentialgeometrie*, Leipzig-Berlin, 1932.

[35] G. Birkhoff and O.D. Kellogg, Invariant points in function space, Trans. Amer. Math. Soc., 23, 1922.

[36] S.R. Bodner, The postbuckling behavior of a clamped circular plate, Quart. Appl. Math., 12:397–401, 1955.

[37] V.V. Bolotin, Applications of probability theory methods in the theory of plates and shells, in: *Proc. IV-th All-Union Conference on Plate and Shell Theory*, pages 15–63, Izd. Akad Nauk Arm. SSR, Erevan, 1964.

[38] V.V. Bolotin, *Probability Theory Methods in the Theory of Reliability in Computations of Structures*, Stroiizdat, Moscow, 1982.

[39] L.E.J. Brouwer, Über Abbildung von Mannigfaltigkeiten, Math. Ann. 71: 158–184, 1912.

[40] F.E. Browder, On the generalization of the Schauder fixed point theorem, Duke Univ. Math. J., 26:116–121, 1954.

[41] F.E. Browder, Nonlinear elliptic boundary value problems, Bull. Amer. Math. Soc. 69:6, 1963.

[42] F.E. Browder, Non-linear maximal monotone operators in Banach spaces, Math. Fund., 175:89–113, 1968.

[43] F.E. Browder, Non-linear monotone and accretive operators in Banach spaces, Proc. Nat. Acad. Sci. USA, 61:388–393, 1968.

[44] I.G. Bubnov, *Ship Construction Mechanics*, Izd. Mor. Minist., St. Petersburg, Part I, 1912, Part II, 1914. See also: *Studies in Plate Theory*, Gostekhizdat, Moscow 1953.

[45] I.G. Bubnov, Review of the paper of Prof. S.P. Timoshenko, "On the stability of elastic systems," in: *Collected Works,* pages 136–139, Gos. Izd. Sudostroit. Promyshl., Leningrad, 1956.

[46] B. Buchberger, D. Collins, R. Loos, eds., *Computer Algebra. Symbolic and Algebraic Computations. A Collection of Papers*, Mir, Moscow, 1986.

[47] B. Budiansky, Theory of buckling and postbuckling behavior of elastic structures, Adv. Appl. Mech., 14:2–66, 1974.

[48] B. Budiansky and J.W. Hutchinson, *Buckling: Progress and Challenge, Trends in Solid Mechanics*, eds. I.F. Besseling and A.M.A. van der Heijden, Proc. of the Symp. dedicated to the sixty-fifth birthday of W.T. Koiter, pages 93–117, the University Press, Delft, 1979.

[49] L.C. Cheo and E.L. Reiss, Unsymmetric wrinkling of circular plates, Quart. Appl. Math., L31:75–91, 1973.

[50] N.G. Chetaev, *Stability of Motion. Studies in Analytical Mechanics*, Fizmatgiz, Moscow, 1962.

[51] W.-Z. Chien, The intrinsic theory of shells and plates, Quart. Appl. Math., 1:297–327, 1943.

[52] W.-Z. Chien, The intrinsic theory of thin shells and plates, Quart. Appl. Math., 2:120–135, 1944.

[53] W.-Z. Chien, The application of perturbation methods to the theory of circular thin plates with large bending, in: [335], pages 56–79.

[54] W.-Z. Chien, The theory of circular plates with large bending under axial symmetry, in: [335], pages 11–38.

[55] W.-Z. Chien and K.-Y.E, Large bending of circular thin plates, in: [335], pages 178–207.

[56] I.D. Chueshov, Existence of statistical solutions of a stochastic system of von Kármán equations in a bounded domain, Mat. Sb., 3:122, 291–312, 1983.

[57] I.D. Chueshov, The Hopf equation for a dynamical system with an infinite-dimensional phase space and the Euclidean field theory, Preprint ITF-84-184P, Inst. Teor. Fiz., Akad. Nauk Ukr. SSR, Kiev, 1985.

[58] P.G. Ciarlet and P. Rabier, *Les equations de von Kármán*, Springer-Verlag, Berlin, 1980.

[59] G. Connor and P. Morin, Perturbation methods in the computation of geometrically nonlinear shells, in: *Numerical Computation of Elastic Structures, Vol. 2*, pages 186–202, Sudostroenie, Leningrad 1974.

[60] R. Courant and D. Hilbert, *Methods of Mathematical Physics*, Gostekhizdat, Moscow–Leningrad, 1951.

[61] D.F. Davidenko, On a new method of numerical solution of systems of nonlinear equations, Dokl. Akad. Nauk SSSR, 88:601-602, 1953.

[62] D. F Davidenko, On the application of the method of variation in parameter to the theory of nonlinear functional equations, Ukr. Mat. Zh., 7:18–28, 1955.

[63] D.F. Davidenko, On the use of the method of variation of parameter in the construction of improved precision iteration formulae in the computation of numerical nonlinear integral equations, Dokl. Akad. Nauk SSSR, 162:499–502, 1965.

[64] D.F. Davidenko, On the construction of improved precision iteration schemes by the method of variation of parameter, in: *Abstracts of Short Communications of the International Congress of mathematicians, Section 14*, page 31, Akad. Nauk SSSR, Moscow, 1966.

[65] P. Destuynder and Ph.G. Ciarlet, A justification of the two-dimensional linear plate model, J. Mécanique, 18:315–344, 1979.

[66] P. Destuynder, Sur l'existence du solution stables pair un modèle coque en élastique non-linéaire, C.R. Acad. Sci., 293:713–716, 1981.

[67] L. Donnell, Stability of thin-walled tubes under torsion, Nat. Adv. Corn. for Aeron. Pep., p. 479, 1934.

[68] L. Donnell, A new theory for the buckling of thin cylinders under axial compression and bending, Trans. ASME, 56:795, 1934.
[69] Yu.A. Dubinskii, On the solvability of a system governing strong deflection of a plate, Dokl. Akad. Nauk SSSR, 175:1026–1029.
[70] Yu.A. Dubinskii, Nonlinear elliptic and parabolic equations, in: *Advances in Science and Technology. Contemporary Problems of Mathematics*, 9:1–131, VINITI, Moscow, 1976.
[71] B.A. Dubrovin, S.G. Novikov, and A.T. Fomenko, *Modern Geometry. Methods and Applications. Part 2*, Springer-Verlag, Berlin, 1985.
[72] M. van Dyke, *Perturbation Methods in Fluid Mechanics*, Academic Press, New York, 1967.
[73] N.V. Efimov, Qualitative questions of the theory of deformable surfaces, Uspekhi Mat. Nauk, 3:47–158, 1948.
[74] L.E. El'sgolts, *Qualitative Methods in Mathematical Analysis*, Gostekhizdat, Moscow, 1955.
[75] L. Euler, *Methodus inveniendi linear curvas maximim minimive proprietate gaudentes (Appendix, De curvis elastics)*, Marcum Michaelem, Bousquet, Lausanne and Geneva, 1744.
[76] L. Euler, *A Method for Finding Curves Having the Maximum or Minimum Property, or the Solution of the Isoperimetric Problem, Understood in the Widest Possible Sense*, ONTI, Moscow–Leningrad, 1934.
[77] V.I. Feodos'ev, On the computation of a buckling membrane, Prikl. Mat. Mekh., 10:295, 1944.
[78] V.I. Feodos'ev, *Elastic Structures in High Precision Device Engineering*, Oborongiz, Moscow, 1949.
[79] S.P. Finikov, *Differential Geometry*, Izd. MGU, Moscow 1961.
[80] A.P. Filin, ed., *Numerical Analysis of Elastic Structure*, Sudostroenie, Leningrad, 1974
[81] K. Forsberg, Evaluation of finite difference and finite element methods applied to the numerical analysis of arbitrary shells, in [80], pages 296–311.
[82] K.O. Friedrichs and R.F. Dressler, A boundary layer theory for elastic plates, Comm. Pure and Appl. Math., 14:1–33, 1961.
[83] K. Friedrichs and J. Stoker, The non-linear boundary value problem of the buckled plate, Proc. Nat. Acad. Sci., 25:535–540, 1939; Amer. J. Math., 63:839–888, 1941; Buckling of the circular plate beyond the critical thrust, J. Appl. Mech., 9:7–14, 1979.
[84] H. Gajewski, K. Groger, and K. Zacharias, *Nichtlineare Operatorgleichungen and Operator Differentialgleichungen*, Academie Verlag, Berlin, 1974.
[85] E. Gagliardo, Ulteriori proprieta di alcune classi di funzioni in piu variabili, Riecerche di Mat., 8:24–51, 1959.
[86] F.D. Gakhov, *Boundary Value Problems*, Pergamon Press, Oxford, 1966.
[87] B.G. Galerkin, Rods and plates, Vestnik Inzhenerov. 1, 19:897–908, 1915.
[88] K.Z. Galimov, A general theory of elastic shells under finite displacements, Izv. KFAN SSSR. Ser. Fiz. Mat. i Tekhn. Nauk, 2:3–38, 1950.
[89] K.Z. Galimov, Towards a general theory of plates and shells under arbitrary displacements and deformations, Prikl. Mat. Mekh., 15:723–742, 1951.
[90] K.Z. Galimov, Some problems of the theory of elasticity for arbitrary displacements, Uch. Zapiski Kazan. Univ., 112:101–119, 1952.
[91] K.Z. Galimov, Surface deformation continuity conditions under arbitrary bending and deformation, Uch. Zapis. Kazan. Univ., 113:161–164, 1953.

[92] K.Z. Galimov, Some problems of shell theory under arbitrary displacements, Izv. KFAN SSSR. Ser. Fiz. Mat. i Tekhn. Nauk, 3:3–16, 1953.
[93] K.Z. Galimov, On the variational methods of solution of problems in nonlinear plate and shell theory, Izv. KFAN SSSR. Ser. Fiz. Mat. i Tekhn. Nauk, 10:3–25, 1956.
[94] K.Z. Galimov, Applications of a variational principle for admissible changes in deformed state to nonlinear theory of shallow shells, Izv. VUZ. Matem., 4:2–11, 1958.
[95] K.Z. Galimov and V.N. Paimushin, *Theory of Solid Shape Shells*, KGU, Kazan', 1985.
[96] E.I. Gokhman, *Introduction to Tensor Analysis*, ONTI NKPT, Khar'kov–Kiev, 1935.
[97] A.L. Gol'denveizer, *Theory of Thin Elastic Shells*, Nauka, Moscow, 1976.
[98] A.A. Gonchar, On the convergence of the generalized Padé approximants of meromorphic functions, Mat. Sb. 98(140):564–577, 1975.
[99] A.A. Gonchar, On the rate of convergence of rational approximants of certain analytic functions, Mat. Sb. 105:147–163, 1975.
[100] A.A. Gonchar and K.N. Lungu, Poles of Padé approximants and analytic continuation of functions, Mat. Sb. 111(153):119–132, 1980.
[101] V.M. Goncharenko, The statistical method in the problem of a pure bending of a cylindrical shell, in: *Proceedings of the All-Union Conference on Plate and Shell Theory*, pages 130–137, Izd. Kazan. Univ., Kazan', 1960.
[102] V.M. Goncharenko, Towards the definition of probability of the loss of stability of a shell, Izv. Akad. Nauk SSSR. OTN. Mekh. Mashin., 1:159–161, 1962.
[103] V.M. Goncharenko, A study of the probability of buckling of an elongated cylindrical panel under random pressure, Prikl. Mat. Mekh., 4:740–744, 1962.
[104] V.M. Goncharenko, Some classes of vector-valued generalized functions and their applications to boundary value problems for random functions, Ukr. Mat. Zh., 2:158–166, 1975.
[105] V.M. Goncharenko, On some spaces of random functions of finite order, Izv. Vyssh. Uchebn. Zaved. Mat., 6:67–69, 1982.
[106] E. Goursat, *A Course of Mathematical Analysis,* Vol. 1, ONTI, Moscow–Leningrad, 1936.
[107] A.E. Green and G.E. Adkins, *Large Elastic Deformations and Nonlinear Continuum Mechanics*, Clarendon Press, Oxford, 1960.
[108] E.I. Grigolyuk, On the question of behavior of a spherical plate after loss of stability, Vest. Inzh. Tekhn. 3:103–106, 1949.
[109] E.I. Grigolyuk and V.I. Shalashilin, On certain versions of the method of continuation in a parameter in problems of nonlinear elasticity, Prikl. Mekh. Teor. Fiz. 5:158–152, 1980.
[110] E.I. Grigolyuk and V.I. Shalashilin, The method of continuation in a parameter in problems of nonlinear deformation of rods, plates and shells, Usp. Mekh. (Warsaw) 4:89–122, 1981.
[111] E.I. Grigolyuk and V.I. Shalashilin, Modified versions of the method of continuation in a parameter in nonlinear problems of mechanics of solid deformable bodies, in: *Problems of Mathematics and Mechanics*, pages 55–70, Nauka, Sib. Otdel, Novosibirsk, 1983.
[112] E.I. Grigolyuk and V.I. Shalashilin, The method of continuation in a parameter in problems of nonlinear deformation of rods, plates and shells, in: *Studies in Plate and Shell Theory, Issue 17, Part I,* pages 3–58, Izd. Kazan' Univ., Kazan', 1984.

[113] M.V. Grosheva, G.V. Efimov, V.A. Brumberg, et al., *Symbolic Computation Systems*, Informator No. 1, Inst. Priklad. Matem. Akad. Nauk SSSR, Moscow, 1983.

[114] V.I. Gulyaev, V.A. Bazhenov, and E.A. Gotsulyak, *Stability of Nonlinear Mechanical Systems*, Vishcha Shkola, L'vov, 1982.

[115] V.I. Gulyaev and G.I. Mel'nichenko, Shapes of postcritical equilibrium of cylindrical and conical shells of elliptic section under an axial load, Izv. Akad. Nauk SSSR, MTT, 5:60–66, 1976.

[116] O.V. Guseva, On boundary value problems for strongly elliptic systems, Dokl. Akad. Nauk SSSR, 102:1069–1070, 1955.

[117] J.S. Hansen, Some two-mode buckling problems and their relation to catastrophe theory, AIAL Journal, 15:1638–1644, 1977.

[118] T.H. Hildebrandt and L.M. Graves, Implicit functions and their differentials in general analysis, Trans. Amer. Math. Soc., 29:, 1927.

[119] I. Hlavaček and I. Naumann, Inhomogeneous boundary value problems for the von Kármán equation, Aplikace Mathematiky, 19:253–269, 1974.

[120] I. Hlavaček and I. Naumann, Inhomogeneous boundary value problems for the von Kármán equations. II., Aplikace Mathematiky, 20:280–297, 1975.

[121] H. Hopf, Vektorfelder in n-dimensionaler Mannigfaltigkeiten, Math. Ann., 96:225–250, 1926.

[122] J.W. Hutchinson and W.T. Koiter, Postbuckling theory, Appl. Mech. Rev., 23:1353, 1970.

[123] V.P. Il'in, On embedding theorems for critical exponent, Dokl. Akad. Nauk SSSR, 96:905–908, 1954.

[124] V.P. Il'in, On embedding theorems, Trudy Mat. Inst. Akad. Nauk SSSR, 53:359–386, 1959.

[125] V.P. Il'in, Certain inequalities in function spaces and their application to the problem of convergence of variational processes, Trudy MIAN, I–III, in: *Studies in Approximation Theory*, pages 64–127, Izd. Akad. Nauk SSSR, Moscow–Leningrad, 1959.

[126] O. John and I. Nečas, On the solvability of von Kármán equations, Apl. Mat., 20:48–62, 1975.

[127] R.I. Kachurovskii, Nonlinear monotone operators in Banach spaces, Uspekhi Mat. Nauk, 23:121–168, 1968.

[128] V.F. Kagan, *Foundations of the Theory of Surfaces. Part I*, Gostekhizdat, Moscow–Leningrad 1947.

[129] E. Kai-Yuan, A study of large deflections of thin annular plates, In [335], pages 79–101.

[130] L.V. Kantorovich and G.P. Akilov, *Functional Analysis* (English transl.), Pergamon, 1982.

[131] von Kármán, J., Festigkeitsproblem in Maschinenbau, Encyclopädia der Mathematischen Wissenschaften, IV:311–385, 1910.

[132] V.V. Karpov and V.V. Petrov, Improvement of the solution in the use of time-stepping methods in the theory of elastic plates and shells, Izv. Akad. Nauk SSSR. MTT, 5:189–191, 1975.

[133] G. Kauderer, *Nonlinear Mechanics*, Inostr. Lit., Moscow 1961.

[134] Ya.F. Kayuk, Analytic continuation of solutions of nonlinear differential equations in a parameter, Ukr. Mat. Zhurn., 19:131–138, 1967.

[135] Ya.F. Kayuk, Deformed state of shallow shells of rotation under large displacements, Izv. Akad. Nauk SSSR. MTT., 5:159–163, 1969.

[136] Ya.F. Kayuk, On the convergence of expansions in a parameter in geometrically nonlinear problems, Prikl. Mekh., 9:83–89, 1973.

[137] Ya.F. Kayuk, *Some Aspects of the Method of Expansion in a Small Parameter*, Naukova Dumka, Kiev, 1980.

[138] M.V. Kel'dysh, On the method of B.G. Galerkin for the solution of boundary value problems, Izv. Akad. Nauk SSSR. Ser. mat., 6:309–330, 1942.

[139] J.B. Keller and S. Antman, eds. *Bifurcation Theory and Nonlinear Eigenvalue Problems*, Benjamin, New York, 1969.

[140] I.Yu. Kharrik, On the approximation of functions that vanish on the boundary together with their partial derivatives, by functions of a particular type, Sib. Mat. Zh., 4:408–425, 1963.

[141] N.A. Kil'chevskii, *Elements of Tensor Analysis and its Applications in Mechanics*, Gostekhizdat, Moscow, 1954.

[142] N.A., Kil'chevskii, *Foundations of Analytical Mechanics of Shells*, Izd. Akad. Nauk SSSR, Kiev, 1963.

[143] G.H. Knightly, An existence theorem for the von Kármán equations, Arch. Rational Mech. Anal., 27:233–242, 1967.

[144] G.H. Knightly and D. Sather, On non-uniqueness of the solutions of the von Kármán equations, Arch. Rational Mech. Anal., 28:65–78, 1968.

[145] I.E. Kochin, *Vector Calculus and Elements of Tensor Analysis*, Nauka, Moscow, 1965.

[146] W.T. Koiter, On the stability of elastic equilibrium, dissertation, Delft, 1945. (Techn. Trans. NASA, 10:833, 1967.)

[147] W.T. Koiter, Elastic stability and postbuckling behavior, *Proc. Symp. Non-linear Prob.*, R.E. Langer, ed., pages 257–275, Univ. of Wisconsin Press, 1963.

[148] W.T. Koiter, On the non-linear theory of thin elastic shells, Proc. Konik. Ned. Akad. Wetensch, 69:1–54, 1966.

[149] W.T. Koiter, The non-linear buckling problem of complete spherical shells under uniform external pressure, Proc. Konik. Ned. Akad. Wetensch Ser. B., 72:40, 1969.

[150] W.T. Koiter, Current trends in the theory of buckling, in: *Buckling Structures*, pages 1–16, Berlin 1976.

[151] W.T. Koiter, Forty years in retrospect, the bitter and the sweet, in: *Buckling: Progress and Challenge, Trends in Solid Mechanics; Proc. Symp. Dedicated to the sixty-fifth birthday of W.T. Koiter*, I.F. Besseling and A.M.A. van der Heijden, eds., pages 237–246, the University Press, Delft, 1979.

[152] W.T. Koiter, The application of the initial postbuckling analysis to shells, in: *Buckling Shells. Proc. State-of-the Art. Colloq. Univ. Stuttgart, May 6-7, 1982*, pages 3–11, Berlin e.a., 1982.

[153] W.T. Koiter and A. van der Neut, Interaction between local and overall buckling of stiffened compression panels, in: *Thin-walled Structures*, I. Rhodes and A.C. Walker, eds., Grenada, London, 1980.

[154] A.N. Kolmogorov, On analytical methods in the theory of probability, Uspekhi Mat. Nauk, 5:5–41, 1938.

[155] A.N. Kolmogorov, Statistical theory of oscillations with a continuous spectrum, in: *The Jubilee Issue, dedicated to 30th Anniversary of the Great October Socialist Revolution*, Vol. 1, pages 245–254, Izd. Akad. Nauk SSSR, Moscow, 1947.

[156] A.N. Kolmogorov and S.V. Fomin, *Elements of Function Theory and Functional Analysis,* Vol. 1: *Metric and Normed Spaces*, Graylock Press, Albany, NY, 1957;

Vol. 2: *Measure. The Lebesgue Integral. Hilbert Space*, Graylock Press, Albany, NY, 1961.

[157] M.A. Koltunov, A higher precision solution of the problem of stability of rectangular panels of elastic shells, Vestnik MGU. Ser. 1. Matem., Mekh., 3:37–46, 1961.

[158] M.A. Koltunov, Stressed stated of elastic shallow shells, Vestnik MGU. Ser. 1. Matem., Mekh., 4:63–69, 1962.

[159] M.A. Koltunov, Stability of a panel of a cylindrical shell, Vestnik MGU. Ser. 1. Matem., Mekh., 6, 1962.

[160] A.I. Koshelev, On the boundedness in L_p of derivatives of solutions of elliptic equations, Mat. Sb., 38(30):278–312, 1956.

[161] A.S. Kosmodamianskii and V.A. Shaldyrvan, *Thick Multiply-connected Plates*, Naukova Dumka, Kiev, 1978.

[162] M.A. Krasnosel'skii, *Topological Methods in the Theory of Nonlinear Integral Equations* (English transl.), Pergamon, 1964.

[163] M.A. Krasnosel'skii, A.I. Perov, A.I. Povolotskii, and P.P. Zabreiko, *Planar Vector Fields,* Fizmatgiz, Moscow, 1963.

[164] M.A. Krasnosel'skii, G.M. Vainikko, P.P. Zabreiko, Ya.B. Rutitskii, and V.Ya. Stetsenko, *Approximate Solution of Operator Equations*, Nauka, Moscow, 1969.

[165] M.A. Krasnosel'skii and P.P. Zabreiko, *Geometrical Methods of Nonlinear Analysis,* (English transl.), Springer-Verlag, Berlin, 1984.

[166] Yu.P. Krassovskii, Study of potentials connected with boundary value problems for elliptic equations, Izv. Akad. Nauk SSSR, Ser. mat., 31:587–640, 1967.

[167] Yu.P. Krassovskii, Determination of the singularity of Green's function, Izv. Akad. Nauk SSSR, Ser. mat., 31:977–1010, 1967.

[168] L. Kronecker, Über die Characteristik von Funktionen Systemen, Monatsbericht. Acad. Wiss. Berlin, 97–121, 1878.

[169] N.M. Krylov, On some directions in the area of approximate solution of problems of mathematical physics, in: *The Jubilee Issue, dedicated to 30th Anniversary of the Great October Socialist Revolution*, Vol. 1, pages 231–241, Izd. Akad. Nauk SSSR, Moscow, 1947.

[170] V.A. Krys'ko, *Nonlinear Statics and Dynamics of Nonhomogeneous Shells*, Izd. Saratov. Univ., Saratov, 1976.

[171] V.D. Kubenko, Yu.N. Nemish, K.I. Sherenko, and I.A. Shul'ga, Perturbation methods in boundary value problems of mechanics of deformable bodies, Prikl. Mekh., 18:3–20, 1982.

[172] V.I. Kublanovskaya, Application of analytic continuation using a change of variables in numerical analysis, Trudy Mat. Inst. Akad. Nauk, 53:13–69, 1959.

[173] A.G. Kurosh, *A Course of Higher Algebra*, Fizmatgiz, Moscow, 1963.

[174] D.I. Kutilin, *Theory of Finite Deformations*, Gostekhizdat, Moscow–Leningrad 1947.

[175] V.V. Kuznetsov, On the use of the method of continuation of the solution in the length of the interval of integration in the computation of circular corrugated plates, Izv. Akad Nauk SSSR, MTT 2:189–191, 1983.

[176] S.G. Lekhnitskii, *Anisotropic Plates*, Gostekhizdat, Moscow, 1957.

[177] L.P. Lebedev, On the equilibrium of a free nonlinear plate, Prikl. Mat. Mekh., 44:162–165, 1980.

[178] J. Leray and J. Schauder, Topology and functional equations, Uspekhi Mat. Nauk, 1:71–95, 1946.

[179] A. Liapunov, Sur les figures d'equilibre peu differentes des ellipsoides d'une masse liquide homogène donnée d'un mouvement de rotation, page 1, Zap. Akad. Nauk, SPB, 1906.

[180] J.-L. Lions, *Quelques Méthodes de Résolution des Problèmes aux Limites Non Linéaires*, Dunod, Paris, 1969.

[181] J.-L. Lions and E. Magenes, *Problèmes aux Limites Non Homogènes et Applications,* Tome 1, Dunod, Paris, 1968.

[182] B.V. Loginov, On invariant solutions in bifurcation theory, Dokl. Akad. Nauk SSSR, 246:1048–1051, 1979.

[183] B.V. Loginov and V.A. Trenogin, On the use of group properties to find multiparameter families of solutions of nonlinear equations, Mat. Sb. 85:440–454, 1971.

[184] A.I. Lur'e, *Nonlinear Elasticity Theory*, Nauka, Moscow, 1980.

[185] A.M. Lyapunov, On nearly ellipsoid equilibrium shapes of rotating homogeneous liquid mass, in: *Collected Works, Vol. 4*, pages 9–209, Izd. Akad. Nauk SSSR, Moscow, 1959.

[186] L.A. Lyusternik, On a class of nonlinear operators in a Hilbert space, Izv. Akad. Nauk SSSR. Ser. mat. , 3:257–264, 1939.

[187] L.A. Lyusternik and L.G. Shnirel'man, *Topological Methods in Variational Problems*, Izd. MGU, Moscow, 1930.

[188] L.A. Lyusternik and V.I. Sobolev, *A Short Course in Functional Analysis*, Vysshaya Shkola, Moscow, 1982.

[189] A.G. McConnell, *Introduction to Tensor Analysis*, Fizmatgiz, Moscow, 1963.

[190] V.I. Mamai, *Interaction of Plates and Shells*, Izd. MGU, Moscow, 1984.

[191] K. Marguerre, Zur theorie der gekrümmten Platte grosser Formänderung, Jahrbuch der deutschen Luftfahrtforschung, 413–418, 1939.

[192] A.I. Markushevich, *The Theory of Analytic Functions,* Hindustani Publishing, Delhi, 1963.

[193] V.G. Maz'ya, Classes of domains and embedding theorems for function spaces, Dokl. Akad. Nauk SSSR, 133:527–530, 1960.

[194] V.G. Maz'ya, Classes of sets and measures connected with embedding theorems, in: *Embedding Theorems and Their Applications. Proceedings of the All-Union Symposium in Embedding Theorems (Baku, 1966)*, pages 142–159, Nauka, Moscow, 1970.

[195] G.K. Mikhailov, Leonard Euler and his contribution to the development of rational mechanics, Usp. Mekh., Warsaw, 8:3–58, 1985.

[196] S.G. Mikhlin, *The Problem of Minimum of a Quadratic Functional*, Holden–Day, San Francisco, 1965.

[197] S.G. Mikhlin, *Variational Methods in Mathematical Physics*, Pergamon Press, Oxford, 1964.

[198] S.G. Mikhlin, *Numerical Implementation of variational Methods*, Fizmatgiz, Moscow, 1966.

[199] G.I. Minty, On the solvability of non-linear functional equations of "monotone" type, Pacific J. Math., 14:??, 1964.

[200] N.F. Morozov, Towards a nonlinear theory of thin plates, Dokl. Akad. Nauk SSSR, 114:968–971, 1957.

[201] N.F. Morozov, Nonlinear problems of the theory of thin plates, Vestnik Leningrad. Univ., 19:100–124, 1958.

[202] N.F. Morozov, Uniqueness of a symmetric solution, Dokl. Akad. Nauk SSSR, 123:417–419, 1958.

[203] N.F. Morozov, Nonlinear problems of the theory of thin anisotropic plates, Izv. Vuzov. Matematika. Kazan', 3:8–12, 1960.

[204] N.F. Morozov, On nonlinear problems of theory of thin plates with axes of symmetry, Dokl. Akad. Nauk BSSR, 7, no. 6, 1963.

[205] N.F. Morozov, *Selected Two-dimensional Problems of Theory of Elasticity*, LGU, Leningrad, 1978.

[206] I.V. Morshneva and V.I. Yudovich, On bifurcations of periodic solutions from equilibria of inversion- and rotation-symmetric dynamical systems, Sib. Mat. Zh., 26:124–133, 1985.

[207] P.P. Mosolov and V.P. Myasnikov, *Mechanics of Rigidly-Plastic Continua,* Nauka, Moscow, 1981.

[208] F.D. Murnagan, *Finite Deformation of an Elastic Solid*, Second Edition, Zonelon, 1967.

[209] Kh.M. Mushtari, Some generalizations of the theory of thin shells, Izv. Fiz. Mat. Obshch. Kazan' Univ. 11:71–150, 1938.

[210] Kh.M. Mushtari, Some generalizations of the theory of thin shells with applications to the solution of problems of stability of an elastic equilibrium, Prikl. Mat. Mekh., 2:439–456, 1939.

[211] Kh.M. Mushtari, On the question of justification of the theory of thin shallow shells, Prikl. Mekh., 5:109–113, 1969.

[212] Kh.M. Mushtari and K.Z. Galimov, *Nonlinear theory of elastic shells*, Tatknigoizdat, Kazan', 1957.

[213] N.I. Muskhelishvili, *Singular Integral Equation*, Noordhoff, Groeningen, 1953.

[214] T.Y. Na and C.E. Turski, Solution of the non-linear differential equations to finite bending of a thin-walled shell by parameter differentiation, J. Aeronaut. Quart., 25:14–18, 1974.

[215] A.H. Nayfeh, *Problems in Perturbation*, John Wiley and Sons, New York, 1985.

[216] J. Nečas and I. Naumann, On a boundary value problem in non-linear theory of thin elastic plates, Aplikace Mathematiky, 19:7–16, 1974.

[217] S.M. Nikol'skii, *Approximation of Functions of Many Variables and Embedding Theorems,* Nauka, Moscow, 1969.

[218] A.P. Norden, *Theory of Surfaces*, Gostekhizdat, Moscow, 1956.

[219] A.P. Norden, On the question of geometrical theory of finite deformations, Izv. KFZN SSSR, Ser. Fiz. Mat. i Tekhn. Nauk, 2:53–61, 1956.

[220] S.G. Novikov, *Lectures in Differential Geometry. Parts I, II*, MGU, Moscow, 1972.

[221] V.V. Novozhilov, *Foundations of Nonlinear Elasticity,* Gostekhizdat, Moscow–Leningrad, 1948.

[222] V.V. Novozhilov, *Theory of Elasticity*, Sudpromgiz, Leningrad, 1958.

[223] J.T. Oden, *Finite Elements in Nonlinear Continua*, McGraw-Hill, 1972.

[224] D. Ortega and W. Reinboldt, *Iterative Methods for Solution of Nonlinear Systems of Equations in Several Variables*, Mir, Moscow, 1985.

[225] R. Osserman, The Plateau problem, Uspekhi Mat. Nauk, 22:55–136, 1967.

[226] V.N. Paimushin, The question of parametrization of the middle surface of a shell with complex geometry, in: *Robustness and Reliability of Complex Systems*, pages 78–84. Naukova Dumka, Kiev, 1979.

[227] D.G. Panov, On the use of the method of B.G. Galerkin to solve certain problems in of the theory of elasticity, Prikl. Mat. Mekh.3:139–142, 1939.

[228] P.F. Papkovich, *Structural Mechanics of Ships*, Part II, Oborongiz, Moscow, 1941.

[229] V.V. Petrov, Study of finite bending of plates and shallow shells by the method of successive loading, in: *Theory of Plates and Shells. Proc. II-nd All-Union Conf. (L'vov, 1961)*, pages 328–331, Izd. Akad. Nauk Ukr. SSR, Kiev, 1962.

[230] V.V. Petrov, *The Method of Successive Loading in Nonlinear Shell Theory*, Izd. Saratov. Univ., Saratov, 1975.

[231] V.V. Petrov, I.V. Neverov, and V.V. Amel'chenko, Some aspects of numerical analysis of shallow shells under large bending using the variational method of Vlasov, Izv. VUZov. Stroi. Arkhitek., 12:22–28, 1968.

[232] V.V. Petrov and P.K. Semenov, Numerical analysis of nonlinearly elastic plates using the generalized Vlasov–Kantorovich method, Izv. VUZov. Stroi. Arkhitek., 2:37–41, 1982.

[233] I.G. Petrovskii, On the analyticity of solutions of systems of partial differential equations. Mat. Sb., 5(47):3–70, 1939.

[234] W. Pietraszkiewicz, Introduction to the non-linear theory of shells, Ruhr-Universität Bochum. Mitteilungen aus dem Institut für Mechanik, 10:1–154, 1977.

[235] W. Pietraszkiewicz, Non-linear theories of thin elastic shells (in Polish), Proc. Polish Symp. Shell Structures. Theory and Applications (p. 25–26, Krakow, 1974), p. 27–50, Polish Sci. Publ., Warsawa, 1978.

[236] W. Pietraszkiewicz, Finite rotations and Lagrangian description in the non-linear theory of shells, Warsawa: Posnan, 1979.

[237] W. Pietraszkiewicz, Niektore problemy nieliniowey teorii powlok, Mechanika teoretyczna i stosowana, 2:169–192, Warsawa, 1980.

[238] W. Pietraszkiewicz, Finite rotations of shells, in: Theory of shells, pp. 445–471, North Holland Publishing Company, 1980.

[239] B.E. Pobedrya, *Lectures on Tensor Analysis*, MGU, Moscow, 1974.

[240] A.V. Pogorelov, *Unique Definition of General Convex Surfaces*, Izd. Akad. Nauk Ukr. SSR, Kiev, 1952.

[241] A.V. Pogorelov, *Infinitely Small Bending of General Convex Surfaces*, Izd. Khar'kov. Univ., Khar'kov, 1959.

[242] A.V. Pogorelov, *Geometrical Methods in Nonlinear Theory of Elastic Shells*, Nauka, Moscow, 1967.

[243] A.V. Pogorelov, *Differential Geometry*, Nauka, Moscow, 1969.

[244] H. Poincaré, *Les Méthodes nouvelles de la mécanique celeste. 1*, Gauthier–Villars, Paris, 1892.

[245] H. Poincaré, On curves defined by differential equations, Gostekhizdat, Moscow–Leningrad, 1947.

[246] M. Poitier-Ferry, Imperfection sensitivity of a nearly double bifurcation points, in: *Stab. Mech. Continua. 2 Symposium, Numbercht Aug. 31–Sept. 4, 1981*, pages 201–214, Berlin, 1982.

[247] P.Ya. Polubarinova-Kochina, On the question of stability of a plate, Prikl. Mat. Mekh., 3:16–22, 1936.

[248] V.S. Pugachev, Random functions, defined by ordinary differential equations, Tr. VVIA im. N.E. Zhukovskii, 118:3–36, 1944.

[249] V.S. Pugachev, *Theory of Random Functions and its Application to Problems of Automatic Control*, Fizmatgiz, Moscow, 1962.

[250] V.S. Pugachev and I.N. Sinitsyn, *Stochastic Differential Systems*, Nauka, Moscow, 1985.

[251] P. Rabier, Résultats d'existence dans les modèles non-linéaires de plaques, C.R. Acad. Sci. Paris, Sér. A., 2:515–518, 1979.

[252] E.A. Rakhmanov, On the convergence of Padé approximations in classes of holomorphic functions, Mat. Sb., 112(154):162–169, 1980.

[253] S.K. Randhamohan, A.V. Setlur, and J.E. Goldberg, Stability of shells by parametric differentiation, J. Struct. Div., Proc. Amer. Soc. Civ. Eng., 97:1775–1790, 1971.

[254] P.K. Rashevskii, *A Course of Differential Geometry*, Gostekhizdat, Moscow, 1956.

[255] P.K. Rashevskii, *Riemannsche Geometrie und Tensoranalysis,* Deutscher Verlag der Wissenschaften, Berlin, 1959.

[256] E. Reissner, On the theory of thin elastic shells, H. Reissner. Anniversary Volume: Contributions to Applied Mechanics, I.W. Edwards, Ann Arbor, Michigan, 231–247, 1949.

[257] E. Reissner, On axisymmetrical deformations of thin shells of revolution, Proceedings of Symposia in Applied Mathematics, 27–52, 1950.

[258] E. Reissner, On the equations for finite symmetrical deflections of thin shells of revolution, Progress in Applied Mechanics, the Prager Anniversary volume, 171–178, 1963.

[259] E. Reissner, On the equations of non-linear shallow shell theory, Studies Appl. Math., 48:171–175, 1969.

[260] E. Reissner, A note on generating generalized two-dimensional plate and shell theories, J. of Appl. Math. and Phys. (ZAMP), 28:633–642, 1977.

[261] E. Riks, Application of Newton's method to the problem of elastic stability, Prikl. Mekh., 4:204–210, 1972.

[262] W. Ritz, Über eine neue Methode zur Lösung gewisser Variationsprobleme der mathematischen Physik, J. Reine Angew. Math., 135:1–61, 1908.

[263] W. Ritz, Theorie der Transversalschwingungen einer quadratischen Platten mit freien Randern, Ann. d. Phys., 28:737–786, 1909.

[264] Ya.A. Roitberg, Elliptic problems with nonhomogeneous boundary conditions and local smoothness increase up to the boundary in generalized solutions, Dokl. Akad. Nauk SSSR, 157:798–801, 1964.

[265] E. Rotte, Zur Theorie der topologische Ordnung und der Vektorfeldes in Banachschen Räumen, Compositio Math. S., 1937.

[266] Yu. Rozhanskaya, Critical points of vector fields, in [245], pages 348–367.

[267] J.L. Sanders, Non-linear theories for thin shells, Quart. Appl. Math., 21:21–36, 1963.

[268] E. Schmidt, Zur Theorie linearen und nichlinearen Integralgleichungen. Theil 3. Über die Auflösungen der nichlinearen Integralgleichungen und die Verzweigung ihrer Lösungen, Math. Ann., 65:370–399, 1910.

[269] J.A. Schouten, *Tensor Analysis for Physicists*, Clarendon Press, Oxford, 1951.

[270] L.I. Sedov, *Continuum Mechanics,* IVth edition, Vol. 3, Nauka, Moscow, 1983; Vol. 4, Nauka, Moscow, 1984.

[271] V.I. Shalashilin, Some algorithms of the method of continuation in parameter in nonlinear problems of the theory of elasticity, in: *Nonlinear Theory of Plates and Shells*, pages 50–51, Izd. Kazan. Univ., Kazan', 1980.

[272] V.I. Shalashilin, Continuation in parameter in problems of stability and of self-oscillations, in: *Proc. All-Union Symp. on Nonlinear Problems of the Theory of Plates and Shells, Vol. III*, pages 29–32, Izd. Saratov. Univ., Saratov, 1981.

[273] V.I. Shalashilin, *Algorithms of the Method of Continuation in Parameter for Nonlinear Equations of Deformable Systems*, MAI, Moscow, 1981.

[274] V.I. Shalashilin, *Algorithms of the Method of Continuation in Parameter in Nonlinear Boundary Value Problems of the Theory of Deformable Systems*, MAI, Moscow, 1981.

[275] V.I. Shalashilin, *Continuation of Solutions through Bifurcation Points*, MAI, Moscow, 1981.

[276] V.I. Shalashilin, Algorithms of the method of continuation in parameter for large axisymmetric bending of shells of revolution, in: *Numerical and Experimental Methods in the Study of Strength, Stability and Oscillations of Aviation Structures*, pages 72–78, MAI, Moscow, 1983.

[277] L.A. Shapovalov, Equations of elasticity of a thin shell under non-axisymmetric deformation, Izv. Akad. Nauk SSSR, MTT, No. 3, 1976.

[278] I. Simmonds and D. Danielson, Non-linear shell theory with a finite rotation vector, Proc. Konik. Ned. Akad. Wetensch. Ser. B., 73:460–478, 1970.

[279] I. Simmonds and D. Danielson, Non-linear shell theory with finite rotation and stress function vectors, Trans. ASME. Ser. E, J. Appl. Mech., 39:1085–1090, 1972.

[280] V.I. Skrypnik, *Nonlinear Elliptic Equations of Higher Order*, Naukova Dumka, Kiev, 1973.

[281] V.I. Skrypnik, Solvability and properties of solutions of nonlinear elliptic equations, in: *Advances in Science and Technology. Contemporary Problems of Mathematics*, 9:131–242, VINITI, Moscow, 1976.

[282] L.N. Slobodetskii, Generalized Sobolev spaces and their applications to boundary value problems in partial derivative equations, Uch. Zapiski A.I. Herzen Leningrad Ped. Inst. 197:54–112, 1958.

[283] V.I. Smirnov, *A Course of Higher Mathematics,* Vol. 5: *Integration and Functional Analysis*, Pergamon Press, N. Y., 1964.

[284] V.I. Sobolev, On eigenelements of certain nonlinear operators, Dokl. Akad. Nauk SSSR, 31:734–736, 1941.

[285] S.L. Sobolev, *Some Applications of Functional Analysis in Mathematical Physics*, LGU, Leningrad, 1950 (NGU, Novosibirsk, 1962).

[286] S.L. Sobolev, *Introduction to the Theory of Cubature Formulae*, Nauka, Moscow, 1974.

[287] S.L. Sobolev and L.A. Lyusternik, *Elements of Functional Analysis*, Second (amended) edition, Nauka, Moscow, 1965.

[288] I.S. Sokolnikov, *Tensor Analysis. Theory and Applications to Geometry and Mechanics of Continua*, New York, 1951.

[289] V.A. Solonnikov, On general boundary value problems that are elliptic in the sense of Douglis-Nirenberg. I, Izv. Akad. Nauk SSSR. Ser. mat. , 28:665–706, 1964.

[290] V.A. Solonnikov, On general boundary value problems that are elliptic in the sense of Douglis-Nirenberg. II, in: *Proc. Steklov Math. Inst. Vol. 92. Boundary Value Problems of Mathematical Physics*, 4:233–296, Nauka, Moscow, 1966.

[291] L.S. Srubshchik, Non-stiffness of a spherical dome, Prikl. Mat. Mekh., 32:435–444, 1968.

[292] L.S. Srubshchik, On the question of non-rigidity in the nonlinear theory of shallow shells, Izv, Akad. Nauk SSSR. Ser. Mat., 4:890–909, 1972.

[293] L.S. Srubshchik, Buckling of elastic shells with initial imperfections in several characteristic forms, Dokl. Akad. Nauk SSSR, 249:808–812, 1979.

[294] L.S. Srubshchik, Influence of initial imperfections on the buckling of elastic shells for multiple critical loads, Prikl. Mat. Mekh., 44:892–904, 1981.

[295] L.S. Srubshchik, *Buckling and Postcritical Behaviour of Shells*, Izd. Rost. Univ., Rostov-on-Don, 1981.

[296] L.S. Srubshchik, Non-axisymmetric buckling and postcritical behavior of elastic spherical shells in the case of a double critical value of the load, Prikl. Mat. Mekh., 47:662–672, 1983.

[297] L.S. Srubshchik and V.I. Yudovich, Asymptotic integration of the system of equations of large bending of symmetrically loaded shells of revolution, Prikl. Mat. Mekh., 26:313–332, 1962.

[298] L.S. Srubshchik and V.I. Yudovich, Asymptotics of equations of large bending of a circular symmetrically loaded plate, Sib. Mat. Zhurn., 4:657–672, 1963.

[299] D.R. Stoutenmyer, Analytical solution of integral equations using computer algebra, SIAM Rev., 18:829, 1971.

[300] G. Strang and G. Fix, *Theory of Finite Elements*, Mir, Moscow, 1977.

[301] G. Strickline (???), Static and dynamical computations of geometrically nonlinear shells of revolution, in [80], pages 272–293.

[302] I.V. Svirskii, On the precision of the Bubnov–Galerkin method, Dokl. Akad. Nauk SSSR, 88:621–624, 1953.

[303] I.V. Svirskii, *Methods of Bubnov–Galerkin Type and of Successive Approximations*, Nauka, Moscow, 1968.

[304] Yu.M. Temis, The method of successive loading with error estimates in geometrically nonlinear elastic problems, in: *Applied Problems of Strength of Materials and Plates, Issue 16*, pages 3–10, Gor'kii, 1980.

[305] G.K. Ter-Grigoryants, On the appearance of doubly periodic convection in the horizontal layer, Prikl. Mat. Mekh., 37:177-184, 1973.

[306] J.M.T. Thompson, The elastic instability of a complete spherical shell, Aeronaut. Quart., 13:295–299, 1962.

[307] J.M.T. Thompson, The rotationally symmetric branching behavior of a complete spherical shell, Proc. Konik. Ned. Akad. Wetensch. ser. B, 67:295–299, 1964.

[308] J.M.T. Thompson, The post-buckling behavior of a spherical shell by computer analysis, in: *World Conference on Shell Structures*, S.I. Medwadowski et al., eds., National Academy of Sciences, Washington, 1964.

[309] G.A. Thurston, Continuation of Newton's method through bifurcation points, Proc. ASME E36:425–430, 1969.

[310] V.A. Trenogin, *Functional Analysis*, Nauka, Moscow, 1980.

[311] H. Triebel, *Interpolation Theory. Function Spaces. Differential Operators*, Mir, Moscow, 1980.

[312] E.S. Tsitlanadze, Some questions of the theory of nonlinear operators and calculus of variations in Banach type spaces, Uspekhi Mat. Nauk, 5:141–176, 1950.

[313] E.S. Tsitlanadze, Minimax points existence theorems in Banach spaces and their applications, Tr. Mosk. Mat. Obshch, 2:235–275, 1952.

[314] S.V. Uspenskii, On embedding theorems for generalized Sobolev classes, Sib. Mat. Zh., 3:418–445, 1962.

[315] S.V. Uspenskii, G.V. Demidenko, and V.G. Perepelkin, *Embedding Theorems and their Applications in Differential Equations*, Nauka, Sib. Otd., Novosibirsk 1984.

[316] D.G. Vainberg and V.I. Gulyaev, Stability of mechanical and physical fields in shells of complex shape, in: *Progress in Mechanics of Deformable Media*, pages 96–104, Nauka, Moscow, 1975.

[317] M.M. Vainberg, On eigenelements of a class of nonlinear operators, Dokl. Akad. Nauk SSSR, 75:609–612, 1950.

[318] M.M. Vainberg, On an unconditional extremum of a functional and on convergence of minimizing sequences, Dokl. Akad. Nauk SSSR, 183:1243–1246, 1968.

[319] M.M. Vainberg, *Variational Methods and the Method of Monotone Operators in the Theory of Nonlinear Equations*, Nauka, Moscow, 1972.

[320] M.M. Vainberg and R.I. Kachurovskii, Towards a variational theory of nonlinear operator equations, Dokl. Akad. Nauk SSSR, 129:1199–1202, 1959.

[321] M.M. Vainberg and V.A. Trenogin, *Branching of Solutions of Nonlinear Equations*, Nauka, Moscow, 1969.

[322] N.V. Valishvili, *Computational Methods for Spherical Shells*, Mashinostroenie, Moscow, 1976.

[323] N.V. Valishvili and V.N. Stegnii, On equilibrium states of hollow spherical shells, MTT, 2:131–134, 1968.

[324] V.V. Vavilov, On the convergence of Padé approximations of meromorphic functions, Mat. Sb. 101(143):44–56, 1976.

[325] I.N. Vekua, *Generalized Analytical Functions*, Pergamon Press, Oxford, 1962

[326] T.V. Vilenskaya and I.I. Vorovich, Asymptotic behavior of solutions of a problem in theory of elasticity for an open hollow sphere of small thickness, in: *Proc. Vth All-Union Conference in Shell and Plate Theory (Moscow, 1965)*, pages 15–16, Izd. Akad. Nauk SSSR, Moscow, 1965.

[327] T.V. Vilenskaya and I.I. Vorovich, Asymptotic behavior of solutions of problems of elasticity theory for a spherical shell of small thickness, Prikl. Mat. Mekh., 30:278–295, 1966.

[328] M.I. Vishik, On general boundary problems for elliptic differential equations, Trudy Mosk. Mat. Obshch., 1:187–246, 1952.

[329] M.I. Vishik, Solution of a system of quasilinear equations in divergence form under periodic boundary conditions, Dokl. Akad. Nauk SSSR, 137:502–505, 1961.

[330] M.I. Vishik, Quasilinear strongly elliptic systems of differential equations in divergence form, Trudy Mosk. Mat. Obshch., 12:125–184, 1963.

[331] M.I. Vishik and A.V. Fursikov, *Mathematical Problems of Statistical Hydrodynamics*, Nauka, Moscow, 1980.

[332] V.Z. Vlasov, The basic differential equations of the general theory of elasticity of shells, Prikl. Mat. Mekh., 8:109–140, 1944.

[333] V.Z. Vlasov, *General Theory of Shells and its Application in Technology*, Gostekhizdat, Moscow–Leningrad, 1949.

[334] L.P. Volevich, Solvability of boundary value problems for general elliptic systems, Mat. Sbornik, 68:373–416, 1965.

[335] A.S. Vol'mir, editor, *Theory of Elastic Plates (translated from Chinese)*, Inostr. Lit., Moscow, 1957.

[336] A.S. Vol'mir, *Stability of Deformable Systems*, Nauka, Moscow, 1967.

[337] A.S. Vol'mir and I.G. Kidil'bekov, Probabilistic characteristics of the behavior of a cylindrical shell under acoustic loading, Prikl. Mekh., 1:1–9, 1965.

[338] I.I. Vorovich, On some direct methods in the nonlinear theory of shallow shells, in: *Abstracts of Presentations at the Symposium on Elasticity, Plasticity and Theoretical Questions of Construction Mechanics (22–25 Dec. 1954)*, pages 21–22, Izd. Akad Nauk SSSR, Moscow–Leningrad, 1954.

[339] I.I. Vorovich, On the existence of solutions in nonlinear shell theory, Izv. Akad. Nauk SSSR. Ser. mat., 19:173–176, 1955.

[340] I.I. Vorovich, On the behavior of a circular plate after loss of stability, Uch. Zap. Rostov Univ. 32:55–60, 1955.

[341] I.I. Vorovich, On some direct methods in nonlinear shellow shell theory, Dokl. Akad. Nauk SSSR, 105:42–45, 1955.

[342] I.I. Vorovich, On some direct methods in nonlinear shallow shell theory, Prikl. Mat. Mekh., 20:449–474, 1956.

[343] I.I. Vorovich, Some problems of nonlinear shell theory, in: *Proc. III All-Union Math. Congress. Section Presentations*, pages 1, 201–202, Izd. Akad. Nauk SSSR, Moscow, 1956.

[344] I.I. Vorovich, On the existence of solutions in nonlinear shell theory, Dokl. Akad. Nauk SSSR, 117:203–206, 1957.

[345] I.I. Vorovich, On the statistical method in the theory of stability of shells, Lecture at the Conference on Theory and Applications of Thin Shells, Tartu, 1957.

[346] I.I. Vorovich, Some questions of shell stability "in the large," Dokl. Akad. Nauk SSSR, 122:37–40, 1958.

[347] I.I. Vorovich, Error estimates in direct methods of nonlinear shell theory, Dokl. Akad. Nauk SSSR, 122:196–199, 1958.

[348] I.I. Vorovich, The statistical method in the theory of stability of shells, Prikl. Mat. Mekh., 23:885–892, 1959.

[349] I.I. Vorovich, Some applications of nonlinear functional analysis in problems of continuum mechanics, in: *Proceedings of the IVth All-Union Mathematical Conference. Section Reports (Leningrad, 1961). Vol. 2*, pages 541–545, Nauka, Leningrad, 1964.

[350] I.I. Vorovich, Some mathematical problems of plate and shell theory, in: *IInd All-Union Congress on Theoretical and Applied Mechanics (Moscow, 29 Jan.–5 Feb., 1964)*, pages 56–57, Nauka, Moscow, 1964.

[351] I.I. Vorovich, Some questions concerning the use of statistical methods in the theory of stability of plates and shells, in: *Proceedings of the IVth All-Union Conference on Plate and Shell Theory*, pages 64–94, Izd. Akad. Nauk Arm. SSR, Erevan, 1964.

[352] I.I. Vorovich, Some mathematical questions of the theory of plates and shells, in: *Proceedings of the IInd All-Union Conference in Theoretical and Applied Mechanics (Moscow, 1964). Survey Lectures, Issue 3. Solid State Mechanics.*, pages 116–136, Nauka, Moscow, 1966.

[353] I.I. Vorovich, General questions of the theory of plates and shells, in: *Proceedings of the VIth All-Union Conference on Plate and Shell Theory (Baku, 1966)*, pages 896–903, Nauka, Moscow, 1966.

[354] I.I. Vorovich, The development of the stability problem in shell theory, in: *Problems of Contemporary Science*, pages 111–126, Izd. Rost. Univ., Rostov-on-Don, 1967.

[355] I.I. Vorovich, Some problems of stress concentration, in: *Stress Concentration. Issue 2. Presentations Made at the II Symposium on Stress Concentration near Openings in Plates and Shells (Kiev, 1967)*, pages 45–53, Naukova Dumka, Kiev, 1968.

[356] I.I. Vorovich, Some estimates of the number of solutions in the von Kármán equations in connection with the problem of stability of plates and shells, in: *Problems of Hydrodynamics and Continuum Mechanics*, pages 111–118, Nauka, Moscow, 1969.

[357] I.I. Vorovich, Uniqueness of solutions of boundary value problems of nonlinear shell theory and the problem of rigidity of a shell, Tr. Tbiliss. Gos. Univ. Inst, Prikl. Mat., Issue 2, pages 49–55, Izd. Tbil. Univ., Tbilisi, 1969.

[358] I.I. Vorovich, On the behavior of plates of arbitrary shape after loss of stability, in: *Problems of Mechanics of Solid Deformable Bodies. A Special Issue to Mark the 60th Birthday of Acad. V.V. Novozhilov*, pages 113–119, Sudostroenie, Leningrad, 1970.

[359] I.I. Vorovich, Nonuniqueness and stability in continuum mechanics, in: *XIII International Congress. Abstracts of Talks*, pages 23–24, Nauka, Moscow, 1972.

[360] I.I. Vorovich, The problem of non-uniqueness and stability in the non-linear mechanics of continuous media, *Applied Mechanics. Proc. Thirteenth Intern. Congr. Theor. Appl. Mech.*, pages 340–357, Springer Verlag, Berlin, Heidelberg, New York, 1973.

[361] I.I. Vorovich, The Bubnov–Galerkin method, its development and role in applied mathematics, in: *Advances in Mechanics of Deformable Media. Dedicated to 100th Anniversary of Acad. B.G. Galerkin*, pages 121–133, Nauka, Moscow, 1975.

[362] I.I. Vorovich, Nonuniqueness and stability in continuum mechanics: Lectures delivered to the Mech. Math. Fac. of Moscow State Univ., in: *Unsolved Problems of Mechanics and Applied Mathematics*, pages 10–47, MGU, Moscow, 1977.

[363] I.I. Vorovich and I.G. Kadomtsev, A qualitative study of the stress-deformed state of a three-layer plate, Prikl. Mat. Mekh., 34:870–876, 1970.

[364] I.I. Vorovich, I.G. Kadomtsev, and Yu.A. Ustinov, Some general properties of the three-dimensional stress-deformed state of a three-layer plate with symmetric structure, in: *Proceedings of the VIth All-Union Conference on Shell and Plate Theory (Leningrad, 1973)*, pages 36–38, Sudostroenie, Leningrad, 1975.

[365] I.I. Vorovich and L.P. Lebedev, On the existence of solutions in nonlinear shallow shell theory, Prikl. Mat. Mekh., 36:691–704, 1972.

[366] I.I. Vorovich, L.P. Lebedev, and Sh.M. Shlafman, On certain direct methods and existence of solutions in the nonlinear theory of non-shallow shells of revolution, Prikl. Mat. Mekh., 38:339–348, 1974.

[367] I.I. Vorovich and O.S. Malkina, The asymptotic method of solution of the problem of the theory of elasticity applied to a thick plate, in: *Materials for the VIth All-Union Conference on Plate and Shell Theory (Baku, 1966)*, pages 277–280, Nauka, Moscow, 1966.

[368] I.I. Vorovich and O.S. Malkina, Stressed state of a thick plate, Prikl. Mat. Mekh., 31:230–241, 1967.

[369] I.I. Vorovich and O.S. Malkina, On the exactness of asymptotic expansions of the problem of the theory of elasticity applied to a thick plate, Inzh. Zh. MTT, 5:92–102, 1967.

[370] I.I. Vorovich and O.S. Malkina, Stress concentration in a thick plate, in: *Stress Concentration. Issue 3*, pages 37–41, Naukova Dumka, Kiev, 1971.

[371] I.I. Vorovich and N.I. Minakova, Stability of a non-shallow spherical dome, Prikl. Mat. Mekh., 32:332–338, 1968.

[372] I.I. Vorovich and N.I. Minakova, A study of the stability of a non-shallow spherical dome using higher approximations, Izv. Akad. Nauk SSSR, MTT, 2:121–128, 1969.

[373] I.I. Vorovich, N.I. Minakova, V.F. Zipalova, L.S. Srubshchik, and V.G. Shepeleva, On the results of an analytical study and of a numerical investigation of stability of certain types of shallow and non-shallow shells, in: *IIIrd All-Union Congress on Theoretical and Applied Mechanics (Moscow, 1968)*, pages 80–83, Izd. Akad. Nauk SSSR, Moscow, 1968.

[374] I.I. Vorovich and Sh.M. Shlafman, On the convergence of the finite element method in nonlinear shell theory, in: *Proceedings of the Xth All-Union Conference on Shell and Plate Theory*, Vol. 1, pages 552–561, Tbilisi, 1975.

[375] I.I. Vorovich and V.G. Shepeleva, A study of nonlinear stability of a shallow doubly curved shell using higher order approximations, Izv. Akad. Nauk SSSR, MTT, 3:69–73, 1969.

[376] I.I. Vorovich and N.A. Shlenev, Plates and shells, in: *Results in Science and Technology. Mechanics*, pages 91–176, Nauka, Moscow, 1963.

[377] I.I. Vorovich and L.S. Srubshchik, Asymptotic analysis of general equations of nonlinear shell theory, in: *Proceedings of the VIIth All-Union Conference on Shell and Plate Theory (Dnepropetrovsk, 1969)*, pages 156–159, Nauka, Moscow 1970.

[378] I.I. Vorovich and M.N. Yatsenko, On a type of loss of stability of a cylindrical panel, in: *Proceedings of the VIIIth All-Union Conference on Shell and Plate Theory (Rostov-on-Don, 1971)*, pages 259–262, Nauka, Moscow, 1973.

[379] I.I. Vorovich and V.F. Zipalova, On the solution of nonlinear boundary value problems of the theory of elasticity by passing to a Cauchy problem, Prikl. Mat. Mekh., 29:150–153, 1965.

[380] I.I. Vorovich and V.F. Zipalova, Analysis of the nonlinear deformation of a spherical dome using higher approximations, Inzh. Zh., MTT, 2:894–901, 1966.

[381] I.I. Vorovich, V.F. Zipalova, N.I. Minakova, M.N. Yatsenko, and V.G. Shepeleva, On some results of a numerical study of nonlinear problems of shell theory, in: *Vth All-Union Conf. on Computers in Construction Mechanics (Tbilisi, 1968). Abstracts of Presentations*, Tbilisi, 1968.

[382] J. Wolkowisky, Existence of buckled states of circular plates, Comm. Pure Appl. Math., 20:549–560, 1967.

[383] J. Wolkowisky, A proof of the existence of buckled states of plates using the Schauder fixed point theorem, in [139], pages 35–45.

[384] I. Yamada and U. Ekoshi, A study of finite deformation of plates and shells, in [80], pages 171–185.

[385] F.S. Yasinskii, *Collected Works on the Stability of Compressed Rods*, supp. IV, Gostekhizdat, Moscow, 1952.

[386] V.I. Yudovich, Free convection and branching, Prikl. Mat. Mekh., 31:101–111, 1967.

[387] O. Zinkiewicz, *The Finite Element Method in Engineering science*, 4th Ed., McGraw-Hill, 1988.

[388] V.F. Zipalova, Stability of a hinged spherical dome, Izv. Akad. Nauk SSSR. MTT, 1:172–177, 1967.

[389] V.F. Zipalova and V.M. Nenast'eva, A study of buckling of a spherical shell under the action of an annular load using higher approximations, in: *Proceedings of VIth All-Union Conference on Plate and Shell Theory*, pages 413–414, Nauka, Moscow, 1966.

[390] V.F. Zipalova and V.M. Nenast'eva, Thermostability of a hinged hollow spherical shell, in: *Proceedings of VIIIth All-Union Conference on Plate and Shell Theory*, pages 289–293, Nauka, Moscow, 1973.

[391] V.F. Zipalova and E.D. Shchepkina, A study of stability of a conical shell using higher approximations, Izv. Akad. Nauk SSSR. MTT, 3:135–139, 1970.

[392] L.M. Zubov, *Methods of Nonlinear Elasticity Theory in the Theory of Shells*, Izd. Rost. Univ., Rostov-on-Don 1982.

Supplementary References for the English Edition

[393] R.A. Adams, *Sobolev Spaces*, Academic Press, NY, 1975.

[394] O. Alexandrescuiosifescu, Existence and regularity of the solution of Koiter nonlinear, 2-dimensional shallow-shell model, Comptes Rendus Acad. Sci., Ser. I, 321:1269–1274, 1995.

[395] S.S. Antman, The influence of elasticity on analysis: Modern developments, Bull. Amer. math. Soc., 9:267–291, 1983.

[396] S.S. Antman, *Nonlinear Problems of Elasticity*, Springer-Verlag, New York, 1996.
[397] V.I. Arnol'd, ed., *Dynamical Systems VI. Singularity Theory,* Encyclopedia of Math. Sciences, vol. 6, Springer-Verlag, Berlin, 1993.
[398] I.J. Bakelman, *Convex Analysis and Nonlinear Geometric Elliptic Equations,* Springer-Verlag, Berlin, 1994.
[399] P.G. Ciarlet, *The Finite Element Method for Elliptic problems,* North Holland, 1978.
[400] P.G. Ciarlet, *Mathematical Elasticity,* Vol. 1: *Three-dimensional Elasticity,* North Holland, 1988.
[401] P.G. Ciarlet, *Plates and Junctions in Elastic Multi-Structures,* Springer-Verlag, Berlin, 1990.
[402] P.G. Ciarlet, *Mathematical Elasticity,* Vol. II, North Holland, 1994.
[403] K. Deimling, *Nonlinear Functional Analysis,* Springer-Verlag, Berlin, 1985.
[404] O. Kavian and B.P. Rao, A remark on the existence of nontrivial solutions to the Marguerre–von Kármán equations, Comptes Rendus Acad. Sci., Ser. I, 317:1137–1142, 1993.
[405] A.M. Khludnev, On a variational inequality for a shallow shell operator with a constraint on the boundary, Prikl. Mat. Mekh., 51:269–272, 1987.
[406] R.S. Kubrusly, On the existence of post-buckling solutions of shallow shells under a certain unilateral constraint, Int. J. Eng. Sci., 20:93–99, 1982.
[407] O.A. Ladyzhenskaya, *The Boundary Value Problems of Mathematical Physics,* Springer-Verlag, Berlin, 1985.
[408] L.P. Lebedev, I.I. Vorovich, and G.M.L. Gladwell, *Functional Analysis; with Applications in Mechanics and Inverse Problems*, Kluwer Acad. Publsihers, 1996.
[409] V.G. Maz'ja, *Sobolev Spaces*, Springer-Verlag, Berlin, 1985.
[410] S. Naomis and P.C.M. Lau, *Computational Tensor Analysis of Shell Structures*, Springer-Verlag, Berlin, 1990.
[411] S.M. Nikol'skii, ed., *Analysis III. Spaces of Differentiable Functions,* Encyclopedia of Math. Sciences, vol. 26, Springer-Verlag, Berlin, 1991.
[412] M. Struwe, *Variational Methods,* 2nd ed., Springer-Verlag, Berlin, 1996.
[413] I.I. Vorovich and L.P. Lebedev, On solvability of the nonlinear problem of equilibrium of a shallow shell, Prikl. Mat. Mekh.52:614–820, 1988.
[414] I.I. Vorovich and L.P. Lebedev, On the finite element method in the nonlinear shell theory, Russian J. Comp. Mech., no. 1, 1–21, 1993.
[415] E. Zeidler, *Nonlinear Functional Analysis and its Applications*. Part 1: *Fixed Point Theorems*, Part 3: *Variational Methods and Optimization*, Part 4: *Applications to Mathematical Physics*, Springer-Verlag, New York, 1985–1988.

List of Symbols

Abbreviations

- BGR ~ Bubnov-Galerkin-Ritz
- KFP ~ Kolmogorov-Fokker-Plank
- LOE ~ linear operator equation
- MIS ~ momentless (membrane) state
- NOE ~ nonlinear operator equation
- PSC ~ piecewise smooth curve
- PSS ~ property shallow shell
- SLOE ~ spectrum of linear operator equation
- SNOE ~ spectrum of nonlinear operator equation
- RNOE ~ regular points of nonlinear operator equation

General Remarks

(1) The letters m, M are positive constants; in expressions where their particular values are of no importance, they are used without numeration.

(2) There is summation over all indices that occur as both sub- and superscripts. As a rule, unless specifically indicated otherwise, summation is over indices that go from one to two. There is no summation over repeated indices m and τ.

(3) $u_{\alpha^k} = \frac{\partial u}{\partial \alpha^k}, \; u_{\alpha^k \alpha^l} = \frac{\partial^2 u}{\partial \alpha^k \partial \alpha^l}.$

(4) $f_{,\alpha^k} = \frac{\partial f}{\partial \alpha^k}$

(5) $\nabla_k u^p$: see (2.3), (2.4).

(6) As a rule, two- and three- dimensional vectors are used.

(7) The coordinates of an arbitrary vector $\mathbf{u}$ in a basis $(\mathbf{e}^1, \mathbf{e}^2, \mathbf{n})$ are denoted by (u_1, u_2, u_3) or by (u_1, u_2, u); in a basis $(\mathbf{e}_1, \mathbf{e}_2, \mathbf{n})$ coordinates are denoted by (u^1, u^2, u^3) or (u^1, u^2, u). Here $u^3 = u_3 = u$.

(8) Dependence of a vector on coordinates is denoted by $\mathbf{u}(u_1, u_2, u)$. Sometimes the same fact is reflected by writing, not entirely consistently, $\mathbf{u} = (u_1, u_2, u)$. Occasionally, we use parentheses to describe the dependence of a vector on the coordinates of the surface, as in $\mathbf{u}(\alpha^1, \alpha^2)$.

(9) From components of a vector $\mathbf{u}$ we construct a new vector $\omega = u_1\mathbf{e}^1 + u_2\mathbf{e}^2$, as a projection onto a plane tangent to the middle surface of the shell; this is indicated in the notation by $\mathbf{u}(\omega, u)$. The various forms of notation described above are here taken to be interchangeable.

(10) δ is used to denote variations.

(11) $[a, b, c]$: for this notation, see (7.25).

Sub- and Superscripts

(1) s (for "stretch") is used for quantities connected with tangential stresses, moduli, etc.

(2) f (for "flexural") is used for quantities connected with bending stresses, moduli, etc.

(3) p stands for a particular solution, a fixed function, etc.

(4) m is used either for a projection of tangential displacements such as ω onto a normal to the boundary of the domain or to denote quantities connected with the normal to the boundary (it is used only to define certain vectors on the boundary of the domain).

(5) τ is used either for a projection of tangential displacements such as ω onto a vector tangent to the boundary of the domain or to denote quantities connected with vectors tangent to the boundary.

(6) n and k (used as subscripts) are used as a rule to index elements in a sequence (in particular, of approximate solutions) or of a basis. The last subscript n is always used to index terms of a sequence.

(7) i, λ are used to denote the components of vectors such as ω.

Description of the Middle Surface of a Shell

(1) S is the middle surface of shell with a boundary Γ.

(2) (α^1, α^2) are curvilinear coordinates of a surface S.

(3) α^3 is the coordinate along a normal to the middle surface of a shell.

(4) $\rho = \rho(\alpha^1, \alpha^2)$ is the position vector of the middle surface S in space.

(5) $\mathbf{e}_k = \rho_{\alpha^k}$, $k = 1, 2$, is a tangent basis at a given point of S.

(6) A_{ij}, A^{ij} are the coefficients of the first quadratic form of the surface.

(7) B_{ij}, B^{ij} are the coefficients of the second quadratic form of the surface.

(8) K is the Gaussian curvature of a surface (7.4).
(9) H is the average curvature of a surface.
(10) Γ^i_{jk} are the Christoffel symbols.
(11) Ω is the image of the surface S in the plane of curvilinear coordinates (α^1, α^2).
(12) D is the Jacobian of this mapping of the middle surface to the plane.
(13) $d\Omega = D d\alpha^1 d\alpha^2$.
(14) Γ_i are parts of the boundary curve.
(15) $\tau = (\tau_1, \tau_2)$ is the tangent to the boundary curve.
(16) $\mathbf{m} = (m_1, m_2)$ is the normal (1.11) to the boundary curve.
(17) s is the arc length of the boundary curve.

Displacements, Deformations

(1) $\mathbf{u} = \mathbf{u}(\omega, w) = \mathbf{u}(w^1, w^2, w) = \mathbf{u}(w^1, w^2, w^3)$; see (3.1).
(2) $\omega = (w^1, w^2)$.
(3) w_m is the projection in the direction of the normal (the sub- or superscript m is always used to denote the projection of ω or of other vectors in the direction m of the normal to the boundary).
(4) w_τ is the projection of ω in the direction of the tangent vector τ to the curve Γ.
(5) $w_4 = \partial w / \partial m$ is the rotation angle of the normal to the middle surface S (normal derivative of w).
(6) φ_k are elements in the basis of the energy space.
(7) ϵ_{ij}, $\overset{0}{\epsilon}_{ij}$, $\overset{1}{\epsilon}_{ij}$ are the components of the strain tensor (3.16), (3.17).
(8) $\gamma_{ii} = \epsilon_{ii}$, $\gamma_{ij} = 2\epsilon_{ij}$, $i \neq j$.
(9) $\delta\mathbf{u} = \mathbf{b} = \chi = (\delta w^1, \delta w^2, \delta w) = (\delta w^1, \delta w^2, \delta w)$ or $(\delta w_1, \delta w_2, \delta w)$ is a vector of "admissible" displacements, a variation of the displacements vector.

Forces and Loads

(1) Arguments in brackets mean that the indicated functions or vector functions are to be substituted instead of the corresponding displacements or stress functions.
(2) M^{ij} are moments (4.11), (4.22).
(3) T^{ij} are tangential stresses (stresses in the plane tangent to the middle surface of the shell) (4.10).
(4) D_{s}^{ijkl} or D_{f}^{ijkl} are the elastic constants of the material.
(5) Ψ is the Airy stress function (7.1).
(6) T_{p}^{ij} is a particular solution of the system (6.17).
(7) k^{ii}, k^{mm}, $k^{\tau\tau}$ are the elastic coefficients of the supports.

Energy and Work

(1) Π is the potential energy.
(2) Π_s is the potential energy of stretching.
(3) Π_f is the bending (flexural) potential energy.
(4) Q_s: see (4.12).
(5) U_{supp} is the elastic energy of the supports (6.4), (6.6), (6.7), (6.10), (6.13), (6.14).
(6) δU is a variation of the potential energy (6.15)
(7) δA is the work done by external forces (6.15), (5.2).
(8) $\mathcal{I}_{t\kappa}$ is the total energy functional (6.38), (12.44).

Function Spaces

General Spaces

(1) A vector function belongs to a function space defined for scalar functions if each of its components belongs to that space.
(2) $\|y\|_X$ is the norm of an element y in a space X.
(3) L^p_Ω is the space of functions the pth powers of which are integrable.
(4) C^k_Ω is the space of functions having k continuous derivatives in Ω.
(5) $H^{k,\lambda}_\Omega$ is the space of Hölder-continuous functions; see (10.2).
(6) $W^{(l)}_{p\Omega}$ is the space of functions having generalized derivatives in L^p_Ω up to order l.

Energy Spaces

(1) H_t is the energy space for the tangential displacemnents ω; the index t corresponds to a type of boundary conditions; the norm is defined by (11.5).
(2) $\overline{H}_t$ is the space of load complexes that describe the boundary conditions for the tangential components of displacement vectors (11.51).
(3) H_κ is the energy space for the normal displacemnents w; the index κ corresponds to a type of boundary conditions; the norm is defined by (12.5).
(4) $\overline{H}_\kappa$ is the space of load complexes that describe the boundary conditions for the normal components of displacement vectors (12.41).
(5) $H_{t\kappa} = H_t \times H_\kappa$ is the energy space for the entire displacement vector (Section 12.5).
(6) $\overline{H}_{t\kappa} = \overline{H}_t \times \overline{H}_\kappa$: see Section 12.8.
(7) H_9 is the space of Airy stress functions; the norm is given by (12.32).
(8) $H_{9\kappa}$ is the space of pairs of elements (Ψ, w); see Section 12.5.

Operators

(1) $\mathbf{G}_{\kappa\kappa\mu}$, $\mathbf{K}_{t\kappa\mu}$ are the homogeneous parts of order of homogeneity μ of the main operators of problems of shell theory (see (13.31), (13.33), (14.1)).

Equilibrium Equations and Boundary Conditions

(1) For the problem in displacements, see (6.34–6.35), (13.5), (13.31), (13.35). The boundary conditions for this case are discussed in Section 6.1.
(2) For the problem with an Airy stress function, see (7.51), (7.60). For the boundary conditions for this case see (7.9), (7.12), (7.13), (7.24).

Index

Applied Mathematical Sciences

(*continued from page ii*)

(continued on next page)

Applied Mathematical Sciences

(continued from previous page)

118. *Godlewski/Raviart:* Numerical Approximation of Hyperbolic Systems of Conservation Laws.
119. *Wu:* Theory and Applications of Partial Functional Differential Equations.
120. *Kirsch:* An Introduction to the Mathematical Theory of Inverse Problems.
121. *Brokate/Sprekels:* Hysteresis and Phase Transitions.
122. *Gliklikh:* Global Analysis in Mathematical Physics: Geometric and Stochastic Methods.
123. *Le/Schmitt:* Global Bifurcation in Variational Inequalities: Applications to Obstacle and Unilateral Problems.
124. *Polak:* Optimization: Algorithms and Consistent Approximations.
125. *Arnold/Khesin:* Topological Methods in Hydrodynamics.
126. *Hoppensteadt/Izhikevich:* Weakly Connected Neural Networks.
127. *Isakov:* Inverse Problems for Partial Differential Equations.
128. *Li/Wiggins:* Invariant Manifolds and Fibrations for Perturbed Nonlinear Schrödinger Equations.
129. *Müller:* Analysis of Spherical Symmetries in Euclidean Spaces.
130. *Feintuch:* Robust Control Theory in Hilbert Space.
131. *Ericksen:* Introduction to the Thermodynamics of Solids, Revised ed.
132. *Ihlenburg:* Finite Element Analysis of Acoustic Scattering.
133. *Vorovich:* Nonlinear Theory of Shallow Shells.
134. *Vein/Dale:* Determinants and Their Applications in Mathematical Physics.
135. *Drew/Passman:* Theory of Multicomponent Fluids.

MIX
Papier aus verantwortungsvollen Quellen
Paper from responsible sources
FSC® C105338
FSC
www.fsc.org

If you have any concerns about our products,
you can contact us on
ProductSafety@springernature.com

In case Publisher is established outside the EU,
the EU authorized representative is:
Springer Nature Customer Service Center GmbH
Europaplatz 3, 69115 Heidelberg, Germany

Printed by Libri Plureos GmbH
in Hamburg, Germany